T0310375

APPLIED MATHEMATICS
FOR THE ANALYSIS OF
BIOMEDICAL DATA

APPLIED MATHEMATICS FOR THE ANALYSIS OF BIOMEDICAL DATA

Models, Methods, and MATLAB®

PETER J. COSTA

WILEY

Published by John Wiley & Sons, Inc., Hoboken, New Jersey.
Published simultaneously in Canada.

For general information on our other products and services or for technical support, please contact our Customer Care Department within the United States at (800) 762-2974, outside the United States at (317) 572-3993 or fax (317) 572-4002.

Wiley also publishes its books in a variety of electronic formats. Some content that appears in print may not be available in electronic formats. For more information about Wiley products, visit our web site at www.wiley.com.

Library of Congress Cataloging-in-Publication Data:

Names: Costa, Peter J., author.
Title: Applied mathematics for the analysis of biomedical data : models, methods, and
    MATLAB / Peter J. Costa.
Description: Hoboken, New Jersey : John Wiley & Sons, 2016. | Includes bibliographical references
    and index.
Identifiers: LCCN 2016017319 | ISBN 9781119269496 (cloth) | ISBN 9781119269519
    (epub)
Subjects: LCSH: Biomathematics. | Bioinformatics–Mathematical models.
Classification: LCC QH323.5 .C683 2016 | DDC 570.1/51–dc23
LC record available at https://lccn.loc.gov/2016017319

Printed in the United States of America

10   9   8   7   6   5   4   3   2   1

IN DEDICATION

Per la mia bella Anne.
L'amore della mia vita,
una luce per la nostra famiglia,
un faro per il mondo.

To William J. Satzer
A great mathematical scientist.
A better friend.

IN MEMORIAM

I was the beneficiary of many superlative educators. Among them the late Professors George Craft, Allen Ziebur (SUNY @ Binghamton), and Edward J. Scott (University of Illinois) are owed more than I can possibly repay. It is impossible for me to adequately thank my dissertation advisor, the late Professor Melvyn S. Berger (University of Massachusetts), for his profound influence on my mathematical and personal development. But thanks are all I can presently offer. My great and avuncular late colleague, Frank P. Morrison, was an enormously positive presence in my life.

Finally, to my long deceased grandparents, I offer my most profound gratitude. You braved an ocean to come to a country where you understood neither the language nor the culture. And yet you persevered, raised families, and lived to see your grandson earn a doctorate in applied mathematics. I hope that this book honors your sacrifices, hardships, and accomplishments. *Molto grazie Nonna e Nonno. Io non ho dimenticato.*

CONTENTS

PREFACE

This is the book I wanted. Or rather I should write, this book would have greatly benefited a considerably younger me.

Just two months after completing my graduate studies, I began my career as a professional mathematician at a prestigious research laboratory. At that time, I was well prepared to prove theorems and to make complicated analytical computations. I had no clue, however, how to model and analyze data. It will come as no surprise, then, that I was not wildly successful in my first job.

But I did learn and mostly that I needed to devise a new approach and new set of tools to solve problems which engineers, physicists, and other applied scientists faced on a day–to–day basis. This book presents some of those tools. It is written with the "new approach" that I often learned the hard way over several decades.

The approach is deceptively simple. Mathematics needs to resemble the world and not the other way around. Most of us learn *iteratively*. We try something, see how well it works, identify the faults of the method, modify, and try the resulting variation. This is how *industrial* mathematics is successfully implemented. It has been my experience that the formula of *data + mathematical model + computational software* can produce insightful and even powerful results. Indeed, this process has been referred to as *industrial strength mathematics*.

This book and its complimentary exercises have been composed to follow this methodology. The reader is encouraged to "play around" with the data, models, and software to see where those steps lead. I have also tried to streamline the presentation, especially with respect to hypothesis testing, so that the reader can locate a technique which "everyone knows" but rarely writes down.

Most of all, I hope that the reader enjoys this book. Applied mathematics is not a joyless pursuit. Getting at the heart of a matter via mathematical principles has proven most rewarding for me. Please have some fun with this material.

ACKNOWLEDGEMENTS

There are 4,632 humans (and a few avians) who found their way into this book. In particular, I wish to thank (at least) the following people.

No two people were more influential and supportive in the development of this work than Anne R. Costa and Dr. William J. Satzer. Anne is my shining light and wife of 30 years. She is aptly named as she greets my typically outlandish suggestions *"sweetheart, I have this idea …"* with grace and aplomb. Her good humor and editorial skills polished the book's presentation and prevented several ghastly errors. Bill Satzer read through the entire manuscript, made *many* insightful recommendations, and helped give the book a coherent theme. He is due significantly more than my thanks.

Dr. Vladimir Krapchev (MIT) first introduced me to the delicate dance of mathematical models and data while Dr. Laurence Jacobs (ADK) showed me the power of computational software. Dr. Adam Feldman (Massachusetts General Hospital) and Dr. James Myrtle (who developed the *PSA* test) greatly enhanced my understanding of *prostate specific antigen* levels and the associated risk of prostate cancer. Dr. Clay Thompson (Creative Creek) and Chris Griffin (St. Jude's Medical) taught me how to program in MATLAB and create numerous data analysis and visualization tools. Thomas Lane and Dr. Thomas Bryan (both of The MathWorks) helped with subtle statistical and computational issues. Professor Charles Roth (Rutgers University) guided me through the mathematical model for real–time polymerase chain reaction. Victoria A. Petrides (Abbott Diagnostics) encouraged the development of outlier filtering and exact hypothesis testing methods. Michelle D. Mitchell helped to develop the *HIV/AIDS* SEIR model. William H. Moore (Northrup Grumman Systems) and Constantine Arabadjis taught me the fundamentals of the *extended Kalman–Bucy filter* and helped me implement an automated outbreak detection method.

Dr. Robert Nordstrom (NIH) first introduced me to and made me responsible for the development of statistical pattern recognition techniques as applied to the detection of cervical cancer. His influence in this work cannot be overstated. Dr. Stephen Sum (infraredx) and Professor Gilda Garibotti (Centro Regional Universitario, Bariloche) were crucial resources in the refinement of pattern recognition techniques. Professor Rüdiger Seydel (Universität Köln) invited me to his department and his home so that I could give lectures on my latest developments. Dr. Cleve Moler (The Math-Works) contributed an elegant M–file (`lambertw.m`) for the computation of the *Lambert W* function. Professor Brett Ninness (University of Newcastle) permitted the use of his team's *QPC* package which allowed me to compute *support vector machine boundaries*. Sid Mayer (Hologic) and I discussed hypothesis testing methods until we wore out several white boards and ourselves. Professor Richard Ellis (University of Massachusetts) provided keen mathematical and personal insight.

For their ontological support and enduring friendship I thank Carmen Acuña, Gus & Mary Ann Arabadjis, Elizabeth Augustine & Robert Praetorius, Sylvan Elhay & Jula Szuster, Alexander & Alla Eydeland, Alfonso & Franca Farina, Vladimir & Tania Krapchev, Stephen & Claudia Krone, Bill & Carol Link, Jack & Lanette McGovern, Bill & Amy Moore, Ernest & Rae Selig, Jim Stefanis & Cindy Sacco, Rüdiger & Friederike Seydel, Uwe Scholz, Clay & Susan Thompson, and many others. To my family, including my parents (Marie and Peter), brothers (MD, Lorenzo, JC, and E), and sisters (V, Maria, Jaki, and Pam), thank you for understanding my decidedly different view of the world. To my nieces (Coral, Jamie, Jessica, Lauren, Natalie, Nicole, Shannon, Teresa, and Zoë the brave hearted) and nephews (Anthony, Ben, Dimitris, Jack, Joseph, Matthew, and Michael) this book explains, in part, why your old uncle is forever giving you math puzzles: Now go do your homework. A hearty *thanks* to my mates at the Atkinson Town Pool in Sudbury, Massachusetts. There is no better way to start the day than with a swim and a laugh. Special thanks to the Department of Mathematics & Statistics at the University of Massachusetts (Amherst) for educating and always welcoming me.

To everyone mentioned above, and a those (4562?) I have undoubtedly and inadvertently omitted, please accept my sincere gratitude. This work could not exist without your support.

All mistakes are made by me and me alone, without help from anyone.

P.J. Costa
Hudson, MA

ABOUT THE COMPANION WEBSITE

This book is accompanied by a companion website:

        www.wiley.com/go/costa/appmaths_biomedical_data/

The website includes:

- MATLAB® code

INTRODUCTION

The phrase, *mathematical analysis of biomedical data*, at first glance seems impossibly ambitious. Does the author assert that inherently irregular biological systems can be described with any consistency via the rigid rules of mathematics? Add to this expression *applied mathematics* and a great deal of skepticism will likely fill the reader's mind. What is the intention of this work?

The answer is, in part, to provide a record of the author's 30-year career in academics, government, and private industry. Much of that career has involved the analysis of biological systems and data via mathematics. More than this, however, is the desire to provide the reader with a set of tools and examples that can be used as a basis for solving the problems he/she is facing. Some uncommon "tricks of the trade" and methodologies rarely broached by university instruction are provided.

Too often, books are written with only an academic audience in mind. This effort is aimed at working scientists and aspiring apprentices. It can be viewed as a combined textbook, reference work, handbook, and user's guide. The program presented here will be example driven. It would be disingenuous to say that the *mathematics* will not be emphasized (the author is, after all, a mathematician). Nevertheless, each section will be motivated by the underlying biology. Each example will contain the MATLAB® code required to produce a figure, result, and/or numerical table.

The book is guided by the idea that applied mathematical models are iterative. Develop a set of equations to describe a phenomenon, measure its effectiveness against data collected to measure the phenomenon, and then modify the model to improve its accuracy. The focus is on solving *real examples* by way of a mathematical method. Sophistication is not the primary goal. A symbiosis between the rigors of mathematical techniques and the unpredictable nature of biological systems is the point of emphasis.

The book reflects the formula that "mathematics + data + scientific computing = genuine insight into biological systems." The computing software of choice in this work is MATLAB. The reader can think of MATLAB as another important mathematical tool, akin to the Fourier transform. It (that is, MATLAB) helps transform data into mathematical forms and vice versa.

The presentation of concepts is as follows.

This introduction gives an overview of the book and ends with a representative example of the "mathematics + data + software" paradigm. The first chapter lists a set of guidelines and methods for obtaining, filtering, deciphering, and ultimately analyzing data. These techniques include data visualization, data transformations, data filtering/smoothing, data clustering (i.e., splitting one collection of samples into two or more subclasses), and data quality/data cleaning. In each case, a topic is introduced along with a data set. Mathematical methods used to examine the data are explained. Specific MATLAB programs, developed for use in an industrial setting, are applied to the data. The underlying assumption of this book is that, unlike most academic texts, data must be examined, verified, and/or filtered before a model is applied.

Following the discussion of data, the second chapter provides a view of the utility of differential equations as a modeling method on three distinct medical issues. The interaction of glucose and insulin levels within a human body is described by way of an elementary *interaction* model. This same approach is applied to the transition of *HIV* to *AIDS* within a patient. The *HIV/AIDS* example portends the *susceptible–exposed–infected–recovered/removed* models detailed in Chapter 3. The renowned *polymerase chain reaction* is presented as a coupled set of differential equations. In all of the cases above, the models are either applied to real clinical data or tested for their predictive value. MATLAB functions and code segments are included.

Chapter 3 focuses on mathematical epidemiology. The approach here is decidedly more involved than the examples provided in Chapter 2. The first section concerns a model, *built on reported clinical data*, that governs the transmission of *HIV/AIDS* through a population. It is, to the author's knowledge, the only such unified approach to the spread of a contagious disease. The second example within Chapter 3 concerns a mathematical method developed to predict the outbreak of a contagious disease based on simulated data (that mimic clinical data) of respiratory infections recorded at Boston Children's Hospital (*BCH*). Due to HIPAA (*Health Insurance Portability and Accountability Act*) laws, the author was unable to include the actual *BCH* data in the analysis. The simulated data, however, very closely resemble the clinical data. In each case, the need to estimate certain parameters crucial to the overall model *reflecting real measurements and recorded populations* is emphasized.

The fourth chapter concerns statistical pattern recognition methods used in the classification of human specimen samples for disease identification. Again, the application of mathematics to clinical data is the central focus of the exposition. *All* of the methods described are applied to data. Numerous figures and MATLAB code segments are included to aid the reader. The chapter ends with a presentation of *support vector machines* and their applications to the classification problem. Special software, developed at the University of Newcastle (Australia), is used with the permission of the design team to calculate the support vector boundaries. This cooperation itself is an example of how things are done in industry: collaborate with other experts; do not reinvent the wheel.

Chapter 5 is dedicated to a key component of biostatistics: hypothesis testing. The mathematical infrastructure is developed to produce the calculations (sample size, test statistic, hypothesis test, p-values) required by review agencies for submissions. This is an encyclopedic chapter listing the most important hypothesis tests and their variations including equivalence, non-inferiority, and superiority tests. As a point of reference, note that the author is on the organizing committee for the annual statistical issues workshop attended by industrial and review agency scientists, statisticians, and policy analysts. Thus, some of the key statistical matters as presented in these workshops are included in the chapter.

The final chapter examines clustered (that is, multi-reader/multi-category) data and the mathematical methods developed to render scientifically justified conclusions. The techniques include hypothesis testing and analysis of variance on clustered data.

0.1 HOW TO USE THIS BOOK

Throughout these chapters, every effort has been made to present the material in as direct and clear a manner as possible. It is assumed that the reader is familiar with elementary differential equations, linear algebra, and statistics. An appendix includes a brief review of the mathematical underpinnings of these subjects. Further, it is hoped that the reader has some familiarity with MATLAB. In order to use the M-files written for this text, the reader must have access to MATLAB *and* the MATLAB Statistics Toolbox. A summary of most of the pertinent M-files and MAT-files which contain the data sets are provided in the *Glossary of MATLAB Functions* located at the end of the book. Also within this section is a recommendation for setting up a workspace to access the M- and MAT-files associated with this text. The reader is *strongly urged* to follow the recommendations contained therein. To gain access to the quadratic programming solver implemented in Chapter 4, the reader must contact Professor Brett Ninness of the University of Newcastle in New South Wales, Australia (http://sigpromu.org/quadprog/index.html). Expertise in any of the aforementioned areas, however, is not crucial to the use and understanding of this work.

The chapters are, by design, independent units. While methods developed in each chapter can be applied throughout the book (especially the chapter on *data*), each topic can be read without reliance on its predecessor. Whenever possible, it is recommended that the reader have MATLAB at the avail so that the examples can be traced along with the provided code.

0.2 DATA AND SOLUTIONS

With these ideas in mind, a few words about the source of all modeling efforts are presented: data. How does an industrial scientist deal with data? As a theoretical physicist once noted, *if the data do not fit your model, change the data.* Naturally, this comment was made with tongue firmly implanted in cheek. Here are some guidelines.

(i) *Data Normalization.* When dealing with time-dependent data, it is often advisable to "center time" based upon the given interval. That is, if t ranges over the discrete values $\{t_0, t_1, \ldots, t_n\}$, then make calculations on times which start at 0 by subtracting off the initial time t_0. More precisely, map t into $t' = t - t_0$: $t_k \mapsto t'_k = t_k - t_0$. Similarly, some large measurements (say, population) can be given as 6,123,000, 6,730,000, etc. Rather than reporting such large "raw" numbers (which can cause overflow errors in computations), compute in smaller units; that is, 6.123 million. Finally, some data are rendered more stable (numerically) when normalized by their mean $\bar{x}$ and standard deviation s: $x_k \mapsto x'_k = \frac{x_k - \bar{x}}{s}$.

(ii) *Data Filtering.* Some collections of measurements are so noisy that they need to be smoothed. This can be achieved via the application of an elementary function (such as the natural logarithm ln). Also, smoothing techniques (e.g., the Fourier transform) can be applied. In other cases, a "distinctly different" measurement or *outlier* from the data can be identified and removed.

(iii) *Think Algorithmically.* A single equation, transformation, or statistic is unlikely to extract a meaningful amount of information from data. Rather, a systematic and consistent algorithmic method will generally be required to understand the underlying information contained within the data. Consequently, the most realistic and hence insightful models will be those that incorporate mathematics and data processing. Some examples of this process are *real-time polymerase chain reaction, the extended Kalman–Bucy filter applied to infectious disease transmission,* and *disease state diagnosis via statistical pattern recognition.*

(iv) *Solution Scale.* Engineers are famous for producing "back of the envelope" calculations. Sometimes, this is all that a particular matter requires. Sometimes, however, it is not. There are (at least) three types of "solutions" the industrial scientist is asked to provide: simple calculation, formal study, and "research effort."

Consider the following example. When processing tissue samples (for pathologists), certain stains are applied to the sample. More specifically, the tissue sample is affixed to a glass slide, the stain is applied, and then the slide is sent off to pathology for review. When this is done on a large scale (say, for cervical cancer screening), the process can be automated. During the automation process, some cells from one slide can migrate to a neighboring slide. Is this "cross-contamination" a serious concern?

The first "solution" could be a basic probability model. What is the probability that a slide sheds a cell? What is the probability that a cell migrates to a neighboring slide? What is the probability that a migrating cell *determines* the outcome of the pathologist's review? Multiply these three probabilities together and a "first-order" estimate of the matter's impact is produced.

The second industrial solution is to design an experiment to test whether the cell migration problem is a cause for cross-contamination from an "abnormal" specimen to a "healthy" one. This entails writing a study protocol complete with sample size justification and the mathematical tools used to determine the outcome of the data analysis. Typically, this is addressed via hypothesis testing. The details can be found in Chapter 5.

The "research effort" solution would require a large-scale mathematical modeling effort. How does a cell migrate from a slide and attach itself to a neighbor? If the stain is applied as a liquid, computation fluid dynamics can come into play. This could easily become a master's thesis or even doctoral dissertation.

The industrial scientist should be aware that all three solutions are actively pursued within the course of a career. It should *never* be the case that only one type of solution is considered. Each matter must be approached openly. Sometimes a problem undergoes all three solution strategies before it is "solved."

Whenever data are part of a modeling effort, these ideas will be kept in mind throughout this book.

0.3   AN EXAMPLE: PSA DATA

The first assignment a scientist is likely to encounter in industry is to "build a model based on these data." Such a straightforward request should have an equally direct answer. This is rarely the case. Examine the set of measurements listed in Table 0.1.

TABLE 0.1   *PSA* measurements

Date	PSA level
16 November 2000	1.7
21 November 2001	1.9
24 October 2003	2.0
6 January 2005	3.6
23 May 2007	2.31
25 July 2008	2.98
28 July 2009	3.45
30 July 2010	3.56
19 August 2011	5.35
27 September 2011	4.51
7 December 2011	5.77
20 March 2012	5.01
9 June 2012	7.19
3 July 2012	6.23
21 July 2012	5.98
21 August 2012	9.11
3 December 2012	6.47
6 April 2013	5.68
7 August 2013	4.7

These are the *prostate specific antigen (PSA)* levels of a man taken over a 13-year span.

To begin the process of modeling these data, it is reasonable to plot them. Even this relatively simple idea, however, is no easy task. First, the dates need to be converted into units that do not introduce numerical errors when fitting curves. Thus, the first date should correspond to time 0 while the final date should be in the units of *years after the initial time*. More precisely, set $t_0 = 0$ and $t_n =$ time (in years) after t_0. Figure 0.1 displays the data in these units, and the MATLAB code used to create the plot is listed immediately thereafter.

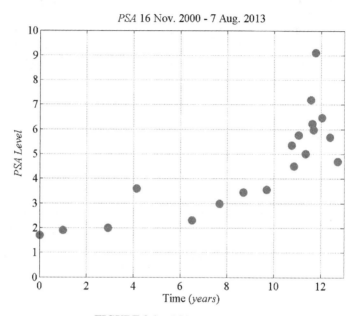

FIGURE 0.1   *PSA* measurements.

MATLAB Commands
(For Figure 0.1)

% Dates measurements taken
```
dates = {'16-Nov-2000','21-Nov-2001','24-Oct-2003','6-Jan-2005','23-
May-2007','25-July-2008','28-July-2009','30-July-2010','19-Aug-
2011','27-Sept-2011','7-Dec-2011','20-Mar-2012','9-June-2012','3-
July-2012','21-July-2012',  '21-August-2012','3-December-2012','6-Apr-
2013','7-Aug-2013'};
```
% *PSA* levels
```
y = [1.7,1.9,2.0,3.6,2.31,2.98,3.45,3.56,5.35,4.51,5.77,5.01,7.19,6.23,
5.98,9.11,6.47,5.68,4.7];
```
% Convert dates to date strings (seconds) …
```
n = cellfun(@datenum,dates);
```
% … and the date strings to years centered at $t_0 = 16$ November 2000
```
t = (n - n(1))/365.25;
```

```
% Plot the scores vs. time
h = plot(t,y,'ro'); grid('on');
set(gca,'FontSize',16,'FontName','Times New Roman');
set(h,'MarkerSize',12,'MarkerFaceColor','r');
xlabel('Time (\ityears\rm)'); ylabel('\itPSA Level');
Tsrt = ...
   ['\itPSA\rm ' strrep(dates{1},'-', ' ') ' - ' strrep(dates{end},
   '-', ' ')];
title(Tsrt);
axis([0,ceil(t(end)),0,ceil(max(y))]);
```

For emphasis, note that time 0 is the 16th of November 2000: $t_0 = 16$ November 2000. The next step in the modeling effort is to postulate what function would best fit these data. A preliminary examination of the shape of the data in Figure 0.1 suggests (at least) two curves: a parabola and an exponential function. These choices are now considered from a mathematical perspective.

Curve 1   Quadratic polynomial $f_2(t) = a_0 + a_1 t + a_2 t^2$
The times and *PSA* levels in Table 0.1 can be mapped into the vectors $t = [t_0, t_1, \ldots, t_n]^T$ and $y = [y_0, y_1, \ldots, y_n]^T$, respectively. Here, $n = 18$ but notice there are 19 measurements, the first occurring at time t_0. At the time points t, the curve takes on the values y. Therefore, the coefficients $a = [a_0, a_1, a_2]^T$ must satisfy the following set of equations

$$\left.\begin{aligned}
y_0 &= a_0 + a_1 \cdot t_0 + a_2 \cdot t_0^2 \\
y_1 &= a_0 + a_1 \cdot t_1 + a_2 \cdot t_1^2 \\
&\;\;\vdots \\
y_n &= a_0 + a_1 \cdot t_n + a_2 \cdot t_n^2
\end{aligned}\right\} \tag{0.1a}$$

The matrix

$$V_2(t) = \begin{bmatrix} 1 & t_0 & t_0^2 \\ 1 & t_1 & t_1^2 \\ \vdots & \vdots & \vdots \\ 1 & t_n & t_n^2 \end{bmatrix} \tag{0.1b}$$

and vectors $a = [a_0, a_1, a_2]^T$, $y = [y_0, y_1, \ldots, y_n]^T$ permit (0.1a) to be written in the compact form

$$V_2(t)a = y \tag{0.2}$$

In the special case of $n = 2$, $V_2(t)$ is called the *Vandermonde matrix*. Equation (0.1b) represents a generalized extension of the Vandermonde system. Typically, the vector a in (0.2) is determined by inverting the matrix $V_2(t)$ and calculating its product with y: $a = V_2(t)^{-1}y$. Observe, however, that $V_2(t)$ is *not* a square matrix so that its inverse must be computed by a generalized method. From this vantage point, it is more efficient to first factor the matrix into a product of other matrices that have easy-to-calculate inversion properties. One such factorization is called the *QR decomposition*. Using this method the matrix, $V_2(t) \in \mathcal{Mat}_{m \times 3}(\mathbb{R})$, can be written as the product of an orthogonal matrix $Q \in \mathcal{Mat}_{m \times 3}(\mathbb{R})$ and a positive upper triangular matrix $R \in \mathcal{Mat}_{m \times 3}(\mathbb{R})$ so that $V_2(t) = QR$. Consequently, $a = R^{-1}Q^{\mathrm{T}}y$. For details, see Trefethen and Bau [5], Demmel [1], or Moler [4]. While MATLAB does indeed have a *QR* decomposition function, it is more direct to fit a second-order polynomial to the data t, y using the MATLAB function polyfit.m as below.

MATLAB Commands
(For Curve 1)

```
% Fit a second-order polynomial to t, y
a = polyfit(t,y,2);
```

This produces the coefficients $a = [0.051, -0.2535, 2.187]^{\mathrm{T}}$. The error obtained from fitting a quadratic polynomial to these data is defined via (0.3a). This is known as the *least squares error* and the coefficients a are selected to minimize (0.3a). Figure 0.2a displays the fitted curve $f_2(t)$ to the data.

$$E[a] = \sum_{k=0}^{n} \left[f_2(t_k) - y_k \right]^2 \tag{0.3a}$$

A variation of the *least squares error*, called the *root mean square error*, is preferred by the engineering community. For a function f defined via the parameters a, the *RMS error* is written as

$$RMS[a] = \sqrt{\frac{1}{n+1} \sum_{k=0}^{n} \left(f(t_k; a) - y_k \right)^2} \tag{0.3b}$$

This error will be more prominently used in succeeding sections.

Observe that, while the fit appears to be "close," there is a cumulative least squares error of 21.7 and a *RMS* error of 1.07. The initial value of the curve is more than 16% larger than the initial measured value: $f_2(t_0) = 1.97 > y_0 = 1.7$, whereas the model produces only 65% of the maximum *PSA* score. Specifically, $f_2(t_{16}) = 5.96$ versus $\max(y) = y_{16} = 9.11$. The terminal value of the quadratic fitted curve, however, is 43% larger than the final *PSA* reading: $f_2(t_n) = 6.73 > y_n = 4.7$. Therefore, this model is at best a crude approximation to the data. This leads to a second model.

Curve 2   Exponential $f(t) = a_0 e^{\alpha(t - t_0)}$
Since it is desirable to have the model equal the data at the initial time point, take $y_0 = f(t_0) = a_0$. How can the exponent α be estimated? One approach is to use calculus to minimize the least squares error functional $E[\alpha] = \sum_{k=0}^{n} [y_0 e^{\alpha(t_k - t_0)} - y_k]^2$. This is

achieved by differentiating $E[\alpha]$ with respect to the exponent α and then setting the derivative to 0. The resulting equation

$$y_0 \sum_{k=0}^{n} (t_k - t_0) e^{2\alpha(t_k - t_0)} = \sum_{k=0}^{n} y_k (t_k - t_0) e^{\alpha(t_k - t_0)}$$

resists a closed form solution for α. Consider, instead, transforming the data via the natural logarithm. In this case, $\ln(f(t)) = \ln(y_0) + \alpha(t - t_0)$. Set $\Delta t = (t - t_0)$, $\Delta z = (z - z_0)$, and $z_k = \ln(y_k)$ for $k = 0, 1, 2, \ldots, n$. Then the transformed error functional becomes $E_{\ln}[\alpha] = \sum_{k=0}^{n} [\Delta z_k - \alpha \cdot \Delta t_k]^2$. To minimize the error with respect to the exponent, differentiate with respect to α, and set the corresponding derivative equal to zero.

$$\frac{dE_{\ln}}{d\alpha}[\alpha] = -2 \sum_{k=0}^{n} [\Delta z_k - \alpha \cdot \Delta t_k] \cdot \Delta t_k \tag{0.4a}$$

$$\frac{dE_{\ln}}{d\alpha}[\alpha] = 0 \Rightarrow \alpha = \frac{\sum_{k=0}^{n} \Delta z_k \cdot \Delta t_k}{\sum_{k=0}^{n} (\Delta t_k)^2} \tag{0.4b}$$

For the data of Table 0.1, the value of the exponent is $\alpha = 0.102$ and the value of the error functional is $E_1[0.102] = 23.27$ with a corresponding *RMS* error of 1.11. These errors are larger than those of the quadratic function. Figure 0.2b shows the fit via the **solid curve**.

Warning: The critical value of α attained via (0.4b) minimizes the error functional $E_{\ln}[\alpha]$ *in natural logarithm space*. It does not necessarily minimize $E[\alpha]$ in *linear space*. To find such a value, use the MATLAB commands below to see that $a \approx 0.106$ minimizes $E[\alpha]$. In this case, the value of $E_2[0.106] = 22.59 < 23.27 = E_1[0.102]$ as determined via equation (0.4b). Figure 0.2b shows this corrected fit via the **dashed curve**. For the exponential model, the parameter vector a corresponds to the values of $a_0 = y_0$ and the estimated exponent α.

MATLAB Commands
(Figure 0.2a)

```
% Evaluate the second-order polynomial at the times t
F = polyval(a,t);
% Compute the least squares error of the difference in-between the polynomial and data y
E = sum( (F-y).^2 )
E =
    21.6916
% Calculate the RMS error as well
RMS = sqrt( (sum( (F-y).^2 )/numel(y) ) )
RMS =
    1.0685
```

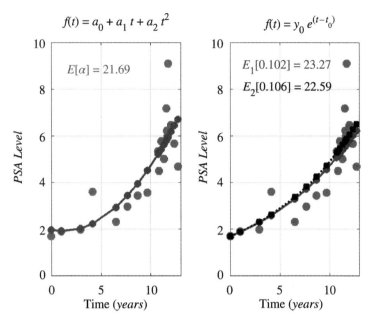

$$f(t) = a_0 + a_1 t + a_2 t^2$$

$$f(t) = y_0 e^{(t - t_0)}$$

$E[\alpha] = 21.69$

$E_1[0.102] = 23.27$

$E_2[0.106] = 22.59$

PSA Level

PSA Level

Time (*years*)

Time (*years*)

FIGURE 0.2   (a) Quadratic curve. (b) Exponential curves.

MATLAB Commands
(Figure 0.2b)

% Compute the values of the transform data $z = \ln(y)$ and the time and data centered at t_0 and z_0, respectively, $\Delta t = (t - t_0)$, $\Delta z = (z - z_0)$

```
Dt = t - t(1); z = log(y); Dz = z - z(1);
```
% Calculate the "optimal" α via equation (0.4b)
```
alpha1 = sum(Dz.*Dt)/sum(Dt.^2);
alpha1 =
      0.1020
```
% Form the function $f(t) = y_0 \exp(\alpha(t - t_0))$ with the value of α_1 above
```
F1 = y(1)*exp(alpha1*Dt);
```
% Find the corresponding *least squares* ...
```
E1 = sum( (F1 - y).^2 );
E1 =
      23.2671
```
% ... and *RMS* errors
```
RMS1 = sqrt( E1/numel(y) )
RMS1 =
      1.1066
```
% Create an in-line error function $E_2[\alpha] = \Sigma_k (y_0 e^{\alpha(t_k - t_0)} - y_k)^2$
```
Dt = t - t(1);
E2 = @(alpha)( sum((y(1)*exp(alpha*Dt) - y).^2) );
```
% Find the value of α that leads to the minimum of $E_2[\alpha]$
% Assume an initial estimate of $\alpha_0 = 0.1$
```
ao = 0.1;
Ao = fminsearch(E2,ao);
```

```
Ao =
    0.1055
% Note the value of the least squares error functional at the minimizing value α₀ = 0.1055
E2(Ao) =
   22.5877
% ... and corresponding RMS error
RMS2 = sqrt( E2(Ao)/numel(y) )
RMS2 =
    1.0903
```

Equations (0.4a) and (0.4b) yield an estimate of the key parameter α for the exponential curve fit $f(t) = a_0 e^{\alpha(t-t_0)}$ by transforming the equation via the natural logarithm ln. An alternate method of achieving such a fit is to *first* transform the data via ln, fit the data in the transformed space, and finally map the fit back into the original space by inverting the transformation. The transformed data can then be fit with a *line*. This can be accomplished via *linear regression*, a common technique used in many data fitting applications. The basic idea of linear regression is that a line $\mathcal{L}(t) = a_0 + a_1 t$ models the data. The parameters $a = [a_0, a_1]^T$ are estimated from the measured information $t = [t_0, t_1, ..., t_n]^T$ and $y = [y_0, y_1, ..., y_n]^T$. This is often written as the paired measurements $\{(t_0, y_0), (t_1, y_1), ..., (t_n, y_n)\}$. In minimizing the error functional

$$E[a] = \sum_{k=0}^{n} [y_k - a_0 - a_1 \cdot t_k]^2 \tag{0.5a}$$

over a, it can be shown (see, e.g., Hastie, Tibshirani, and Friedman [3]) that the parameters are estimated as

$$\left.\begin{aligned}\hat{a}_1 &= \frac{\displaystyle\sum_{k=0}^{n} (t_k - \bar{t})(y_k - \bar{y})}{\displaystyle\sum_{k=0}^{n} (t_k - \bar{t})^2} \\[2mm] \hat{a}_0 &= \bar{y} - \hat{a}_1 \cdot \bar{t}\end{aligned}\right\} \tag{0.5b}$$

$$\bar{t} = \frac{1}{n}\sum_{k=0}^{n} t_k, \quad \bar{y} = \frac{1}{n}\sum_{k=0}^{n} y_k \tag{0.5c}$$

This constitutes a least squares solution of (0.5a) for $a = [a_0, a_1]^T$. It should be observed, however, that *simple linear regression* assumes that modeling errors are compensated only in the regression coefficients a. More specifically, the errors occur only with respect to the x-axis (which, in this case, corresponds to the time measurements $t_0, t_1, ..., t_n$).

Deming regression takes into account errors in both the x-axis $(t_0, t_1, \ldots, t_n)$ *and the* y-axis $(y_0, y_1, \ldots, y_n)$. It comprises a *maximum likelihood* estimate for the coefficients $\mathbf{a}$ and is summarized in the formulae below.

$$
\left.
\begin{aligned}
\hat{a}_1 &= \frac{s_y^2 - s_t^2 + \sqrt{\left(s_y^2 - s_t^2\right)^2 + 4 s_{t,y}^2}}{2\, s_{t,y}} \\[2mm]
\hat{a}_0 &= \bar{y} - \hat{a}_1 \cdot \bar{t}
\end{aligned}
\right\}
\tag{0.6a}
$$

$$
s_y^2 = \frac{1}{n}\sum_{k=0}^{n}(y_k - \bar{y}), \; s_t^2 = \frac{1}{n}\sum_{k=0}^{n}(t_k - \bar{t}), \; s_{t,y} = \frac{1}{n}\sum_{k=0}^{n}(t_k - \bar{t})(y_k - \bar{y}) \tag{0.6b}
$$

Note: Since there are $n + 1$ measurements, the factor normalizing the sample variance s_y^2 and sample covariance $s_{t,y}$ is $1/n$. If only n measurements were available (i.e., $\{(t_1, y_1), \ldots, (t_n, y_n)\}$), then the normalizing factor would be $1/(n-1)$.

Figures 0.3a and 0.3b show the result of computing a Deming regression fit $L(t)$ to the transformed data $\{(t_0, \ln(y_0)), (t_1, \ln(y_1)), \ldots, (t_n, \ln(y_n))\}$ and then mapping the line back into *linear space* via the inverse transformation $f(t, \mathbf{a}) = \exp(L(t)) = A_0\, e^{\hat{a}_1(t-t_0)}$ where $A_0 = e^{\hat{a}_0}$. As can be seen from the figure, the error in the transformed logarithm space is 0.83, whereas the error in linear space is 22.89. As before, a "good" linear fit in logarithm space does not guarantee the best fit in the original space.

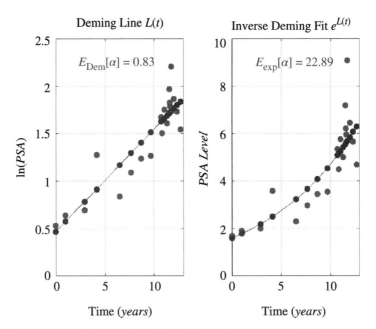

FIGURE 0.3   Data and model in (a) *logarithm space* and (b) *linear space*.

MATLAB Commands
(Figure 0.3a)

```
% Map the PSA scores into ln space
z = log(y);
% Set the significance level a and find the Deming fit in ln space
alpha = 0.05;
S = deming(t,z,alpha);
S =
          m: 0.1079
          b: 0.4689
        rho: 0.8959
        CIm: [0.0815 0.1343]
        CIb: [0.2347 0.7031]
      CIrho: [0.7448 0.9596]
% Create the Deming line and compute the corresponding least squares ...
L = S.m*t + S.b;
ELd = sum( (z - L).^2 );
ELd =
      0.8329
% ... and RMS errors
Lrms = sqrt( ELd/numel(z) );
Lrms =
      0.2093
% Map the Deming fit from ln space back into linear space via the exponential function
eL = exp(L);
% Compute the least squares ...
Ed = sum( (y - eL).^2 );
Ed =
     22.8850
% ... and corresponding RMS errors
Drms = sqrt( Ed/numel(y) )
Drms =
      1.0975
```

Again, a visual inspection of any of the curves exhibited in Figures 0.2a, 0.2b, and 0.3b reveals that the maximum value of the *PSA* scores is *not* well fit. Also, the right end points of the models tend to be lower than the data values. Thus, a better model is required. Now is the time to introduce the *biology* of *prostate-specific antigens* so that the model reflects the underlying science.

There are three issues to consider.

(i) What are the maximum and *most recent PSA levels*? Since these readings *can* dictate medical treatment, it is a crucial portion of the model to estimate accurately.

(ii) What is the value of the first measured *PSA* level? This is important to set as a baseline. For example, if a 25-year-old man had an initial *PSA* reading of 5.3, this would not necessarily indicate prostate cancer. The standard of care will require a biopsy. The absence of cancer upon a pathologist's review will indicate that the patient has an abnormally high *PSA* level. If the level does not increase significantly over time and if there is no family history of prostate cancer, then continued monitoring alone could be sufficient.

(iii) What is the rate of increase in *PSA* level in-between measurement periods? A large *derivative* in *PSA* level can indicate rapid tumor growth.

With these considerations in mind, a more complicated model is proposed.

Curve 3 Combined quadratic and skewed Gaussian $f(t; a; b) = f_2(t; a) \cdot \mathcal{I}_{[t_0, t_q]}(t) + F_{skew}(t; b) \cdot \mathcal{I}_{[t_q, t_n]}(t)$

This model is used to satisfy conditions (i) and (ii). Here, as with *Curve 1*, $f_2(t; a) = a_0 + a_1 t + a_2 t^2$ is a quadratic that fits select points (t_0, y_0), (t_m, y_m), and (t_q, y_q), $q < n$. The *indicator function* $\mathcal{I}_{[ta, tb]}(t)$ is 1 on the interval $[t_a, t_b]$ and 0 otherwise.

$$\mathcal{I}_{[t_a, t_b]}(t) = \begin{cases} 1 & \text{for } t_a \leq t \leq t_b \\ 0 & \text{otherwise} \end{cases} \tag{0.7}$$

The points $(t_0, y_0) = (0, 1.7)$, $(t_m, y_m) = (6.5, 2.31)$, and $(t_q, y_q) = (10.86, 4.51)$ are selected as they appear to most sensibly form a parabola. There is nothing inherently obvious about the choice of a second-order polynomial modeling this portion of the data. Rather, this choice reflects the *art* of applied mathematical modeling. The coefficients $a = [a_0, a_1, a_2]$ that fit the points (t_0, y_0), (t_m, y_m), and (t_q, y_q) to the quadratic $f_2(t; a)$ can readily be computed via the matrix (0.1b) with elements t_0, t_m, and t_q:

$$V_2([t_0, t_m, t_q]) = \begin{bmatrix} 1 & t_0 & t_0^2 \\ 1 & t_m & t_m^2 \\ 1 & t_q & t_q^2 \end{bmatrix}. \text{ Solving } a = V_2^{-1} y \text{ with } y = [y_0, y_m, y_q]^{\mathrm{T}} \text{ yields the}$$

values $a = [1.7, -0.154, 0.038]$.

The second portion of the curve $F_{skew}(t; b)$ is the *translated skewed Gaussian* function defined via (0.8).

$$\left. \begin{aligned} F_{skew}(t; b) &= b_0 + \Phi_{skew}(t; b) \\ \Phi_{skew}(t; b) &= \frac{1}{b_1} \phi\left(\frac{t - t_m}{b_1}\right) \cdot \left(1 + erf\left(b_2 \left[\frac{t - t_m}{b_1}\right]\right)\right) \\ \phi(z) &= \frac{1}{\sqrt{2\pi}} e^{-\frac{1}{2}z^2}, \quad erf(z) = \frac{2}{\sqrt{\pi}} \int_0^z e^{-x^2} \, dx \end{aligned} \right\} \tag{0.8}$$

Here, ϕ is the *standard normal probability density function*, *erf* is the *error function*, and $\Phi_{skew}(t; b)$ is the *skewed Gaussian*. The skewed Gaussian is implemented in this setting since it is, by construction, asymmetrical with respect to a peak or maximum value. For the "non-quadratic" portion of the *PSA* measurements (i.e., for $t > t_q$), there is neither evidence nor reason to suggest readings should be symmetric about the maximum y_M. Some numerical experimentation suggests that $b = [4.5, 0.29, 0.1]$

makes for a reasonable fit to the data over the time range $t \in (t_q, t_n]$. Figure 0.4 illustrates the model against the data.

Observe that the least squares error $E_{q,s}$ of the quadratic-skewed Gaussian function with respect to the parameter vectors $\boldsymbol{a} = [a_0, a_1, \alpha]^T = [1.7, -0.154, 0.038]^T$ and $\boldsymbol{b} = [4.5, 0.29, 0.1]^T$ is 22.4 with corresponding *RMS* error of 1.085. This is comparable to the least squares errors of the quadratic, exponential, or Deming regression curves previously examined.

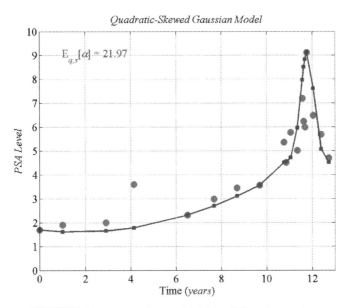

FIGURE 0.4   Data and *quadratic-skewed Gaussian* model.

MATLAB Commands
(Figure 0.4)

```
% Find the index of the maximum PSA scores ...
,~,ii] = max(y);
% ... and note the time and value of the maximum PSA
tM = t(ii); yM = y(ii);
% Mark the index of the right end of the quadratic fit ...
k = 8;
% ... along with the associated time and PSA value
tq = t(k); yq = y(k);
% Initial and terminate times and PSA values
to = t(1); tn = t(end); yo = y(1); yn = y(end);
% Midpoint of the quadratic function
tm = t(5); ym = y(5);
% Time range
N = 100; T = linspace(to,tn,N);
```

% Define the *skewed Gaussian* $F_{skew}(t; \boldsymbol{b}) = \phi((t - t_m)/b_2)(1 - \text{erf}(b_3(t - t_m)/\sqrt{2b_2})/b_2$

```
Fskew = ...
  @(t,b) (normpdf(t,tM,b(2)).*(1+erf(b(3)*(t-tM)/(sqrt(2)*b(2))))/b(2) );
```
% Skewed Gaussian coefficient values $\boldsymbol{b} = [b_1, b_2, b_3]$
```
b(2) = 1/3.4; b(3) = 0.1; b(1) = yM - Fskew(tM,b);
```
% Indicator function $\mathcal{I}_{[ta, tb]}(t) = 1$ for $t_a \leq t \leq t_b$ and 0 otherwise
```
Ind = @(t,ta,tb) (ta <= t & t <= tb);
```
% Quadratic function $f_2(t; \boldsymbol{a})$ which fits (t_0, y_0), (t_5, y_5), and (t_q, y_q)
```
f2 = @(t,a) (a(1) + a(2)*t + a(3)*t.^2);
```
% Compute the coefficients via the *Vandermonde* matrix
```
A = fliplr(vander([to,tm,tq]));
a = inv(A)*[yo;ym;yq];
```

% Combined quadratic skewed Gaussian function
```
f = @(t,a,b) (f2(t,a).*Ind(t,to,tq) + (b(1)+Fskew(t,b)).*Ind(t,t(k+1),tn));
```
% Compute the *least squares* ...
```
Eqs = sum( (f(t,a,b) - y).^2 );
Eqs =
   21.9672
```
% ... and the corresponding *RMS errors*
```
LSrms = sqrt( Eqs/numel(y) );
LSrms =
    1.0753
```
% Parameter values $\boldsymbol{a} = [a_0, a_1, a_2]$ and $\boldsymbol{b} = [b_0, b_1, b_2]$
```
a =
    1.7000    -0.1068     0.0308
b =
    4.4982     0.2941     0.1000
```

Though the *quadratic-exponential* model gives a comparable least squares error and an improved fit at the endpoints with respect to the aforementioned models, it does not cover the variations in the data. Indeed, from 19 August 2011 onward, the data appear to oscillate. This may be due to the device on which the blood samples were measured. It may also be due to the nature of the *prostate-specific antigen* itself (see Walsh [6]). Therefore, the introduction of a model error to compensate for the vacillating measurements is realistic. This is achieved by adding a variability of 1.5% of the value to each measurement. Figure 0.5 illustrates the consequence of this supplement to the *quadratic-skewed Gaussian* model. As can be seen, the envelope (denoted by the dotted lines) now includes most of the *PSA* values. Also observe that the measurement at time $t_3 = 6$ January 2005 is excluded (Feldman [2]) as it was viewed as an outlier due to normal prostate activity.

A summary of the models, their estimated parameters, least squares, and *RMS* errors is provided in Table 0.2. While the most naïve model, the quadratic $f(t) = a_0 + a_1 t + a_2 t^2$, has the smallest least squares and *RMS* errors, it is also the least satisfying from a biological perspective. Would any physician believe that an untreated patient's *PSA* level would increase as the square of time? This would be unlikely. The most intuitively appealing model is the combined quadratic-skewed Gaussian with error envelope, $f(t; \boldsymbol{b}) = F_{skew}(t; \boldsymbol{b}) \cdot (1 + \varepsilon \cdot (1 - t))$ where $\varepsilon \in (0, 1)$ is the expected error in *PSA* measurement. This curve, illustrated in Figure 0.5 with an error of 1.5% or

TABLE 0.2   Summary of models and fitting errors

Model	Parameter estimates	Least squares error	RMS error
Quadratic	$a = [0.05, -0.25, 2.19]$	21.69	1.068
Exponential	$\alpha = 0.102$	23.27	1.107
	$\alpha = 0.1055$	22.29	1.09
Linear fit after $z = \ln(y)$	$m = 0.108$	22.89	1.0975
transformation $\Leftrightarrow y = \exp(z)$	$b = 0.47$		
Quadratic + Skewed Gaussian	$a = [1.7, -0.15, 0.04]$	22.4	1.085
	$b = [4.5, 0.29, 0.1]$		

$\varepsilon = 0.015$, mimics the measurements and introduces the most realistic aspect of any of the presented models. Most notably, this model captures the nature of *PSA* levels within a given patient as intrinsically *variable*.

Other models, naturally, can be developed. One variation on the *quadratic-skewed Gaussian* model would be to remove the constraint that the *maximum* value of the *PSA* range fit *exactly*. This idea and any others which come to mind are left for the interested reader to investigate.

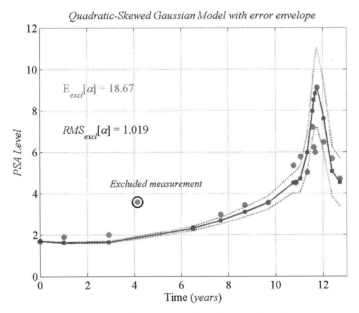

FIGURE 0.5   Model with uncertainty band and excluded measurement.

MATLAB Commands
(Figure 0.5)

```
% Note the time point which should be excluded
ie = strcmp('6-Jan-2005',dates);
% Remove the unwanted date ...
Dates = dates; Dates(ie) = [];
% ... and corresponding datum
Y = y; Y(ie) = [];
clear n
% Convert dates to date strings (seconds)
n = cellfun(@datenum,Dates);
% Convert date strings to years
T = (n - n(1))/365.25;
```
% Define the indicator function $\mathcal{I}_{[ta,\,tb]}(t) = 1$ for $t_a \le t \le t_b$ and 0 otherwise
```
Ind = @(t,ta,tb) (ta <= t & t <= tb);
```
% Combined quadratic skewed Gaussian function (with one point removed)
```
f = @(t,a,b) (f2(t,a).*Ind(t,to,tq) + (b(1)+Fskew(t,b)).*Ind
(t,t(k),tn));
```
% Calculate the values of the *quadratic-skewed Gaussian* curve model over the filtered time
```
F = f(T,a,b);
```
% Add in the 1.5% per year uncertainty/variability factor into the *PSA* model
```
Nf = numel(F);
In = 0:0.015:0.015*(Nf-1);
```
% Upper and lower bounds on the model
```
Fup = F.*(1 + In); Flow = F.*(1 - In);
```
% *Least squares* ...
```
E = sum( (Y - F).^2 );
E =
    18.6738
```
% ... *RMS errors*
```
RMS = sqrt( sum( (Y - F).^2 )/numel(Y) );
RMS =
    1.0185
```

EXERCISES

0.1 Develop a mathematical model in which the *maximum PSA* value is *not* fit *exactly*. What are the consequences of this model?

0.2 Refine the *quadratic-skewed Gaussian* model so that the quadratic fit on $t_0 \le t \le t_q$ is *not* restricted to any pre-selected points. Moreover, use non-linear optimization methods to find the parameters $\boldsymbol{b} = [b_0, b_1, b_2]$ for the *translated skewed Gaussian* that minimize the *least squares error* on $t_q < t \le t_n$.

0.3 Develop a mathematical model in which the *PSA* values level off and are asymptotically constant. This can occur if the tumor is *indolent* (that is, does not grow significantly, has low malignancy Gleason score, and does not spread outside of the prostate capsule). *Hint*: Consider an affine transformation of $f(x) = tanh(x)$.

REFERENCES

[1] Demmel, J. W., *Applied Numerical Linear Algebra*, SIAM Books, Philadelphia, PA, 1997.

[2] Feldman, A., *Private Communication*, March 2012, Department of Urology, Massachusetts General Hospital, Boston, MA.

[3] Hastie, T., R. Tibshirani, and J. Friedman, *The Elements of Statistical Learning, Data Mining, Inference, and Prediction*, Second Edition, Springer, New York, 2009.

[4] Moler, C. B, *Numerical Computing with MATLAB*, SIAM Books, Philadelphia, PA, 2004.

[5] Trefethen, L. N., and D. Bau III, *Numerical Linear Algebra*, SIAM Books, Philadelphia, PA, 1997.

[6] Walsh, P. C., *Dr. Patrick Walsh's Guide to Surviving Prostate Cancer*, Hachette Book Group, New York, 2012.

1

DATA

Arguably, the most important role of the applied mathematical scientist is to model and analyze data. While this statement may appear to be obvious, just how to proceed given any set of data remains as much art as it does science. Should the data be calibrated, normalized, transformed, smoothed, filtered, separated into subclasses? These are just a few of the approaches one can take to data before beginning any substantial analysis. Therefore, this chapter will be dedicated to a select few methods: Data visualization, data transformation, data filtering, data clustering (not to be confused with clustered data of Chapter 4), and data quality.

1.1 DATA VISUALIZATION

Yogi Berra, the Baseball Hall of Fame catcher for the New York Yankees 1946–1965 and accidental metaphysician once remarked [2; https://en.wikipedia.org/wiki/Yogi_Berra]

You can observe a lot by watching.

Fanciful or not, this comment takes to heart the power of modern computing and the MATLAB software system in particular. In the Glossary of MATLAB Functions written for this text, the reader will find a number of specialized visualization M-files designed to plot data that have been processed or transformed. This includes

Applied Mathematics for the Analysis of Biomedical Data: Models, Methods, and MATLAB®, First Edition. Peter J. Costa.
© 2017 Peter J. Costa. Published 2017 by John Wiley & Sons, Inc.
Companion website: www.wiley.com/go/costa/appmaths_biomedical_data

principal component axes and *discriminant analysis feature extraction coordinates* for multivariate data. For single variable matched-pairs data, *Bland–Altman plots* of *allowable total difference zones* can be constructed.

More practically, the question this section will focus on is "What can be inferred or deduced from these data?"

Consider a question raised in Chapter 5. That is, height, weight, and age data for every 2014–2015 season active National Basketball Association (NBA) and National Hockey League (NHL) player are available from www.nba.com and www.nhl.com, respectively. For the sake of convenience, these data are summarized in two separate MAT files contained in the software associated with this text. How do the heights of the players from both leagues compare? Figures 5.1a and 5.1b of Chapter 5 provide the weighted histograms for the player height data. This is certainly one way to examine these data. Alternately, a point-by-point plot of the data along with a means of each collection of measurements is presented in Figure 1.1. The dots (•) on the upper graph represent the heights of the active 2014–2015 NHL players and the thick line (——) through the center of these data is the average height (indicated by the symbol μ_{NHL}). In a similar manner, the dots (•) on the lower portion of the graph represent the active 2014–2015 NBA players with average height (——) denoted by μ_{NBA}. Instinctively, the sense of the figure is that (on average) NBA players are taller than NHL players. This is verified in Chapter 5 on hypothesis testing.

While the *average* player heights are statistically different (see Table 5.1 of Chapter 5), the question of *classification* remains. More specifically, from the triple of height, weight, and age, can such information be used to determine whether a particular player measurement (*Ht, Wt, Age*) indicates to which *class* (namely, NBA or

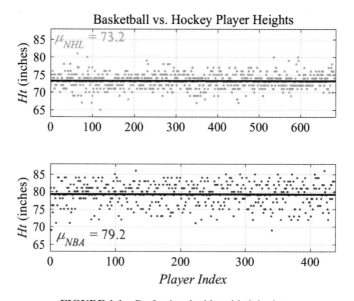

FIGURE 1.1 Professional athletes' height data.

Basketball vs. Hockey Players

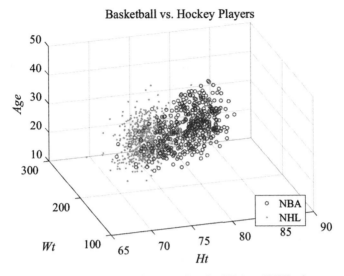

FIGURE 1.2   Height, weight, age data for NBA and NHL players.

NHL) the player belongs? The answer is discussed in Chapter 4 on classification. The raw data, obtained from www.nhl.com and www.nba.com, for the active players in the 2014–2015 season can be viewed in Figure 1.2. What is the best approach to be used in separating these groups? This will be the concern of the next section.

EXERCISES

1.1   Make plots of the NBA vs. NHL weight data. Do the same for the *age* data. Are the differences plain from the graphs? These data are located in C:\Database\ Math_Biology\Chapter_1\Data as BasketballData.mat and Hockey Data.mat. They are obtained via the MATLAB commands

```
Ddir = 'C:\Database\Math_Biology\Chapter_1\Data';
load(fullfile(Ddir,'BasketballData.mat'));
load(fullfile(Ddir,'HockeyData.mat'));
```

   Hint: The weight data are B.Wt and D.Wt while the age data are B.Ages and D.Ages.

1.2   DATA TRANSFORMATIONS

Once data have been recorded, how should they be treated so that the *content* of the information contained therein is most plainly revealed? This question is unanswerable, as it presupposes there is any discriminatory information within the measurements. There are, however, standard approaches that can be applied to make the data more regular. These methods are listed sequentially.

1.2.1   Normalization

The basic idea behind *normalization* is to collect a representative data set and then, from each measurement within the set, subtract the mean and divide by the standard deviation. If the data are multidimensional, this can be achieved by *subtracting off* the mean vector and multiplying by the inverse of the associated covariance matrix. This notion is made precise in Chapter 4, equation (4.5). As a review, suppose that $X \in \mathcal{M}at_{n \times p}(\mathbb{R})$ is the data matrix of n p-dimensional measurements from the same source. For the NHL (*Ht*, *Wt*, *Age*) data, this means that $n = 684$ and $p = 3$. The *standard normalization* of the data matrix X is

$$Z = (X - \mathbf{1}_{n \times 1} \cdot \boldsymbol{m}) \cdot S^{-1}. \tag{1.1}$$

Here $\mathbf{1}_{n \times 1} = [1, 1, \ldots, 1]^T$ is the n-dimensional *column vector* of 1's, $\boldsymbol{m} = [m_1, m_2, \ldots, m_p]$ is the p-dimensional row vector whose every element is the column mean of the data matrix X. Finally, $S = diag(s_1, s_2, \ldots, s_p)$ is the $p \times p$ diagonal matrix whose nonzero entries are the column standard deviations of X (e.g., see Johnson and Wichern [8] for details about normalization). Applying this transformation to the data matrices for the (*Ht*, *Wt*, *Age*) NBA and NHL triplets results in the normalized data displayed in Figure 1.3.

The reader can see that the normalized data provide a comparable display to the unnormalized data.

Rather than simply normalizing the data, projecting these measurements into coordinates that amplify differences is now prescribed.

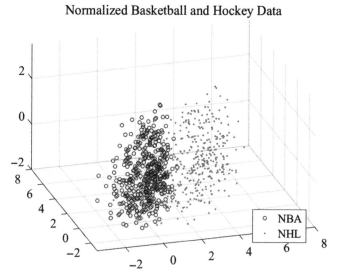

FIGURE 1.3   Normalized NBA and NHL player (*Ht*, *Wt*, *Age*) data.

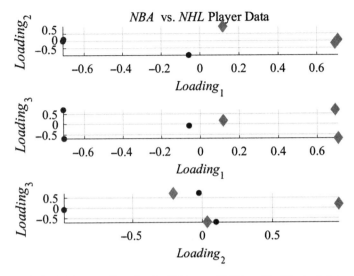

FIGURE 1.4 Loadings for the NBA • and NHL ◆ (*Ht, Wt, Age*) data.

1.2.2 Principal Components (Karhunen–Loève Transform)

In Section 4.3 (Chapter 4), the notion of principal components as an orthogonal set of axes along the descending amounts of variation in a data matrix X is detailed. In particular, equations (4.9a)–(4.10d) present the principal component axes and projection into *PC space*. The projection, defined in Chapter 4, equation (4.10d) is sometimes referred to as the *Karhunen–Loève* transform. As noted in Chapter 4, the projection matrix is a product of the first ρ singular values[1] of the data matrix X and the associated loading matrix. Here, ρ is the desired number of principal components (e.g., the number of *PCs* required to attain say 99.9% of the total data variation). Figure 1.4 illustrates the loadings for the NBA/NHL (*Ht, Wt, Age*) data analysis.

To see *how many PCs* are required to meet a desired percentage of the total variance, a *screeplot* can be created. This is a depiction of the percentage of the total variance captured by the number of principal components used as a function of the normalized eigenvalues.

The data visualized in this graphic are taken from a four class, multivariate collection of measurements in which one level of a disease is normal (no-disease) and the remaining three are of increasingly serious levels. The screeplot in Figure 1.5 gives the following information. For levels 0–2, it is seen that 5–6 *PCs* provide the vast majority of the variance information. At level 3, there are only 10 non-trivial *PCs* with the bulk of the information contained in the first 5 *PCs*. The data have been normalized as per (1.1).

---

[1] For more information about linear algebra and singular values, see Appendix Section A.2.

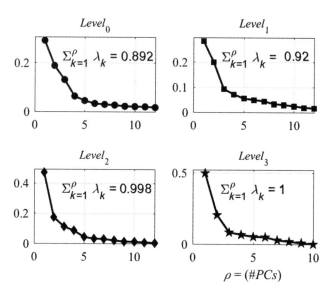

FIGURE 1.5 Screeplot for up to 12 *PCs* for multivariate data.

1.2.3 DAFE Coordinates

Discriminant analysis feature extraction (DAFE) coordinates are independent axes along the direction of maximal discriminant information. The DAFE axes projection mapping Π_r into the first r *DAFE coordinates* is composed of the first r columns of the orthogonal matrix from the singular value decomposition of the Fisher discriminant matrix. This matrix $F = C_{pool}^{-1} \cdot C_{btwn}$ is described in equation (4.13b) of Section 4.4 (Chapter 4). The projection matrix Π_r is defined via equation (4.10d) of Section 4.3 (Chapter 4). Figures 4.10 and 4.11 of Chapter 4 illustrate DAFE coordinates and the corresponding weightings on each class projection.

The exercises will give the reader some experience in reduction of dimension via DAFE coordinates and how such axes act to separate data classes.

EXERCISES

1.2 Figure 1.4 illustrates the PCA loadings for the NBA vs. NHL (*Ht, Wt, Age*) data. Use the M-file `scorecompare.m` to plot the PCA scores for these data. These data are obtained via the MATLAB commands

```
Ddir = 'C:\Database\Math_Biology\Chapter_1\Data';
load(fullfile(Ddir, 'BasketballData.mat'));
load(fullfile(Ddir, 'HockeyData.mat'));
```

1.3 How many DAFE coordinates are required to recover 99.9% of the discriminant information contained in the NBA vs. NHL (*Ht, Wt, Age*) data? Use the M-files `dafe.m` and `dafeplot.m` and the commands in Exercise 1.2 to project the data into DAFE space and then plot it, respectively.

1.3   DATA FILTERING

Data are by their very nature noisy. Indeed, data are measurements that are recorded either via device or human beings. No machine can perfectly register a series of measurements without some manner of error. And we humans are legendary for our inability to accurately record and repeat measurements. It is this portion of our humanity that motivates the design and development of machines to perform repetitive tasks. Consequently, any set of measurements must be viewed as inherently imprecise.

Consider the simulated respiratory infection data from Chapter 3 (Figure 3.10b). While these data follow what appears to be a regular sinusoidal pattern, it is evident that there is plenty of "jitter" in the plot. One approach to filter the data is to smooth the signal. This can be achieved in a number of ways. Two methods will be discussed: Convolution and Fourier transforms.

1.3.1   Convolution and Smoothing

Again, referring to the respiratory infections data of Chapter 3, the question of how to smooth this signal arises. Smoothing by convolution is one approach. If $f(t)$ is an integrable function, then the *convolution* of f with the function g is defined as

$$(f \odot g)(t) = \int_{-\infty}^{\infty} f(s) \cdot g(t - s)\, ds. \tag{1.2}$$

If $g(t)$ is a "square wave" as indicated in Figure 1.6, then by convolving against the function f, the nonzero portion of g smooths those portions of f of equal length

Square Wave

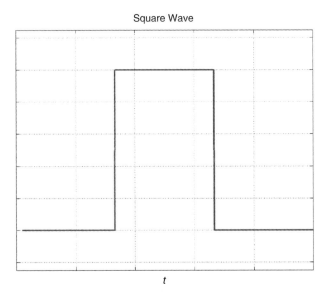

t

FIGURE 1.6   Square wave smoothing function.

to g. Andrews and Shivamoggi [1] and Costa [3] contain discussions on convolution and Fourier transforms (which will also be used to smooth signals).

The numerical algorithm used for smoothing a signal $f(t)$ via a smoothing function $g(t)$ is presented in the next section. For the sake of simplicity, it will be assumed that the nonzero portion of g consists of 2^p points.

1.3.1.1   Convolution Smoothing Algorithm

(i) Convolve f with g via (1.2). Obtain $\phi(t) = (f \odot g)(t)$.

(ii) Select the smoothing function g to have 2^p elements. Remove the first $2^{p-1} - 1$ and the last $2^{p-1} - 1$ elements from the new convolution vector $\phi(t)$.

(iii) Normalize $\phi(t)$ by the number of nonzero points in the smoothing function g.

Figure 1.7 illustrates the results of this smoothing approach. Observe that as the number of elements in the smoothing vector increases so does the amount of smoothing. As the amount of smoothing increases, however, the results exhibit two deficiencies: Loss of amplitude and increase in "ghosting." Ghosting or *aliasing* is the artificial suppression of the smoothed data near the initial and terminal portions of the signal. The first (loss of amplitude) is evident from the plot. At the end of the smoothed functions, moreover, tails trail off to zero. This behavior largely ignores the original data trend. This phenomenon is called *aliasing* (see, e.g., Costa [3] for more details). The MATLAB code used to generate the smoothed functions displayed in Figure 1.7 is presented below.

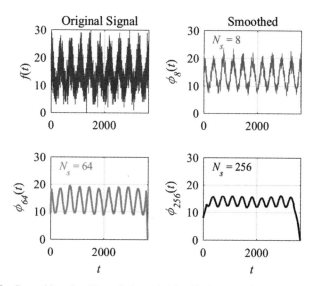

FIGURE 1.7   Smoothing for $N_s = 8$ (top right), 64 (lower left), and 256 (lower right) elements.

MATLAB Commands
(*Smoothing via Convolution*)

```
% Data directory ...
Ddir = 'C:\Database\Math_Biology\Chapter_1\Data';
% ... for the respiratory infection data.
data = load(fullfile(Ddir,'SEIR_Simulated_Data.mat'));
t = load(fullfile(Ddir,'SEIR_Time.mat'));
% Place the data and time into MATLAB vectors
D = data.D; T = t.T; clear data t
% Number of measurements (infected patients)
N = numel(D);
% Smoothing vectors of various lengths
s{1} = ones(1,2^3); s{2} = ones(1,2^6); s{3} = ones(1,2^8);
% Number of elements per smoothing vector
Ns = cellfun(@numel,s);

% Convolve the data with the smoothing vectors s
for j = 1:numel(s);
    % indices to be removed from the right and left portion of the convolution
    iR = [1:(-1+Ns(j)/2),N-(-1+Ns(j)/2):N];
    % Convolve D and s: (D ⊙ s)(t)
    tmp = conv(D,s{j});
    % Remove the points at the undesired indices
    tmp(iR) = [];
    % Normalize the convolution
    Ds{j} = tmp/Ns(j);
end
```

1.3.2   Fourier Transform and Smoothing

Rather than using convolution to smooth a signal, the *Fourier transform* can be utilized. Definitions for the transform and its inverse are provided in equations (1.3a) and (1.3b). The basic idea is that the Fourier transform maps a time (or space)-based function into frequency space. By eliminating high frequencies in the transformed function, the inverse transform will smooth the frequency-truncated data. This is formalized via the algorithm and Figure 1.8.

$$\mathcal{F}[f(t)](\omega) = \frac{1}{\sqrt{2\pi}} \int_{-\infty}^{\infty} f(t)\, e^{-it\omega}\, dt \tag{1.3a}$$

$$\mathcal{F}^{-1}[F(\omega)](t) = \frac{1}{\sqrt{2\pi}} \int_{-\infty}^{\infty} F(\omega)\, e^{it\omega}\, d\omega \tag{1.3b}$$

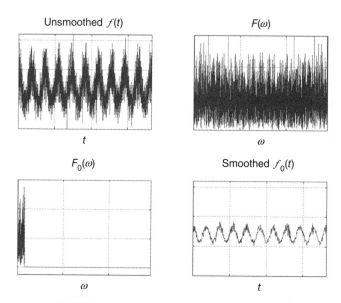

FIGURE 1.8   Smoothing via Fourier transform.

Remarks:

(1) The notation $F(\omega)$ is often used in place of $\mathcal{F}[f(t)](\omega)$. Also, the transform and its inverse are frequently defined in the alternate forms $F(\omega) = \int_{-\infty}^{\infty} f(t)\, e^{-2\pi it\omega}\, dt$ and $f(t) = \int_{-\infty}^{\infty} F(\omega)\, e^{2\pi it\omega}\, d\omega$.

(2) The *Heaviside* function is the distribution whose values are 1 on the positive real line. That is, $H(x) = \begin{cases} 1 & \text{for } x \geq 0 \\ 0 & \text{otherwise} \end{cases}$. Thus, $S(x, 1) = H(x) - H(x - 1)$ is a *square wave*. More generally, $S(x, x_0) = H(x) - H(x - x_0)$ is the rectangular wave defined over $[0, x_0]$ and $S(x, [x_a, x_b]) = H(x - x_a) - H(x - x_b)$ is defined over $[x_a, x_b]$.

1.3.2.1   Smoothing Algorithm Fourier Transform

(i) Apply the Fourier transform to the function $f(t)$ and obtain $F(\omega)$.

(ii) Multiply $F(\omega)$ by $S(\omega, \omega_0)$ to eliminate any frequencies greater than ω_0. $F_0(\omega) = F(\omega) \cdot S(\omega, \omega_0)$.

(iii) Apply the inverse Fourier transform to $F_0(\omega)$.

(iv) Obtain the smoothed function $f_0(t) = \mathcal{F}^{-1}[F_0(\omega)](t)$.

These steps are illustrated in Figure 1.8 as upper left (step i), upper right (step ii), lower left (step iii), and lower right (step iv). Since the function $f(t)$ is obtained as the vector of measurements $f = [f_1, f_2, \ldots, f_n]$, the *discrete Fourier transform* (DFT) is used. The DFT and its inverse are provided via equations (1.4a) and (1.4b).

$$DFT[f] = \left[\sum_{j=1}^{n} f_j, \sum_{j=1}^{n} f_j(e^{-2\pi i \cdot (j-1)}), \sum_{j=1}^{n} f_j(e^{-2\pi i \cdot 2(j-1)}), \cdots, \sum_{j=1}^{n} f_j(e^{-2\pi i \cdot (n-1) \cdot (j-1)}) \right]$$

(1.4a)

$$IDFT[F] = \frac{1}{n} \left[\sum_{j=1}^{n} F_j, \sum_{j=1}^{n} F_j(e^{2\pi i \cdot (j-1)}), \sum_{j=1}^{n} F_j(e^{2\pi i \cdot 2(j-1)}), \cdots, \sum_{j=1}^{n} F_j(e^{2\pi i \cdot (n-1) \cdot (j-1)}) \right]$$

(1.4b)

As with the smoothing via convolution, the greater the amount of smoothing, the smaller the amplitude envelop on the smoothed function. The *Fourier* method does not appear to present the problem of aliasing. The MATLAB code below is used to compute the smoothed vector f_0.

MATLAB Commands
(*Smoothing via Fourier Transform*)

```
% Data directory ...
Ddir = 'C:\Database\Math_Biology\Chapter_1\Data';
% ... for the respiratory infection data.
data = load(fullfile(Ddir,'SEIR_Simulated_Data.mat'));
t = load(fullfile(Ddir,'SEIR_Time.mat'));
% Place the data and time into MATLAB vectors
D = data.D; T = t.T; clear data t
% Number of measurements (infected patients)
N = numel(D);
% Take the (discrete) Fourier of the Respiratory data
F = fft(D);
% Form the square wave by removing all but the first 200 wavelengths.
S = ones(1,N); w = T/(2*pi); wo = w(200);
ind = w > wo; S(ind) = 0;
% Compute the Fourier transform of the truncated frequency data and its inverse
transform
Fo = F.*S; fo = ifft(Fo);
% Plot the results step-by-step
figure;
% Unsmoothed function f(t)
subplot(2,2,1); plot(T,D,'b'); grid('on');
set(gca,'FontSize',16,'FontName','Times New Roman');
title('Unsmoothed \itf\rm(\itt\rm)');
axis([0,T(end),0,max(ceil(D))]);
% Discrete Fourier transform of f(t), F(ω)
```

```
subplot(2,2,2); plot(w,abs(F),'b'); grid('on');
set(gca,'FontSize',16,'FontName','Times New Roman');
title('\itF\rm(\it\omega\rm)'); axis([0,w(end),0,600]);
% Truncated wavelength transform F_0(ω)
subplot(2,2,3); plot(w,abs(Fo),'b'); grid('on');
set(gca,'FontSize',16,'FontName','Times New Roman');
title('\itF_{o}\rm(\it\omega\rm)'); axis([0,w(end),-50,600]);
% Inverse transform of F_0(ω) ⇒ smoothed function f_0(t)
subplot(2,2,4); plot(T,abs(fo),'b'); grid('on');
set(gca,'FontSize',16,'FontName','Times New Roman');
title('Smoothed \itf_{o}\rm(\itt\rm)'); axis([0,T(end),0,
max(ceil(D))]);
```

1.3.3 Outlier Filtering

Smoothing alters *every* element of the data set under consideration. What happens if there are only a few "extraordinary" members of the collection of measurements? How can such exceptional measurements be identified?

The first course of action is to define what "extraordinary" or "exceptional" means. Such measurements are referred to as *outliers*. An *outlier* is a measurement that is markedly different from all others in a sample. The notion of an outlier is simultaneously obvious and difficult to formalize. Indeed, outliers can best be described by a phrase made famous by the US Supreme Court Associate Justice Potter Stewart [12].

> *I shall not today attempt further to define the kinds of material I understand to be embraced within that shorthand description; and perhaps I could never succeed in doing so. But I know it when I see it.*

To enhance the presentation, let $x = \{x_1, x_2,..., x_n\}$ be a sample of a random variable X. One approach to identify sample outliers is to normalize each element of the sample by the mean and standard deviation. That is, set $z_j = \dfrac{x_j - \bar{x}}{s}$ for $j = 1, 2,...,$ n, $\bar{x} =$ the sample mean, and $s =$ the sample standard deviation. The z_j comprise the *normalized sample* and, in limit, follow a standard normal distribution (for more information concerning probability distributions, see Appendix Section A.3). Consequently, measurements at the extreme margins (e.g., the 0.1% or 99.9% quantiles) can be considered *outliers*. If ω is the selected quantile, then $\mathcal{I}(\omega) = [\bar{x} - s \cdot \omega, \bar{x} + s \cdot \omega]$ is the *inclusion interval*. Thus, $x_k \notin \mathcal{I}(\omega)$ means that x_k is an *outlier*. This method can be reviewed in Johnson and Wichern [8] and with limiting values suggested by Tholen et al. [13]. For non-symmetric quantile limits (i.e., the lower limit is a 0.5% quantile while the upper limit is the 99.99% quantile), the inclusion interval can be generalized to $\mathcal{I}(\boldsymbol{\omega}) = [\bar{x} - s \cdot \omega_\ell, \bar{x} + s \cdot \omega_u]$, where $\boldsymbol{\omega} = [\omega_\ell, \omega_u]$ is the vector of the lower and upper quantiles ω_ℓ and ω_u, respectively.

This section will provide a different method for identifying outliers. The details of this development can be viewed in Costa [4].

Rather than normalizing the sample as above, replace the minimum and maximum of the sample by its nearest neighbor. Specifically, let $x_{order} = \{x_{(1)}, x_{(2)}, \ldots, x_{(n)}\}$ be the ordered values of x. That is, $\min\limits_{1 \leq j \leq n}\{x_j\} \equiv x_{(1)} \leq x_{(2)} \leq x_{(3)} \leq \cdots \leq x_{(n)} \equiv \max\limits_{1 \leq j \leq n}\{x_j\}$. Next remove $x_{(1)}$ and $x_{(n)}$ from the sample and replace them by $x_{(2)}$ and $x_{(n-1)}$, respectively. The *truncated* sample $x_T = \{x_{(2)}, x_{(2)}, x_{(3)}, x_{(4)}, \ldots, x_{(n-1)}, x_{(n-1)}\}$ is then used to normalize the sample via the truncated mean $\bar{x}_T$ and standard deviation s_T.

$$z_j = \frac{x_j - \bar{x}_T}{s_T} \tag{1.5a}$$

$$\bar{x}_T = \frac{1}{n}\left(x_{(2)} + \sum_{j=2}^{n-1} x_{(j)} + x_{(n-1)}\right) \tag{1.5b}$$

$$s_T = \sqrt{\frac{1}{n-1}\left((x_{(2)} - \bar{x}_T)^2 + \sum_{j=2}^{n-1}(x_{(j)} - \bar{x}_T)^2 + (x_{(n-1)} - \bar{x}_T)^2\right)} \tag{1.5c}$$

For a selected quantile ω the *truncated inclusion interval* is

$$\vartheta_T(\omega) = [\bar{x}_T - s_T \cdot \omega, \bar{x}_T + s_T \cdot \omega] \tag{1.5d}$$

If $x_k \notin \vartheta_T(\omega)$, then x_k is an outlier with respect to the *truncated outlier filtering method*. To see how the truncated outlier filter contrasts with conventional outlier identification, consider a sample formed by selecting 100 draws from a uniform distribution over $[0, 1]$ and two draws from $\mathfrak{U}[0, 10]$. That is, $x = \{x_1, x_2, x_3, \ldots, x_{102}\}$ where $x_{i_1}, x_{i_2}, \ldots, x_{i_{100}} \overset{i.i.d}{\sim} \mathfrak{U}[0, 1]$ and $x_{i_{101}}, x_{i_{102}} \overset{i.i.d}{\sim} \mathfrak{U}[0, 10]$ for some set of indices $\{i_1, i_2, \ldots, i_{102}\} \subset \{1, 2, \ldots, 102\}$. The conventional outlier detection method normalizes the data with respect to the sample mean and sample standard deviation taken with respect to the entire data set. The inclusion interval is then calculated with respect to a selected quantile ω. The truncated outlier filter uses equations (1.5a)–(1.5d) to calculate the inclusion interval. Figure 1.9 provides an illustration of conventional versus truncated outlier filtering. As can be seen, the conventional method identifies only one of the samples from $\mathfrak{U}[0, 10]$ as an outlier while the truncated method indicates that both samples from $\mathfrak{U}[0, 10]$ are outliers. Moreover, the inclusion interval produced by the conventional method (illustrated along the y-axis) is considerably wider than the truncated filter method.

Among the virtues of the truncated outlier filter is its extensibility to higher dimensions. Indeed, rather than n one-dimensional samples $x = \{x_1, x_2, \ldots, x_n\}$, suppose there are n p-dimensional samples contained in the data matrix X.

$$X = \begin{bmatrix} x_{1,1} & x_{1,2} & \cdots & x_{1,p} \\ x_{2,1} & x_{2,2} & \cdots & x_{2,p} \\ \vdots & \vdots & \ddots & \vdots \\ x_{n,1} & x_{n,2} & \cdots & x_{n,p} \end{bmatrix} \equiv \begin{bmatrix} x_1 \\ x_2 \\ \vdots \\ x_n \end{bmatrix} \tag{1.6}$$

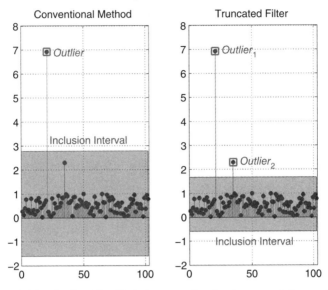

FIGURE 1.9   Outliers identified via conventional and truncated filter methods.

Now, order the data matrix column-wise to obtain

$$
X_{order} = \begin{bmatrix}
x_{(1),1} & x_{(1),2} & \cdots & x_{(1),p} \\
x_{(2),1} & x_{(2),2} & \cdots & x_{(2),p} \\
\vdots & \vdots & \ddots & \vdots \\
x_{(n),1} & x_{(n),2} & \cdots & x_{(n),p}
\end{bmatrix}
\tag{1.7}
$$

where $x_{(1),j} \le x_{(2),j} \le x_{(3),j} \le \cdots \le x_{(n),j}$ are the order statistics of the jth column of X. In each column, the minimum $x_{(1),j}$ is replaced by the second smallest value $x_{(2),j}$. Similarly, the maximum in each column $x_{(n),j}$ is replaced by the penultimate order statistic $x_{(n-1),j}$. The result is the ordered, truncated sample matrix.

$$
X_{T,order} = \begin{bmatrix}
x_{(2),1} & x_{(2),2} & \cdots & x_{(2),p} \\
x_{(2),1} & x_{(2),2} & \cdots & x_{(2),p} \\
x_{(3),1} & x_{(3),2} & \cdots & x_{(3),p} \\
\vdots & \vdots & \ddots & \vdots \\
x_{(k),1} & x_{(k),2} & \cdots & x_{(k),p} \\
\vdots & \vdots & \ddots & \vdots \\
x_{(n-2),1} & x_{(n-2),2} & \cdots & x_{(n-2),p} \\
x_{(n-1),1} & x_{(n-1),2} & \cdots & x_{(n-1),p} \\
x_{(n-1),1} & x_{(n-1),2} & \cdots & x_{(n-1),p}
\end{bmatrix}
\tag{1.8}
$$

From this matrix, compute the truncated sample mean vector $\bar{x}_T$ and the truncated sample standard deviation matrix Σ_T.

$$
\left.
\begin{aligned}
\bar{x}_T &= [\bar{x}_{T,1}, \bar{x}_{T,2}, \ldots, \bar{x}_{T,p}] \in \mathbb{R}^p \\
\bar{x}_{T,\ell} &= \frac{1}{n}\left(x_{(2),\ell} + x_{(n-1),\ell} + \sum_{j=2}^{n-1} x_{(j),\ell}\right), \quad \ell = 1, 2, \ldots, p
\end{aligned}
\right\} \tag{1.9}
$$

$$
\left.
\begin{aligned}
\Sigma_T &= \begin{bmatrix} s_{T,1}^2 & 0 & \cdots & 0 \\ 0 & s_{T,2}^2 & \cdots & 0 \\ \vdots & \vdots & \ddots & \vdots \\ 0 & 0 & \cdots & s_{T,p}^2 \end{bmatrix} \\
s_{T,\ell}^2 &= \frac{1}{n-1}\left((x_{(2),\ell} - \bar{x}_{T,\ell})^2 + (x_{(n-1),\ell} - \bar{x}_{T,\ell})^2 + \sum_{j=2}^{n-1}(x_{(j),\ell} - \bar{x}_{T,\ell})^2\right)
\end{aligned}
\right\} \tag{1.10}
$$

The metric used to determine the utility of a measurement is the *Mahalanobis distance* with respect to a *weighting matrix M*.

$$
d_M^2(x_k, \bar{x}_T) = (x_k - \bar{x}_T) \cdot M^{-1} \cdot (x_k - \bar{x}_T)^T \tag{1.11}
$$

If the number of samples n is sufficiently large (e.g., $n \geq 2 + \frac{1}{2}\, p(p + 1)$), then the sample truncated covariance matrix $\mathrm{cov}(X_T)$ of (1.12) can be used in place of the weighting matrix M in (1.11).

$$
\mathrm{cov}(X_T) = \frac{1}{n-3}(X_T - \mathbf{1}_{n\times 1} \cdot \bar{x}_T)^T (X_T - \mathbf{1}_{n\times 1} \cdot \bar{x}_T) \tag{1.12}
$$

Otherwise, set $M = \Sigma_T$. Observe that the normalization factor in the covariance matrix formula (1.12) is $1/(n-3)$ rather than the usual $1/(n-1)$ since the first and last rows of X_T are duplicates of rows 2 and $n-1$. Thus, rather than n independent measurements, there are only $n-2$.

The Mahalanobis distance is distributed as a χ^2-random variable with p degrees of freedom regardless of choice of weighting matrix. If $\chi_p^2(\gamma)$ is the $\gamma \cdot 100\%$ quantile of a χ_p^2 distribution, then any measurement x_k whose Mahalanobis distance exceeds the value of $\chi_p^2(\gamma)$ is characterized as an outlier. That is, $d_M^2(x_k, \bar{x}_T) > \chi_p^2(\gamma)$ implies that x_k is an outlier. In parallel to the example illustrated in Figure 1.9, an *independent identically distributed (i.i.d.)* sample of 100 taken from $\mathcal{N}(0, 1)$ along with two *i.i.d.* samples taken from $\mathcal{N}(0, 100)$ are collected as the vector x. Similarly, an additional *i.i.d.* sample of 100 taken from $\mathcal{N}(0, 1)$ along with two *i.i.d.* samples taken from $\mathcal{N}(0, 100)$ are collected as the vector y. These data are examined for outliers using the

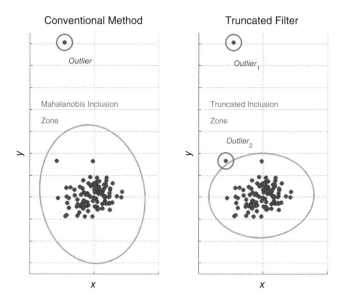

FIGURE 1.10   Two-dimensional outliers identified via conventional and truncated filter methods.

conventional and truncated filter methods. The conventional outlier method calculates the Mahalanobis distance (1.11) using the entire data matrix X to form either the covariance matrix (1.12) or diagonal matrix (1.10). The conventional method is only able to identify one of the two distinct measurements while the truncated filter method identifies both $\mathcal{N}(0, 100)$ samples as outliers. Moreover, the (Mahalanobis) inclusion zone $\mathcal{I}_M(\gamma) = \{x \in \mathbb{R}^P | d_M^2(x, \bar{x}_T) \leq \chi_2^2(\gamma)\}$ is more compact for the truncated filter than it is for the conventional method. In this case, $\gamma = 0.999$. The results are presented in Figure 1.10.

MATLAB Commands
(*Two-Dimensional Outliers*)

```
% Data directory
Ddir = 'C:\Database\Math_Biology\Chapter_1\Data';
% Load the 2–dimensional outlier data
load(fullfile(Ddir,'X_2D_Outlier.mat'));
load(fullfile(Ddir,'Y_2D_Outlier.mat'));
% Compute the data covariance matrix M and mean vector m
A = [x,y]; M = cov(A); m = mean(A);
% χ²–quantile
c = chi2inv(0.999,2);
% Mahalanobis distance
d = mahalanobis(A,m,M);
```

```
% Indices ...
ii = (d.^2 > c);
% ... and list of (conventional) outliers
A(ii,:) =
    -2.2761    14.1121
% Compute the truncated filter outliers
[Aout,iOut,S] = toutlier2(A,c);
A(iOut,:) =
    -2.8610     3.3224
    -2.2761    14.1121
```

EXERCISES

1.4 Use the data contained in the MAT-files X_2D_Outlier.mat and Y_2D_Outlier.mat along with the MATLAB commands contained in the table above to determine the outliers for the two-dimensional data set $[x, y]$ using different quantiles. Do the computations above for $\chi_2^2(\gamma)$ with $\gamma = 0.9$, 0.975, and 0.99. How does this choice of quantile affect the outlier selection for the conventional and truncated filter method? These data are obtained via the MATLAB commands

```
Ddir = 'C:\Database\Math_Biology\Chapter_1\Data';
load(fullfile(Ddir,'X_2D_Outlier.mat'));
load(fullfile(Ddir,'Y_2D_Outlier.mat'));
```

1.4   DATA CLUSTERING

There are collections of measurements from a particular class of objects that can be separated into two or more distinct subclasses. For example, the large (and imprecise) class of *automobiles* can be separated into sedans, station wagons, sport utility vehicles, and crossover vehicles. Each subclass, in turn, can be further refined via brand (e.g., Ford, VW, Toyota, etc.). This class distillation is particularly prominent in biology. Disease states, viral strains, and common food crops are often lumped into large and ill-defined categories. For the purposes of automated identification, more precise and well-defined subclasses can produce better classification.

Thus, this section will be concerned with the mathematical approach used to separate multivariate data of a single class into distinct subclasses.

Suppose a measurement $v_0 \in \mathbb{R}^2$ is compared against two classes *Class*$_1$ and *Class*$_2$ as represented by the data matrices X and Y and illustrated in Figure 1.11. If the *Mahalanobis distance* (see equations (1.11) and (4.4) of Chapter 4) is the metric of proximity with weight matrix M equal to the class covariance, then the vector v_0 is closer to *Class*$_1$ (dots •) than it is to *Class*$_2$ (diamonds ♦) since its Mahalanobis distance to *Class*$_1$ is smaller than the comparable distance to *Class*$_2$. An examination

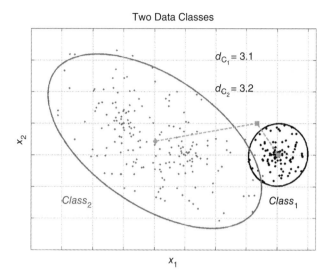

FIGURE 1.11   Mahalanobis distances for two distinct data classes.

of Figure 1.11, however, suggests that $Class_2$ has two concentrations or *clusters* of data within the error ellipse. If a mathematical process known as *data clustering* is applied to that data matrix Y for $Class_2$, it is seen that two distinct subclasses $Class_{2,1}$ (diamonds ◆) and $Class_{2,2}$ (stars ∗) arise. Moreover, these classes form a different partition of the classification space so that v_0 has a smaller Mahalanobis distance to $Class_{2,1}$ than to either $Class_{2,2}$ or $Class_1$. Figure 1.12 demonstrates these remarks.

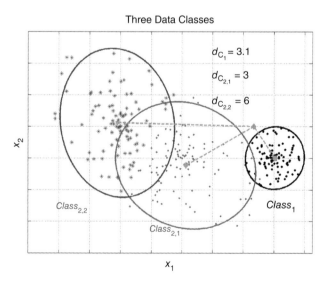

FIGURE 1.12   Mahalanobis distances for three distinct data classes.

What is this process of *data clustering* and how can it be applied to data matrices? The answer to this question is the focus of the remainder of this section.

1.4.1 Distances

The first step is to define the notion of a *distance*. Plainly the usual *Euclidean* distance from the selected point v_0 appears closest to the *Class$_1$* center. For a non-identity weighting matrix M, the Mahalanobis distance d_M can produce a different class assignment. Thus, the selection of a distance metric is critical to the classification process.

Definition 1.1   A *distance d* is a function mapping *p*-dimensional vectors onto the positive real line $\mathbb{R}^+$ so that

   (i) $d : \mathbb{R}^p \otimes \mathbb{R}^p \rightarrow \mathbb{R}^+$

  (ii) $d(x,y) \geq 0$ for all $x, y \in \mathbb{R}^p$ *(non-negativity)*

 (iii) $d(x,y) = 0 \Leftrightarrow x = y$ *(reflexivity)*

 (iv) $d(x,y) = d(y,x)$ *(commutivity)*

  (v) $d(x,y) \leq d(x,z) + d(z,y)$ for any $x, y, z \in \mathbb{R}^p$.   *(triangle inequality)*

The following are examples of popular distances.

Euclidean: $d(x,y) = \sqrt{\sum_{j=1}^{p} (x_j - y_j)^2}$

City Block (also called the *Manhattan* distance): $d(x,y) = \sum_{j=1}^{p} |x_j - y_j|$

Minkowski (or ℓ_q-norm): $d_q(x,y) = \left(\sum_{j=1}^{p} |x_j - y_j|^q \right)^{1/q}$

Mahalanobis: $d_M(x,y) = \sqrt{(x - y) \cdot M^{-1} \cdot (x - y)^T}$

Maximum (or Chebyshev): $d(x,y) = \max_{1 \leq j \leq p} |x_j - y_j|$

If two measurements have a "small" distance with respect to a data class, they can be thought of as *similar*. As with *distance* this notion can be formalized.

Definition 1.2   A *similarity measure s* is a metric that gauges the proximity of two measurements with respect to a distance *d*. Such a metric has the following properties.

   (i) $s: \mathbb{R}^p \otimes \mathbb{R}^p \rightarrow [0, 1]$

  (ii) $0 \leq s(x,y) \leq 1$ for all $x, y \in \mathbb{R}^p$

 (iii) $s(x,x) = 1$ and $s(x,y) = s(y,x)$

This formalism is a way to codify the idea that the smaller the distance in-between two measurements, the more "similar" they are. That is, $d(x, y) = 0$ should imply $s(x, y) = 1$. The connection in-between distance and similarity measure is summarized in the following results.

Theorem 1.1   If d is any distance metric, then $s(x, y) = \dfrac{|1 - d(x, y)|}{1 + d^2(x, y)}$ is a similarity measure.

Proof: Since d is a distance, then $s(x, y) = \dfrac{|1 - d(x, y)|}{1 + d^2(x, y)} = \dfrac{|1 - d(y, x)|}{1 + d^2(y, x)} = s(x, y)$.

Also, $s(x, x) = \dfrac{|1 - d(x, x)|}{1 + d^2(x, x)} = \dfrac{|1 + 0|}{1 + 0^2} = 1$. Thus, condition *iii* of Definition 1.2 is satisfied. Clearly, $s(x, y) \geq 0$ as it is defined as the ratio of a non-negative numerator and a positive denominator. Therefore, to show conditions *i* and *ii* of the definition, it remains to show that $s(x, y) \leq 1$ for any $x, y \in \mathbb{R}^p$. There are two cases to consider.

Case 1. $d(x, y) \geq 1$. In this case, $|1 - d(x, y)| = d(x, y) - 1$ so that $0 \leq \dfrac{|1 - d(x, y)|}{1 + d^2(x, y)} = \dfrac{d(x, y) - 1}{1 + d^2(x, y)} = \dfrac{1 - 1/d(x, y)}{d(x, y) + 1/d(x, y)}$. But this is the ratio of a number that is less than 1 (namely, $1 - 1/d(x, y)$) and a number larger than 1 ($d(x, y) + 1/d(x, y)$). Hence, the ratio is less than 1 and therefore, $s(x, y) \leq 1$.

Case 2. $d(x, y) < 1$. In this case, $0 \leq \dfrac{|1 - d(x, y)|}{1 + d^2(x, y)} = \dfrac{1 - d(x, y)}{1 + d^2(x, y)} < \dfrac{1}{1 + d^2(x, y)} \leq 1$.

In either case, $0 \leq s(x, y) \leq 1$ for any $x, y \in \mathbb{R}^p$ so that conditions *i* and *ii* are satisfied.   ■

Remarks:

1. If $d(x, y) < \varepsilon \ll 10^{-n}$ for "large" n, then $s(x, y) \approx \dfrac{1 - \varepsilon}{1 + \varepsilon^2} \approx 1$. Thus, small distances indicate a similarity near 1.

2. If $d(x, y) > N \gg 1$, then $s(x, y) \approx \dfrac{N}{1 + N^2} \approx \dfrac{1}{N} \to 0$ as $N \to +\infty$. Hence, large distances indicate a similarity near 0.

The next example combines the ideas of *distance* and *similarity metrics* to associate a measurement with a set of data classes.

Example 1.1:   Suppose $v_0 = [-0.65, 1]$ is a measurement, X is the data class matrix from $Class_1$, and Y is the data matrix from $Class_2$ of Figure 1.11. Table 1.1 summarizes the distances and similarity measures from v_0 to each class with respect to

TABLE 1.1   Distance and similarity measures for Figure 1.11

Metric	$Class_1$	$Class_2$	Measure	Association
$d(v, Class)$	0.34	0.64	Euclidean	1
$s(v, Class)$	0.59	0.254		1
$d(v, Class)$	3.12	3.20	Mahalanobis	1
$s(v, Class)$	0.1975	0.1001		1
$d(v, Class)$	0.298	0.51	Maximum	1
$s(v, Class)$	0.6445	0.39		1
$d(v, Class)$	0.4238	0.8992	City Block	1
$s(v, Class)$	0.4885	0.0557		1

several of the distances presented at the beginning of this section. Notice that, in each case, the distance from v_0 to the $Class_1$ data matrix X is smallest (simultaneously, the similarity of v_0 to X is greatest) so that v_0 would be associated with $Class_1$. Once a clustering method is applied to $Class_2$, however, the association will be reversed. This will be detailed in the next section.

1.4.2   The K-Means Method

Figures 1.11 and 1.12 illustrate the idea of taking a single data class and "splitting" it into two distinct subclasses. In the example mentioned earlier, the data from $Class_2$ were divided into the subclasses $Class_{2,1}$ and $Class_{2,2}$. How is this division achieved?

Suppose $X = \begin{bmatrix} x_{1,1} & x_{1,2} & \cdots & x_{1,p} \\ x_{2,1} & x_{2,2} & \cdots & x_{2,p} \\ \vdots & \vdots & \ddots & \vdots \\ x_{n,1} & x_{n,2} & \cdots & x_{n,p} \end{bmatrix}$ is the data matrix for a particular class of

objects. If $x_k = [x_{k,1}, x_{k,2}, \ldots, x_{k,p}]$ represents the kth row of X, then the data matrix can be viewed as a set of p-dimensional measurements: $X = \{x_1, x_2, \ldots, x_p\}$. A *partition* of X is a collection of distinct subsets whose union is X. More precisely, $P = \{P_1, P_2, \ldots, P_K\}$ is a partition of X provided:

(i) $P_k \subset X$ for each $k = 1, 2, \ldots, K$

(ii) $X = \bigcup_{k=1}^{K} P_k$

(iii) $P_j \cap P_k = \varnothing$ for any $j \neq k$.

Let $v_k = |P_k|$ be the cardinality of the set P_k; that is, v_k is the number of elements in the set P_k. The *attractor z_k for the kth member of a partition P_k* is the "average" of the elements within the partition element.

$$z_k = \frac{1}{v_k} \sum_{x \in P_k} x \qquad (1.13)$$

The *K-means clustering method* requires the construction of a set of K *attractors* $Z = \{z_1, z_2, \ldots, z_K\} \subset X$ so that the associated partition $P = \{P_1, P_2, \ldots, P_K\}$ provides the maximum separation in-between the subclasses P_k with respect to a distance d. Therefore, the *K-means* algorithm is implemented to minimize the function

$$E(X) = \sum_{k=1}^{K} \sum_{x \in P_k} d(x, z_k) \tag{1.14}$$

with respect to a given partition P of X. Thus, if $\mathcal{P}_X$ is the collection of all partitions of X, the K-means method is to find the partition P so that $E(X)$ is minimized.

$$\min_{P \in \mathcal{P}_X} E(X) = \min_{P \in \mathcal{P}_X} \sum_{k=1}^{K} \sum_{x \in P_k} d(x, z_k) \tag{1.15}$$

Example 1.1 (*Continued*):   After splitting *Class*$_2$ into two subclasses *Class*$_{2,1}$ and *Class*$_{2,2}$, calculate the distance of $v_0 = [-0.65, 1]$ to these data classes along with the original *Class*$_1$. Now the choice of distance metric is crucial as v_0 is closer to *Class*$_{2,1}$ than either *Class*$_{2,2}$ or *Class*$_1$ with respect to the Mahalanobis distance. Otherwise, v_0 is associated with *Class*$_1$. Table 1.2 summarizes the computations.

The selection of distance metric is crucial to object classification. It also plays a crucial role in data clustering. This is the *art* of applied mathematics: The selection of methods and techniques that best respond to the data at hand. The general rule of thumb *for classification* is to use the *Mahalanobis distance provided the weight matrix M is well conditioned.* The Euclidean distance, whose weight matrix is the identity I, is recommended for the K-means clustering algorithm. For a well-conditioned weight matrix M, the Mahalanobis distance will separate classes along the eigenvectors of M. The Euclidean metric separates classes along the standard basis in which the data operates ($\mathbb{R}^p$). Care then should be taken in understanding

TABLE 1.2   Distance and similarity measures for Figure 1.12

Metric	$Class_1$	$Class_{2,1}$	$Class_{2,2}$	Measure	Association
$d(v, Class)$	0.34	0.64	2.81	Euclidean	1
$s(v, Class)$	0.59	0.254	0.204		1
$d(v, Class)$	3.12	3.045	5.985	Mahalanobis	2
$s(v, Class)$	0.1975	0.199	0.135		2
$d(v, Class)$	0.298	0.5086	2.62	Maximum	1
$s(v, Class)$	0.6445	0.39	0.206		1
$d(v, Class)$	0.424	0.8992	3.4	City Block	1
$s(v, Class)$	0.4885	0.0557	0.191		1

how the data are to be classified/clustered so that the end goal of providing the best categorization of the data is achieved. This requires some understanding of the underlying biological phenomenon being studied.

The example illustrated in Figures 1.11 and 1.12 is produced via the MATLAB commands shown below. In particular, kmeans.m (from the Statistics Toolbox) and kcluster.m are the MATLAB functions that perform the clustering via the *K*-means method.

MATLAB Commands
(*K-Means Clustering*)

```
% Data directory
Ddir = 'C:\PJC\Math_Biology\Chapter_1\Data';
% Retrieve the two-dimensional clustered data classes
load(fullfile(Ddir,'Clustered_Data_Classes1_2.mat'));
X = C{1}; Y = C{2};

% Split Class₂ into two subclasses
S = kcluster(Y,2);
% Identify the split classes as Z₁ and Z₂
Z1 = S.C{1}; Z2 = S.C{2};
```

1.4.3   Number of Clusters

As seen previously, the *K*-means clustering algorithm partitions a single data class into a specified number (K) of subclasses. What is the optimal number of subclasses? That is, is there a "best" K? This question is generally approached by way of *cluster validity indices*. These indices are a measure of how effective the cluster size is in reorganizing the data. There are *many* validity indices and the reader is referred to Gan et al. [7] for a thorough presentation of this topic. Instead, two indices with relatively straightforward interpretation are described and implemented on the data listed in the table "MATLAB Commands" from Section 1.4.2.

Before these indices are described, however, some notation must be established. The *data class matrix* $X \in \mathcal{Mat}_{n \times p}(\mathbb{R})$ will be treated as a collection of p-dimensional vectors $\{x_1, x_2, \ldots, x_n\}, x_k \in \mathbb{R}^p$. The partition $P = \{P_1, P_2, \ldots, P_K\}$ of X with each $P_j \in \mathcal{Mat}_{n_j \times p}(\mathbb{R})$ and $\sum_{k=1}^{K} n_k = n$ is a *K-partition cluster* with *centroid element* $\mu_k = \dfrac{1}{n_k} \sum_{x \in P_k} x \in \mathbb{R}^p$. Observe that $\mu_k = [\mu_{k,1}, \mu_{k,2}, \ldots, \mu_{k,p}]$ implying the *mean of the centroid* is $\bar{\mu}_k = \dfrac{1}{n_k} \sum_{\ell=1}^{n_k} \mu_{k,\ell}$. Finally, the *mean center of the data matrix* X is defined to be $\bar{X} = [\bar{x}_1, \bar{x}_2, \ldots, \bar{x}_p]$ with $\bar{x}_k = \dfrac{1}{n} \sum_{j=1}^{n} x_{j,k} \equiv$ the column mean for $k = 1, 2, \ldots, p$.

1.4.3.1   The RMSSTD Validity Index   The *root mean square standard deviation* (RMSSTD) index measures the variance of elements of a data class collection from its centroid elements. The optimal partition of the class collection occurs at the value j_0 at which the plot of the points $(j, RMS_{std}(j))$ attains a "knee," $j = 1, 2,..., K$ (see Gan et al. [7, p. 310]. For any given partition $P = \{P_1, P_2,..., P_K\}$ of X, let $\Pi_k = P_k - \mathbf{1}_{n_k \times 1} \cdot \mu_k$ where $\mathbf{1}_{n_k \times 1} = [1, 1,..., 1]^T$ is the column vector of n_k ones. Then $\Pi_k \in \mathcal{Mat}_{n_k \times p}(\mathbb{R})$ for each k, $\pi_{i,j}^{(k)} = p_{i,k}^{(k)} - \mu_{k,j}$, and

$$\Pi_k = \begin{bmatrix} \pi_{1,1}^{(k)} & \pi_{1,2}^{(k)} & \cdots & \pi_{1,p}^{(k)} \\ \pi_{2,1}^{(k)} & \pi_{2,2}^{(k)} & \cdots & \pi_{2,p}^{(k)} \\ \vdots & \vdots & \ddots & \vdots \\ \pi_{n_k,1}^{(k)} & \pi_{n_k,2}^{(k)} & \cdots & \pi_{n_k,p}^{(k)} \end{bmatrix}$$

for each member of the partition.

$$RMS_{std} = \sqrt{\frac{\sum_{k=1}^{K} \sum_{i=1}^{n_k} \sum_{j=1}^{p} \left(\pi_{i,j}^{(k)} \right)^2}{p(n-K)}} \tag{1.16}$$

As is indicated in Figure 1.13, the "knee" of the validity curve for the RMS_{std} index occurs at $K = 2$.

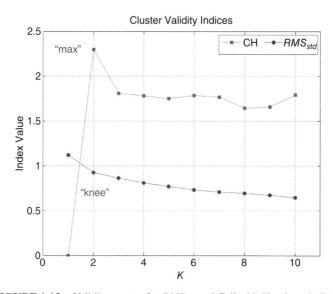

FIGURE 1.13   Validity curves for RMS_{std} and Calinski–Harabasz indices.

1.4.3.2   The Calinski–Harabasz Index   This index is the ratio of the traces of the in-between-cluster scatter matrix and the within-cluster scatter matrix. The *Calinski–Harabasz* or *CH* index provides an optimal partition $P = \{P_1, P_2, \ldots, P_K\}$ of the data set X at the maximum value of the index.

$$CH = \frac{(n - K) \sum\limits_{k=1}^{K} n_k(\boldsymbol{\mu}_k - \bar{X})(\boldsymbol{\mu}_k - \bar{X})^T}{(K - 1) \sum\limits_{k=1}^{K} \sum\limits_{x \in P_k} (x - \boldsymbol{\mu}_k)(x - \boldsymbol{\mu}_k)^T} \qquad (1.17)$$

An inspection of Figure 1.13 reveals that the maximum value of the *CH* index occurs at $K = 2$. Therefore, using either the *CH* or RMS_{std} indices, splitting $Class_2$ into two subclasses provides the optimal clustering effect.

EXERCISES

1.5 Try computing the *SD* and *SDbw* indices via the MATLAB functions `sdindex.m` and `sdbwindex.m` using the Mahalanobis distance options (see Gan et al. [7] for more details). Plot the results. What conclusions can be drawn from these indices? These data are obtained via the MATLAB commands

```
Ddir = 'C:\Data\Math_Biology\Chapter_1\Data';
load(fullfile(Ddir,'Clustered_Data_Classes1_2.mat'));
X = C{1}; Y = C{2};
```

1.6 Using the clustered data class contained in the MAT-file sited in the command table at the end of 1.4.2 (`Clustered_Data_Classes1_2.mat`), compute the distance and similarity metrics of the Minkowski measure using $q = 1$, 3, and 4. The commands listed in Exercise 1.5 will yield the desired data.

1.5   DATA QUALITY AND DATA CLEANING[2]

Data quality is the most significant and uncontrolled matter the data analyst/ mathematical modeller must address. What is *data quality*? First, *data quality* refers to the state of the data under consideration. How were the data recorded? Were the data reviewed? What are the sources for data entry error? Can the data be edited and corrected for errant entries? These are *some* of the questions that the *quality control* specialist confronts. Hence, *data quality* should be thought of as a process in which the integrity of data can be ensured and safeguarded.

The issues that arise with respect to data quality vary considerably depending on the sources and kinds of measurements made. A significant portion of current interest

---

[2] The author wishes to thank Dr. William J. Satzer who inspired and co-wrote this section.

is focused on data mining, and thus often very large databases. There have been cases of passengers improperly flagged on "no-fly" lists due to spelling or transcription errors. If "John Doe" is a notorious terrorist while "Jon Dough" is an amiable pastry salesman, how can screening officials clearly identify friend from foe? And what to do with "J. Doe," "John B. Doe," and "John Deo?" With such a large list of names from which to flag an "undesirable" passenger, the accuracy of the data entry is crucial. Data quality problems can also arise, however, in other smaller scale settings. These need to be addressed in a systematic fashion.

Most students encounter only well managed or artificial data sets. Too often, textbooks or instructors provide "cleaned data" so that readers can apply various mathematical/statistical techniques to replicate examples. These data are close to perfect and do not represent the kinds of information that data modellers will actually encounter. No data are missing, no values are obvious typographical errors, and all the data are in one tidy file.

Indeed, one feature of this book is that, whenever possible, data sets are drawn from actual reports, studies, or genuine measurements.

The focus of the *quality control* specialist is to ensure data quality by *data cleaning*. This includes, but is not limited to, verifying data sources, reviewing data entry procedures, correcting misspellings, inversions, or inaccurate entries, missing information, invalid measurements, and identifying unlikely values. For example, if the heights and weights of dancers are presented as a data set, does the entry *Zoë Ballerina*, height $7'\ 2''$, weight 97 pounds make any sense?

Data cleaning aims to detect and remove errors and inconsistencies from data in order to improve its quality and prevent problems in subsequent data analyses. Sometimes data corruption occurs because of human error in data recording, and other times because of software or instrument error. When multiple data sources need to be integrated, the need for data cleaning is even more significant. In situations like this, data can be redundant, contradictory, and have multiple and inconsistent representations.

So far there is no science of data cleaning, only a collection of methods selected to meet needs of a particular environment. Dasu and Johnson [5], McCallum [10], Osborne [11], and Maydanchik [9] provide guidelines and insights into data cleaning. Eldén [6] examines the mathematics of *data mining*. A set of basic tenets for data extraction software is listed below. That is, when writing software that reads data from a set of measurements, the following recommendations are offered:

(i) Flag/identify any missing entry. If there is a vacant or missing data entry, replace the empty "value" with a NaN (an *IEEE* arithmetic representation for an object which is *not a number*).

(ii) Collect heading identifiers and indicate duplications. For example, if *Ht.*, *Height*, *H*, and *Hght* are column headers used to describe height measurements, note this naming disparity. Select a single column identifier and replace all "synonyms" with the representative.

(iii) Ensure consistent units. Again, if one column of data specified by *Height* provides measurements in *meters* while another column *H* lists measurements in *feet*, select a single unit and convert *feet* to *meters* (or vice versa).

(iv) Write code that verifies the data file being read and from which data are extracted. If there are two files, `DataFile1` and `DataFile2`, be certain that the extraction program is reading in the file that contains the requisite measurements. If reading data from Excel spreadsheets, check that the *sheet name* (i.e., the particular sheet/lower tab of the entire spreadsheet which contains data) is the one containing the desired information.

(v) Flag any column/heading that *should* be present in the data file but is not located via the extraction code.

(vi) Check for inconsistent information. If one column of data contains color information (e.g., red, green, or blue) and a second column contains RGB coordinates ([1, 0, 0] → red, [0, 1, 0] → green, and [0, 0, 1] → blue), check that the color (red) and coordinate ([1, 0, 0]) are consistent. If the (color, coordinate) pair is (green, [0, 0, 1]), then the measurement is *inconsistent*.

No exhaustive list of recommendations/guidelines can be given. The intention here is only to draw attention to the issue, identify some approaches, and offer select references to the developing literature in the area. The most important message of this section is the need to recognize that *data cleaning needs to precede data analysis*. No matter how insightful or complete is the model, the application of mathematics to poorly gathered data will result in nonsense or worse.

Data cleaning is variously known as "data scrubbing," "data wrangling," "data janitor work," as well as several other disparaging terms. It is regarded as a thankless job, yet interviews and expert estimates suggest that data cleaning can consume 50–80% of a data analyst's time. While considerable effort is being devoted to developing software that automates part of the task, much of it is essentially manual labor.

The aforementioned represent a few key ideas. They merit repetition:

(i) Inspect the data carefully before performing any analysis.

(ii) When possible, perform a visual inspection of the data files. To that end, this book provides a set of data visualization plotting functions (written in MATLAB).

(iii) Review the data (plotted or otherwise) to uncover anomalies such as missing data, mismatched data types, obvious outliers, data out-of-time sequence, and perhaps inconsistent scaling or mismatched measurement units within the data.

(iv) Document each data cleaning operation along the way.

This is the second crucial message of this section. *Document everything*: Merged files, obvious irregularities, dropped values, data transformations, and any new data derived from more raw forms. Further, always retain previous versions of data files;

never discard the original information. Otherwise, an error in data cleaning could effectively destroy the original data.

These recommendations listed are forged from many years of (often bitter) experience. They are offered in the hope that the reader will adopt a healthy skepticism toward data files and their processing. *Caveat emptor.*

REFERENCES

[1] Andrews, L. C. and B. K. Shivamoggi, *Integral Transforms for Engineers and Applied Mathematicians*, Macmillan Publishing Company, New York, 1988.

[2] Berra, Y., *The Yogi Book*, Workman Publishing, New York, 1998.

[3] Costa, P. J., *Bridging Mind and Model*, St. Thomas Technology Press, St. Paul, MN, 1994.

[4] Costa, P. J., "Truncated outlier filtering," *Journal of Biopharmaceutical Statistics*, vol. 24, pp. 1115–1129, 2014.

[5] Dasu, T. and T. Johnson, *Exploratory Data Mining and Data Cleaning*, John Wiley & Sons, Inc., Hoboken, NJ, 2003.

[6] Eldén, L., *Matrix Methods in Data Mining and Pattern Recognition*, SIAM Books, Philadelphia, PA, 2007.

[7] Gan, G., C. Ma, and J. Wu, *Data Clustering: Theory, Algorithms, and Applications*, ASA–SIAM Series on Statistics and Applied Probability, SIAM Books, Philadelphia, PA, ASA, Alexandria, VA, 2007.

[8] Johnson, R. A. and D. W. Wichern, *Applied Multivariate Statistical Analysis*, Fourth Edition, Prentice–Hall, Inc., Upper Saddle River, NJ, 1998.

[9] Maydanchik, A., *Data Quality Assessment*, Technics Publications, Bradley Beach, NJ, 2007.

[10] McCallum, Q. E., *The Bad Data Handbook, Cleaning Up The Data So You Can Get Back To Work*, O'Reilly Media, Sebastopol, CA, 2013.

[11] Osborne, J. W., *Best Practices in Data Cleaning: A Complete Guide to Everything You Need to do Before and After Collecting Your Data*, Sage Publications, Los Angeles, CA, 2013.

[12] Stewart, Potter, *United States Supreme Court Decision Jacobellis versus Ohio 1964*, 378 U.S. 184, https://en.wikipedia.org/wiki/Jacobellis_v._Ohio.

[13] Tholen, D.W., A. Kallner, J. W. Kennedy, J. S. Krouwer, and K. Meier, *Evaluation of Precision Performance of Quantitative Measurement Methods: Approved Guideline*, Second Edition, CLSI Document EP5–A2, vol. 24, no. 25, NCCLS, Wayne, PA, ISBN 1–56238–542–9.

2

SOME EXAMPLES

In this chapter, the interaction of *rates* used to govern the processing of caloric intake, disease progression, and *DNA* identification is explored using models based on differential equations. Diabetes is a disease caused by a metabolism disorder in our body that leads to excess sugar levels in the blood. This can be due to the inability of the pancreas to produce a sufficient level of the hormone insulin, which acts to lower blood sugar. Glucose is a sugar that the body produces via the digestion of food. Plainly, understanding the interaction of glucose and insulin rates is important in determining the causes and treatments of diabetes.

Acquired immunodeficiency syndrome (*AIDS*) is a serious auto-immune disease caused by infection with the human immunodeficiency virus (*HIV*). The present standard of care for *AIDS* patients is to administer *highly active antiretroviral therapy* (*HAART*) medications. While long-term (i.e., decades long) effects of *HAART* treatments remain unclear, these medications have proved successful in treating *HIV-positive* (*HIV*⁺) populations and delay, perhaps indefinitely, the onset of *AIDS*.

The *polymerase chain reaction* (*PCR*) is the method currently used to amplify and ultimately identify *deoxyribonucleic acid* (*DNA*) samples. This process has been used forensically within the US judicial system and to great fanfare within popular culture. Real-time *PCR* is presented here in the context of use to identify various biomarkers as indication of the presence of disease.

All of these examples show how differential equations relay the interactions of rates and dissipations of concentrations within a particular biological feedback system.

Applied Mathematics for the Analysis of Biomedical Data: Models, Methods, and MATLAB®, First Edition. Peter J. Costa.
© 2017 Peter J. Costa. Published 2017 by John Wiley & Sons, Inc.
Companion website: www.wiley.com/go/costa/appmaths_biomedical_data

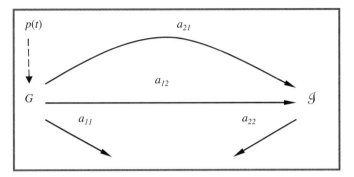

FIGURE 2.1   Glucose/insulin interaction.

2.1 GLUCOSE–INSULIN INTERACTION

As we humans ingest food, the blood sugar concentration in our bodies changes. This sugar or *glucose* is a primary source of nutrition for our bodies' cells. The fats and oils found in our blood are called *lipids* and are also a source of energy. Glucose is transported from the liver to cells by way of the bloodstream. It is made ready for cellular absorption via the hormone *insulin* that is produced by the pancreas. See, for example, Lewin et al. [4]; http://en.wikipedia.org/wiki/Insulin, for details. Consequently, there is a dynamic interaction in-between blood glucose levels and insulin production.

Yeargers et al. [12] propose a simple model of glucose/insulin concentrations as illustrated in Figure 2.1.

In this setting $G(t)$ and $\mathcal{I}(t)$ are the glucose and insulin concentrations, respectively, in a system, a_{ij} are the rate proportions, and $p(t)$ is the glucose infusion rate. This schematic leads to the set of coupled linear ordinary differential equations (2.1) with the general initial conditions $G_0 = G(t_0)$ and $\mathcal{I}_0 = \mathcal{I}(t_0)$.

$$\left.\begin{aligned}
\frac{dG}{dt} &= -a_{11}\,G(t) - a_{12}\,\mathcal{I}(t) + p(t) \\
\frac{d\mathcal{I}}{dt} &= a_{21}\,G(t) - a_{22}\,\mathcal{I}(t)
\end{aligned}\right\} \tag{2.1}$$

By setting $\boldsymbol{u}(t) = [G(t), \mathcal{I}(t)]^T$, $\boldsymbol{u}_0 = \boldsymbol{u}(t_0) = [G(t_0), \mathcal{I}(t_0)]^T$, $\mathcal{A} = \begin{bmatrix} -a_{11} & -a_{12} \\ a_{21} & -a_{22} \end{bmatrix}$, and $\boldsymbol{p}(t) = [p(t), 0]^T$, then (2.1) can be written as

$$\left.\begin{aligned}
\frac{d\boldsymbol{u}}{dt} &= \mathcal{A}\,\boldsymbol{u} + \boldsymbol{p}(t) \\
\boldsymbol{u}(t_0) &= \boldsymbol{u}_0
\end{aligned}\right\}. \tag{2.2}$$

In this environment, the concentrations can be treated as proportions. It can be shown that the solution of (2.2) is $\boldsymbol{u}(t) = e^{\mathcal{A}(t-t_0)}\,\boldsymbol{u}_0 + \int_{t_0}^t e^{\mathcal{A}(t-s)}\,\boldsymbol{p}(s)\,ds$. This will be left as an exercise for the reader.

Example 2.1: Yeargers et al. [12] suggest that each $a_{ij} = 1$, $u_0 = [1, 0]^T$, and $p(t) \equiv 0$ for all $t \geq t_0 = 0$ (i.e., no glucose infusion). That is,

$$\frac{dG}{dt} = -G(t) - \mathcal{I}(t), G(0) = 1$$

$$\frac{d\mathcal{I}}{dt} = G(t) - \mathcal{I}(t), \mathcal{I}(0) = 0$$

or simply $\frac{du}{dt} = Au$, $u(0) = [1, 0]^T$.

This version of (2.1)–(2.2), namely $p(t) \equiv 0$, is referred to as a *homogeneous ordinary differential equation*. When the glucose infusion rate $p(t)$ is *not* identically zero, the ordinary differential equations (2.1)–(2.2) are called *inhomogeneous*.

The matrix $A = \begin{bmatrix} -1 & -1 \\ 1 & -1 \end{bmatrix}$ has *eigenvalues* $\lambda_1 = -1 + i$ and $\lambda_2 = -1 - i$, along with *eigenvectors* $v_1 = \frac{\sqrt{2}}{2} \begin{bmatrix} 1 \\ -i \end{bmatrix}$ and $v_2 = \frac{\sqrt{2}}{2} \begin{bmatrix} 1 \\ i \end{bmatrix}$. Hence, the solution of $\frac{du}{dt} = Au$ is $u(t) = e^{A(t-t_0)} u(0)$. Observe that $e^{A(t-t_0)} = V e^{\Lambda(t-t_0)} V^T$, where $V = [v_1 \; v_2] = \frac{\sqrt{2}}{2} \begin{bmatrix} 1 & 1 \\ -i & i \end{bmatrix}$, $\Lambda = [\lambda_1, \lambda_2]^T$, and $e^{\Lambda(t-t_0)} = \begin{bmatrix} e^{\lambda_1(t-t_0)} & 0 \\ 0 & e^{\lambda_2(t-t_0)} \end{bmatrix} = e^{-t} \begin{bmatrix} e^{it} & 0 \\ 0 & e^{-it} \end{bmatrix}$. Therefore, since $t_0 = 0$, $e^{A(t-t_0)} = V e^{\Lambda t} V^T = e^{-t} \begin{bmatrix} \cos(t) & \sin(t) \\ \sin(t) & -\cos(t) \end{bmatrix}$.

For $u_0 = [1, 0]^T$, $e^{A(t-t_0)} u_0 = e^{-t} \begin{bmatrix} \cos(t) \\ \sin(t) \end{bmatrix}$. For more detailed information about matrices, eigenvalues, and eigenvectors, see Appendix Section A.2.

Consequently, $u(t) = e^{-t} \begin{bmatrix} \cos(t) \\ \sin(t) \end{bmatrix}$ so that $G(t) = e^{-t} \cos(t)$ and $\mathcal{I}(t) = e^{-t} \sin(t)$.

Figure 2.2 illustrates a graph of the solution over the time frame $0 \leq t \leq 5$.

The model yields a glucose concentration that is negative on the (approximate) time interval $[1.6, 4.7]$. This appears to defy both biology and common sense. Thus, it is reasonable to reconsider the contents of the model.

Example 2.2: Reconsider equation (2.1) so that the glucose infusion rate is *not* identically zero. That is, set $a_{ij} = 1$ for $i, j = 1, 2$ as before but use the exponentially declining infusion rate of $p(t) = B_0 e^{-k(t-t_0)}$ as per Fisher [2]. Then for $u_0 = [1, 0]^T$, a direct computation can show

$$u(t) = e^{-t} \begin{bmatrix} \cos(t) \\ \sin(t) \end{bmatrix} + \frac{B_0}{1 + (1-k)^2} e^{-t} \begin{bmatrix} \sin(t) - (1-k)\cos(t) + (1-k) \cdot e^{(1-k)t} \\ -\cos(t) - (1-k)\sin(t) + e^{(1-k)t} \end{bmatrix}.$$
$$\text{(2.3)}$$

For $B_0 = 0.5$ and $k = 0.05$, Figure 2.3 illustrates the solution. Here the interaction appears to be more physiological than the model without glucose infusion (i.e., $p(t) \equiv 0$). The insulin level is non–negative throughout the time interval $[0, 10]$.

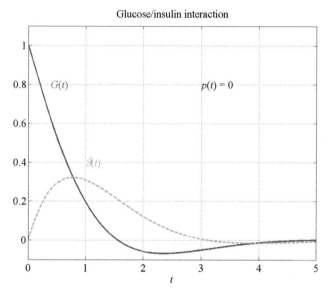

FIGURE 2.2   Glucose/insulin interaction without glucose infusion.

Moreover, the concentrations of glucose and insulin appear to reach an equilibrium at time 4 (approximately). This is sensible as the body does not need to produce *more* insulin as glucose levels stabilize. Hence, the model better reflects the biology. More complex paradigms will be discussed in the chapter on *SEIR* (*Susceptible, Exposed, Infected, Recovered/Removed*) models.

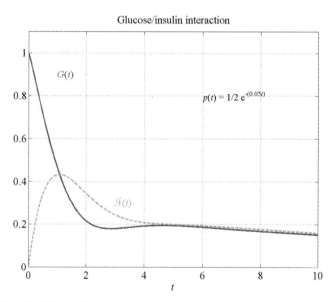

FIGURE 2.3   Glucose/insulin interaction with glucose infusion (exponential decay).

EXERCISES

2.1   Show that $A = \begin{bmatrix} -1 & -1 \\ 1 & -1 \end{bmatrix}$ does in fact have *eigenvalues* $\lambda_1 = -1 + i$ and $\lambda_2 = -1 - i$, and *eigenvectors* $v_1 = \frac{\sqrt{2}}{2}\begin{bmatrix} 1 \\ -i \end{bmatrix}$ and $v_2 = \frac{\sqrt{2}}{2}\begin{bmatrix} 1 \\ i \end{bmatrix}$. Verify calculations that result in the formulae $G(t) = e^{-t}\cos(t)$ and $\mathscr{G}(t) = e^{-t}\sin(t)$.

2.2   Reconsider equation (2.3). What happens to the concentrations $G(t)$ and $\mathscr{G}(t)$ for $k = 0.1, 0.5, 0.8$?

2.3   Show that $u(t) = e^{A(t-t_0)} u_0 + \int_{t_0}^{t} e^{A(t-s)} p(s)\, ds$ is the solution of (2.2). *Hint:* Use the fact that $\frac{d}{dt}e^{At} = Ae^{At}$ and apply the *Fundamental Theorem of Calculus* to the integral term $\int_{t_0}^{t} e^{A(t-s)} p(s)\, ds$.

2.4   Show that (2.3) is the solution of (2.1) with $a_{ij} = 1$ for each $i, j = 1, 2$, $u_0 = [1, 0]^T$, and $p(t) = B_0 e^{-kt}$.

2.2   TRANSITION FROM *HIV* TO *AIDS*

Antiretroviral medications have significantly reduced the number of people who have contracted *HIV* from developing *AIDS*. Still, significant populations of disease carriers[1] are largely unaware of their *HIV* status. Consequently, a reliable and reasonable mathematical model of the transition from being HIV^+ to the development of *AIDS* is of value to public health agencies in setting national policy. Consider the following very simple model.

Let $H(t)$ be the fraction of a population that is HIV^+ and $A(t)$ be that fraction of the HIV^+ subpopulation who develop *AIDS*. Since H and A account for all people who are HIV^+, then these quantities can be treated as proportions. Hence, $H(t) + A(t) = 1$ for all $t \geq t_0$. The symbol t_0 represents some initial reference time. For the purposes of this presentation, t_0 is assumed to be zero: $t_0 = 0$. The proposed relationship in-between H and A is that the rate of change in H is negatively proportional to itself while the rate of change in A is directly proportional to H by the same factor. Further, it is assumed that, at time zero, no one has yet developed *AIDS* and the entire infected population is HIV^+. As a differential equation, this is written as

$$\left.\begin{aligned} \frac{dH}{dt}(t) &= -v(t) \cdot H(t), \quad H(0) = 1 \\ \frac{dA}{dt}(t) &= v(t) \cdot H(t), \quad A(0) = 0 \end{aligned}\right\} . \tag{2.4}$$

Plainly these initial assumptions can be altered to reflect more realistic conditions of a given population. The model presented in (2.4) is most appropriately applied to a population in which *HIV* has just been introduced.

---

[1] For example, sexually active urbanites in Africa and Asia.

TABLE 2.1   *AIDS* **onset rates following** *HIV* **infection**

$t^\dagger$	0	0.5	1.0	1.5	2.0	2.5	3.0	3.5	4.0	4.5	5.0	5.5
$\dfrac{dA}{dt}$	0	0.05	0.08	0.175	0.155	0.09	0.155	0.04	0.04	0.08	0.02	0.04

†Time t is measured as years following *HIV* infection.

The differential equations in (2.4) are uncoupled and can therefore be integrated directly. By setting $V(t) = \int_0^t v(s)\,ds$, it is seen that $H(t) = e^{-V(t)}H(0) = e^{-V(t)}$. It is left as an exercise to show that $A(t) = \int_0^t v(s) \cdot H(s)\,ds = 1 - e^{-V(t)} \equiv 1 - H(t)$. This is, of course, an obvious result of the assumption $H(t) + A(t) = 1$. How, then, should $v(t)$ be modeled? The most sensible approach is to look at some data. Table 2.1 gives *AIDS* rates over half-year intervals as reported by Peterman et al. [9]. These data will be supplemented by the assumption that the rate at time zero will be zero: $\dfrac{dA}{dt}(0) = 0$.

The *HIV*$^+$ proportion $H(t)$ is determined from the differential equations (2.4) to be an exponential function. One of the simplest forms for the growth/decay rate $v(t)$ is a linear function. That is, assume $v(t) = \alpha\,t$ for an as yet to be determined constant α. In this case, $V(t) = \int_0^t v(s)\,ds = \int_0^t \alpha\,s\,ds = \frac{1}{2}\,\alpha\,t^2$ so that $H(t) = e^{-\frac{1}{2}\alpha t^2}$ and $\dfrac{dA}{dt}(t) = v(t)H(t) = \alpha\,t\,e^{-\frac{1}{2}\alpha t^2}$. Recalling the idea from the prostate specific antigen (*PSA*) data of Section 3 of the Introduction, an error functional with respect to the unknown constant α can be formed.

$$E[\alpha] = \sum_t \left(\frac{dA}{dt}(t) - v(t)\,H(t) \right)^2 = \sum_t \left(\frac{dA}{dt}(t) - \alpha\,t\,e^{-\frac{1}{2}\alpha t^2} \right)^2 \qquad (2.5)$$

The minimum of $E[\alpha]$ can be determined from the discrete set of points provided in Table 2.1. It can be seen that the function $E[\alpha]$ attains a minimum at $\alpha = a_{min} = 0.042$ and $E[\alpha_{min}] = E[0.042] = 0.05$. The natural question to ask then is, *How well does this model fit the data?* The rate data, model fit, and corresponding error functional are illustrated in Figure 2.4. As can be seen, while the value of the error functional is relatively small, the model is not visually satisfying.

What do the values of the model expressed in (2.4) with $v(t) = \alpha t$ reveal? To gauge the effectiveness of the model, consider Figure 2.5. First observe that at $t_e \approx$ 4.38 years, $H(t_e) = A(t_e)$. That is, at this time t_e, the proportion of the *HIV*$^+$ population that has converted to *AIDS* is $\frac{1}{2}$. Moreover, at $t \approx 8.0$ years, virtually 100% of the *HIV*$^+$ population has converted to *AIDS*.

This is a naive model. The notion that *every person infected with* HIV *develops* AIDS *after approximately 8 years* seems less than reliable. What are the alternative models? One candidate is the *hazard function*.

The hazard function, also called the *failure rate function*, is the probability of lifetime failure conditioned on survival at time t. In mathematical terms, let $\mathcal{T} : \mathbb{R}^+ \to \mathbb{R}^+$ be a continuous random variable with probability density function (*pdf*) $f_{\mathcal{T}}$ and cumulative density function (*cdf*) $F_{\mathcal{T}}$. Then, $F_{\mathcal{T}}(t) = \text{Prob}(\mathcal{T} < t) = \int_0^t f_{\mathcal{T}}(\tau)\,d\tau$. The *survivor function* ($S(t)$), also called the *reliability function*) at time t, is the probability of

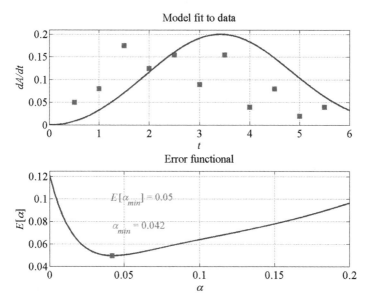

FIGURE 2.4   *AIDS* onset rates and error functional.

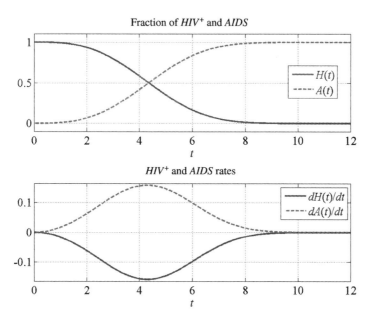

FIGURE 2.5   *HIV$^+$/AIDS* fractions and onset rates for $v(t) = \alpha t$.

survival until at least time t: $S(t) = \text{Prob}(\mathcal{T} \geq t) = 1 - F_{\mathcal{T}}(t)$. The *hazard function* is the ratio of the *pdf* to the *survivor function* at time t.

$$Haz(t) = \frac{f_{\mathcal{T}}(t)}{S(t)} = \frac{f_{\mathcal{T}}(t)}{1 - F_{\mathcal{T}}(t)} \tag{2.6}$$

Using a *modified Weibull* probability density function $W(t; a, b, c) = c \cdot p_w(t; a, b)$, a more appealing fit to the *AIDS* onset rate data can be attained. The *Weibull pdf* p_w is determined by the shape parameters a, b, and c. That is,

$$p_w(t; a, b) = \frac{b}{a^b} t^{(b-1)} \exp\left(-\left(\frac{t}{a}\right)^b\right). \tag{2.7}$$

Fitting a *modified Weibull pdf* product (i.e., $dA(t)/dt = c \cdot p_w(t; a, b) \cdot H(t)$) to the data in Table 2.1 yields parameter values $a = 3.0354$, $b = 3.5$, and $c = 0.497$ with a squared error of 0.0252: $E[a, b, c] = \Sigma_t[dA(t)/dt - c \cdot p_w(t; a, b) \cdot H(t)]^2 = 0.0252$. How does the selection of $v(t) = {}^1/_2 p_w(t; 3.04, 3.5)$ reflect upon the original model of (2.4)?

First notice that unlike the *linear model* $v(t) = \alpha t$, which results in *every HIV⁺* patient transitioning to *AIDS* after approximately 4.4 years, the *hazard function model* indicates that the fraction of *HIV⁺* cases levels out at $t \approx 5.0$ years. Indeed, for $t \geq$ 4 years, the proportion of patients who are *HIV⁺* (only) and those who have developed *AIDS* becomes constant at 0.61 and 0.39, respectively. Although this may be a more realistic model of the disease progression, it is plainly inadequate. In the absence of antiretroviral medication, the model implies that nearly 40% of *HIV⁺* cases will develop full-blown *AIDS* after approximately 5 years. See Figures 2.6–2.7.

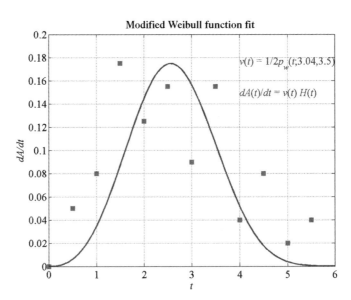

FIGURE 2.6   Hazard function fit to the *AIDS* onset rates.

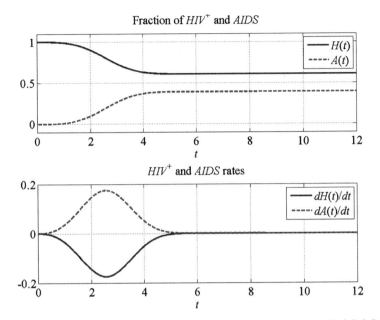

FIGURE 2.7   *HIV⁺/AIDS* fractions and onset rates for $v(t) = W(t; 3.04, 3.5, 0.5)$.

Therefore, additional models should be considered. This pursuit will be left as an exercise for the interested reader.

EXERCISES

2.5 By solving the first ordinary differential equation in (2.4), it was seen that $H(t) = e^{-V(t)}$, where $V(t) = \int_0^t v(s)\,ds$. Show that the second equation $dA(t)/dt = v(t)H(t)$ has the solution $A(t) = 1 - e^{-V(t)}$. *Hint*: Since $A(t) = \int_0^t v(s)\,e^{-V(s)}\,ds$, let $u = V(s)$ and integrate directly.

2.6 Using the methodology described via the *hazard function*, model the scaling term by an exponential: $v(t) = a\,e^{-\frac{1}{b^2}(t-t_c)^2}$. Solve for the parameters a, b, and t_c and construct plots parallel to Figures 2.3 and 2.4. What is the size of the error $E[a,b,t_c] = \sum_t \left[\dfrac{dA}{dt}(t) - a\,e^{-\frac{1}{b^2}(t-t_c)^2} \right]^2$ for the "optimal" values of a, b, and t_c?

2.3 REAL-TIME POLYMERASE CHAIN REACTION

The *polymerase chain reaction* (*PCR*) is a renowned methodology in molecular biology. Its primary use is to amplify a small (or even single strand) sample of *DNA* so that thousands of copies of the *DNA* sequence can be obtained. The applications of *PCR* include the analysis of genes, the diagnosis of hereditary diseases, genetic identification (forensic matching), and the detection of infectious diseases. This seminal method was developed by Kary Mullis in 1983 [10, 11] for which he shared the 1993 Noble Prize in chemistry.

The basic notion of *PCR* is to heat the original *DNA* sample to approximately 95°C. This temperature range [94°C, 97°C] causes the double helix structure of *DNA* to break apart into two complementary strands called *templates*. The two templates are referred to as the *associative* and *dissociative* template pairs. The process of breaking the original *DNA* sample into two template strands is known as *denaturing*. The denatured template strands are cooled to approximately 60°C and then attached to pre-specified *DNA* fragments called *primers*. The cooling is referred to as the *annealing* step. The *primer–template strand* combinations are known as *hybrids*.

An intermediate set of reactions (equilibrium and kinetic) follow the annealing phase. Once these reactions stabilize, the temperature is again raised to approximately 70°C and a *Taq* polymerase enzyme is introduced. This is an enzyme that was initially extracted from the bacterium called *Thermus aquaticus*: Hence, the eponymous contraction *Taq*. An *enzyme* is a protein produced by (living) cells that catalyze specific biochemical reactions such as digestion. A *polymerase enzyme* is one that catalyzes the formation of *DNA*.

The *Taq* enzyme binds to the hybrid *DNA* strands and enlarges or *amplifies* the hybrids. This amplification stage is called *synthesis* and marks the completion of the first stage of the *PCR* cycle. The synthesis set is also referred to as the *extension stage*. Using the amplified hybrids from the first cycle as the "input *DNA*" into the denaturing/annealing/synthesis process results in an even larger amplification of the original *DNA* sample in the second cycle. Continue this process n times and the *PCR* method is completed.

How large does n need to be until a sufficient quantity of amplified *DNA* is produced to attain the desired effect (e.g., disease diagnosis, forensic *DNA* match)? That is the dominion of *real-time polymerase chain reaction* (*rtPCR*). A mathematical model based on the work of Gevertz et al. [3] is presented for *rtPCR*. The outcomes of the *rtPCR* process are fluorescence intensities. The greater the level of *DNA* in a sample, the higher the fluorescence. The fluorescence, in turn, is directly proportional to the efficiencies of the amplified *DNA* concentrations at each cycle in the *rtPCR* process. The overall *PCR* efficiency, at each cycle, is a product of the efficiencies at each of the three stages: Denaturation, annealing, and synthesis.

Before the mathematical model of the *rtPCR* method is described, some notation must be established. Then each stage is developed and the model is exercised.

2.3.1 Nomenclature

PCR	*polymerase chain reaction*
rtPCR	real-time *PCR*
N	number of *PCR* cycles
$\varepsilon_{PCR}^{(n)}$	overall *PCR* efficiency at cycle n
$\varepsilon_{den}^{(n)}$	efficiency at cycle n during denaturation
$\varepsilon_{ann}^{(n)}$	efficiency at cycle n during annealing

$\varepsilon_{syn}^{(n)}$	efficiency at cycle n during synthesis
t_d	time for denaturation (nominally $t_d = 5$ seconds)
t_a	time for annealing (nominally $t_a = 20$ seconds)
t_s	time for synthesis (nominally $t_s = 15$ seconds)
F_n	fluorescence level at cycle n
$[X]$	concentration of species X
P_1	associative primer
P_2	dissociative primer
T_1	associative template
T_2	dissociative template
H_1	associative hybrid
H_2	dissociative hybrid
U	template hybrid
D	primer hybrid (primer–dimer)

The overall efficiency is the product of the stage efficiencies.

$$\boxed{\varepsilon_{PCR}^{(n)} = \varepsilon_{den}^{(n)} \cdot \varepsilon_{ann}^{(n)} \cdot \varepsilon_{syn}^{(n)}}$$

2.3.2   Fluorescence

The fluorescence at the conclusion of a *PCR* cycle is proportional to the overall *PCR* efficiency at the cycle index n.

$$F_n = \alpha \cdot [DNA]_0 \cdot \prod_{j=1}^{n} \left(1 + \varepsilon_{PCR}^{(j)}\right) \tag{2.8}$$

The symbol $[DNA]_0$ corresponds to the initial concentration of *DNA*.

2.3.3   Efficiency During Denaturation

The assumption is that denaturation results in 100% efficiency at each cycle n.

$$\varepsilon_{den}^{(n)} = 1, \quad \text{for all } n \tag{2.9}$$

2.3.4   Efficiency During Annealing

The annealing efficiency is determined by the *PCR* kinetics and total template concentrations at each cycle. It is the ratio of the hybrid concentrations (at each step) over the sum of the total template concentrations. That is,

$$\varepsilon_{ann}^{(n)} = \frac{[H_1]^{(n)} + [H_2]^{(n)}}{[T_1]_T^{(n)} + [T_2]_T^{(n)}} \tag{2.10}$$

$$[T_j]_T^{(n)} = [T_j]^{(n)} + [H_j]^{(n)} + [U]^{(n)}, \quad j = 1, 2. \tag{2.11}$$

To see how the hybrids H_j and total template T_j concentrations change over the annealing step, let $k = [k_{aH_1}, k_{dH_1}, k_{aH_2}, k_{dH_2}, k_{aU}, k_{dU}, k_{aD}, k_{dD}]^T \equiv [k_1, k_2, k_3, k_4, k_5, k_6, k_7, k_8]^T$ be the vector of association and dissociation rates with $k_1 \equiv k_{aH_1}$, $k_2 \equiv k_{dH_1}$, and so forth. Define the *kinetics vector* as $x = [[P_1], [P_2], [T_1], [T_2], [H_1], [H_2], [U], [D]]^T \equiv [x_1, x_2, x_3, x_4, x_5, x_6, x_7, x_8]^T \in \mathbb{R}^8$, where $x_1 = [P_1]$, $x_2 = [P_2]$, $x_3 = [T_1]$, $x_4 = [T_2]$, $x_5 = [H_1]$, $x_6 = [H_2]$, $x_7 = [U]$, and $x_8 = [D]$. The dynamical flow of the various concentrations is governed by the non-linear system of ordinary differential equations (2.12).

$$\frac{dx}{dt} = f(x, k; t) \tag{2.12}$$

The right-hand side of (2.12) is the vector of quasilinear terms as displayed below.

$$f(x, k; t) = \begin{bmatrix} -(k_1 x_3 + k_7 x_2)x_1 + k_2 x_5 + k_8 x_8 \\ -(k_3 x_4 + k_7 x_1)x_2 + k_4 x_6 + k_8 x_8 \\ -(k_1 x_1 + k_6 x_4)x_3 + k_2 x_5 + k_6 x_7 \\ -(k_3 x_2 + k_6 x_3)x_4 + k_4 x_6 + k_6 x_7 \\ k_1 x_1 x_3 - k_2 x_5 \\ k_3 x_2 x_4 - k_4 x_6 \\ k_5 x_3 x_4 - k_6 x_7 \\ k_7 x_1 x_2 - k_8 x_8 \end{bmatrix}$$

$$\equiv \begin{bmatrix} -(k_{aH_1}[T_1] + k_{aD}[P_2])[P_1] + k_{dH_1}[H_1] + k_{dD}[D] \\ -(k_{aH_2}[T_2] + k_{aD}[P_1])[P_2] + k_{dH_2}[H_2] + k_{dD}[D] \\ -(k_{aH_1}[P_1] + k_{aU}[T_2])[T_1] + k_{dH_1}[H_1] + k_{dU}[U] \\ -(k_{aH_2}[P_2] + k_{aU}[T_1])[T_2] + k_{dH_2}[H_2] + k_{dU}[U] \\ k_{aH_1}[P_1][T_1] - k_{dH_1}[H_1] \\ k_{aH_2}[P_2][T_2] - k_{dH_2}[H_2] \\ k_{aU}[T_1][T_2] - k_{dU}[U] \\ k_{aD}[P_1][P_2] - k_{dD}[D] \end{bmatrix}$$

At the first *PCR* cycle (i.e., $n = 1$), integrate system (2.12) for $t \in [t_d, t_d + t_a]$ to obtain $[T_1]^{(1)}, [T_2]^{(1)}, [H_1]^{(1)}, [H_2]^{(1)}$, and $[U]^{(1)}$. These concentrations, in turn, yield $[T_j]_T^{(1)}$ for $j = 1, 2$ by (2.11) and hence $\varepsilon_{ann}^{(1)}$ by (2.10). Continue in this fashion to calculate $\varepsilon_{ann}^{(n)}$ for any cycle index $n = 1, 2, \ldots, N$.

2.3.5 Efficiency During Synthesis

The efficiency of the synthesis stage of the real-time *PCR* process is tied to the concentration of the *Taq* polymerase $[E]$ at cycle n and the primer–template hybrids referred to as substrate concentration $[S]$ during synthesis. First, note as a function of cycle index n the *Taq* enzyme concentration at each stage of the *rtPCR* process evolves according to the kinetics in (2.13a).

$$[E]^{(n+1)} = [E]^{(n)} \exp(-k_{deg} t_d) \tag{2.13a}$$

Here, $k_{deg} = 1.9 \times 10^{-4}$ per second is the *Taq* degradation constant (the rate at which the enzyme loses its concentration per second) and t_d is the denaturation time. For $t_d = 5$ seconds, $\exp(-k_{deg}t_d) = 0.9991$. Consequently, the *Taq* polymerase is assumed to lose only 0.1% of its enzyme activity at each step in the *rtPCR*. Subsequently, (2.13a) is replaced by (2.13b).

$$[E]^{(n)} = [E]^{(0)}(0.9990)^n \tag{2.13b}$$

The initial *Taq* concentration $[E]^{(0)}$ is related to the ratio V_{max}/k_{cat}. Indeed, the time varying version of the Michaelis–Menten kinetics [5] states that $V_{max} = k_{cat}[E]^{(n)}$. By (2.13b), however, $V_{max} = k_{cat}[E]^{(0)}(0.9990)^n$. Therefore, V_{max} must be treated as a function of the cycle index. That is, $V_{max} = V_{max}(n)$. The value of $V_{max}(0)$ is temperature-dependent. During the annealing process, at 72°C, the estimate $V_{max}(0) = 2.22 \times 10^{-4}$ mol/m^3/s is provided by Dijksman and Pierik [1]. The primer–template substrate concentration $[S]$ evolves according to the first-order ordinary differential equation as established by Michaelis and Menten [5],

$$\frac{d[S]}{dt} = -\frac{V_{max}(n) \cdot [S]}{K_m + [S]} \tag{2.14}$$

subject to the initial conditions $[S]|_{t=t_0} = [S]_T$. The *total substrate concentration* at each cycle level n, $[S]_T^{(n)}$, is computed as a sum of the hybrid concentrations $[S]_T^{(n)} = [H_1]^{(n)} + [H_2]^{(n)}$. Integrating (2.14) directly and applying the initial conditions yields the implicit algebraic equation

$$\left(\frac{[S]}{K_m}\right) \cdot \exp\left(\frac{[S]}{K_m}\right) = \frac{[S]_T}{K_m} \exp\left(\frac{[S]_T - V_{max}(t - t_0)}{K_m}\right). \tag{2.15a}$$

By letting $w = [S]/K_m$, $w_0 = [S]_T/K_m$, $\eta_0 = V_{max}(n)/K_m$, and $\theta(\tau, w_0, \eta_0) \equiv w_0 \cdot \exp(w_0 - \eta_0\tau)$, then the right-hand side of (2.15a) can be written as $\theta(t - t_0; w_0, \eta_0) = w_0 \cdot e^{w_0} \cdot e^{-\eta_0(t-t_0)}$. Therefore, (2.15a) can be simplified to the equation

$$we^w = \theta(t - t_0; w_0, \eta_0). \tag{2.15b}$$

The solution of (2.15b) is attained via a special function called the *Lambert w-function* λ_w. Formally, the Lambert *w-function* is the inverse of the exponential function $f(w) \equiv we^w$. Thus, if $z = we^w$, then $w = \lambda_w(z)$. Olver et al. [8] discuss the Lambert *w-function* in some detail. Moler [6,7] has developed a short and effective MATLAB program `lambertw.m` using Halley's method (see http://blogs. mathworks.com/cleve/2013/09/02/the-lambert-w-function/ for more details). The solution of (2.15b) is then

$$w = \lambda_w(\theta(t - t_0; w_0, \eta_0)). \tag{2.15c}$$

As stated earlier, the estimate $V_{max}(0) = 2.22 \times 10^{-4}$ mol/m^3/s for the *Taq* polymerase is provided by Dijksman and Pierik [1]. Thus, $k_{cat} [E]^{(0)} = V_{max}(0) \approx 2.22 \times 10^{-4}$. The dimensionless constant $K_m = 1.5 \times 10^{-9}$ is provided by Gevertz et al. [3]. Solve (2.15c) at $t = t_s$ to obtain $[S]_{t_s}$.

The *synthesis efficiency* at each cycle index n is the normalized difference inbetween the total and synthesis substrate concentrations as indicated by (2.16).

$$\varepsilon_{ext}^{(n)} = \frac{[S]_T^{(n)} - [S]_{t_s}^{(n)}}{[S]_T^{(n)}} \tag{2.16}$$

2.3.6 Overall PCR Efficiency

Combining (2.9), (2.10), and (2.15c) yields the overall *PCR efficiency* ε_{PCR} and so by (2.8) the cycle fluorescence.

$$\varepsilon_{PCR}^{(n)} = \varepsilon_{den}^{(n)} \cdot \varepsilon_{ann}^{(n)} \cdot \varepsilon_{ext}^{(n)} \tag{2.17}$$

2.3.7 Real-Time PCR Process

The method for determining the overall *PCR* efficiency proceeds in the three stages outlined earlier: Denaturation, annealing, and synthesis. The fourth stage yields fluorescence.

Stage 1: Denaturation
It is assumed that the efficiency *at every PCR* level is 100% so that $\varepsilon_{den}^{(n)} = 1$ for *each n* and all $t \in [0, t_d] = [0, 5]$.

Stage 2: Annealing
The efficiency $\varepsilon_{ann}^{(n)}$ is computed at each *PCR cycle* n by first integrating the ordinary differential equation (2.12) over $t \in [t_d, t_d + t_a] = [5, 25]$ with respect to the initial conditions $x_0^{(n)}$. For $n = 1$, $x_0^{(1)} = [[P_1]_T^{(1)}, [P_2]_T^{(1)}, [T_1]_T^{(1)}, [T_2]_T^{(1)}, 0, 0, 0, 0]^T$; for $n = 2$, $x_0^{(2)}$ will take on the same form as $x_0^{(1)}$ *following* the extension/synthesis step. Since the denaturing temperature is so high (95–97°C), the model assumption is that the total concentrations of all primer–template hybrids, template hybrids, and primer–dimers after denaturation is zero. That is, $[H_1]_T^{(n)} = [H_2]_T^{(n)} = [U]_T^{(n)} = [D]_T^{(n)} = 0$ for each n. At each *PCR* cycle index n, $\varepsilon_{ann}^{(n)}$ is computed via (2.10)–(2.11).

Stage 3: Synthesis
The efficiency $\varepsilon_{ext}^{(n)}$ is more complicated logistically to compute. Step 1 is to integrate the ordinary differential equation (2.12) at *PCR* cycle level n over $t \in [t_d, t_d + t_a] = [5, 25]$ and obtain the solution $x^{(n)}(t_d + t_a)$. Select the requisite elements of $x^{(n)}(t_d + t_a)$ (namely elements 5 and 6) to calculate the initial condition $S_0 \equiv [S]_T^{(n)} = [H_1]^{(n)} + [H_2]^{(n)}$ for the ordinary differential equation (2.14). Solve for $[S]_{t_s}^{(n)}$ via (2.15c) at $t = t_d + t_a + t_s$ to obtain the value $[S]_{t_s}^{(n)} = K_m \cdot \lambda_w(\theta(t - t_0), [S]_T^{(n)}/K_m, V_{max}(n)/K_m)$ at each *PCR* cycle level n. Here, $t_0 = t_d + t_a$. Then use (2.16) to compute $\varepsilon_{ext}^{(n)}$. Now $\varepsilon_{PCR}^{(n)}$ is computed from (2.17). Next set the

TABLE 2.2   Constants associated with the 18S complementor DNA (as template)

$[P_1]_T^{(1)}$	$[P_2]_T^{(1)}$	$[T_1]_T^{(1)}$	$[T_2]_T^{(1)}$	K_{H_1}	K_{H_2}	K_U
10^{-6} mol	10^{-6} mol	10^{-10} mol	10^{-10} mol	5.5531×10^{-13} mol	8.1493×10^{-11} mol	≈ 0
K_D	k_{aH_1}	k_{aH_2}	k_{aU}	k_{aD}	k_{dH_1}	k_{dH_1}
10^{-2} mol	$10^6 \cdot$1/mol/s	$10^6 \cdot$1/mol/s	$10^6 \cdot$1/mol/s	$10^6 \cdot$1/mol/s	5.5531×10^{-7} 1/s	8.1493×10^{-11} 1/s
k_{dU}	k_{dD}	K_m	k_{cat}	k_{deg}	t_{den}	t_{syn}
≈ 0	$10^4 \cdot$1/mol/s	1.5×10^{-2} mol	$10^4 \cdot$1/mol/s	$1.9 \times 10^{-4} \cdot$1/s	5 s	15 s

initial condition $x_0^{(n)}$ via equations (2.18a)–(2.18d).

$$x_0^{(n)} = \left[[P_1]_T^{(n)}, [P_2]_T^{(n)}, [T_1]_T^{(n)}, [T_2]_T^{(n)}, 0, 0, 0, 0 \right]^T \tag{2.18a}$$

$$\left. \begin{aligned} [P_1]_T^{(n+1)} &= [P_1]_T^{(n)} - \varepsilon_{syn}^{(n)} [H_1]^{(n)} \\ [P_2]_T^{(n+1)} &= [P_2]_T^{(n)} - \varepsilon_{syn}^{(n)} [H_2]^{(n)} \end{aligned} \right\} \tag{2.18b}$$

$$\left. \begin{aligned} [T_1]_T^{(n+1)} &= [T_1]_T^{(n)} + \varepsilon_{syn}^{(n)} [H_2]^{(n)} \\ [T_2]_T^{(n+1)} &= [T_2]_T^{(n)} + \varepsilon_{syn}^{(n)} [H_1]^{(n)} \end{aligned} \right\} \tag{2.18c}$$

$$[H_1]_T^{(n)} = [H_2]_T^{(n)} = [U]_T^{(n)} = [D]_T^{(n)} \tag{2.18d}$$

Stage 4: Fluorescence
Equation (2.8) gives the desired result.

Using the constants supplied by Gevertz et al. [3] (and listed in Table 2.2), computations were made assuming the initial amount of *DNA* is 1 mole and $\alpha = 1$ as per

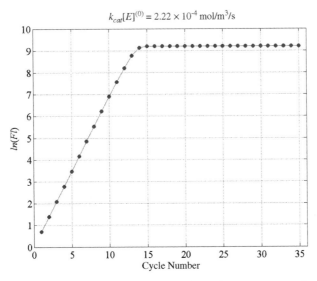

$k_{cat}[E]^{(0)} = 2.22 \times 10^{-4}$ mol/m³/s

FIGURE 2.8   Fluorescence for 72°C annealing substrate concentration rate.

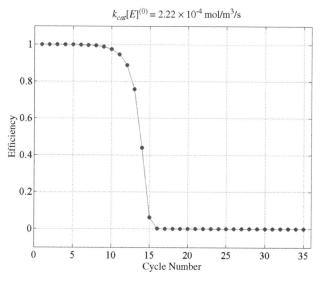

FIGURE 2.9   *PCR* efficiency for 72°C annealing substrate concentration rate.

equation (2.8). The corresponding fluorescence and *PCR* efficiency are displayed in Figures 2.8 and 2.9.

MATLAB Commands
(Real-Time PCR Simulation and Plotting Results)

```
%Load the structure of constants
P = load('C:\PJC\Math_Biology\Chapter_2\Data\Ktrpcr.mat'));
% Extract the structure for input
K = P.K;
% Run the real-time PCR simulation
tic; F = rtpcr(35,K,1); toc
Elapsed time is 383.260878 seconds.
% Plot the fluorescence ...
figure;
h = plot(F.cycles,log(F.fluorescence),'bo-'); grid('on');
set(h,'MarkerFaceColor','b');
set(gca,'FontSize',16,'FontName','Times New Roman');
xlabel('\itCycle Number'); ylabel('\itln\rm(\itFl\rm)');
title('\itk_{cat}\rm[\itE\rm]^{(0)} = 2.22\bullet10^{-4}
  \itmol/m^3/s');
axis([0,36,0,10]);
% ... and PCR efficiency
figure;
```

```
h = plot(F.cycles,F.Eff.pcr,'bo-'); grid('on');
set(h,'MarkerFaceColor','b');
set(gca,'FontSize',16,'FontName','Times New Roman');
xlabel('\itCycle Number'); ylabel('\itEfficiency');
title('\itk_{cat}\rm[\itE\rm]^{(0)} = 2.22\bullet10^{-4}
\itmol/m^3/s');
axis([0,36,-0.1,1.1]);
```

REFERENCES

[1] Dijksman, J. F. and A. Pierik, "Mathematical analysis of the real time PCR (RTA PCR) process," *Chemical Engineering Science*, vol. 71, pp. 496–506, 2012.

[2] Fisher, M. E., "A semiclosed–loop algorithm for the control of blood glucose levels in diabetics," *IEEE Transactions on Biomedical Engineering*, vol. 38, no. 1, pp. 57–61, 1991.

[3] Gevertz, J. L., S. M. Dunn, and C. M. Roth, "Mathematical model of real–time PCR kinetics," *Biotechnology and Bioengineering*, vol. 92, no. 3, pp. 346–355, November 2005.

[4] Lewin, B., L. Cassimeris, V. R. Lingappa, and G. Plopper, editors, *Cells*, Jones and Bartlett, Sudbury, MA, 2007.

[5] Michaelis, L. and M. Menten, "Die kinetik der invertinwirkung," *Biochemische Zeitschrift*, vol. 49, pp. 333–369, 1913.

[6] Moler, C. B., *Private Communication*, August 2013.

[7] Moler, C. B., "Cleve's corner," *MathWorks Technical Newsletter*, vol. 2, September 2013, http://blogs.mathworks.com/cleve/2013/09/02/the-lambert-w-function/

[8] Olver, F. W. J., D. W. Lozier, R. F. Boisvert, and C. W. Clark, *NIST Handbook of Mathematical Functions*, Cambridge University Press, 2010.

[9] Peterman, T. A., D. P. Drotman, and J. W. Curran, "Epidemiology of the acquired immunodeficiency syndrome (AIDS)," *Epidemiological Reviews*, vol. 7, pp. 7–21, 1985.

[10] Saiki, R., S. Scharf, S. Faloona, K. Mullis, G. Horn, J. Erlich, and N. Arnheim, "Enzymatic amplification of *beta–goblin* genomic sequences and restriction site analysis for diagnosis of sickle cell anemia," *Science*, vol. 230, no. 4732, pp. 1350–1354, 1985.

[11] Saiki, R., D. Gelfand, S. Stoffel, S. Scharf, R. Higuichi, G. Horn, K. Mullis, and J. Erlich, "Primer–directed enzymatic amplification of DNA with a thermostable DNA polymerase," *Science*, vol. 239, no. 4839, pp. 487–491, 1988.

[12] Yeargers, E. K., J. V. Herod, and R. W. Shonkwiler, *An Introduction to the Mathematics of Biology*, Birkhäuser Boston, 1996.

FURTHER READING

Demmel, J. W., *Applied Numerical Linear Algebra*, SIAM Books, Philadelphia, PA, 1997.

Moler, C. B., *Numerical Computing with MATLAB*, SIAM Books, Philadelphia, PA, 2004.

Trefethen, L. N. and D. Bau III, *Numerical Linear Algebra*, SIAM Books, Philadelphia, PA, 1997.

3

SEIR MODELS

One of the fundamental ideas within the sub-specialty of *mathematical epidemiology* is to model the outbreak of an infectious disease through a population. This can be achieved via *Susceptible–Exposed–Infected–Removed/Recovered (SEIR)* models. Figure 3.1 illustrates a typical *SEIR* scenario in which members of the susceptible population $S(t)$ become part of the exposed population $E(t)$ at the rate proportional to a_1. The exposed, in turn, transition to the infected populace $I(t)$ at a rate proportional to a_2. The infected *recover* at a rate proportional to a_3 and those who recover $R(t)$ from the disease return to the population of susceptibles at a rate proportional to a_4.

This diagram can be converted into a mathematical format via the set of differential equations (3.1).

$$\left.\begin{array}{l} \dfrac{dS}{dt}(t) = -a_1 S(t) + a_4 R(t) \\[2mm] \dfrac{dE}{dt}(t) = a_1 S(t) - a_2 E(t) \\[2mm] \dfrac{dI}{dt}(t) = a_2 E(t) - a_3 I(t) \\[2mm] \dfrac{dR}{dt}(t) = a_3 I(t) - a_4 R(t) \end{array}\right\} \qquad (3.1)$$

If $Z(t) = S(t) + E(t) + I(t) + R(t)$, then observe $dZ(t)/dt = 0$ and hence, $Z(t) \equiv$ constant. This is because Figure 3.1 and its subsequent model equation (3.1) describe

Applied Mathematics for the Analysis of Biomedical Data: Models, Methods, and MATLAB®, First Edition. Peter J. Costa.
© 2017 Peter J. Costa. Published 2017 by John Wiley & Sons, Inc.
Companion website: www.wiley.com/go/costa/appmaths_biomedical_data

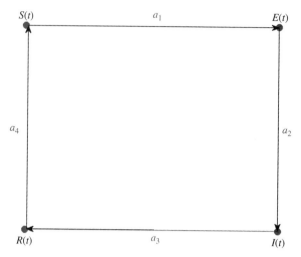

FIGURE 3.1 A generic *SEIR* schema.

a *closed system*. No new members are introduced into the population of susceptibles and no one fails to recover from the disease. Real-life situations are plainly quite different.

Consider the simple model of a viral interaction within the body in-between an invading virus and the immune system. The model comes courtesy of Britton [3]. A virus particle is called a *viron*. It infects the body by latching onto a healthy cell and extracting nutrition from the healthy cell's cytoplasm. This permits the viron to replicate, infect the healthy cell, and then move on to a new cell to infect. If $V(t)$ is the population of virons at time t, $H(t)$ the population of healthy cells, $I(t)$ the population of infected cells, and R the repository of dead cells, then the *SEIR* diagram for this virus/immune system interaction is represented by Figure 3.2.

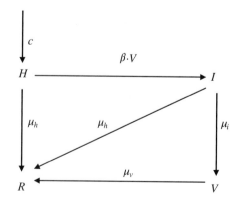

FIGURE 3.2 The *SEIR* diagram for virus/immune system interaction.

Here, c is the rate at which the body produces new healthy cells, μ_h is the rate at which healthy cells are removed (i.e., die), μ_i is the rate at which infected cells are converted to virons, and μ_v is the rate at which virons are removed. If β is the contact rate of a viron with a healthy cell, then $\beta \cdot V$ is the rate at which healthy cells are infected. The upper left portion of Figure 3.2 indicates that the rate of change of the healthy cell population $H(t)$ is proportional to the input rate c and the outgoing rates $\beta \cdot V(t)$ and μ_h. That is, $dH(t)/dt = c - \beta \cdot V(t) \cdot H(t) - \mu_h H(t)$. Hence, Figure 3.2 leads to the set of nonlinear ordinary differential equations (3.2).

$$
\left.
\begin{aligned}
\frac{dH}{dt} &= c - (\beta \cdot V(t) - \mu_h) \cdot H(t) \\
\frac{dI}{dt} &= \beta \cdot V(t) \cdot H(t) - (\mu_i + \mu_h) \cdot I(t) \\
\frac{dV}{dt} &= \mu_i \cdot I(t) - \mu_v \cdot V(t)
\end{aligned}
\right\}
\tag{3.2}
$$

To exercise this model, estimates of the parameters c, β, μ_h, μ_i, and μ_v are required along with knowledge of the initial populations $H_0 = H(t_0)$, I_0, and V_0. How can such parameters be estimated? Are there literature references or other resources available to the mathematical scientist to guide such a process? That is the key component of this chapter. Two examples, (i) the spread of human immunodeficiency virus (*HIV*) and its conversion to acquired immune deficiency syndrome (*AIDS*) and (ii) the transmission of a respiratory infection through a population, are examined in detail. *All of the parameters are either estimated from actual measurements or modeled after reported behavior.* With respect to the transmission of an infectious respiratory disease, measurements from the Boston Children's Hospital are mimicked to form the basis of the model. Since the measurements came from Boston, the various parameters were estimated from data published by the Commonwealth of Massachusetts. Similarly, the *HIV/AIDS* model uses parameters estimated by data reported in studies of infection rates, seroconversion rates, and death rates due to the acquisition of the disease.

The respiratory model has a collection of reported infections as a function of time. This is commonly called a *time series*. These measurements along with the corresponding *SEIR* model contain all of the components required for one of the twentieth century's most celebrated estimation methods: The *Kalman filter*. The filter combines the time series data and mathematical model to produce an optimal estimate of the model "state" along with a probabilistic error. Some "tricks of the trade" and guidelines into the "tuning" of a Kalman filter will also be discussed. Historically, the Kalman filter has been used in radar applications. Here, it is utilized to help predict the outbreak of a contagious disease.

From the generic *SEIR* scenario, realistic models of disease outbreak can be developed. The crucial components of these models are data. This chapter details the elements of an effort used to estimate the model parameters and develop a realistic result.

3.1 PRACTICAL APPLICATIONS OF *SEIR* MODELS

3.1.1 Transmission of *HIV/AIDS*

Consider the transmission of *HIV* and *AIDS* through a certain restricted population; say a city or large university campus. The following notation will be used to construct an *SEIR* model for the transmission of *HIV* and conversion to *AIDS*.

$S(t)$	the population of people susceptible to the transmission of *HIV* at time t
$X(t)$	the population of people who are *HIV* positive (HIV^+) at time t
$X_q(t)$	the population of people who are HIV^+ but are no longer infectious at time t
$Y(t)$	the population of people who have developed *AIDS* at time t
$Y_q(t)$	the population of people who have developed *AIDS* but are no longer infectious at time t
B	the number of new susceptibles introduced into the population per time unit
μ_D	the rate at which people are removed from the entire population due to "non-*AIDS*" causes
μ_A	the rate at which people with *AIDS* die from *AIDS*
p	the proportion of the infectious population who are not "quarantined"
$I(t) = X(t) + Y(t)$	the total *infectious* population at time t
$N(t) = S(t) + I(t)$	the total *active* at-risk population at time t
$\beta(t)$	the probability of *HIV* transmission at time t
$\lambda(t) = \beta(t)\frac{I(t)}{N(t)}$	the probability of *HIV* transmission (infection) per contact at time t
$c(t)$	the number of contacts (partners) at time t
$v(t)$	the seroconversion rate from *HIV* to *AIDS* ($1/v(t)$ = incubation period)
$\boxed{\text{Births}}$	the number of people entering the susceptible population (births, immigration)
$\boxed{D_N}$	the number of people leaving the susceptible population (non-*AIDS* deaths, emigration)
$\boxed{D_A}$	the number of deaths due to *AIDS*

Remarks: The total *infectious* population $I(t) = X(t) + Y(t)$ *does not* include either $X_q(t)$ or $Y_q(t)$ since these "quarantined" populations are no longer viewed as vectors for disease transmission.

The word "quarantined" does not necessarily mean physically isolated from the community at large. Rather, it should be understood to mean the subpopulation of the *HIV*- or *AIDS-infected* populations who no longer shed the virus due to (i) abstention from activities that can spread the disease, (ii) a regimen of antiretroviral medications, or (iii) other medical interventions. New *susceptibles* are introduced into the

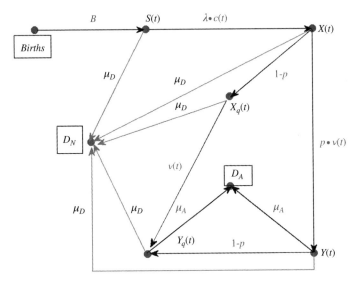

FIGURE 3.3   The *SEIR* model for the transmission of *HIV* and seroconversion to *AIDS*.

population $S(t)$ via "births" (i.e., immigration, being born, or otherwise moving into an area).

The *SEIR* diagram for the transmission of *HIV* and seroconversion to *AIDS* is presented in Figure 3.3. Observe that the first (upper) portion of the diagram indicates that the rate of change of the susceptible population is *increased* by the number of "births" B (per time unit) and *decreased* by the rate of infection $\lambda \bullet c(t)$ and "death rate" μ_D. The parameter μ_D is the rate at which people die (of non-*AIDS* related causes) or emigrate out of the population.

This leads to the differential equation $dS(t)/dt = B - (\lambda \bullet c(t) + \mu_D) S(t)$. To see that this is sensible, consider the case in which $\lambda = 0 = \mu_D$. Then $dS(t)/dt = B$ so that $S(t) - S(t_0) = B \bullet (t - t_0) \Leftrightarrow S(t) = S(t_0) + B \bullet (t - t_0)$. That is, the population (with no deaths or emigration) grows linearly in time according to the number of "births."

The complete model, based on the work of Gupta et al. [15], as illustrated in Figure 3.3, leads to the set of coupled, nonlinear ordinary differential equations (3.3).

$$
\left.
\begin{aligned}
\frac{dS}{dt}(t) &= B - \left(\mu_D + \lambda \bullet c(t)\right) S(t) \\
\frac{dX}{dt}(t) &= \lambda \bullet c(t)S(t) - \left(\mu_D + (1-p) + p\bullet v(t)\right) X(t) \\
\frac{dX_q}{dt}(t) &= (1-p)X(t) - \left(\mu_D + v(t)\right) X_q(t) \\
\frac{dY}{dt}(t) &= p\bullet v(t)X(t) - \left(\mu_D + \mu_A + (1-p)\right) Y(t) \\
\frac{dY_q}{dt}(t) &= v(t)X_q(t) - (\mu_D + \mu_A)Y_q(t) + (1-p) Y(t)
\end{aligned}
\right\}
\qquad (3.3)
$$

Now, the task at hand is to find estimates of the assorted rate parameters B, $\beta(t)$, $c(t)$, p, μ_A, μ_D, and $v(t)$. To construct a realistic model, a set of assumptions must be made. First, it is assumed that *everyone* is susceptible to infection and no one recovers from the disease. Recent medical studies, however, have indicated that some HIV^+ patients may become *seronegative* (i.e., virus free) after treatment with *highly active antiretroviral therapy* (HAART). This has not been confirmed for all HIV^+ recipients of HAART; see Finzi et al. [13] and Dornadula et al. [12]. For the purpose of this analysis, it will be assumed that there is no recovery once a person is infected with *HIV*. Hence, "infecteds" are removed from the population only via "death." Plainly, if an HIV^+ person develops *AIDS* and then dies from the disease (at a rate of μ_A), then this person is removed from the population.

When modeling the microcosm of a city, for example, entry into and exit from the population of *susceptibles* becomes more problematic. Births and deaths from natural (i.e., non-*AIDS*) causes are the obvious manners in which the susceptible population grows and declines, respectively. People also move into (immigrate) as well as move out of (emigrate) communities. Hence, the *SEIR* diagram labels $\boxed{\text{Births}}$ and $\boxed{D_N}$ represent the sources of the susceptible population's growth and decline. The parameters B and μ_D will be treated as constants. The label $\boxed{D_A}$ represents the decline in the *AIDS* population due to death by the disease.

Consider the *greater Boston area* to be defined as towns and cities within a 15-mile radius of Boston, Massachusetts. Renski et al. [27] report that, for the period 2005–2010, the number of *births*, *domestic immigration*, and *international immigration* into the greater Boston area were 122,374, 526,510, and 153,105, respectively. In this setting, the "births per year" is the sum of these populations divided by the number of years: $B = (122{,}374 + 526{,}510 + 153{,}105)/5 = 160{,}397$ *people per year*. Since the non-*AIDS* population depletion rate μ_D is in proportion to the susceptible population, it is estimated from the number of deaths, domestic emigration, and international emigration *out of* the greater Boston area. Again, following Renski et al. [27], it is seen that from 2005 to 2010, $\mu_D \approx$ [#(deaths) + #(domestic emigrants) + #(international emigrants)]/population(2010)/#(years) $= (71{,}114 + 547{,}465 + 154{,}198)/1{,}975{,}155/5 = 0.0782$. The symbol #($\Theta$) means the number of people in the population defined by Θ while #(years) means the number of years over which the data have been gathered. The population in 2005 was estimated as 1,945,942, while the population in 2010 was listed as 1,975,155. The *Centers for Disease Control* (CDC) [6] note that, in 2011, the number of people with HIV^+ and *AIDS* per $N_0 = 100{,}000$ were $H_0 = 314.6$ and $A_0 = 187.8$, respectively.

These numbers can be used to provide initial estimates of the populations of *susceptibles* $S(t_0)$, HIV^+ $X(t_0)$, HIV^+ but not infectious $X_q(t_0)$, and *AIDS* $Y(t_0)$. To that end, take t_0 to be the year 2010. Then an estimate of the initial susceptible population is $S(t_0) = 1{,}975{,}155$. The CDC [6] infection rates mentioned above imply that the initial population of HIV^+ *who still shed the virus* is $X(t_0) = p \cdot S(t_0) \cdot H_0/N_0$. Here, p is the proportion of the HIV^+ population who remain infectious. The remaining initial population estimates are given directly as $X_q(t_0) = (1 - p) \cdot S(t_0) \cdot H_0/N_0 = (1 - p) \cdot X(t_0)/p$, $Y(t_0) = p \cdot S(t_0) \cdot A_0/N_0$, and $Y_q(t_0) = (1 - p) \cdot S(t_0) \cdot A_0/N_0 = (1 - p) \cdot Y(t_0)/p$. This development is summarized in Table 3.1.

TABLE 3.1   Initial values for the various *SEIR* populations as a function of the proportion infectious *p*

		Initial values, t_0 occurs at year 2010		
$S(t_0)$	$X(t_0)$	$X_q(t_0)$	$Y(t_0)$	$Y_q(t_0)$
1,975,155	$p \cdot 6{,}214$	$(1-p) \cdot 6{,}214$	$p \cdot 3{,}709$	$(1-p) \cdot 3{,}709$

Next consider the transmission parameter β. Hyman and Stanley [18, 19] note that the probability of transmission of *HIV* from male to female β_{MF} during sexual intercourse per contact is 0.001, while the reverse (probability of transmission from female to male β_{FM}) is believed to be substantially less at 0.000025. The probability of transmission from male to male is $\beta_{MM} = 0.1$. Wawer et al. [31] estimate an average rate of *HIV* transmission of 0.008 per contact. Consequently, estimating the probability of transmission is very much dependent upon the composition of the susceptible population. Petersen et al. [26], however, have published data on transmission probabilities as a function of contact (with an infected carrier) and *time*. These values are provided in Table 3.2. Plotting these data in Figure 3.4 gives rise to a mathematical model for the probability of transmission.

TABLE 3.2   Petersen et al. [26] *HIV* transmission data

Probability of *HIV* transmission $\hat{\beta}(t)$	Time t (days)
0.63	40
0.58	60
0.5	80
0.43	100
0.36	120
0.3	140
0.275	160
0.26	180
0.24	200
0.23	220
0.2	240
0.17	260
0.14	280
0.12	300
0.11	320
0.12	340
0.125	360
0.12	380
0.12	400
0.11	420
0.095	440
0.08	460
0.07	480
0.08	500

The model for the probability of *HIV* transmission per contact at time t (years) is an exponential. Since Petersen et al. [26] presented no measurements from time $t = 0$ to $t_1 = 40$ days $(= 0.1095$ years), a linear ramp will be assumed. Thus, the model is

$$\beta(t;c) = \begin{cases} \dfrac{\beta(t_1;c)}{t_1}t & \text{for } 0 \leq t \leq t_1 \\ c_1 \exp[-c_2(t - t_1)] + c_3 & \text{for } t > t_1. \end{cases} \tag{3.4}$$

A nonlinear regression least-squares fit to the data provided in Table 3.2, as per Seber and Wild [29], indicates that $c = [c_1, c_2, c_3] = [0.5745, 2.9466, 0.0708]$ for $t_1 = 0.1095$ years. That is, the functional $E[c] = \sum_{j=1}^{N}\left(c_1 \exp[-c_2(t_j - t_1)] + c_3 - \hat{\beta}(t_j)\right)^2$ is optimized by setting its gradient to the zero vector $\mathbf{0}$ and solving (numerically) for c. The vector c which solves the equation $\nabla E[c] = \left[\dfrac{\partial E}{\partial c_1}, \dfrac{\partial E}{\partial c_2}, \dfrac{\partial E}{\partial c_3}\right] = \mathbf{0}$ is the best fit to the data. Subsequently, $\boxed{\beta(t_1;c) = 0.6453}$ so that $\beta(t_1;c)/t_1 = 5.8924$. The root mean square error $RMS_{err} = \sqrt{E[c]/N}$ for this fit is 0.0151.

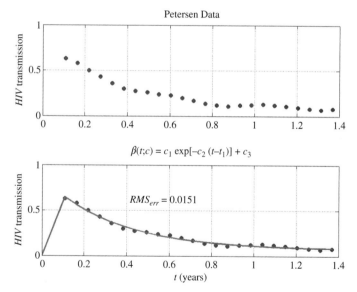

FIGURE 3.4   Least-squares fit to the Petersen et al. [26] data.

**MATLAB Commands
(Probability of *HIV* Transmission)**

```
% HIV transmission probabilities
Hp = [0.63,0.58,0.5,0.43,0.36,0.3,0.275,0.26,0.24,0.23,0.2,0.17,
0.14,0.12, 0.11,0.12,0.125,0.13,0.12,0.11,0.095,0.08,0.07,0.08];
% Time (in days) converted to years
T = 40:20:500; td = T/365.25;
```

```
% Transmission model ...
f = @(c,t)(c(1)*exp(-c(2)*(t-td(1))) + c(3));
% ... and nonlinear least square parameter fit
c = nlinfit(td,Hp,f,[Hp(1),3,Hp(end)])
c =
      0.5745      2.9466      0.0708
% Error functional and RMS error
Ec = sum((f(b,td) - Hp).^2); RMS = sqrt(Ec/numel(Hp))
RMS =
      0.0151
```

Plainly, once a value for β is established, then the probability of *HIV* transmission (infection) per contact at time t, $\lambda(t) = \beta(t; c) \bullet [X(t) + Y(t)]/[S(t) + X(t) + Y(t)]$ can be computed.

The remaining parameters $c(t) =$ the number of contacts at time t, $v(t) =$ the seroconversion rate at time t, and $\mu_A =$ the death rate due to *AIDS* are considerably more challenging to estimate. First, consider seroconversion. Bacchetti and Moss [1] provide estimates of the probability of developing *AIDS*, once a person is HIV^+, as a function of time. *Seroconversion* is the process of transitioning from *seronegative* (i.e., being infected with *HIV* but having no detectable antibodies) to *seropositive* (*HIV* with detectable antibodies). The estimates that Bacchetti and Moss [1] list are of the *hazard function*: The probability of developing *AIDS* in $[t, t + \Delta t]$ given that the person does not have *AIDS* at time t. If $F(t)$ is the cumulative density function (*cdf*) of the seroconversion process, then $Haz(t) = \dfrac{F'(t)}{1 - F(t)} = \dfrac{f(t)}{1 - F(t)}$, where $f(t)$ is the *probability density function* (*pdf*). An elementary discussion of density functions and probability is provided in the Appendix. Solving the first-order separable differential equation $Haz(t) = \dfrac{F'(t)}{1 - F(t)}$ for $F(t)$ in terms of the *hazard function* yields

$$F(t) = 1 - \exp\left(-\int Haz(t)dt\right) \tag{3.5}$$

so that the discrete estimate is $\hat{F}(t) = 1 - \exp(-\sum_{\tau \le t} \widehat{Haz}(\tau))$. See Table 3.3.

The strategy here is to fit a mathematical model $F(t)$ to the estimated *cdf* $\hat{F}(t)$, then use the resulting *pdf* and definition of the hazard function to form $Haz(t)$ for all $t \ge 0$. The work of Chiarotti et al. [7] examines two *cdf* models: The *Weibull distribution* $F_W(t; a) = 1 - \exp[-(t/a_2)^{a_1}]$ and the *generalized exponential distribution* $F_G(t; b) = (1 - \exp[-t/b_2])^{b_1}$. The following MATLAB commands demonstrate that the generalized exponential distribution has the smallest *root mean square* error. For both distributions, the *RMS* errors are relatively "small." As Figure 3.5 exhibits, both distributions appear to fit the data in column 3 of Table 3.3 quite well.

How do these models, as identified by the hazard function $Haz(t)$, fit the data from column 2 of Table 3.3? By examining Figure 3.6 and the MATLAB commands table, the generalized hazard function $Haz_G(t)$ yields a smaller *RMS* error as well as a more

TABLE 3.3   Estimated *hazard function* and *cdf* for transition to *AIDS*

Time after seroconversion (years)	Estimated probability of developing *AIDS*: $\widehat{Haz}(t)$	Cumulative density function: $\hat{F}(t) =$ $1 - \exp\left(-\sum_{\tau \leq t} \widehat{Haz}(\tau)\right)$
0	0	0
1	0.002	0.002
2	0.007	0.009
3	0.022	0.0305
4	0.043	0.0713
5	0.061	0.1263
6	0.073	0.1878
7	0.082	0.2517
8	0.081	0.31
9	0.074	0.3592
10	0.067	0.4007

visually appealing fit to the data. Consequently, the function in (3.6) will be used to model the seroconversion rate.

$$
\left.
\begin{aligned}
v(t; \boldsymbol{b}) &= Haz_G(t; \boldsymbol{b}) = \frac{f_G(t; \boldsymbol{b})}{1 - F_G(t; \boldsymbol{b})} \\
f_G(t; \boldsymbol{b}) &= \frac{b_1}{b_2} \exp(-t/b_2)[1 - \exp(-t/b_2)]^{b_1 - 1} \\
\boldsymbol{b} &= [b_1, b_2] = [2.85, 7.52]
\end{aligned}
\right\}
\tag{3.6}
$$

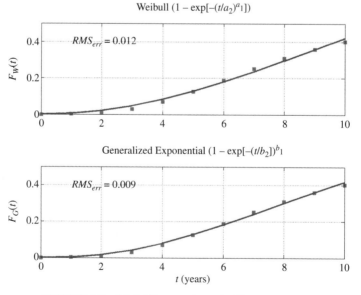

FIGURE 3.5   Model fits to estimated *cdf* measurements.

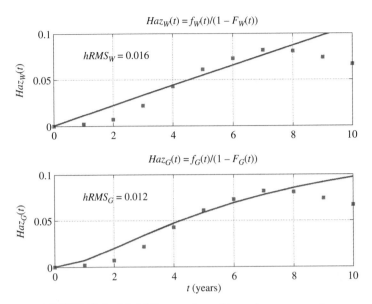

FIGURE 3.6   Model fits to estimated hazard measurements.

The last parameter that can be modeled from the available information is the death rate μ_A due to the onset of *AIDS*. Hawala [16] used a Weibull-based function $S_W(t; c) = \exp[-(t/c_2)^{c_1}]$ to represent the probability that a patient would live for *at least* t days after developing *AIDS*. He reported $\hat{c}_2 = 845.317361$ days and $\hat{c}_1 = 1.996985$ as estimates of the model parameters. To ensure that units are consistent with the estimates produced for $\beta(t; c)$ and $v(t; b)$, set $\lambda_2 = \hat{c}_2/365.25$ days/year and $\lambda_1 = \hat{c}_1$. Then λ_2 is measured in *years* and models the rate at which (untreated) patients die from *AIDS*.

$$\mu_A(t; \lambda) = \exp[-(t/\lambda_2)^{\lambda_1}] \tag{3.7}$$

MATLAB Commands
(*AIDS* Seroconversion *cdf*)

```
% Hazard function measurements
haz = [0,0.002,0.007,0.022,0.043,0.061,0.073,0.082,0.081,
0.074,0.067];
% Time (years)
T = 0:10;
% Estimated cdf
F = 1 - exp(-cumsum(haz));
% Create candidate cdfs via MATLAB "anonymous" functions.
Fw = @(a,t)(1 - exp(-(t/a(2)).^a(1)));  % Weibull
Fg = @(b,t)((1-exp(-t/b(2))).^b(1));
% Generalized exponential
```

```
% Estimate the parameter values for the Weibull ...
aa = nlinfit(T,F,Fw,[10,2])
aa =
     1.9857    13.5710
% RMS error for the Weibull distribution fit
Ew = sum((Fw(aa,T) - F).^2); RMSw = sqrt(Ew/numel(F))
RMSw =
     0.0122
% Generalized exponential cdfs
bb = nlinfit(T,F,Fg,[5,4])
bb =
     2.8478     7.5237
% RMS error for the Generalized exponential distribution fit
Eg = sum((Fg(bb,T) - F).^2); RMSg = sqrt(Eg/numel(F))
RMSg =
     0.0086
% Weibull pdf: fW(t) = dFW(t)/dt
fw = @(c,t)((c(1)/c(2)).*((t/c(2)).^(c(1)-1)).
   *exp(-(t/c(2)).^c(1)));
% Hazard function HazW(t) = fW(t)/[1 - FW(t)]
hazw = @(c,t)(fw(c,t)./(1 - Fw(c,t)));
% RMS error for HazW(t)
HrmsW = sqrt(sum((hazw(aa,T) - haz).^2)/numel(haz))
HrmsW =
     0.0161
% Generalized exponential pdf: fG(t) = dFG(t)/dt
fg = @(c,t)((c(1)/c(2))*((1 - exp(-t/c(2))).^ (c(1)-1)).*
   exp(-t/c(2)));
% Hazard function HazG(t) = fG(t)/[1 - FG(t)]
hazg = @(c,t)(fg(c,t)./(1 - Fg(c,t)));
% RMS error for HazG(t)
HrmsG = sqrt(sum((hazg(bb,T) - haz).^2)/numel(haz))
HrmsG =
     0.0122
```

The remaining model parameters p = the proportion of the population not "quarantined" and $c(t)$ = the number of contacts at time t will be varied over a range of plausible values. Table 3.4 lists a summary of the parameter values.

Remarks: Modeling the behavior of sexual partners is a daunting and difficult task. Hyman and Stanley [18,19] have produced some probabilistic models based on cohort groups. Laumann et al. [23] have published a statistical abstract of sexual practices. Murray [25] uses constants to model the contact profile $c(t)$. Gupta et al. [15] employ a similar approach.

TABLE 3.4   Parameter values for *HIV/AIDS SEIR* transmission model (3.1)

Parameter	Value
B	160,397 (people/year)
μ_D	0.0782 (dimensionless)
$\mu_A(t)$	$(e^{-t/2.3144})^{1.997}$ for $t \geq 0$ (t in years)
$\beta(t)$	$\begin{cases} 5.8924\,t & \text{for } 0 \leq t \leq 0.1095 \text{ (years)} \\ 0.5745\,e^{-2.95(t-0.1095)} + 0.07 & \text{for } t > 0.1095 \text{ (years)} \end{cases}$
$v(t)$	$\dfrac{0.38\,e^{-t/7.52}\,(1 - e^{-t/7.52})^{1.85}}{1 - (1 - e^{-t/7.52})^{2.85}}, \quad t \geq 0 \text{ (years)}$
p	1, 0.80, 0.50
$c(t) \equiv Cprofile$	Contact profiles
$S(t_0), X(t_0), X_q(t_0), Y(t_0), Y_q(t_0)$	See Table 3.1

A set of *contact profiles* are presented here as a genesis to examine model behavior. Three such profiles are considered: Lifelong monogamy, multiple simultaneous partners, and piecewise monogamy. It is assumed that *HIV* is transmitted via sexual contact, although other vectors (e.g., intravenous drug use) can be included in these profiles. The following are the mathematical descriptions of these distinct "interactions." Notice that $c(t)$ is the number of *cumulative* contacts at time t. It can be the case that $c(t) = 0$ for $t \in [t_a, t_b]$ since this describes a period of abstinence; hence, there are no contacts in that time frame.

$$\text{Lifelong monogamy: } c(t) = 1 \quad \text{for all } t \geq t_0 \qquad (3.8a)$$

That is, each member of the population has one partner for life (*single* contact profile).

$$\text{Multiple simultaneous partners: } c(t) = 3 \quad \text{for all } t \geq t_0 \qquad (3.8b)$$

Each member of the population has three distinct partners simultaneously (*multiple* contact profile).

$$\text{Piecewise monogamy: } c(t) = \begin{cases} 0 & \text{for } t_0 \leq t < t_1 \\ 1 & \text{for } t_1 \leq t < t_2 \\ 0 & \text{for } t_2 \leq t < t_3 \\ 2 & \text{for } t_3 \leq t < t_4 \\ 0 & \text{for } t_4 \leq t < t_5 \\ 3 & \text{for } t \geq t_5 \end{cases} \qquad (3.8c)$$

In this situation, all members of the population have no partners (or do not engage in activities which could transmit disease) on the time interval $[t_0, t_1)$; one partner on $[t_1, t_2)$; no partners on $[t_2, t_3)$; a new partner (distinct from the first contact) on $[t_3, t_4)$; no partners on $[t_4, t_5)$; and a lifetime partner after time t_5. The *piecewise monogamy* function will also be referred to as the *compound contact profile*.

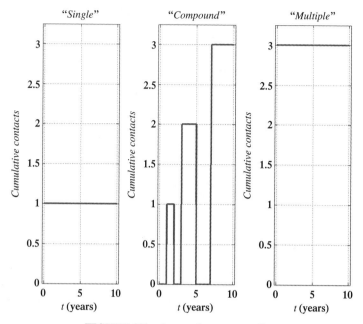

FIGURE 3.7   Assorted contact profiles.

These profiles are by no means exhaustive or representative. They are presented solely as a starting point. Figure 3.7 illustrates the behavior of each profile.

The assumption of a homogeneous contact profile for a large population is an exceptionally poor one. Nevertheless, *qualitative* behavior of a population can be studied from these simple models. Multi-city transmission model by Hyman and LaForce [17] can be adapted to model a subgroup-to-subgroup interaction.

Figures 3.8a (*"single"* partner profile), 3.8b (*"compound"* partner profile), and 3.8c (*"multiple"* partners profile) illustrate the population evolution over a 10-year time frame for each of the contact profiles. A suite of MATLAB functions has been developed for this model. They are included in this text.

MATLAB Commands
(*HIV/AIDS* Transmission Greater Boston)

% "Single" partner profile: $c(t) = 1$ for all $t \geq t_0 = 0$
```
hivseirplot('single',0);
```
% "Compound" partner profile: $c(t) = 0$ for $t_0 \leq t < t_1$; $c(t) = 1$ for $t_1 \leq t < t_2$;
% $c(t) = 0$ for $t_2 \leq t < t_3$, $c(t) = 2$ for $t_3 \leq t < t_4$; $c(t) = 0$ for $t_4 \leq t < t_5$; $c(t) = 3$
for $t \geq t_5$
```
hivseirplot('compound',[0,1,2,3,5,7]);
```
% "Multiple" partners profile: $c(t) = 3$ for all $t \geq t_0 = 0$
```
hivseirplot('multiple',0);
```

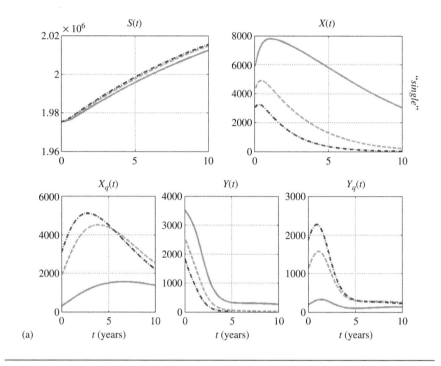

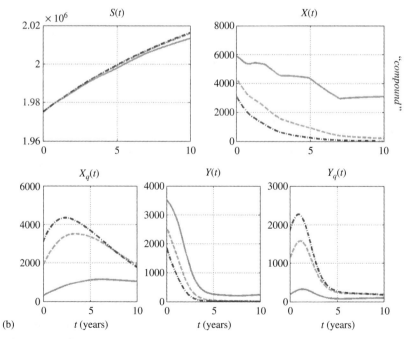

FIGURE 3.8 *HIV/AIDS SEIR* model for (a) *"single partner profile,"* (b) *"compound part-ner profile,"* (c) *"multiple partners profile"* with $p = 0.95$ (—), 0.70 (- -), 0.50 (- • -).

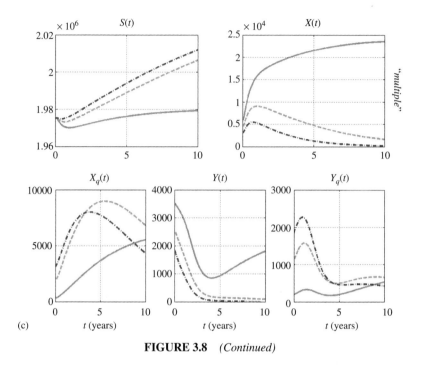

FIGURE 3.8 *(Continued)*

3.1.1.1 Quantitative Analysis The primary test of any mathematical model is whether its consequences produce sensible results. An inspection of Figures 3.8a–3.8c yields the following observations.

For the *"single"* (i.e., lifelong monogamy) and *"compound"* (piecewise monogamy) partner profiles, the proportion p of *HIV/AIDS* infected who remain infectious (i.e., are not quarantined) has little impact on the evolution of the susceptible population $S(t)$. Furthermore, for all partner profiles, the larger the rate p, the larger the populations of *HIV/AIDS* infecteds $X(t)$ and $Y(t)$. The opposite occurs with respect to the "quarantined" *HIV/AIDS* populations $X_q(t)$ and $Y_q(t)$. This, of course, makes sense. As p approaches 1, $X(t)$ and $Y(t)$ will increase while $X_q(t)$ and $Y_q(t)$ will decrease.

The *"multiple"* partners profile has distinctly different behavior for "large" p. Indeed, for $p = 0.95$, the susceptibles $S(t)$ exhibit a "flat" curve with almost no growth. All other populations of infecteds $X(t)$, $X_q(t)$, $Y(t)$, and $Y_q(t)$ increase with respect to their *"single"* or *"compound"* profile counterparts. For *AIDS* patients, this appears to happen near the $t = 5$ mark. Again, this is quite reasonable: As high-risk (for infection) behavior increases throughout the population, so should infections.

These comments reflect only a perfunctory review of the system output. The results, nevertheless, lend credibility to the model (3.1), the parameter functions (3.4)–(3.7), the initial condition estimates in Tables 3.1 and 3.4, and the various assumptions. Therefore, this entire system can serve as a foundation for prediction and model enhancement.

EXERCISES

3.1 Determine what happens when p approaches 0. That is, what is the effect of high levels of reduction in infection ("quarantine") on the various subpopulations? This may be due to prophylactic measures (i.e., widespread abstinence or the use of condoms) or medical intervention (antiretroviral drugs).

3.2 Develop other contact profiles. This could be achieved by conducting an anonymous survey to determine patterns of behavior. See, for example, Laumann et al. [23].

3.1.1.2   Project   Use the multi-city model of Hyman and LaForce [17] to construct a multi-contact profile model for the transmission of the disease throughout a heterogeneous population.

3.1.2   Modeling the Transmission of a Respiratory Epidemic

In the previous section, the mathematical structure required to model the transmission of a disease through a population was developed. The model uses measurements of population patterns in order to find mathematical functions that describe this behavior. What approach should be taken when direct observations of the population are available?

That is, suppose the set of differential equations (3.9a) describes the evolution of a population *and* a corresponding set of measurements (3.9b) describes the population at regular time intervals. Can the measurements be incorporated into the mathematical model? The answer is *yes*. Indeed, this is an informal description of the *filtering problem*.

$$\frac{dx}{dt}(t) = f(x, t) \tag{3.9a}$$

$$m(t) = h(x, t) \tag{3.9b}$$

In the special case of linearity, $f(x, t) = Fx(t)$ and $h(x, t) = Hx(t)$ with F, H matrices, Kalman [21] established the seminal filter which bears his name. Shortly thereafter, Kalman and Bucy [22] extended the result for general functions f and h. These methods are known as the *Kalman filter* and the *Extended Kalman–Bucy filter (EKBF)*, respectively.

This section is dedicated to the application of the *EKBF* to respiratory infection data and an *SEIR* model constructed to reflect the spread of these viruses. The recent concern centering about a global outbreak of an infectious respiratory disease such as *avian flu*, *H1N1* ("swine") *flu*, or *SARS* (severe acute respiratory syndrome) has highlighted the need for an "early warning system." One approach is to predict the outbreak of an epidemic respiratory disease based upon a limited reporting of data along with a mathematical model. Such a system will be the focus of the remainder of this chapter. In order to develop an algorithmic predictor, a set of assumptions must be made.

The primary assumption is that daily measurements of the infectious population are recorded and made available to epidemiologists. Furthermore, it is assumed that the infection is sufficiently fast acting that death among the infected population is

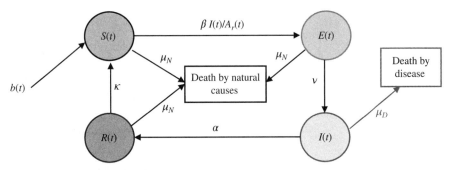

FIGURE 3.9   The *SEIR* diagram for infectious respiratory disease.

caused exclusively by the disease. Finally, a portion of those exposed to the disease are considered infectious but not necessarily infected. The infected population comprises those who are suffering from the disease. The resulting *SEIR* diagram is presented in Figure 3.9.

The nomenclature of the diagram is as follows.

$b(t)$	the rate at which new *susceptibles* enter the population at time t
$S(t)$	the population of those *susceptible* to the disease at time t (*susceptibles*)
$E(t)$	the population of those *exposed* to the disease at time t
$I(t)$	the population of those *infected* by the disease at time t (*infecteds*)
$R(t)$	the population of those who *recovered* from the disease at time t
$A_r(t) = S(t) + E(t) + I(t)$	the active *at risk* population at time t
β	the probability of disease transmission
v	the rate of seroconversion
α	the recovery rate
κ	the rate at which the *recovered* return to the *susceptible* population (loss of immunity)
μ_N	the death rate due to natural causes
μ_D	the death rate due to the disease

Records of visits to the Boston Children's Hospital emergency department for respiratory distress were provided by Dr. Kenneth Mandl in 2005 [24]. Due to HIPAA (*Health Insurance Portability and Accountability Act*) restrictions, however, these data cannot be reported beyond their initial publication in Costa and Dunyak [8]. As a consequence, simulated data which accurately mimic the Children's Hospital records will be utilized. The contrast in-between these data sets can be seen in Figures 3.10a and 3.10b. As will be seen from the output of the *EKBF*, the filter acts as a combined smoother and outlier identifier. Therefore, the data pre-processing detailed in Chapter 1 will not be applied to the simulated data. Some preliminary checks, however, are in order. The minimum and maximum numbers of reported infections (i.e., simulated emergency room visits) are 0 and 32, respectively. This is well within

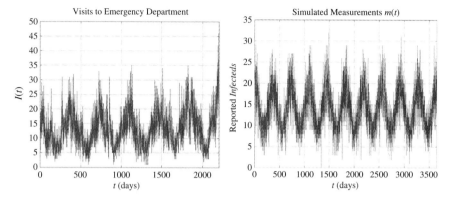

FIGURE 3.10   (a) Boston Children's Hospital data. (b) Simulated respiratory infections.

both the reported Children's Hospital data and the realm of common sense. Plainly, if the data indicated that 500 emergency room visits occurred in a single day at Children's (or any other) Hospital, there would be the need for *data cleaning*.

Proceeding again as in the first section of this chapter, a set of ordinary differential equations can be derived from the *SEIR* diagram in Figure 3.9.

$$
\left.
\begin{aligned}
\frac{dS}{dt}(t) &= b(t) - \left(\beta \cdot \frac{I(t)}{A_r(t)} + \mu_N \right) \cdot S(t) + \kappa \cdot R(t) \\
\frac{dE}{dt}(t) &= \beta \cdot \frac{I(t)}{A_r(t)} \cdot S(t) - (v + \mu_N) \cdot E(t) \\
\frac{dI}{dt}(t) &= v \cdot E(t) - (\mu_D + \alpha) \cdot I(t) \\
\frac{dR}{dt}(t) &= \alpha \cdot I(t) - (\mu_N + \kappa) \cdot R(t)
\end{aligned}
\right\}
\tag{3.10}
$$

Equation (3.10) is called the *state model* or *state dynamics* and the symbol $x(t) = [S(t), E(t), I(t), R(t)]^T$ is known as the *state vector*. By setting $b(t) = b(t) \bullet [1, 0, 0, 0]^T$, $N(x(t)) = b \bullet (I(t)/A_r(t)) \bullet S(t) \bullet [-1, 1, 0, 0]^T$, A equal to the 4×4 matrix

$$
A = \begin{bmatrix}
-\mu_N & 0 & 0 & \kappa \\
0 & -(\mu_N + v) & 0 & 0 \\
0 & v & -(\mu_D + \alpha) & 0 \\
0 & 0 & \alpha & -(\mu_N + \kappa)
\end{bmatrix}
$$

and $f(x, t) = b(t) + Ax(t) + N(x(t))$, then (3.10) can be written in the form of (3.9a). Since the measurements are of the population infected at time t, $I(t)$, then these data take the form (3.9b) when $m(t) = h(x, t) = Hx(t)$ and $H = [0, 0, 1, 0]$.

The intent of any good mathematical model should be a realistic representation of the phenomenon under examination. While considerable effort is expended in taking all influences into account, this model is not perfect. To account for unforeseen or hidden model parameters, the state vector $x(t)$ can be considered as the probabilistic expectation of a stochastic random vector X: $x(t) = E[X(t)]$. With this subtlety in mind, the model is modified by adding "noise" to the *state equation* (3.10) to form the *stochastic differential equation*

$$\frac{dX}{dt}(t) = b(t) + AX(t) + N(X(t)) + B\,w(t) \equiv f(X, t) + B\,w(t) \qquad (3.11a)$$

in which B is a vector of weights and $w(t)$ is *Gaussian white noise.*[1] Letting $W(t) = Bw(t)$ implies that the *process noise covariance matrix* $Q(t) \equiv E[W(t)W^T(t)] = B \cdot \text{var}(w(t)) \cdot B^T$. The weighting vector B can be comprised of distinct elements. For this application, the process noise is taken to be uniform across all states so that $B = b_0 \cdot [1, 1, 1, 1]^T$. In this sense, b_0 is the *plant* or *process noise*.

Just as the model cannot be perfect, the measurements will inevitably contain errors. It is not difficult to understand that improper recording, misdiagnosis, or the general chaos present in any emergency room can contribute to measurement error $u(t)$. Thus, the measurement model becomes

$$m(t) = HX(t) + u(t) \qquad (3.11b)$$

with a corresponding noise variance $U(t) = \text{var}(u(t))$.

Observe that $m(t)$ is a sequence of *integers* since this measurement records the number of people infected with a disease at a given time t. The simulated infections are modeled as $m(t) = [\![5 \cdot (\cos(2\pi t/365.25) + w_t)]\!]$ where $w_t \sim \mathcal{N}(0, 1/2)$ is white noise and $[\![\bullet]\!]$ is the nearest integer function. Moreover, since infections are the only state variable measured, then the variance $U(t)$ is merely the variance on the measured infections $\sigma^2_{m,I}(t)$. Observe that the notation $\sigma^2_{m,I}(t)$ refers to the variance of the *measurement* variable I. This differs from the variance on the modeled state element $\sigma^2_{M,I}(t)$. Equations (3.11a)–(3.11b) comprise the *nonlinear filtering problem*. As will be discussed, one solution to this problem is the *EKBF*. The fundamental idea of the *EKBF* is to blend the predictions of the mathematical model and system measurements to form a probabilistically weighted, corrected estimate. If the mathematical model is more precise than the measurements, then the model process noise $Q(t)$ is small and the prediction is weighted more heavily. Conversely, if the measurements are better, then the measurement noise $U(t)$ is small and the measurements are more influential. The weighting is achieved via the *Kalman gains matrix*. This helps incorporate the measurements into the model prediction (i.e., integration of the state dynamics) in a manner that is "optimized."

For a purely linear system with Gaussian noise, the Kalman filter is indeed the optimal estimate of the state. The *EKBF* is a first-order approximation and therefore only accurate to that extent (i.e., suboptimal).

The steps outlining the implementation of the *EKBF* are listed sequentially. For more details, see Kalman [21], Kalman and Bucy [22], Bucy and Joseph [4], Costa [9], Gelb [14], Jazwinski [20], Satzer [28], and Sontag [30].

$$\text{State vector: } x(t) \in \mathbb{R}^n$$

$$\text{State dynamics: } \frac{dx}{dt}(t) = f(x(t)) \qquad (3.12a)$$

$$\text{Measurements: } m(t) = Hx(t)$$

---

[1] That is, a stochastic process which has the standard normal probability density function.

A measurement at time t_k will be designated as m_k so that $m_k = m(t_k)$. The collection of measurements from time t_1 through time t_k will be denoted as M_k. That is, $M_k = \{m_1, m_2, \ldots, m_k\}$. At time t_0, M_k is the empty set: $M_0 = \emptyset$.

$$\text{State prediction: } x_p(t_k; M_{k-1}) = \int_{t_{k-1}}^{t_k} f\left(x(\tau; M_{k-1})\right) d\tau \qquad (3.12b)$$

The state prediction is the solution of the state dynamics (3.12a) over the time interval $[t_{k-1}, t_k]$, which is influenced by the measurements up to time t_{k-1} by way of M_{k-1}. Here the "integration" is purely symbolic. Typically, the state dynamics are advanced in time via a numerical method used to solve nonlinear ordinary differential equations.

$$\text{Predicted measurement: } m_p(t_k; M_{k-1}) = Hx_p(t_k; M_{k-1})$$

Once the state prediction has occurred, the predicted (i.e., *modeled*) measurement is calculated.

$$\text{Measurement residual: } \Delta m(t_k; M_k) = m_k - m_p(t_k; M_{k-1}) = m_k - Hx_p(t_k; M_{k-1})$$
$$(3.12c)$$

This is the difference in-between the actual and predicted measurement at time t_k.

$$\text{State residual error: } \Delta x(t_k; M_{k-1}) = x_p(t_k; M_{k-1}) - \hat{x}(t_{k-1}; M_{k-1})$$

Once an estimate of the state vector at time t_{k-1} has been determined and a prediction at time t_k is made, then the residual error is calculated. The error covariance is then the expected value of the residual error outer product.

$$\text{State covariance: } P(t_k; M_{k-1}) = E[\Delta x(t_k; M_{k-1})\Delta x^T(t_k; M_{k-1})]$$

The state covariance evolves according to the *degenerate Ricatti equation*[2]

$$\frac{dP}{dt} = P(t)F^T(t) + F(t)P^T(t) + Q(t).$$

Here, $F(t) = F(x, t)$ is the state Jacobian matrix $F(x, t) = \dfrac{\partial f(x, t)}{\partial x} = \left[\dfrac{\partial f_j}{\partial x_k}\right]_{j=1, k=1}^{n, n}$ and $Q(t) = B \bullet \text{var}(w(t)) \bullet B^T$ is the process noise covariance matrix. The state covariance update is achieved via the *transition matrix* $\Phi(t, s)$ and the equation (3.12d). Note that the diagonal elements of the state covariance P represent the variance of the state elements. Thus, $p_{11}(t_k, M_{k-1}) = \sigma_{M,S}^2(t_k; M_{k-1})$, $p_{22}(t_k, M_{k-1}) = \sigma_{M,E}^2(t_k; M_{k-1})$, $p_{33}(t_k, M_{k-1}) = \sigma_{M,I}^2(t_k; M_{k-1})$, and $p_{44}(t_k, M_{k-1}) = \sigma_{M,R}^2(t_k; M_{k-1})$. The symbol $\sigma_{M,X}^2(t; M)$ is the variance of the modeled state element X estimated at time t with respect to the ensemble of measurements M.

$$\text{Transition matrix: } \Phi(t, s)$$

---

[2] The full quadratic Ricatti equation is $\frac{dP}{dt} = PF^T + FP^T + PH^T UHP^T$. The quadratic term is dropped from the *EKBF* as it is a first-order method.

This is a solution of the matrix differential equation $\dfrac{d\Phi}{dt}(t, s) = F(t)\,\Phi(t, s)$ with initial condition $\Phi(t_0, t_0) = I$ (i.e., the $n \times n$ identity matrix).

Covariance update:

$$P(t_k; M_{k-1}) = \Phi(t_k, t_{k-1})\, P(t_{k-1}; M_{k-1})\, \Phi^T(t_k, t_{k-1}) + \int_{t_{k-1}}^{t_k} \Phi(t_k, s)\, Q(s)\, \Phi^T(t_k, s)\, ds$$

(3.12d)

Once the state vector has been "integrated" to time t_k, then the next step in the filter process is to incorporate the influence of the measurements up to time t_k. This is accomplished via the *Kalman gains matrix K* and information matrix $\mathcal{I}$.

Kalman gains matrix: $K(t_k) = P(t_k; M_{k-1})H^T(t_k)\mathcal{I}(t_k; M_{k-1})$ (3.12e)

Information matrix: $\mathcal{I}(t_k; M_{k-1}) = [H(t_k)P(t_k; M_{k-1})H^T(t_k) + U(t_k)]^{-1}$ (3.12f)

Measurement Jacobian: $H(t) = \dfrac{\partial m(t)}{\partial x} = \dfrac{\partial}{\partial x}(Hx) = H$ (3.12g)

Now the measurement incorporation at time t_k can occur.

State correction: $x_c(t_k; M_k) = K(t_k)\Delta m(t_k; M_k) = K(t_k)[m_k - Hx_p(t_k; M_{k-1})]$

(3.12h)

State update (integration and measurement): $\hat{x}(t_k; M_k) = x_p(t_k; M_{k-1}) + x_c(t_k; M_k)$

(3.12i)

The only remaining step in the *EKBF* is to incorporate the measurement at time t_k into the covariance matrix. This is achieved via the Joseph form [4].

Covariance update:

$$P(t_k; M_k) = [I - K(t_k)H(t_k)]P(t_k; M_{k-1})[I - K(t_k)H(t_k)]^T + K(t_k)U(t_k)K^T(t_k)$$

(3.12j)

Together, equations (3.12a)–(3.12j) comprise the extended Kalman–Bucy filter. Table 3.5 summarizes the *EKBF* for the system (3.11a)–(3.11b) with state Jacobian F.

$$F(x) = \begin{bmatrix} -\left(\beta\dfrac{-(E+I)\cdot I}{(S+E+I)^2} + \mu_N\right) & \beta\dfrac{S\cdot I}{(S+E+I)^2} & -\beta\dfrac{(S+E)\cdot S}{(S+E+I)^2} & \kappa \\[3mm] \beta\dfrac{(E+I)\cdot I}{(S+E+I)^2} & -\beta\dfrac{S\cdot I}{(S+E+I)^2} - (v+\mu_N) & \beta\dfrac{(S+E)\cdot S}{(S+E+I)^2} & 0 \\[3mm] 0 & v & -(\alpha+\mu_D) & 0 \\[3mm] 0 & 0 & \alpha & -(\kappa+\mu_N) \end{bmatrix}$$

3.1.2.1 Parameter Estimation and Modeling

Equations (3.11a)–(3.11b) establish the mathematical models for the disease transmission and measurement of infections. Table 3.5 summarizes the *EKBF* which will incorporate the model and measurements so that the state can be estimated. Before proceeding with the implementation of the filter, the model parameters β (probability of disease

TABLE 3.5   Application of the *EKBF* to the *SEIR* system (3.11a) and (3.11b)

State vector $x(t)$	$[S(t), E(t), I(t), R(t)]^T$

State dynamics $\dfrac{dx}{dt}(t) = f(x, t)$

$$f(x, t) = \begin{bmatrix} b(t) - \left(\beta \cdot \dfrac{I(t)}{A_r(t)} + \mu_N\right) \cdot S(t) + \kappa \cdot R(t) \\ \beta \cdot \dfrac{I(t)}{A_r(t)} \cdot S(t) - (\nu + \mu_N) \cdot E(t) \\ \nu \cdot E(t) - (\mu_D + \alpha) \cdot I(t) \\ \alpha \cdot I(t) - (\mu_N + \kappa) \cdot R(t) \end{bmatrix}$$

Measurements $m(t) = Hx(t)$	$H = [0, 0, 1, 0]$
State Jacobian $F(x; t)$	$F(x)$ (see formula above)
Measurement Jacobian $H(t)$	$H = [0, 0, 1, 0]$

Covariance matrix P

$$\begin{bmatrix} P_{1,1} & P_{1,2} & P_{1,3} & P_{1,4} \\ P_{1,2} & P_{2,2} & P_{2,3} & P_{2,4} \\ P_{1,3} & P_{2,3} & P_{3,3} & P_{3,4} \\ P_{1,4} & P_{2,4} & P_{3,4} & P_{4,4} \end{bmatrix}$$

Error values $U(t) = \sigma_{m,I}^2(t), p_{3,3}(t_k; M_{k-1}) = \sigma_{M,I}^2(t_k; M_{k-1})$

Information matrix $\Im(t_k; M_{k-1})$ $[\sigma_{M,I}^2(t_k; M_{k-1}) + \sigma_{m,I}^2(t_k)]^{-1}$

Kalman gains matrix $K(t)$

$$\frac{1}{\left(\sigma_{M,I}^2(t_k; M_{k-1}) + \sigma_{m,I}^2(t_k)\right)} \begin{bmatrix} P_{1,3}(t_k; M_{k-1}) \\ P_{2,3}(t_k; M_{k-1}) \\ \sigma_{M,I}^2(t_k; M_{k-1}) \\ P_{4,3}(t_k; M_{k-1}) \end{bmatrix}$$

transmission), ν (rate of seroconversion), α (recovery rate), κ (rate of loss of immunity), μ_N (death rate by natural causes), and μ_D (death rate by disease) must be modeled. This is accomplished by examining the available data.

Influenza is the most commonly reported and treated respiratory infection. Therefore, the aforementioned parameters will be estimated from data gathered on flu infections. The Centers for Disease Control (*CDC*) estimate the duration of the flu to be in-between 3 and 10 days. The investigation of Cauchemez et al. [5] produced a duration of 3.6 days. Hence, the duration and recovery rates are estimated to be $\delta_r = 1/3.6$ day^{-1} and $\alpha = 1 - \delta_r = 0.7222$ day^{-1}, respectively. In addition, Cauchemez et al. [5] estimate the probability of instantaneous risk of infection to be $\beta = 0.0056$ day^{-1}. A summary of all of the parameters used in the computations related to this model are presented in Table 3.6.

Mandl [24] suggested a seroconversion rate of 1.5 year^{-1}. The death rates due to disease (μ_D) and natural causes (μ_N) are estimated from Massachusetts population data [27]. These data and estimates are exhibited in Table 3.7. The loss of immunity rate κ and measurement noise fraction b_0 are *system adjustable parameters* (*SAP*). That is, their values are to be "tuned" to the model. The selection of *SAP* values, however, is restricted by model consistency and common sense. For example, κ *must* be a percentile, $\kappa \in [0, 1]$. As can be seen from Table 3.7, the rates μ_D and μ_N are in the units *per year*. Hence, division by the scale factor $d_y = 365.25$ days/year so that μ_D/d_y and μ_N/d_y are in units *per day*. This is consistent with the measurement model.

TABLE 3.6   Parameters used in the *EKBF* calculations on model (3.10)

System adjustable parameters		Initial conditions	
v	$1.5/d_y$	$S(t_0)$	65,000
β	0.0056	$E(t_0)$	$S(t_0) \cdot 0.05$
α	0.7222	$I(t_0)$	$m(t_0)$
μ_D	$0.00032/d_y$	$R(t_0)$	$m(t_0) \cdot 0.5$
μ_N	$0.0087/d_y$	$m(t)$ = simulated measurements at time t	
κ	0.95		
b_0	$0.03 \cdot d_y \cdot I(t_0)$		
d_y	365.25 days/year		
$U(t)$	$\text{var}(m(t))/10$		

The plant noise (b_0) following equation (3.11a) is another *SAP*, as are the initial conditions. These parameters are selected as per Table 3.6. It remains to find a reasonable model for $b(t)$ = the rate at which new susceptibles enter the population. Assuming that everyone is susceptible to influenza, $b(t)$ is the rate of change in the population at time t. The data presented in columns 1 and 2 of Table 3.7 are plotted and fit with a third-order polynomial $p_3(\tau) = a_0 + a_1\tau + a_2\tau^2 + a_3\tau^3$, where τ is "normalized time," $\tau = (t - \mu_t)/\sigma_t$, $\mu_t = 2000$ is the mean year, and $\sigma_t = 6.2$ is the standard deviation of time (in years). The population is converted to millions so that numerical overflow problems are alleviated.

TABLE 3.7   Population and respiratory disease data for Massachusetts [27]

Year	Population	Total number of deaths (all causes)	Total number of deaths due to flu/pneumonia	Estimated μ_N	Estimated μ_D
1990	6016425	53008	2279	0.0088	0.00037
1991	6018101	53010	2385	0.0088	0.00037
1992	6037065	53804	2421	0.0089	0.00038
1993	6063539	55557	2778	0.0092	0.00039
1994	6106792	54914	2636	0.0090	0.00040
1995	6142813	55296	2710	0.0090	0.00045
1996	6184354	55187	2815	0.0089	0.00043
1997	6227622	54634	2677	0.0088	0.00044
1998	6291263	55204	1524	0.0088	0.00045
1999	6360236	55763	1506	0.0088	0.00042
2000	6349097	56591	1585	0.0089	0.00024
2001	6407631	56733	1475	0.0089	0.00024
2002	6431788	56881	1536	0.0088	0.00025
2003	6438510	56194	1517	0.0087	0.00023
2004	6433676	54419	1360	0.0085	0.00024
2005	6429137	53776	1398	0.0084	0.00024
2006	6434389	53293	1226	0.0083	0.00021
2007	6449755	52690	1212	0.0082	0.00022
2008	6497967	53341	1174	0.0082	0.00019
2009	6593587	N/A	N/A	Mean(μ_N)	Mean(μ_D)
2010	6547629	N/A	N/A	0.0087	0.00032

MATLAB Commands
(Massachusetts Population Data, Figure 3.11)

```
% Time (in years) and Population
t = 1990:2010;
Pop = [6016425, 6018101, 6037065, 6063539, 6106792, 6142813,
6184354, 6227622, 6291263, 6360236, 6349097, 6407631,
6431788, 6438510, 6433676, 6429137, 6434389, 6449755,
6497967, 6593587, 6547629];
% Convert to "normalized time" and "per million" population
Mut = mean(t); St = std(t); T = (t - Mut)/St;
Y = Pop/1e+06;
% Fit the polynomial p3(τ) = a0 + a1 τ + a2 τ² + a3 τ³ to the "per million"
population
a3 = polyfit(T,Y,3);
p3 = polyval(a3,T);
% Coefficients for b(t) = dp3(τ)/dt
b2 = [3,2,1].*a3(1:end-1)/St;
% RMS error
E3 = sqrt(sum((Y - p3).^2)/numel(Y));
E3 =
    0.0330
```

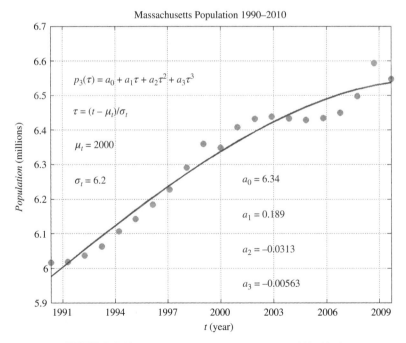

FIGURE 3.11   Massachusetts population data 1990–2010.

To determine the "precision" of the fit, calculate the *root mean square (RMS)* error. As a function of the vector of parameters $\boldsymbol{a} = [a_0, a_1, a_2, a_3]$, the *RMS* error of the model $p_3(\tau)$ with respect to the normalized data $\boldsymbol{y} = [y_1, y_2, \ldots, y_n]$ is

$$RMS\left[p_3(\tau; \boldsymbol{a})\right] = \sqrt{\frac{1}{n} \sum_{k=1}^{n} \left(y_k - p_3(\tau_k)\right)^2}. \tag{3.13}$$

Recall that the first term $b(t)$ of the differential equation for $S(t)$ in (3.10) is the population's *rate of change*. Therefore, $b(t)$ can be modeled as the derivative of the population function $p_3(\tau)$. Observe, however, $p_3(\tau)$ has been computed for normalized time $\tau = (t - \mu_t)/\sigma_t$, so that $b(t) = \dfrac{dp_3(\tau)}{dt} = \dfrac{dp_3(\tau)}{d\tau} \cdot \dfrac{d\tau}{dt} = \dfrac{1}{\sigma_t} p_3'(\tau) = $
$\dfrac{1}{\sigma_t}(a_1 + 2a_2\tau + 3a_3\tau^2)$. That is, $b(t) = b_0 + b_1\tau + b_2\tau^2$, where $b_0 = 0.03$, $b_1 = -0.01$, and $b_2 = -0.003$. All of the parameters listed after Figure 3.9 have either been estimated or treated as a system adjustable parameter. The idea now is to run the filter of (3.12a)–(3.12j) and Table 3.5 over the model (3.11a)–(3.11b) with the parameter values from Table 3.6 and $b(t)$. The results of this modeled effort are captured in Figures 3.12a and 3.12b.

MATLAB Commands
(EKBF)

```
% Load in the simulated measurements and associated times
Ddir = 'C:\PJC\Math_Biology\Chapter_3\Data';
data = load(fullfile(Ddir,'SEIR_Simulated_Data.mat'));
D = data.D;
t = load(load(fullfile(Ddir,'SEIR_Time.mat'))); T = t.T;
clear data t
% Initial conditions for the nonlinear ODE (state dynamics)
So = 6.5e+04; IC = [So,So*0.05,D(1),D(1)*0.5];
% Structure of system adjustable parameters (SAP)
dy = 365.25;
Sap.v = 1.5/dy; Sap.beta = 0.0056; Sap.a = 0.7222;
Sap.d = 0.0003224/dy; Sap.kappa = 0.95; Sap.nd = 0.008724/dy;
Sap.bo = 0.03*dy*IC(3); Sap.Mnoise = 0.1;
% Run the filter
tic; F = seirekbf(IC,Sap,D,T); toc
Elapsed time is 28.743055 seconds.
% Plot the results
seirekbfplot(F,T,D);
% Mean and range of the innovation measurement residuals Δm(tₖ) = m(tₖ)
     − Hx(tₖ, Mₖ₋₁)
mean(F.In) =
   -0.0262
[min(F.In), max(F.In)] =
-17.0378    15.5723
```

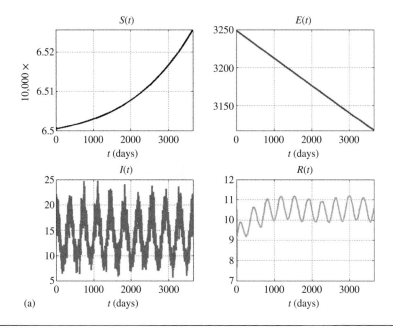

(a)

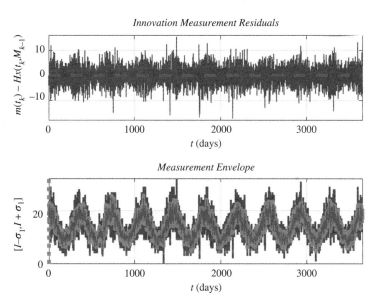

FIGURE 3.12   (a) *EKBF* state estimates. (b) Filter tuning results. (c) Distribution of the innovation process $\vartheta_p(t) = (m(t) - Hx(t)) \bullet (HP(t)H^T + U)^{-1}$.

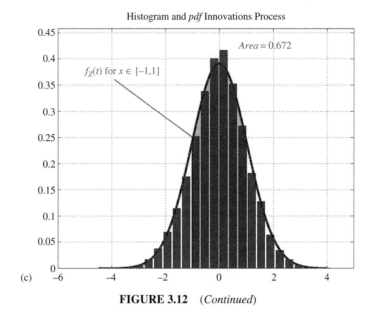

(c)

FIGURE 3.12 *(Continued)*

The specialized MATLAB functions `seirekbf.m` and `seirekbfplot.m` are used to apply the filter and plot the results.

As can be seen from the lower left-hand subplot of Figure 3.12a and the measurement envelope of Figure 3.12b, the estimate of the infecteds $I(t)$ is both a good qualitative and quantitative representation of the measured data (namely, respiratory infections). In addition, the mean value of the innovations is -0.03 with range $[-17.04, 15.57]$. This resembles white noise as is predicted by the Kalman–Bucy theory. Moreover, the behavior of the estimated measurement residuals $\vartheta_p(t) = (m(t) - Hx(t)) \cdot (HP(t)H^T + U)^{-1}$ should be approximately standard normal $\mathcal{N}(0, 1)$. Examining Figure 3.12c gives strong indication that this is indeed the case. The computations show that the area subtended by the probability density function $f_Z(x)$ for $x \in [-1, 1]$ is close to that of the standard normal distribution. Specifically, $\text{Prob}(-1 \leq \vartheta_p \leq 1) = \int_{-1}^{1} f_Z(x)\,dx = 0.6721$, whereas $\text{Prob}(-1 \leq Z \leq 1) = \int_{-1}^{1} \varphi(x)\,dx = 0.6827$ when $Z \sim \mathcal{N}(0, 1)$ with standard normal *pdf* $\varphi(x)$. See the Appendix for greater detail concerning normal distributions. Subsequently, the white noise nature of the *innovation measurement residuals* $\Delta m(t) = m(t) - Hx(t)$ and the standard normal behavior of the *innovations process* $\vartheta_p(t) = \Delta m(t) \cdot (HP(t)H^T + U)^{-1}$ indicate that the tuning of the *EKBF* is reasonable. These results are summarized in Table 3.8.

TABLE 3.8 Filter tuning summary results

Process $\omega(t)$	Mean $\omega(t)$	Random variable Z	Prob$(-1 \leq Z \leq 1)$
Innovation measurements			
$\omega(t) = m(t) - Hx(t)$	-0.0262	$Z = \vartheta_p(t)$	0.6721
Gaussian white noise $\omega(t)$	0	$Z \sim \mathcal{N}(0, 1)$	0.6827

This outcome provides confidence that the model can be used to predict the outbreak of a respiratory epidemic. The next section undertakes this challenge.

Remarks: The discussion of how to select the *system adjustable parameters* so that the filter accurately responds to the model and data is referred to as the "tuning" process. It is not uncommon for those who implement Kalman filters to refer to tuning as a "black art." Groucho Marx is attributed as the apocryphal source of the statement, *you can tune a piano but you can't tuna fish*. Kalman filter practitioners, conversely, have often lamented, *it is easier to tune a fish than it is to tune a Kalman filter*. Consequently, a set of tuning guidelines is provided. It is emphasized that these guidelines are more recommendations than inviolable rules for tuning accuracy.

3.1.2.2 Kalman Filter Tuning Guidelines

(i) *Validate the state dynamics.* "Integrate" the state dynamics (3.12a) from time t_0 to some time $t_f > t_0$. Formally, $x(t) = x(t_0) + \int_{t_0}^{t} f(x(s))\, ds$. If the resulting state estimate $x(t_f)$ is wildly divergent from the initial condition, then the model is only valuable for small time increments. That is, $\|x(t_f)\| \gg \|x(t_0)\|$ implies that the model is unsuitable for long timescale. This requires the use of short timescale measurements (i.e., $\Delta t_j = t_j - t_{j-1}$ should be "small") and "large" process noise covariance $Q(t)$.

(ii) *Test the covariance propagation without noise.* Set $Q(t) = O_{n \times n}$ and integrate the Ricatti equation $dP/dt = P(t)F^T(t) + F(t)P^T(t)$ from t_0 to t_f as in step (i). If trace $(P(t_f)) \gg$ trace$(P(t_0))$, then the covariance will require the influence of process noise to control stability. Unless *a priori* information gives an estimate of the process covariance, assume that it is a diagonal matrix

$$Q(t) = \begin{bmatrix} q_{11} & 0 & \cdots & 0 \\ 0 & q_{22} & \cdots & 0 \\ \vdots & \vdots & \ddots & \vdots \\ 0 & 0 & \cdots & q_{nn} \end{bmatrix}.$$

(iii) *Check that the measurement residuals are Gaussian white noise.* The *innovation measurement residuals* $\Delta m(t) = m(t) - H \cdot x(t, M_t)$ should form a $\mathcal{N}(0, \sigma^2)$ distribution. When the measurements are multidimensional (i.e., $m(t) \in \mathbb{R}^N$ for $N > 1$), take $U(t_0) = (m(t_0) - \bar{m})^T (m(t_0) - \bar{m})$ where $\bar{m}$ is the average value of the elements of $m(t_0)$. If this is *not* the case, alter the process noise $Q(t)$ and/or measurement noise $U(t)$ so that the mean of the innovation residuals approaches 0. In the case of the filter tuning as discussed earlier, the mean and variance of the innovation residuals are $\mu_{In} = -0.026$ and $\sigma_{In}^2 = 14.85$, respectively. The distribution of $\Delta m(t)$ illustrated in Figure 3.13 indicates a good approximation to a $\mathcal{N}(0, 19)$ random variable.

(iv) *Check the measurement envelope.* The *innovation process* $\vartheta_p(t) = (m(t) - Hx(t)) \bullet (HP(t)H^T + U)^{-1}$ should form a standard normal distribution. If not,

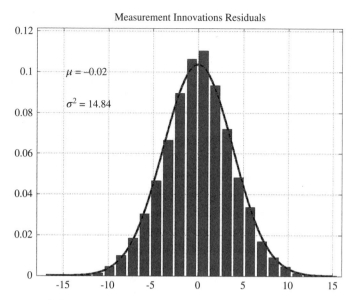

FIGURE 3.13 Distribution of the innovation residuals.

adjust the process and measurement noise to compensate. In the case afore-mentioned, the mean and variance of $\vartheta_p(t)$ are $\mu = -0.007$ and $\sigma^2 = 1.045$, respectively. Moreover, a band of one standard deviation about the mean of $\vartheta_p(t)$ yields an area of 0.6721. For a standard normal distribution, a one standard deviation band about 0 produces an area of 0.6827. This information, coupled with the histogram in Figure 3.12c, lends credibility to the notion of a properly tuned filter. Therefore, the tuning above yields the desired results.

There are many more factors that can be considered in filter tuning. These are mostly *ad hoc* methods. The interested reader is referred to Costa [10, 11] for additional details about filter tuning.

MATLAB Commands
(Tuning Results)
% Determine the tuning results via display as a normal distribution on
$Z = (m(t) - Hx(t, M_t)) \bullet (HP(t, M_t)H^T + U(t))^{-1}$
% Normalized state estimated *infecteds* with respect to measurement residuals
z = F.In.*F.Info;
% Mean and standard deviation of the estimated measurement residuals
mu = mean(z); s = std(z);
% Area under $f_Z(z)$ for $x \in [-1, 1]$: Prob$(-1 \leq Z \leq 1)$, Z the *innovations process*
Area = normcdf(1,mu,s) - normcdf(-1,mu,s)

```
Area =
    0.6721
```
% Area under $\phi(z)$ for $x \in [-1, 1]$: Prob$(-1 \le Z \le 1)$, $Z \sim \mathcal{N}(0,1)$
```
Area = normcdf(1,0,1) - normcdf(-1,0,1)
Area =
    0.6827
```
% Plot the distribution for $f_Z(x)$
```
histplot(z,25);
```

EXERCISES

3.3 Rather than modeling the population of Massachusetts by a third-order polynomial $p_3(t)$, employ the *logistics model*

$$\frac{dN}{dt} = rN\left(1 - \frac{1}{k}N\right), \quad N(t_0) = N_0.$$

Show that $N(t) = \dfrac{kN_0\, e^{r(t-t_0)}}{k - N_0 + N_0\, e^{r(t-t_0)}}$. Find a "best least-squares" fit to the data with respect to the parameters k and r. Calculate the *RMS* error for these parameters. The constants k and r are sometimes referred to as the *steady-state population* and *net per capital growth rate*, respectively. *Hint*: $k \approx 7$ (million) and $r \approx 0.02$. See Britton [3].

3.4 Vary the *system adjustable parameters* κ and U of Table 3.6. Also, vary the proportion q_0 of the initial measurements $m(t_0)$ that comprise the initial number of *recovered* (i.e., infectious people who have recovered from the disease): $R(t_0) = m(t_0) \cdot q_0$. How does the adjustment of these parameters affect the outcome of the model, *EKBF* tuning parameters, and filter behavior?

3.1.2.3 Project Combine the respiratory disease model (3.10) along with the *HIV/AIDS* transmission model (3.1) to produce a paradigm for the transmission of the Ebola virus. This requires estimating the rate of infectiousness, rate of recovery, disease death rate, and determining whether a person infected with the virus is permanently immune thereafter.

3.1.3 Predicting the Outbreak of a Respiratory Epidemic

The advent of highly contagious and lethal ailments such as *SARS* (severe acute respiratory syndrome 2002–2003), H1N1 (swine flu 2009), or H5N1 (avian flu 2008) influenza give impetus to the development of an "automated" epidemic alert system. Indeed, the combination of the *SEIR* model (3.10) and the *EKBF* (3.12a)–(3.12j) is the foundation of an algorithmic approach in predicting the outbreak of an infectious disease. In this section, a simulated outbreak is injected into the data used in Section 3.1.2 that mimics historical respiration infection counts.

The simulated data illustrated in Figure 3.10b were generated to reflect the true cyclical nature of respiratory infections in the northeastern United States. Notably, the infection peak occurs from December to February of a given year, while the ebb of such infections coincides with the summer months (June–August). Therefore, to simulate an outbreak of a disease, the history of the disease corresponding to the time of a significant change in reported infections must be taken into account.

Consider the following simple scenario. The simulated data depicted in Figure 3.10b are constructed to have initial time from 1 January 2000. That is, $t_0 = 1$ January 2000. These data are simulated for 10 years so that, from 1 January 2000 through 1 January 2010, the number of infections is recorded. Could an outbreak beginning on [say] 3 February 2005 be determined from the data previously recorded on 3 February 2000, 3 February 2001, 3 February 2002, 3 February 2003, and 3 February 2004? The answer is "yes." To see how, proceed as follows.

1. Let $t_{out} = 3$ February 2005 be the date of the start of the outbreak and τ_1, $\tau_2, \ldots, \tau_n$ be the "historic" times corresponding to previously recorded data at these dates. In this example, $\tau_1 = 3$ February 2000, $\tau_2 = 3$ February 2001, $\tau_3 = 3$ February 2002, $\tau_4 = 3$ February 2003, and $\tau_5 = 3$ February 2004 so that $n = 5$.

2. Set the length of the outbreak at d days.

3. Run the *EKBF* on the data from $t_0 = 1$ January 2000 until $t_f = t_{out} + 2 \cdot d$. This is to ensure that the filter has run through the outbreak period and into the "recovery" time.

4. Obtain estimates of the state vector x and covariance matrix P at the times $t \in [t_0, t_f]$ and for measurements $M = \{m_0, m_1, \ldots, m_f\}$, where m_k is the measurement at time t_k.

5. Collect the state vectors and covariance matrices from the historic time intervals $\vartheta_1 = [\tau_1, \tau_1 + 2 \cdot d]$, $\vartheta_2 = [\tau_2, \tau_2 + 2 \cdot d], \ldots, \vartheta_n = [\tau_n, \tau_n + 2 \cdot d]$. That is,

$$
\begin{array}{ll}
x(\tau_1), x(\tau_1 + 1), \ldots, x(\tau_1 + 2 \cdot d) & P(\tau_1), P(\tau_1 + 1), \ldots, P(\tau_1 + 2 \cdot d) \\
x(\tau_2), x(\tau_2 + 1), \ldots, x(\tau_2 + 2 \cdot d) & P(\tau_2), P(\tau_2 + 1), \ldots, P(\tau_2 + 2 \cdot d) \\
\quad \vdots & \quad \vdots \\
x(\tau_n), x(\tau_n + 1), \ldots, x(\tau_n + 2 \cdot d) & P(\tau_n), P(\tau_n + 1), \ldots, P(\tau_n + 2 \cdot d)
\end{array}
$$

6. Average the state vectors and covariance matrices over the times $[\tau_1, \tau_2, \ldots, \tau_n]$, $[\tau_1 + 1, \tau_2 + 1, \ldots, \tau_n + 1]$, $[\tau_1 + 2, \tau_2 + 2, \ldots, \tau_n + 2], \ldots$, and $[\tau_1 + 2 \cdot d, \tau_2 + 2 \cdot d, \ldots, \tau_n + 2 \cdot d]$, respectively, to obtain

$$
\bar{x}(\delta\tau_1) = \frac{1}{n} \sum_{j=1}^{n} x(\tau_j) \qquad\qquad \bar{P}(\delta\tau_1) = \frac{1}{n} \sum_{j=1}^{n} P(\tau_j)
$$

$$
\bar{x}(\delta\tau_2) = \frac{1}{n} \sum_{j=1}^{n} x(\tau_j + 1) \qquad\qquad \bar{P}(\delta\tau_2) = \frac{1}{n} \sum_{j=1}^{n} P(\tau_j + 1)
$$

$$
\vdots \qquad\qquad\qquad\qquad \vdots
$$

$$
\bar{x}(\delta\tau_{2d+1}) = \frac{1}{n} \sum_{j=1}^{n} x(\tau_j + 2 \cdot d) \qquad \bar{P}(\delta\tau_{2d+1}) = \frac{1}{n} \sum_{j=1}^{n} P(\tau_j + 2 \cdot d)
$$

Observe that these vectors and matrices are in one-to-one correspondence with the outbreak times $t_1^* = t_{out}$, $t_2^* = t_{out} + 1$, $t_3^* = t_{out} + 2, \ldots, t_{2d+1}^* = t_{out} + 2 \cdot d$.

7. From the collection of mean state vectors $\bar{x}(\delta\tau_k)$ and mean covariance matrices $\bar{P}(\delta\tau_k)$, $k = 1, 2, \ldots, 2d+1$, form the sequence of *inclusion intervals of width* ω, $IZ_k(\omega)$, by projecting these estimates into measurement space. That is, apply the measurement projection matrix $H = [0, 0, 1, 0]$ to $\bar{x}(\delta\tau_k)$ and the corresponding similarity transformation to the $\bar{P}(\delta\tau_k)$ to obtain the estimated measurement and measurement variance at each time t_k^*. Specifically, $IZ_k(\omega) = H\bar{x}(\delta\tau_k) \pm \omega \cdot H\bar{P}(\delta\tau_k)H^T$.

8. Inject an outbreak into the measurements at time t_{out} via an outbreak model function f_{out}. These models are described via equations (3.20a)–(3.20d). Over the outbreak time interval $\vartheta_{out} = [t_{out}, t_{out} + d]$ with time partition $\{t_1^*, t_2^*, \ldots, t_{2d+1}^*\}$, obtain state estimates $x_{out}(t_1^*), x_{out}(t_2^*), \ldots, x_{out}(t_{2d+1}^*)$ via the *EKBF* using the measurements $m^* = \{m(t_1^*) + f_{out}(t_1^*), m(t_2^*) + f_{out}(t_2^*), \ldots, m(t_{2d+1}^*) + f_{out}(t_{2d+1}^*)\}$.

9. Observe that the projection matrix $H = [0, 0, 1, 0]$ maps $\bar{P}(\delta\tau_k)$ into $\bar{\sigma}_I^2(t_k^*)$, the variance of the estimated infectious population. Select $\omega = 3$ so that $IZ_k(\omega)$ is a three-*standard deviation* or 3-σ band inclusion zone, $IZ_k(\omega) = H\bar{x}(\delta\tau_k) \pm 3 \cdot \bar{\sigma}_I(t_k^*)$. While the notion of using $(1 - \alpha) \times 100\%$ confidence intervals $CI_k = H\bar{x}(\delta\tau_k) \pm z_{1-\alpha/2} \cdot \bar{\sigma}_I(t_k^*)$ as the inclusion zone is appealing, these intervals do not fully cover the data extent. Here, $z_{1-\alpha/2}$ is the $(1 - \alpha/2) \times 100\%$ quantile of the standard normal distribution. For $a = 0.05$, $z_{1-\alpha/2} = 1.96$ which is substantially smaller than $\omega = 3$. That this "confidence interval" extent $z_{1-\alpha/2}$ is too small to fully cover the data range is illustrated in Figures 3.14a and 3.14b. Therefore, the 3-σ band inclusion zones will be implemented.

10. If the number of reported infections from an outbreak "escapes" the inclusion zone (i.e., exceeds 3-σ upper limit) for a specified number of consecutive days, then an outbreak is declared. If no such "escape" is observed, then the outbreak is undetected. Figure 3.15 illustrates inclusion zones, mean state vector estimates, and filter returns for an outbreak starting on 3 February 2005.

The process outlined in steps 1–10 will yield a *probability of outbreak detection* estimate. An example of an inclusion zone and an injected outbreak is presented in Figure 3.15. The dots (•) represent the simulated data $m(t)$. The squares (■) are *EKBF* estimates of the state vector projected into the measurement space $Hx(t, M)$. The dotted bands (- - -) about the filter estimates illustrate a 3-σ inclusion zone for the first historic data intervals $\vartheta_1, \vartheta_2, \ldots, \vartheta_n$. Finally the diamonds ◆ depict the *EKBF estimates* of the outbreak. As presented in Figure 3.15, the outbreak would be detected over three consecutive days (e.g., at times t_6^*, t_7^*, and t_8^*).

This measure of outbreak detection, however, does not consider a natural or cyclic surge in a disease. Consequently, a method to determine a real detection from a false detection is required. Therefore, to gauge the traditional metrics *sensitivity* and *specificity*, a more sophisticated measure is required. The *sensitivity* of a procedure is the ratio of the *true positives* to all positives, while the *specificity* is the ratio of the

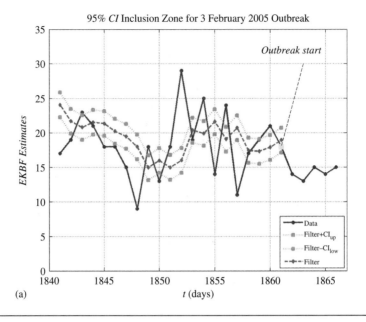

(a)

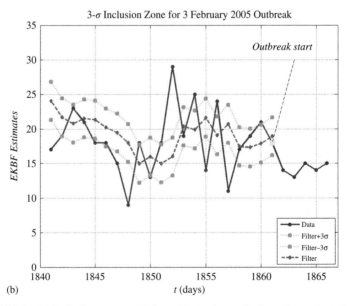

(b)

FIGURE 3.14   (a) Inclusion zone as 95% confidence intervals about the estimated measurements. (b) Inclusion zone as 3-σ band about the estimated measurements.

Inclusion Zone and Outbreak over 3–18 February 2005

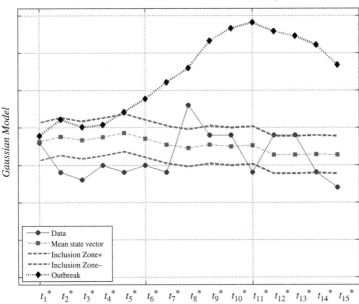

FIGURE 3.15 Inclusion zone and outbreak injection.

true negatives to all negatives. A *true positive* T_p is the detection of an anticipated event (such as an infection of a patient or the outbreak of a communicable disease) when the event actually occurs. Failure to detect such an event (i.e., failing to detect a true positive) results in a *false negative* F_n. Similarly, the "detection" of a positive event in the absence of a genuine positive is a *false positive* F_p. Declaring "no detection" when no positive event has occurred is called a *true negative* T_n.

These metrics are generally reported as percentages and are defined via (3.14a) and (3.14b).

$$\text{Sen} = \frac{T_p}{T_p + F_n} \cdot 100\% \tag{3.14a}$$

$$\text{Spec} = \frac{T_n}{T_n + F_p} \cdot 100\% \tag{3.14b}$$

How then are outbreaks modeled? How can the *sensitivity* and *specificity* of an outbreak detection algorithm be calculated? This is achieved via the *EKBF* estimates and the *Mahalanobis distance*. The *Mahalanobis distance* of a vector $x \in \mathbb{R}^p$ to a center x_0 with respect to the $p \times p$ weighting matrix M is defined as

$$D_M(x, x_0) = \sqrt{(x - x_0) \cdot M^{-1} \cdot (x - x_0)^T}. \tag{3.15}$$

On the historic time intervals $\vartheta_j = [\tau_j, \tau_j + d]$ that parallel the interval $\vartheta_{out} = [\tau_{out}, \tau_{out} + d]$ under which a potential outbreak is tested, filter estimates of the state vector and covariance matrix can be made at each measurement within ϑ_j, $j = 1, 2, \dots, n$. More specifically, suppose $\vartheta_j = [\tau_j, \tau_j + d] = \{s_{1j}, s_{2j}, \dots, s_{\ell j}\}$, where the s_{hj} are times with associated measurements. Subsequently, for each interval ϑ_j, ℓ measurements are available. Over the interval ϑ_j, the *EKBF* yields estimates of the state vectors $\hat{x}(s_{1j}, M_{1j}), \hat{x}(s_{2j}, M_{2j}), \dots, \hat{x}(s_{\ell j}, M_{\ell j})$ and covariance matrices $\hat{P}(s_{1j}, M_{1j}), \hat{P}(s_{2j}, M_{2j}), \dots, \hat{P}(s_{\ell j}, M_{\ell j})$. Consequently, an ℓ-dimensional vector of (squared) Mahalanobis distances can be computed by the components

$$
\left.
\begin{aligned}
D_{\hat{M}}^2(\hat{x}_{hj}, m_{hj}) &= (\hat{x}_{hj} - m_{hj}) \cdot \hat{M}^{-1} \cdot (\hat{x}_{hj} - m_{hj})^T \\
\hat{x}_{hj} &= \hat{x}(s_{hj}, M_{hj}), \quad m_{hj} = m(s_{hj}) \\
\hat{M} &= H\hat{P}(s_{hj}, M_{hj})H^T + U
\end{aligned}
\right\}, \quad h = 1, 2, \dots, \ell. \quad (3.16)
$$

That is, the $\delta_{h,j} \equiv D_{\hat{M}}^2(\hat{x}_{hj}, m_{hj})$ form an $n \times \ell$ matrix for all Mahalanobis distances computed over the intervals $\vartheta_1, \vartheta_2, \dots, \vartheta_n$. This is exhibited in (3.17).

$$
\Delta =
\begin{bmatrix}
\delta_{1,1} & \delta_{2,1} & \cdots & \delta_{\ell,1} \\
\delta_{1,2} & \delta_{2,2} & \cdots & \delta_{\ell,2} \\
\vdots & \vdots & \ddots & \vdots \\
\delta_{1,n} & \delta_{2,n} & \cdots & \delta_{\ell,n}
\end{bmatrix}
\quad (3.17)
$$

In a similar fashion, a vector of ℓ Mahalanobis distances can be calculated for the outbreak data and corresponding *EKBF* estimates associated with ϑ_{out}. Thus, if $f_{out}(t)$ is an outbreak model and $\mu_h = m(s_h) + f_{out}(s_h)$ are the measurements obtained from the outbreak interval $\vartheta_{out} = [\tau_{out}, \tau_{out} + d] = \{s_1, s_2, \dots, s_\ell\}$, then the *EKBF* estimates of the state vector $\tilde{x}$ and covariance matrix $\tilde{P}$ yield the desired result.

$$
\left.
\begin{aligned}
D_{\tilde{\Omega}}^2(\tilde{x}_h, \mu_h) &= (\tilde{x}_h - \mu_h) \cdot \tilde{\Omega}^{-1} \cdot (\hat{x}_h - \mu_h)^T \\
\tilde{x}_h &= \tilde{x}(s_h, M_h), \quad \mu_h = m(s_h) + f_{out}(s_h) \\
\tilde{\Omega} &= H\tilde{P}(s_h, M_h)H^T + U
\end{aligned}
\right\}, \quad h = 1, 2, \dots, \ell \quad (3.18)
$$

The aforementioned vector is $d_\mu = [D_{\tilde{\Omega}}^2(\tilde{x}_1, \mu_1), D_{\tilde{\Omega}}^2(\tilde{x}_2, \mu_2), \dots, D_{\tilde{\Omega}}^2(\tilde{x}_\ell, \mu_\ell)]$. Observe that the square of the Mahalanobis distance of (3.15), D_M^2, is distributed as a χ^2 random variable with p-degrees of freedom: $D_M^2 \sim \chi_p^2$.

For every measurement μ_h within a surveilled interval ϑ_{out}, a corresponding Mahalanobis distance $D_{\tilde{\Omega}}^2(\tilde{x}_h, \mu_h)$ calculated from the *EKBF* estimates can be compared to a threshold θ. If the Mahalanobis distance exceeds this threshold for a specified number of consecutive days (say 3 or 4), then an outbreak can be declared. If this occurs over a surveilled interval, then an outbreak is declared when an outbreak genuinely is occurring. This will be counted as a *true positive* T_p. Conversely, if the vector of Mahalanobis distances d_μ does *not* contain a consecutive sequence of elements that exceed the threshold θ, then the outbreak is not detected. This will be recorded as a *false negative* F_n. This combination of *true positives* and *false negatives* will be combined via (3.14a) to determine the *sensitivity* of the outbreak detection algorithm.

To test the *specificity* of the outbreak detection method, compute the *column means* of the historic data Mahalanobis distances from Δ of (3.17).

$$\left. \begin{aligned} d_\Delta &= [\bar{\delta}_1, \bar{\delta}_2, \dots, \bar{\delta}_\ell] \\ \bar{\delta}_h &= \frac{1}{n} \sum_{j=1}^{n} \delta_{h,j} \end{aligned} \right\} \tag{3.19}$$

If the $\bar{\delta}_h$ exceed the threshold θ for the select number of consecutive days, then an outbreak is declared and a *false positive* F_p is registered. Otherwise, a *true negative* T_n is the determination. From this logic, the *specificity* of the outbreak detection procedure is calculated.

Now the question becomes, *What is a sensible model for a disease outbreak?* As with most problems within the sphere of mathematical biology, the answer is rooted in both art and science. Consider the following transmission behaviors.

(i) A disease that mimics common maladies (such as a cold or influenza) begins a slow influx into a population. There is a steady increase in the number of infections. Public health officials identify the new infection and either neutralize it via medical intervention (i.e., vaccine) or via policy intervention (isolating/eliminating the source).

(ii) A disease has a steady migration into a population and then saturates.

(iii) A disease has a steady rise and fall within a population.

These scenarios can be modeled by four elementary functions. In fact, behavior (i) parallels the *SARS* outbreak in Hong Kong. For a timeline and containment strategies, see Timeline of the *SARS* outbreak, https://en.wikipedia.org/wiki/SARS_outbreak. Behavior (ii) is modeled by a linear ramp, while scenario (iii) can be mimicked by a linear hat function or a Gaussian distribution. For each function, m represents the *rate of infection* and d is the *number of days* in the outbreak.

Linear increase and collapse

$$L_m(t) = \begin{cases} m \cdot (t - t_0) & t_0 \leq t \leq t_0 + d \\ 0 & \text{otherwise} \end{cases} \tag{3.20a}$$

Linear ramp and saturation

$$L_r(t) = \begin{cases} m \cdot (t - t_0) & t_0 \leq t \leq t_0 + d \\ m \cdot d & t > t_0 + d \\ 0 & t < t_0 \end{cases} \tag{3.20b}$$

Linear growth and decay

$$L_{hat}(t) = \begin{cases} m \cdot (t - t_0) & t_0 \leq t \leq t_0 + d \\ m \cdot d - m(t - t_0 - d) & t > t_0 + d \\ 0 & t < t_0 \end{cases} \tag{3.20c}$$

Exponential growth and decay

$$\varphi(t) = m \cdot d \cdot \exp\left(-\frac{(t - t_0 - d)^2}{2(d/z_{0.975})^2}\right) \tag{3.20d}$$

For the Gaussian function $\varphi(t)$, the "variance" is normalized by the 97.5% quantile of the standard normal distribution. This quantity is included to ensure the right "spread" in the distribution of the outbreak. It is part of the *art of mathematical biology*. Illustrations of these outbreak models are provided in Figure 3.16 with $m = 3$ infections/day and $d = 7$ days.

With this process in mind, a series of simulations using various models, infection rates m, and number of days d are processed. In particular, the outbreak length in each simulation will be $d = 7$ days; the infection rates will vary over the discrete interval $m \in \{1, 2, 3, 4\}$; and the number of *consecutive days* in which the estimated infections must exceed a 3-σ band will be $N_{days} = 3$ days. For each outbreak model, 200 trials are injected into the data as follows. Two years of the simulated respiratory data were used to tune and train the *EKBF*. That is, from time t_0 (1 January 2000) to time t_{train} (1 January 2002), run the *EKBF* on the simulated data as discussed in Section 3.1.2. Then, at a random time $t_1' > t_{train}$, inject an outbreak of length d days

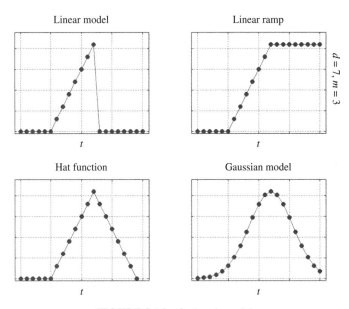

FIGURE 3.16   Outbreak models.

and run the filter to time $t'_{1,f} = t'_1 + 2 \cdot d$. Over this time interval $[t'_1, t'_1 + 2 \cdot d]$, find the corresponding dates (from other years) to use as "historical data" as shown in Figure 3.15. From these historical data, state estimates and corresponding inclusion zones are computed via the *EKBF* of equations (3.12a)–(3.12j) (Section 3.1.2) applied to the model (3.10). Each day yields a "yes/no" result as to whether the state estimates exceed the 3-σ inclusion zone computed via the *EKBF* on historical data. Moreover, a true positive, true negative, false positive, or false negative is recorded for every day via the Mahalanobis distance (3.18). The Mahalanobis threshold mentioned in the comments following equation (3.18) is selected as $\theta = \chi_1^2(1 - \tau) =$ the $(1 - \tau)$ 100% quantile of the χ^2-distribution with 1 degree of freedom (for more details about the χ^2-distribution, see Section A.3.2 in the Appendix). If the Mahalanobis distance of an injected outbreak exceeds the threshold for three or more consecutive days, then the outbreak is correctly detected: $d_M^2(outbreak) \geq \chi_1^2(1 - \tau) \Rightarrow$ *true detection* or *true positive*. Similarly, if the Mahalanobis distance *does not exceed* the threshold over three consecutive days, then the outbreak is incorrectly undetected. That is, if $d_M^2(outbreak) < \chi_1^2(1 - \tau)$, then there is a failure to detect an outbreak and a *false negative* is recorded. For the outbreak-free data run through the *EKBF* on the historic times $\tau_1, \tau_2, \ldots, \tau_n$, compute the mean Mahalanobis distances as indicated via equation (3.16). Compare these distances against the threshold $\chi_1^2(1 - \tau)$. If the distances exceed the threshold for three consecutive days, then an outbreak is detected where no outbreak exists. This results in a *false positive*. Finally, if the distances do not exceed the threshold for three consecutive days, then no outbreak is detected where none occurs. This yields a *true negative*. Continue this process for the randomly selected times $t'_2, t'_3, \ldots, t'_N$, where N is the number of outbreaks injected into the method. Next, count up the number of true negatives T_n, true positives T_p, false negatives F_n, and false positives F_p accumulated from the N-injected outbreaks.

A probability of *outbreak detection* p_d as well as a *sensitivity* and *specificity* can now be calculated from this simulation. Tables 3.9a–c summarize the simulated outbreak detection method results. Figure 3.17 displays the graphs of $[1 - specificity, sensitivity]$ coordinates. The functions connecting the measured $[1 - spec, sen]$ points are referred to as *receiver–operator characteristic* (*ROC*) curves. The *ROC* curves and the area they subtend can be used to assess algorithm performance. This will be

TABLE 3.9 **(a) Probabilities of detection, (b) Sensitivities, and (c) Specificities for simulated outbreak models**

	(a) Outbreak model: p_d			
M	Linear	Ramp	Hat	Gaussian
0.5	0.402	0.835	0.61	0.747
1	0.852	0.995	0.975	0.995
1.5	0.98	1	1	1
2	1	1	1	1
2.5	1	1	1	1
3	1	1	1	1
3.5	1	1	1	1
4	1	1	1	1

TABLE 3.9 (*Continued*)

M	Linear	Ramp	Hat	Gaussian
		(b) Outbreak model: Sensitivities		
0.5	0.372	0.365	0.34	0.364
1	0.357	0.298	0.359	0.308
1.5	0.538	0.322	0.354	0.325
2	0.616	0.379	0.487	0.449
2.5	0.875	0.518	0.636	0.618
3	0.95	0.717	0.795	0.765
3.5	0.965	0.843	0.853	0.889
4	0.995	0.925	0.919	0.92
		(c) Outbreak model: Specificities		
0.5	0.668	0.725	0.74	0.742
1	0.75	0.717	0.672	0.702
1.5	0.688	0.704	0.692	0.725
2	0.707	0.717	0.729	0.768
2.5	0.69	0.653	0.727	0.764
3	0.72	0.692	0.69	0.699
3.5	0.724	0.731	0.69	0.707
4	0.704	0.734	0.712	0.724

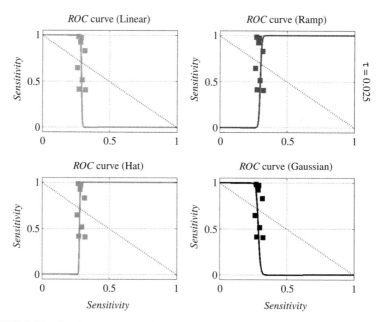

FIGURE 3.17 *Receiver–operator characteristic* curves for the outbreak detection method.

greatly amplified in Chapter 4 on statistical pattern recognition. The point at which line $y = 1 - x$ intersects the *ROC* curve corresponds to the theoretical equality of *Sensitivity* and *Specificity*. This is called the *Q-point* of the *ROC* curve and will be further detailed in Chapter 4.

For modest "infection rates" $m \geq 1.5$, the method does quite well at detecting the outbreak of an "epidemic" infection. The models, however, do not report reliable sensitivities until m exceeds 3. This may be due to the strictness of the declaration protocol; namely, requiring the reported infections to exceed their historic *inclusion zones* for at least three consecutive days. It is left as an exercise to determine whether using two consecutive days or four consecutive days produces better outbreak performance.

EXERCISE

3.5 Run the outbreak detection model using $N_{days} = 2$ and then $N_{days} = 4$ to see the difference in performance. *Hint*: Use the function `seirekbfout.m` and MATLAB commands listed at the end of this section.

MATLAB Commands
(Simulated Outbreak Results)

```
% Load the data and corresponding times
Ddir = 'C:\PJC\Math_Biology\Chapter_3\Data';
data = load(fullfile(Ddir,'SEIR_Simulated_Data.mat'));
D = data.D;
t = load(fullfile(Ddir,'SEIR_Time.mat')); T = t.T;
% Eliminate unnecessary variables
clear data t
% Set the vector of initial conditions
So = 6.5e+04; IC = [So,So*0.05,D(1),D(1)*0.5];
% Structure of system adjustable parameters
dy = 365.25; Sap.v = 1.5/dy; Sap.beta = 0.0056; Sap.a = 0.7222;
Sap.d = 0.0003224/dy; Sap.kappa = 0.95; Sap.nd = 0.008724/dy;
Sap.bo = 0.03*dy*IC(3); Sap.Ndays = 3; Sap.thres = 0.025;
Sap.Mnoise = 0.1; Sap.IZwidth = 3;
% Range of disease rates and number of injected outbreaks Nout
Rates = 0.5:0.5:4;
Nr = numel(Rates);   Nout = 200;
% Linear increase and collapse after peak (vaccine, isolation) SL
% Linear increase and saturation SR, Hat outbreak SH, Gaussian outbreak SG
tic;
    for j = 1:Nr
        SL{j} = seirekbfout(IC,Sap,D,T,'alin',Nout,7,Rates(j));
        SR{j} = seirekbfout(IC,Sap,D,T,'lramp',Nout,7,Rates(j));
        SH{j} = seirekbfout(IC,Sap,D,T,'hat',Nout,7,Rates(j));
        SG{j} = seirekbfout(IC,Sap,D,T,'Gauss',Nout,7,Rates(j));
    end
toc
Elapsed time is 790.042206 seconds.
```

MATLAB Commands
(*Recover the Assessment Metrics*)

```
% Probabilities of detection
pdl = cellfun(@(x)(x.Pd),SL); pdr = cellfun(@(x)(x.Pd),SR);
pdh = cellfun(@(x)(x.Pd),SH); pdg = cellfun(@(x)(x.Pd),SG);
Pd = {pdl,pdr,pdh,pdg}; PrD = (vertcat(Pd{:})).';
% Sensitivities
Senl = cellfun(@(x)(x.Sen),SL);
Senr = cellfun(@(x)(x.Sen),SR);
Senh = cellfun(@(x)(x.Sen),SH);
Seng = cellfun(@(x)(x.Sen),SG);
Sen = {Senl,Senr,Senh,Seng};
SEN = (vertcat(Sen{:})).';
% Specificities
Specl = cellfun(@(x)(x.Spec),SL);
Specr = cellfun(@(x)(x.Spec),SR);
Spech = cellfun(@(x)(x.Spec),SH);
Specg = cellfun(@(x)(x.Spec),SG);
Spec = {Specl,Specr,Spech,Specg};
SPEC = (vertcat(Spec{:})).';
```

REFERENCES

[1] Bacchetti, P. and A. R. Moss, "Incubation period of AIDS in San Francisco," *Nature*, vol. 338, pp. 251–253, 16 March 1989.

[2] Banks, H. T. and C. Castillo–Chavez, *Bioterrorism: Mathematical Modeling Applications in Homeland Security*, Frontiers in Applied Mathematics, vol. 28, SIAM Books, Philadelphia, PA, 2003.

[3] Britton, N. F., *Essential Mathematical Biology*, Springer–Verlag, London, 2003.

[4] Bucy, R. S. and P. D. Joseph, *Filtering for Stochastic Processes with Applications to Guidance*, Chelsea Publishing, New York, 1968.

[5] Cauchemez, S., F. Carrat, C. Viboud, A. J. Valleron, and P. Y. Boelle, "A Bayesian MCMC approach to study transmission of influenza: application to household longitudinal data," *Statistics in Medicine*, vol. 23, no. 22, pp. 3469–3484, November 30, 2004.

[6] CDC, HIV Surveillance Report, *Statistics epidemiology of infection through 2011*, http://www.cdc.gov/hiv/library/reports/surveillance/index.html#panel, 2011.

[7] Chiarotti, F., M. Palombi, N. Schinaia, A. Ghirardini, and R. Bellocco, "Median time from seroconversion to AIDS in Italian HIV–positive haemophiliacs: different parametric estimates," *Statistics in Medicine*, vol. 13, pp. 136–175, 1994.

[8] Costa, P. J. and J. P. Dunyak, "Models, prediction, and estimation of outbreaks of an infectious disease," in Proceedings of the IEEE SoutheastCon, edited by Yair Levy (IEEE Catalog Number 05CH37624), pp. 174–178, April 8–9, 2005.

[9] Costa, P. J., *Bridging Mind and Model*, St. Thomas Technology Press, St. Paul, MN, 1994.

[10] Costa, P. J., *A Compendium on Track Filter Designs, Development, and Analyses*, Raytheon Company Technical Memorandum, May 1992.

[11] Costa, P. J., "An estimation method for the outbreak of an infectious disease," MITRE Technical Report, MTR05B0000093, November 2005.

[12] Dornadula, G., H. Zhang, B. Van Uitert, J. Stern, L. Livornese Jr., M. Ingerman, J. Witek, R. Kedanis, J. Natkin, J. DeSimone, and R. Pomerantz, "Residual HIV–1 RNA blood plasma of patients taking suppressive highly active antiretroviral therapy," *JAMA*, vol. 282, no. 17, pp. 1627–1632, November 3, 1999.

[13] Finzi, D., J. Blankson, J. Siliciano, J. Margolick, K. Chadwick, T. Pierson, K. Smith, J. Lisziewicz, F. Lor, C. Flexner, T. Quinn, R. Chaisson, E. Rosenberg, B. Walker, S. Grange, J. Gallant, and R. Siliciano, "Latent infection of CD4+ T cells provides a mechanism for lifelong persistence of HIV–1, even in patients on effective combination therapy," *Nature Medicine*, vol. 5, no. 5, pp. 512–517, May 1999.

[14] Gelb, A., *Applied Optimal Estimation*, MIT Press, Cambridge, MA, 1974.

[15] Gupta, S., R. Anderson, and R. May, "Mathematical models and the design of public health policy: HIV and antiviral therapy," *SIAM Review*, vol. 35, no. 1, pp. 1–16, March 1993.

[16] Hawala, S., "Chi–squared goodness of fit test for lifetime data," Ph.D. Dissertation, University of California at Davis, September 1994.

[17] Hyman, J. M. and T. LaForce, "Modeling the spread of influenza among cities," in *Bioterrorism, Mathematical Modeling Applications in Homeland Security*, edited by H. T. Banks and C. Castillo–Chavez, SIAM Books, Philadelphia, PA, pp. 211–236, 2003.

[18] Hyman, J. M. and E. A. Stanley, "A risk–based heterosexual model for the AIDS epidemic with biased sexual partner selection," in *Modeling the AIDS Epidemic: Planning, Policy, and Prediction*, edited by E. H. Kaplan and M. L. Brandeau, Raven Press, 1994.

[19] Hyman, J. M. and E. A. Stanley, "The effects of social mixing patterns on the spread of AIDS," in *Mathematical Approaches to Problems in Resource Management and Epidemiology*, Lecture Notes in Biomathematics, vol. 81, edited by C. Castillo–Chavez, S. A. Levin, and C. A. Shoemaker, Springer–Verlag, Berlin, 1989.

[20] Jazwinski, A. H., *Stochastic Processes and Filtering Theory*, Dover Publications, Mineola, NY, 1970.

[21] Kalman, R. E., "A new approach to linear filtering and prediction problems," *Journal of Basic Engineering (ASME)*, vol. 82D, pp. 35–45, March 1960.

[22] Kalman, R. E. and R. S. Bucy, "New results in linear filtering and prediction theory," *Journal of Basic Engineering (ASME)*, vol. 83, pp. 95–108, 1961.

[23] Laumann, E. O., J. H. Gagnon, R. T. Michael, and S. Michaels, *The Social Organization of Sexuality: Sexual Practices in the United States*, University of Chicago Press, 1994.

[24] Mandl, K., *Private Communication*, Department of Biomedical Information and Population Health, Boston Children's Hospital, November 2005.

[25] Murray, J. D., *Mathematical Biology*, Springer–Verlag, Berlin, 1989.

[26] Petersen, L. R., G. A. Satten, M. Busch, S. Kleinman, A. Gridon, B. Lenes, and the HIV Seroconversion Study Group II, "Duration of time from onset of human immunodeficiency virus type–I infectiousness to development of detectable antibody," *Transfusion*, vol. 34, no. 4, pp. 283–289, 1994.

[27] Renski, H., L. Koshgarian, and S. Strate, "Long–term population projections for Massachusetts regions and municipalities," *UMass Donahue Institute Report*, Prepared for

the Office of the Secretary of the Commonwealth of Massachusetts, http://pep.donahue institute.org/, November 2013.

[28] Satzer, W. J., "Track reconstruction and data fusion using optimal smoothing," in *Bridging Mind and Model*, edited by P. J. Costa, St. Thomas Technology Press, St. Paul, MN, pp. 64–90, 1994.

[29] Seber, G. A. F. and C. J. Wild, *Nonlinear Regression*, Wiley–Interscience, 1989.

[30] Sontag, E. D., *Mathematical Control Theory: Deterministic Finite Dimensional Systems*, Springer–Verlag, New York, 1998.

[31] Wawer, M. J., R. H. Gray, N. K. Sewankambo, D. Serwadda, X. Li, O. Laevendecker, N. Kiwanuka, G. Kigozi, M. Kiddugavu, T. Lutalo, F. Nalugoda, F. Wabwire–Mangen, M. P. Meehan, and T. C. Quinn, "Rates of HIV–1 transmission per coital act by stage of HIV–1 infection in Rakai, Uganda," *Journal of Infectious Diseases*, vol. 191, no. 9, pp. 1403–1409, 2005.

FURTHER READING

Becker, N. G. and L. R. Egerton, "A transmission model for HIV," *Mathematical Biosciences*, vol. 199, pp. 205–224, 1994.

Celum, C. L., R. W. Coombs, M. Jones, V. Murphy, L. Fisher, C. Grant, L. Corey, T. Inui, M. H. Werner, K. K. and Holmes, "Risk factors for repeatedly reactive HIV–1, EIA, and indeterminate western blots," *Archive of Internal Medicine*, vol. 154, pp. 1129–1137, May 1994.

Cox, D. R., R. M. Anderson, and H. C. Hillier, "Epidemiological and statistical aspects of the AIDS epidemic," *Philosophical Transactions of the Royal Society of London*, vol. 325, pp. 37–187, 1989.

Cristea, A. and J. Strauss, "Mathematical model for the AIDS epidemic evolution in Romania," *Revue Romaine de Virologie*, vol. 44, no. 1–2, pp. 21–47, 1993.

de Vincenzi, I., "A longitudinal study of human immunodeficiency virus transmission by heterosexual partners," *The New England Journal of Medicine*, vol. 331, no. 6, pp. 341–346, August 1994.

Goedart, J. J., "Prognostic markers for AIDS," *Annals of Epidemiology*, vol. 1, no. 2, pp. 129–139, December 1990.

Jewell, N. P., K. Dietz, and V. T. Farewell, editors, *AIDS Epidemiology: Methodological Issues*, Birkhäuser, Boston, 1992.

Kim, M. Y. and S. W. Lagakes, "Estimating the infectivity of HIV from partner studies," *Annals of Epidemiology*, vol. 1, no. 2, pp. 117–128, December 1990.

Kroner, B. L., P. S. Rosenberg, L. M. Aledort, W. G. Alvord, and J. J. Goedart for the Multi–Center Hemophilia Cohort Study, "HIV–1 infection incidence among persons with hemophilia in the United States and Western Europe, 1978–1990," *Journal of Acquired Immune Deficiency Syndromes*, vol. 7, pp. 279–286, 1994.

Malice, M–P and R. J. Kryscio, "On the role of variable incubation periods in simple epidemic models," *IMA Journal of Mathematics Applied in Medicine and Biology*, vol. 6, pp. 233–242, 1989.

Mode, C. J., H. E. Gollwitzer, M. A. Salsburg, and C. K. Sleeman, "A methodological study of nonlinear stochastic model of an AIDS epidemic with recruitment," *IMA Journal of Mathematics Applied in Medicine and Biology*, vol. 6, pp. 179–203, 1989.

Padian, N. S., S. C. Shiboski, and N. P. Jewell, "Female–to–male transmission of human immunodeficiency virus," *Journal of the American Medical Association*, vol. 266, no. 12, pp. 1664–1667, September 1991.

Pedersen, C., "Infection with HIV type–1," *Danish Medical Bulletin*, vol. 41, no. 1, pp. 12–22, February 1994.

Peterman, T. A., D. P. Drotman, and J. W. Curran, "Epidemiology of the acquired immunodeficiency syndrome (AIDS)," *Epidemiological Reviews*, vol. 7, pp. 7–21, 1985.

Philips, A. N., C. A. Sabin, J. Elford, M. Bofil, G. Janossy, and C. A. Lee, "Acquired immunodeficiency syndrome (AIDS) risk in recent and long–standing human immunodeficiency virus type–1 (HIV–1) infected patients with similar CD4 lymphocyte counts," *American Journal of Epidemiology*, vol. 138, no. 10, pp. 870–878, November 1993.

Taylor, J. M. and Y. Chon, "Smoothing grouped bivariate data to obtain incubation statistics in medicine," *Statistics in Medicine*, vol. 13, no. 9, pp. 969–981, May 1994.

4

STATISTICAL PATTERN RECOGNITION AND CLASSIFICATION

When pathologists or radiologists examine a slide containing a tissue sample or an image obtained via *X-ray*, *MRI*, or *CAT* scan, they are asked to *interpret* what they see. In particular, when a diagnostician views the results of a biopsy or mammogram, he/she views the slide/image and makes a determination as to disease state. Does the patient have cancer? A radiologist tries to ascertain, among other conditions, whether a patient has a broken bone or a torn ligament. For these devices, diagnosis is achieved by the most proficient pattern recognition system presently known: Human sensory analysis. The eyes examine a slide (containing a tissue sample or cellular swab) or a radiological image and send sensory information to the brain. Then the specialist's catalogue of *previously viewed cases* guides her/him to a diagnostic conclusion.

This process can be summarized mathematically as *statistical pattern recognition* (*SPR*). The focus of this chapter will be on a general description of *SPR* methods along with the mathematical techniques used to determine a *classification*. Consider the following simplification. Suppose a person falls and feels a sharp pain in his arm. An *X-ray* of the injured area is taken and the developed film is sent to a radiologist. There are three possible diagnoses: (1) The bone is *not* broken, (2) the bone is broken, and (3) the bone is fractured. In the case of a broken bone, the *X-ray* image will show a separation in-between two sections of the bone. A *fracture* (Case 3) will show a line or crack in the bone. The *diagnosis* of the fracture/break will be a consequence of the experience and training of the radiologist and the accuracy of the *X-ray* machine.

Applied Mathematics for the Analysis of Biomedical Data: Models, Methods, and MATLAB®, First Edition. Peter J. Costa.
© 2017 Peter J. Costa. Published 2017 by John Wiley & Sons, Inc.
Companion website: www.wiley.com/go/costa/appmaths_biomedical_data

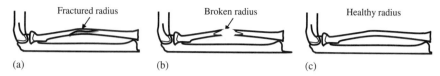

FIGURE 4.1   (a) Fractured radius. (b) Broken radius. (c) Healthy (unbroken) radius.

What is at play here? It is assumed that the radiologist has seen many examples of *unbroken, fractured*, and *broken* bones. These three categories or *classes* are the sets which partition the "injured bone" space. Using this reservoir of examples from known classes (i.e., broken, fractured, or unbroken bones), the radiologist views a film and, in essence, says *this film shows a fractured ulna because it most closely resembles other images of fractured ulnae*. Figures 4.1a–4.1c provide example images of fractured, broken, and unbroken radius bones.

Plainly not every fracture is identical to the one depicted in Figure 4.1a. The radiologist, however, will determine a "bone state" by comparing the image under review against the (mental) catalogue of other fractures already diagnosed. Whichever class (broken, fractured, unbroken) the review image is "closest to" will be the diagnosis rendered.

How can this process be described mathematically? Generically, a reviewer (pathologist, radiologist) inspects an image (i.e., measurement) and computes a "distance" of the measurement to one of the diagnostic classes. The class which has the minimum distance to the image will be the diagnosis determined by the reviewer.

Furthermore, it is rare that a reviewer is looking for a single "feature" on an image. For pathology or *MRI*, morphology (shape), size, cellular structure, continuity (tears, lacerations), and a host of other quantities are taken into consideration before a diagnosis is established. Consequently, an image is viewed heuristically as a multivariate measurement. The outline of this chapter is to

1. formalize the notion of a data class
2. review data preparation procedures
3. detail the process of transforming the data into lower dimensional coordinate systems (which most efficiently extract the data's information content)
4. present various classification methods
5. explain how the methods are "tuned" to select a desired performance level
6. demonstrate a method by example and
7. discuss a nonlinear technique.

4.1   MEASUREMENTS AND DATA CLASSES

As mentioned in the introductory remarks to this chapter, the image/slide/scan a diagnostician is asked to review is treated as a multivariate measurement. If the reviewer needs to consider p features, then the measurement takes the form of a *row vector*

$x = [x_1, x_2, \ldots, x_p]$. Suppose there are g diagnostic categories and n_k samples in each category, $k = 1, 2, \ldots, g$. The *data matrices*

$$X_k = \begin{bmatrix} x_{1,1}^{(k)} & x_{1,2}^{(k)} & \cdots & x_{1,p}^{(k)} \\ x_{2,1}^{(k)} & x_{2,2}^{(k)} & \cdots & x_{2,p}^{(k)} \\ \vdots & \vdots & \ddots & \vdots \\ x_{n_k,1}^{(k)} & x_{n_k,2}^{(k)} & \cdots & x_{n_k,p}^{(k)} \end{bmatrix} \tag{4.1}$$

provide the mathematical description of each category or *data class*. These categories are identified via the symbols $\mathscr{C}_1, \mathscr{C}_2, \ldots, \mathscr{C}_g$. The data matrices X_k, in turn, produce *class mean vectors* $\bar{x}_k \in \mathbb{R}^p$ and *class covariance matrices* $C_k \in \mathcal{Mat}_{p \times p}$ as defined via (4.2a) and (4.2b).

$$\bar{x}_k = \left[\frac{1}{n_k} \sum_{j=1}^{n_k} x_{j,1}^{(k)}, \frac{1}{n_k} \sum_{j=1}^{n_k} x_{j,2}^{(k)}, \ldots, \frac{1}{n_k} \sum_{j=1}^{n_k} x_{j,p}^{(k)} \right], \quad k = 1, 2, \ldots, g \tag{4.2a}$$

$$\left. \begin{aligned} C_k &= \frac{1}{n_k-1}(X_k - \mathbf{1}_{n_k} \cdot \bar{x}_k)^T (X_k - \mathbf{1}_{n_k} \cdot \bar{x}_k) \\ \mathbf{1}_{n_k} &= [1, 1, \ldots, 1]^T \in \mathbb{R}^{n_k} \end{aligned} \right\}, \quad k = 1, 2, \ldots, g \tag{4.2b}$$

Figure 4.2 illustrates four distinct data classes of multivariate measurements with respect to a set of wavelengths $\Omega = \{\omega_1, \omega_2, \ldots, \omega_p\}$. These data are hyperspectral measurements of fluorescence scans of tissue with varying disease states. The data classes consist of normal (i.e., no evidence of disease) and three increasing levels of

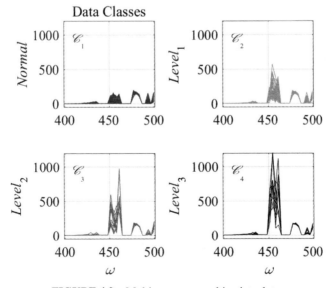

FIGURE 4.2   Multi-category, multivariate data.

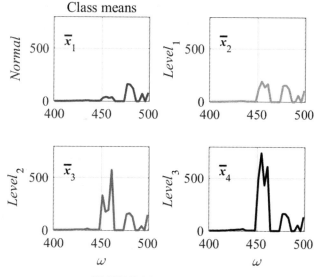

FIGURE 4.3   Class means.

disease. Disease *level* 3 is considered the most severe. These data are contained in the Chapter 4 data directory as the MAT-file `Multivariate_MultiClass_Data.mat`.

Observe that Figure 4.3 depicts the four class means $\bar{x}_1, \bar{x}_2, \bar{x}_3$, and $\bar{x}_4$ as computed via (4.2a).

What insight into disease status can be gleaned from these data? First, it is evident that there are wavelengths at which very little information can be gained from the measurement. Also, the "middle wavelengths" (i.e., $\omega \in [440, 460]$) seem to contain the bulk of the distinguishing features. A manner to formally validate these observations will be provided in the next section.

The total number of samples across all categories is $n = \sum_{k=1}^{g} n_k$. In addition to the class mean and covariance, three other measures are introduced; namely the *total data matrix* X_T, the *overall (sample) mean vector* $\bar{x}_T$, and the *(sample) variance across a selected feature* s_k^2. The word "sample" is used to reinforce the notion that the mean and variance are *estimated* from the class data matrix. The "exact" mean vector or covariance matrix is only known theoretically. These measures are defined via equations (4.3a)–(4.3c).

$$X_T = \begin{bmatrix} X_1 \\ X_2 \\ \vdots \\ X_g \end{bmatrix} \equiv \begin{bmatrix} x_{1,1} & x_{1,2} & \cdots & x_{1,p} \\ x_{2,1} & x_{2,2} & \cdots & x_{2,p} \\ \vdots & \vdots & \ddots & \vdots \\ x_{n,1} & x_{n,2} & \cdots & x_{n,p} \end{bmatrix} \tag{4.3a}$$

$$\left. \begin{array}{l} \bar{x}_T = \left[\bar{x}_{T,1}, \bar{x}_{T,2}, \dots, \bar{x}_{T,p} \right] \\ \bar{x}_{T,k} = \dfrac{1}{n} \sum\limits_{j=1}^{n} x_{j,k}, \, k = 1, 2, \dots, g \end{array} \right\} \tag{4.3b}$$

$$s_k^2 = \frac{1}{n-1} \sum_{j=1}^{n} \left(x_{j,k} - \bar{x}_{T,k}\right)^2, \ k = 1, 2, \ldots, p$$

$$s = [s_1, s_2, \ldots, s_p]$$

$$(4.3c)$$

Here, s is the vector of feature (*sample*) *standard deviations*. How can these concepts be utilized to quantify the idea of a distance to a data class? The answer is via the *Mahalanobis distance* $d_M(m, \bar{x}_k)$ of a measurement $m = [m_1, m_2, \ldots, m_p]$ to the kth data class with respect to a *weighting matrix M*.

$$d_M(m, \bar{x}_k) = \sqrt{(m - \bar{x}_k) \cdot M^{-1} \cdot (m - \bar{x}_k)^T} \qquad (4.4)$$

If $d_M(m) = [d_M(m, \bar{x}_1), d_M(m, \bar{x}_2), \ldots, d_M(m, \bar{x}_g)]$ is the vector of Mahalanobis distances from the measurement m to each data class and $d_M(m, \bar{x}_\ell) = \min(d_M(m)) \equiv \min_{1 \leq k \leq g} d_M(m, \bar{x}_k)$, then m is closest to class $\mathscr{C}_\ell$. There are many possible candidates for the weighting matrix M. One choice is $M = C_k$ at each distance $d_M(m, \bar{x}_k)$. The selection of the weighting matrix M is crucial to the separation of the g data classes. This is the concern of the next section.

It is emphasized that the Mahalanobis distance depends on the weighting matrix M. In the special case in which $M = I_{p \times p} =$ the identity matrix, then the Mahalanobis distance is equivalent to the *Euclidean distance*. Figure 4.4 gives an example in which $M = C_k$ and the *Mahalanobis* distance d_M of a select point $m = [0.8, 3.5]$ to a select class (*Class*₁) is smaller than it is to the alternate class (*Class*₂). The opposite, however, occurs with respect to the *Euclidean* distance d_I (namely, $M = I_{2 \times 2}$). That is, when the weighting matrix is the data class covariance C_k, the *Mahalanobis distance* assigns the point m to *Class*₁ as its distance to this class is smallest. Under

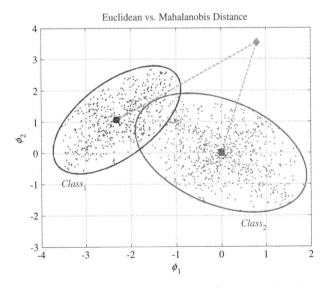

FIGURE 4.4   Mahalanobis and Euclidean distances to data classes.

TABLE 4.1 ***Mahalanobis (d_M) versus Euclidean (d_I) distances***

$m = [0.8, 3.53]$						
Class	μ	M	$d_M(m, \mu)$	*Class association*	$d_I(m, \mu)$	*Class association*
1	[−2.32, 1.1]	$\begin{bmatrix} 0.34 & 0.26 \\ 0.26 & 0.51 \end{bmatrix}$	5.33	1	3.95	2
2	[−0.0085, 0.022]	$\begin{bmatrix} 0.57 & -0.21 \\ -0.21 & 0.58 \end{bmatrix}$	5.48		3.6	

the Euclidean distance (i.e., weighting matrix is the identity I), $\boldsymbol{m}$ is closest to $Class_2$. Table 4.1 summarizes the results.

EXERCISE

4.1 Let $C_{pool} = \frac{1}{n}[(n_1 - 1) \cdot C_1 + (n_2 - 1) \cdot C_2]$ be the *pooled within-class covariance* matrix. Using the data from Table 4.1, compute the Mahalanobis distances $d_{C_{pool}}(\boldsymbol{m}, \boldsymbol{\mu}_j)$ for $j = 1, 2$. How does this compare to the Euclidean distance (i.e., $M = I_{2\times 2}$)? The data are located in the Chapter 4 data directory under the MAT-file names `Mahal_Example_X.mat` (for data $Class_1$) and `Mahal_Example_Y.mat` (for data $Class_2$).

4.2 DATA PREPARATION, NORMALIZATION, AND WEIGHTING MATRIX

Assume, as in Section 4.1, there are g p-dimensional data classes $\mathscr{C}_1, \mathscr{C}_2, \ldots, \mathscr{C}_g$. The data matrix X_k corresponding to $\mathscr{C}_k$ is $n_k \times p$ matrix of the form (4.1). Each column of X_k represents the measurement of a specific variable throughout the class. Each variable or *feature* may be in different units. Therefore, the data need to be transformed into a *dimensionless* (i.e., unit-free) coordinate frame. This can be accomplished by data normalization. In particular, each data matrix X_k is mapped into *standard normal form* via the transformation (4.5).

$$\left.\begin{aligned} & X_k \mapsto Z_k \equiv (X_k - \mathbf{1}_{n_k} \cdot \bar{\boldsymbol{x}}_T) \cdot S^{-1} \\ & S = \text{diag}(\boldsymbol{s}) = \begin{bmatrix} s_1 & 0 & \cdots & 0 \\ 0 & s_2 & \cdots & 0 \\ \vdots & \vdots & \ddots & \vdots \\ 0 & 0 & \cdots & s_p \end{bmatrix} \\ & \mathbf{1}_{n_k} = [1, 1, \ldots, 1]^T \in \mathbb{R}^{n_k}, k = 1, 2, \ldots, g \end{aligned}\right\} \tag{4.5}$$

Observe that this normalization of a data matrix uses the data *across all classes.* Instead of this global normalization, a *within-groups normalization* can be applied. For this approach, the class feature sample variance is defined via the data matrix X_k and equation (4.6).

$$\bar{x}_j(k) = \frac{1}{n_k} \sum_{\ell=1}^{n_k} x_{\ell,j}^{(k)}$$

$$s_j^2(k) = \frac{1}{n_k - 1} \sum_{\ell=1}^{n_k} \left(x_{\ell,j}^{(k)} - \bar{x}_j(k) \right)^2 \quad \Bigg\} \quad (4.6)$$

$$s(k) = \left[s_1(k), s_2(k), \dots, s_p(k) \right]$$

The symbols $\bar{x}_j(k)$ and $s_j^2(k)$ are the *class sample mean* and *sample variance* of the jth feature. If $S_k = \text{diag}(s(k))$ is the diagonal matrix with elements $s_1(k), s_2(k), \dots, s_p(k)$, then the *within-groups normalization* is

$$X_k \mapsto \tilde{Z}_k \equiv (X_k - \mathbf{1}_{n_k} \cdot \bar{x}_k) \cdot S_k^{-1}. \quad (4.7)$$

Here, $\bar{x}_k$ is the class mean vector from (4.2a).

Once the class data (i.e., data matrices) are normalized, then any measurement $m = [m_1, m_2, \dots, m_p]$ must be normalized in a comparable manner. When using the standard normal form, the measurement is mapped to $m_{norm} = (m - \bar{x}_T) \bullet S^{-1}$. Conversely, if a within-groups normalization is utilized, then g transformations need to be implemented in order to compute the Mahalanobis distance to each class: $m \mapsto m_{norm}(k) \equiv (m - \bar{x}_k) \cdot S_k^{-1}, k = 1, 2, \dots, g$.

With these developments in mind, data normalization can proceed. Before racing into this computation, however, a measure of caution is advised. As noted for the data illustrated in Figure 4.2, there are wavelength subregions over which the measurements appear to contain little useful information. These data may well have a sample variance close to 0. This would cause numerical instability in the inversion of the diagonal matrix S of sample variances from equation (4.5). Therefore, it is crucial to indicate unnecessary or dependent variables (i.e., wavelengths). How can these be identified? One approach is to determine which of the variables are *correlated* to others and then eliminate unnecessary wavelengths.

The *correlation* of two random variables X and Y is defined as the ratio of their covariance to the square root of the product of their variances. More specifically, $cor(X, Y) = \frac{\text{cov}(X,Y)}{\sqrt{\text{var}(X) \cdot \text{var}(Y)}} \equiv \frac{\sigma_{XY}}{\sqrt{\sigma_X^2 \cdot \sigma_Y^2}} = \frac{\sigma_{XY}}{\sigma_X \cdot \sigma_Y}$. Generally, the variance and covariance of random variables are not known *a priori*. Hence, a sample of a variable will be required to estimate the mean and variance. The sample mean and sample variance of a feature have already been defined via (4.6). The *sample covariance* is defined as $\widehat{\text{cov}}(x, y) \equiv s_{x,y} = \sum_{\ell=1}^{m} (x_\ell - \bar{x}) \cdot (y_\ell - \bar{y})/(m - 1)$. The *sample correlation* of the two variables is the ratio of the sample covariance to the product of the sample standard deviations

$$\rho(x, y) \equiv \rho_{x,y} = \frac{s_{x,y}}{s_x \cdot s_y}. \tag{4.8}$$

Here, s_x is the *sample standard deviation* of the variable x: $s_x = \sqrt{\frac{1}{m-1} \sum_{\ell=1}^{m} (x_\ell - \bar{x})^2}$.

How is correlation used to gauge the information content of the data presented in Figure 4.2? First, observe that for any two variables, $-1 \leq \rho(x_j, x_k) \leq 1$. Next, set a threshold τ so that $|\rho(x_j, x_k)| \geq \tau$ implies that the two features are correlated. In this sense, "correlated" means that the two variables are linearly related. Namely, $x_j = m x_k + b$. The correlation is a measure of how well the sample of the features fit a linear model. In the special case in which $s_{x_j} = s_{x_k}$, then the slope m of the line which "best fits" the samples of (x_j, x_k) is equal to the correlation in-between x_j and x_k. Also, if the variables x_j and x_k are normally distributed and independent, then their correlation is 0. In this sense, correlation measures the linear dependence of two variables.

Consider the collection of measurements made at the wavelengths $\omega_3 = 406.45$ and $\omega_6 = 416.13$. The sample correlation of the data collected at these wavelengths is $\rho_{3,6} = 0.916$. The "best fit" linear model for the measurements made at ω_3 and ω_6 is $X(\omega_6) = 4.8 \cdot X(\omega_3) - 0.37$. The method for obtaining a "best linear fit" to a collection of data pairs (x_j, y_j) is detailed in Section 1 of the Appendix. *Deming regression* is used to determine the linear model and associated parameters (slope m, intercept b, and correlation). A discussion of regression is given in Chapter 5 on biostatistics and hypothesis testing. This correlation value of 0.916 lends credence to the notion of linear dependence in-between the measurement values at ω_3 and ω_6. Figure 4.5 illustrates the relationship of these measurements.

Continuing in this vein, all wavelength pairs (ω_j, ω_k) are compared via correlation. Setting a threshold of 75%, this degree of "linear dependence" will be considered sufficient for wavelength consolidation. That is, $|\rho_{j,k}| = |cor(\omega_j, \omega_k)| \geq \tau = 0.75$ implies ω_j and ω_k are linearly dependent. As computations show, the wavelengths enumerated in Table 4.2 are considered mutually dependent and therefore can be replaced by a single member of the set. For example, the measurements obtained at the nine wavelengths $\omega_1, \omega_2, \omega_3, \ldots, \omega_9$ can be replaced by the measurements obtained solely at ω_9. Notice that *any one* of the nine wavelengths can be selected as the correlation group's representative. The last wavelength is selected as a convenience. By eliminating 15 wavelengths, the data matrices are now of size 237×17 (normal), 45×17 ($Level_1$), 14×17 ($Level_2$), and 10×17 ($Level_3$). Furthermore, notice that the wavelength sets indexed by [13, 14, 15, 16, 21] and [14, 15, 16, 23] have a non-empty intersection. They are represented by the wavelengths ω_{21} and ω_{23}, respectively. The corresponding measurements have correlation well below the 0.75 threshold and are thereby *not* considered linearly dependent: $\rho_{21,23} = 0.4663$. Table 4.2 summarizes the sets of correlated wavelengths, their associated indices, and waveset representative.

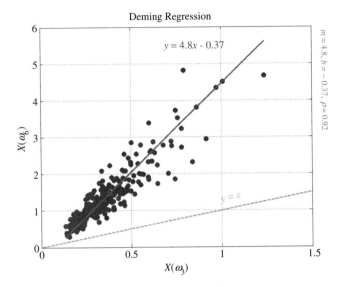

FIGURE 4.5   Correlated variables ω_3 and ω_6.

The normalization of the (wavelength reduced) data can now be actualized via the application of formula (4.5). Figure 4.6 shows the resulting dimensionless (i.e., unitless) measurements. By removing correlated variables (wavelengths), the condition number of the total data matrix X_T is reduced by 39% (from 2759.1 to 1684.9). The *condition number* of a matrix is the ratio of the largest to smallest nonzero singular values: $\kappa = \sigma_{max}/\sigma_{min}$. The nonzero singular values of a matrix A are the square roots of the nonzero eigenvalues of the matrix $A^T A$. A brief discussion of *matrices* and linear algebra is provided in Section 2 of the Appendix. The closer the condition number of a matrix is to 1, the more stable (numerically) is the matrix. A stable matrix is one which can be readily inverted without concern for round-off error or numerical overflow. Thus, anything which can be done to reduce the condition matrix of the total data matrix will enhance the numerical stability and hence reliability of the ensuing analysis.

TABLE 4.2   Correlated variables

	Correlated variables	
Indices	Values	Representative
[1, 2, 3, 4, 5, 6, 7, 8, 9]	[400, 403.2, 406.5, 409.7, 412.9, 416.1, 419.4, 422.6, 425.8]	$\omega_9 = 425.8$
[13, 14, 15, 16, 21]	[438.7, 441.9, 445.2, 448.4, 464.5]	$\omega_{21} = 464.5$
[14, 15, 16, 23]	[441.9, 445.2, 448.4, 471]	$\omega_{23} = 471$
[26, 27, 28, 29]	[480.6, 483.9, 487.1, 490.3]	$\omega_{29} = 490.3$

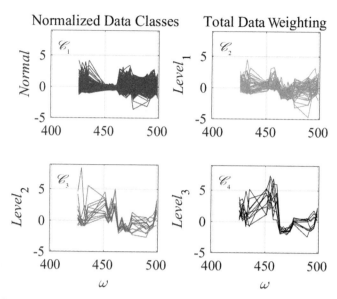

FIGURE 4.6   Normalized data using total data matrix (following correlated variable reduction).

In the subsequent section, other methods for variable reduction will be detailed. In summary, methods for the elimination of redundant information (correlated variables) and data normalization have been developed. The weighting matrix used in normalization can be one of two choices. The first is the diagonal matrix of the

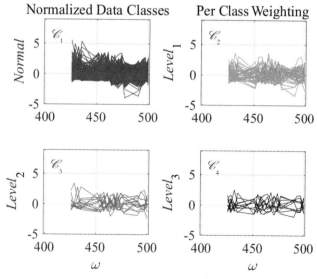

FIGURE 4.7   Normalized data using each data matrix (following correlated variable reduction).

variable standard deviations across the entire data set as defined in equation (4.5). Alternately, each data class can be normalized with respect to standard deviations of its corresponding data matrix as per equation (4.7). Such a normalization is exhibited in Figure 4.7. Notice that this technique tends to "smear out" the data and makes the classes less distinctive. This is counter to the notion of class separation or *classification* which will be explored later in this chapter. The following are the MATLAB commands used to generate the figures.

MATLAB Commands
(*Correlated Variables*; Figure 4.5)

```
% Load in the multivariate, multi-category data
Ddir = 'C:\PJC\Math_Biology\Chapter_4\Data';
load(fullfile(Ddir,'Multivariate_MultiClass_Data.mat'));
% Set of wavelengths, number of wavelengths, significance (α), correlation threshold (ρ)
w = D.w; p = numel(w); alpha = 0.05; rho = 0.75;
% Total data matrix
X = vertcat(D.X{:});
% Artificial set of variable names, significance level α, and correlation level ρ
for j = 1:p; Feature{j} = ['omega_' num2str(j)]; end
alpha = 0.05; rho = 0.75;
% Values of the wavelengths at ω₃ and ω₆.
w([3,6]) =
  406.4516   416.1290
% Compute the correlation in-between the data measured at ω₃ and ω₆
x = X(:,3); y = X(:,6);
rho36 = corr(x,y)
rho36 =
  0.9161
% Plot the Deming regression and scatter of the data measured at ω₃ and ω₆
S = deming(x,y,alpha, 'plot');
text(1,1.35,'\ity\rm = \itx','Rotation',15,'FontSize',15, 'FontName','Times
New Roman','Color','g');
text(0.75,5.5,'\ity\rm = 4.8\itx\rm - 0.37','FontSize',15,'FontName','Times
New Roman','Color','r');
S =
        m: 4.8349
        b: -0.3719
      rho: 0.9161
      CIm: [4.7907 4.8792]
      CIb: [-0.3872 -0.3567]
    CIrho: [0.8960 0.9324]

% Correlated variables
[R,F] = corrvars(D.X,Feature,rho,alpha);
R =
     1     2     3     4     5     6     7     8     9
     2     3     4     5     6     7     0     0     0
    13    14    15    16    21     0     0     0     0
    14    15    16    23     0     0     0     0     0
    26    27    28    29     0     0     0     0     0
    28    29     0     0     0     0     0     0     0
```

```
% Compute the correlation in-between the data measured at ω21 and ω23
x = X(:,21); y = X(:,23);
rho2123 = corr(x,y)
rho2123 =
    0.4663
```

```
% Groups of correlated variables
```

F{1} =	F{3} =	F{5} =
1 omega_1	13 omega_13	26 omega_26
2 omega_2	14 omega_14	27 omega_27
3 omega_3	15 omega_15	28 omega_28
4 omega_4	16 omega_16	29 omega_29
5 omega_5	21 omega_21	
6 omega_6		
7 omega_7		
8 omega_8		
9 omega_9		

4.3   PRINCIPAL COMPONENTS

As mentioned in the previous section, multivariate data can contain redundant and/or unnecessary information. Section 4.2 discussed the notion of reducing the dimension of the data by removing correlated variables. Another approach is to transform the data into a lower dimensional space by capturing *most* of the essential information. Reviewing Figure 4.2 indicates that there are regions along the ω-axis in which the data have almost no variation. One approach then is to transform the data into coordinates which point in the direction of maximal data variance. This is accomplished via *principal components*.

Fundamentally, the *principal components* of an $n \times p$ matrix are the projections of its columns onto its weighted singular vectors. More specifically, let $X \in \mathcal{M\!at}_{n \times p}(\mathbb{R})$ with *singular value decomposition* $X = UDV^T$, $U \in \mathcal{M\!at}_{n \times n}(\mathbb{R})$, $V \in \mathcal{M\!at}_{p \times p}(\mathbb{R})$ *orthogonal matrices*, and $D = \begin{bmatrix} D_{q \times q} \\ 0_{m-q \times q} \end{bmatrix}$, $q = \min[n, p]$, $m = \max[n, p]$. A square (i.e., $n \times n$) matrix M is *orthogonal* provided $MM^T = I_{n \times n} = M^T M$. Hence, for the SVD of X, $UU^T = I_{n \times n} = U^T U$ and $VV^T = I_{p \times p} = V^T V$. The notation $D_{q \times q}$ signifies the upper $q \times q$ sub-matrix of the larger $m \times q$ matrix D. This is also referred to as the *primary cofactor* of D. Finally, the diagonal elements of $D_{q \times q}$ are the nonzero singular values of X in *decreasing* order. A more complete discussion of the *singular value decomposition* is contained in Section 2 of the Appendix.

Since it is usually the case that there are more variables than measurements with respect to a data class, assume that $n > p$. Recall the definitions of the *total data matrix* $X_T \in \mathcal{M\!at}_{n \times p}(\mathbb{R})$ and the *mean total data vector* $\bar{x}_T \in \mathbb{R}^p$ via equations (4.3a) and (4.3b), respectively. For the vector of n ones $\mathbf{1}_n = [1, 1, ..., 1] \in \mathbb{R}^n$, the matrix whose every row is a copy of the mean total data vector is defined as $\overline{X}_T = \mathbf{1}_n^T \cdot \bar{x}_T \in$

$\mathcal{Mat}_{n\times p}(\mathbb{R})$. Set $Y = X - \overline{X}_T \in \mathcal{Mat}_{n\times p}(\mathbb{R})$. The *sample covariance matrix* for Y is $S = \frac{1}{n-1} Y Y^T \in \mathcal{Mat}_{p\times p}(\mathbb{R})$. Suppose $U\Sigma V^T = Y$ is the singular value decomposition of Y with orthogonal matrices U and V as noted earlier. Since $n > p$, then Σ has the form

$$\Sigma = \begin{bmatrix} \Sigma_{p\times p} \\ 0_{n-p\times p} \end{bmatrix} \text{ and } \Sigma_{p\times p} = \begin{bmatrix} \sigma_{1,1} & 0 & \cdots & 0 \\ 0 & \sigma_{2,2} & \cdots & 0 \\ \vdots & \vdots & \ddots & \vdots \\ 0 & \cdots & 0 & \sigma_{p,p} \end{bmatrix}.$$

Let $\Gamma_{p\times p} = \frac{1}{n-1}\Sigma^2_{p\times p}$ and $\Lambda_{p\times p} = \frac{1}{tr(\Gamma_{p\times p})}\Gamma_{p\times p}$. The *trace* of a matrix A, $tr(A)$, is the sum of the diagonal elements. Then $\Lambda_{p\times p}$ is a diagonal matrix whose (nonzero) elements $\lambda_{j,j}$ are the decreasing fraction of the total variance. That is, $\lambda_{1,1} \geq \lambda_{2,2} \geq \lambda_{3,3} \geq \cdots \geq \lambda_{p,p}$ and $\sum_{j=1}^{p} \lambda_{j,j} = 1$.

Therefore, if [say] 95% of the total variance of the centered data matrix Y is desired, find the smallest r so that $\sum_{j=1}^{r} \lambda_{j,j} \geq 0.95$. If $\Lambda_{r\times r}$ is the diagonal matrix whose nonzero entries are $\lambda_{1,1}, \lambda_{2,2},..., \lambda_{r,r}$, and $V_r \in \mathcal{Mat}_{n\times r}(\mathbb{R})$ is the matrix comprised of the first r columns of the *SVD* factorization of Y, then $\Pi_r = \Lambda_{r\times r} \cdot V_r^T \in \mathcal{Mat}_{r\times n}(\mathbb{R})$ is the *projection matrix* onto the first r *principal components*. The matrix Π_r is also referred to as the *Karhunen–Loéve transform* [4]. The matrix $V_r \equiv L_d(r)$ is referred to as the *loading* of the first r *principal components*. Finally, $S_c(r) = \Lambda_{r\times r} \cdot L_d(r) \bullet Y^T$ are the *scores* of the first r *principal components*. Principal component analysis will now be applied to the multivariate data set of Section 4.2.

To garner 99% of the total variance, $r = 6$ principal components (*PC*s) are required. As Figure 4.8 illustrates, the $g = 4$ data classes (normal, *level*$_1$, *level*$_2$, *level*$_3$) appear to be readily separated with respect to the first three *PC*s. The scores of the $r = 6$

Normalized Principal Components

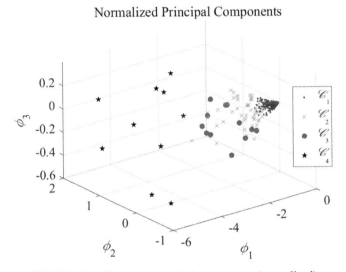

FIGURE 4.8 First three principal components (normalized).

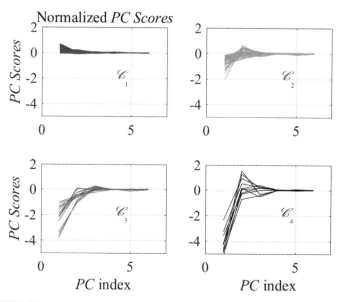

FIGURE 4.9   Normalized principal component scores with respect to $r = 6$ PCs.

principal components are presented in Figure 4.9. The *scores* can be interpreted as the projections of the data classes into the r-dimensional *principal component axes*. By using principal components, a more direct manner for dimension reduction *along the axes of greatest data variation* can be calculated. Indeed the scores in Figure 4.9 give a lower dimensional representation of the multivariable data originally presented in Figure 4.2. Moreover, the *PC* scores appear to preserve the *qualitative* content of the original data. That is, the *normal* disease state data are tightly grouped with little variation while *levels* 1, 2, and 3 data have increasing variation and data range. It is also apparent from Figure 4.9 that *PC*s 4–6 have scores which do not appear to contribute significantly to class separation.

The *Principal Component Axes, Scores, Loading,* and *Projection* are summarized in equations (4.9a)–(4.10d). For details of this development see Johnson and Wichern [8] or Jolliffe [9].

$$\text{Data matrix } X \in \mathcal{M}at_{n \times p}(\mathbb{R}) \tag{4.9a}$$

$$\text{Mean matrix } \bar{X} \in \mathcal{M}at_{n \times p}(\mathbb{R}) \tag{4.9b}$$

$$\text{Centered matrix } Y = X - \bar{X} \in \mathcal{M}at_{n \times p}(\mathbb{R}) \tag{4.9c}$$

$$\text{Singular value decomposition } Y = U\Sigma V^T \tag{4.9d}$$

$$U \in \mathcal{M}at_{n \times n}(\mathbb{R}), V \in \mathcal{M}at_{p \times p}(\mathbb{R}) \text{ orthogonal}, \Sigma = \begin{bmatrix} \Sigma_{p \times p} \\ 0_{n-p \times p} \end{bmatrix}$$

$$\left. \begin{array}{l} \Lambda_{p \times p} = \frac{1}{tr(\Gamma_{p \times p})} \Gamma_{p \times p} \\ \\ \Gamma_{p \times p} = \frac{1}{n-1} \Sigma^2_{p \times p} \end{array} \right\}$$

Normalized singular values $\qquad\qquad$ (4.9e)

Number of required/desired principal components $r \leq p \qquad$ (4.10a)

The value r is selected so that $\sum_{j=1}^{r} \lambda_{jj} \geq t_{frac}$ (the desired fraction of the total variation).

$$r \text{ Principal component loading } L_d(r) = V_{p \times r} \in \mathcal{M\!at}_{p \times r}(\mathbb{R}) \qquad (4.10b)$$

$$r \text{ Principal component scores } S_c(r) = \Sigma_{r \times r} \cdot L_d^T(r) \cdot Y^T \in \mathcal{M\!at}_{r \times n}(\mathbb{R}) \qquad (4.10c)$$

$$\text{Projection onto } r \text{ principal components } \Pi_r = \Sigma_{r \times r} \cdot L_d^T(r) \qquad (4.10d)$$

Principal component projection provides a systematic way to reduce data dimensionality. As Figure 4.8 suggests, it also provides a method to identify or *classify* data. This will be the focus of the next section.

Again, it is noted that from upper left to lower right the four classes of Figure 4.9 represent $\mathcal{C}_1$ = normal (i.e., no disease), $\mathcal{C}_2$ = *disease level$_1$*, $\mathcal{C}_3$ = *disease level$_2$*, and $\mathcal{C}_4$ = *disease level$_3$* diagnoses.

EXERCISE

4.2 Using the notation developed in the beginning of the section $\Lambda_{p \times p} = \frac{1}{tr(\Gamma_{p \times p})} \Gamma_{p \times p}$, show (i) $\Lambda_{p \times p} = \frac{1}{\sum_{j=1}^{p} \sigma_{jj}^2} \cdot \begin{bmatrix} \sigma_{11}^2 & 0 & \cdots & 0 \\ 0 & \sigma_{22}^2 & \cdots & 0 \\ \vdots & \vdots & \ddots & \vdots \\ 0 & 0 & \cdots & \sigma_{pp}^2 \end{bmatrix}$, (ii) $\lambda_{1,1} \geq \lambda_{2,2} \geq \lambda_{3,3} \geq \cdots \geq \lambda_{p,p}$, and (iii) $\sum_{j=1}^{p} \lambda_{jj} = 1$.

4.4 DISCRIMINANT ANALYSIS

This chapter, thus far, has been dedicated to the discussion of data matrices, correlation, data normalization, principal components, and reduction of dimension. What has only been alluded to are (i) the coordinates which yield *optimal* data information and (ii) a method of *classifying* a measurement as a member of one of the prescribed data classes $\mathcal{C}_1, \mathcal{C}_2, \ldots, \mathcal{C}_g$. In this section, attention is paid to the first of these issues.

Principal components project the data into a lower dimensional *orthogonal* coordinate space in the *direction of greatest variation*. Indeed, Duda et al. [2] note that while "*PCA seeks directions that are efficient for representation, discriminant analysis seeks directions that are efficient for discrimination.*" Plainly, if one can readily discriminate in-between classes, then the *classification* of a measurement follows directly. It is now sensible to ask "Are there coordinates which map the data into the direction of *greatest information?*" The answer is a qualified "yes" and the coordinates are called *discriminant analysis feature extraction* (*DAFE*). Before *DAFE* coordinates are established, two important matrices are described.

The *pooled within-groups covariance matrix* C_{pool} is the weighted average of the class covariance matrices $C_1, C_2,...,C_g$ defined in equation (4.2b). This matrix is the weighted average of the covariance matrices across all data classes.

$$C_{pool} = \frac{1}{n - g} \sum_{k=1}^{g} (n_k - 1) \cdot C_k \in \mathcal{Mat}_{p \times p}(\mathbb{R}), n = \sum_{k=1}^{g} n_k \qquad (4.11a)$$

The *between-class covariance matrix* C_{btwn} is the weighted sum of the mean outer products. It measures the total variance across each data class.

$$C_{btwn} = \frac{1}{g} \sum_{k=1}^{g} n_k (\bar{x}_k - \bar{x}_T)^T (\bar{x}_k - \bar{x}_T) \in \mathcal{Mat}_{p \times p}(\mathbb{R}) \qquad (4.11b)$$

Note: The vectors $\bar{x}_k$ and $\bar{x}_T$ are $1 \times p$ so that their outer product, as expressed in (4.11b), produces a $p \times p$ matrix. Moreover, the matrices C_{pool} and C_{btwn} are positive and symmetric as they are the sums of other positive symmetric matrices. Consequently, the eigenvalues of these matrices are all positive. This, in turn, implies that $F = C_{pool}^{-1} \cdot C_{btwn}$ is also a positive matrix. The *Exercises* continue these ideas. See, for example, Gnanadesikan [5] or Fukunaga [4].

How do the aforementioned *within-* and *between-class covariance matrices* aid in discrimination? Consider the problem of a set of p-dimensional measurements taken from just two classes $\mathscr{C}_1$ and $\mathscr{C}_2$. If $\bar{x}_k \in \mathbb{R}^p$ are the data class means for $k = 1, 2$, and $w \in \mathbb{R}^p$ is a weight which projects any measurement x onto the real line, then $m_k = w^T \cdot \bar{x}_k$ are the projected class means. The projected sample scatter is $s_k^2 = \sum_{y \in \mathscr{Y}_k} (y - m_k)^2$, where $\mathscr{Y}_k$ is the projection of the class $\mathscr{C}_k$ onto $\mathbb{R}$. Maximizing the *Fisher linear discriminant* $J(w) = \frac{(m_1 - m_2)^2}{s_1^2 + s_2^2}$ with respect to all such weight vectors w will produce the best separation (in $\mathbb{R}$) for the two sets $\mathscr{Y}_1$ and $\mathscr{Y}_2$. For the general g-class problem, the Fisher discriminant becomes $J(W) = \frac{|W^T \cdot C_{btwn} \cdot W|}{|W^T \cdot C_{pool} \cdot W|}$, where $W \in \mathcal{Mat}_{m \times p}(\mathbb{R})$ is a rectangular matrix. Finding such a matrix W that provides a maximum value for $J(W)$ is equivalent to finding the generalized eigenvectors that correspond to the descending (from largest to smallest) eigenvalues of $F = C_{pool}^{-1} \cdot C_{btwn}$. More precisely, find v so that

$$Fv = \lambda v, \qquad (4.12)$$

where $\lambda \in \{\lambda_{(1)} \geq \lambda_{(2)} \geq \cdots \geq \lambda_{(p)}\}$.

Observe that (4.12) is equivalent to $C_{pool}^{-1} \cdot C_{btwn} v = \lambda v \Leftrightarrow C_{btwn} v = \lambda C_{pool} v$. As with principal components, these generalized eigenvectors can be recovered from the singular value decomposition of $F = C_{pool}^{-1} \cdot C_{btwn}$. The eigenvector v_{max} corresponding to the largest eigenvalue $\lambda_{max} = \lambda_{(1)}$ comprises the first column of W; the eigenvector v_2 corresponding to $\lambda_{(2)} = $ the next largest eigenvalue comprises column 2 of W; etc. In parallel to Section 4.3, the *DAFE* method is listed in sequence. For the complete development of *DAFE* coordinates, see Duda et al. [2] or Gnanadesikan [5]. The normalized singular values $\lambda_{jj} = \gamma_j / \sum_{k=1}^{p} \gamma_k$ defined via (4.13d) represent the fraction of the *discriminant informa-tion* contained in the *j*th *DAFE* coordinate. The sum $\sum_{j=1}^{r} \lambda_{jj}$ is the *total discrimi-nant information* contained in the first *r DAFE* coordinates. Again, as with principal components, if $\iota_{frac} \bullet 100\%$ is the percent of the total discriminant information desired and $\sum_{j=1}^{r} \lambda_{jj} \geq \iota_{frac}$, then *r-DAFE* coordinates yield the desired information content.

$$\text{Data matrix } X_k \in \mathcal{M}at_{n_k \times p}(\mathbb{R}) \tag{4.13a}$$

$$\text{Fisher linear discriminant matrix } F = C_{pool}^{-1} \cdot C_{btwn} \in \mathcal{M}at_{p \times p}(\mathbb{R}) \tag{4.13b}$$

$$\text{Singular value decomposition } F = U\Gamma V^T \tag{4.13c}$$

$$U \in \mathcal{M}at_{p \times p}(\mathbb{R}), \ V \in \mathcal{M}at_{p \times p}(\mathbb{R}) \text{ orthogonal}, \quad \Gamma = \text{diag}(\gamma_1, \gamma_2, \ldots, \gamma_p)$$

$$\text{Normalized singular values } \Lambda_{p \times p} = \frac{1}{tr(\Gamma)} \text{diag}(\Gamma) = \frac{1}{\sum\limits_{j=1}^{p} \gamma_j} \begin{bmatrix} \gamma_1 & 0 & \cdots & 0 \\ 0 & \gamma_2 & \cdots & 0 \\ \vdots & \vdots & \ddots & \vdots \\ 0 & 0 & \vdots & \gamma_p \end{bmatrix} \tag{4.13d}$$

$$\text{Number of required/desired } DAFE \text{ coordinates } r \tag{4.14a}$$

The value r is selected so that $\sum_{j=1}^{r} \lambda_{jj} \geq \iota_{frac}$.

$$\text{r-Dimensional DAFE projections} = X_{DAFE}^r(k) = X_k \bullet U_{p \times r} \in \mathcal{M}at_{n_k \times r}(\mathbb{R}) \tag{4.14b}$$

$$\text{r-Dimensional (right) DAFE projection matrix } \Pi_r = U_{p \times r} \in \mathcal{M}at_{p \times r}(\mathbb{R}) \tag{4.14c}$$

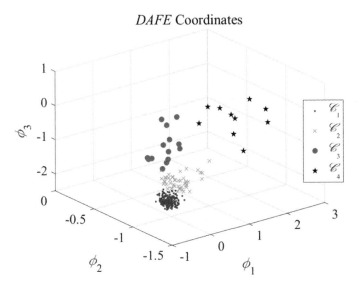

FIGURE 4.10   *DAFE* coordinates for $\iota_{frac} = 0.99$.

Observe via (4.14b) that the projection $\Pi_r = U_{p \times r}$ must be applied on the right-hand side of the data matrix X_k.

This procedure was applied to the data first exhibited in Figure 4.2. It turns out that the projections of the data classes onto the three *DAFE* coordinates account for 99% of the discriminant information as presented in Figure 4.10. This means that $r = 3$ and

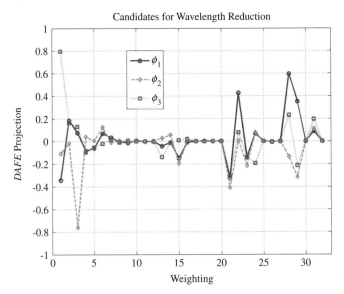

FIGURE 4.11   *DAFE* coordinates and wavelength weightings.

Wavelength Reduction   Via *DAFE* Projection

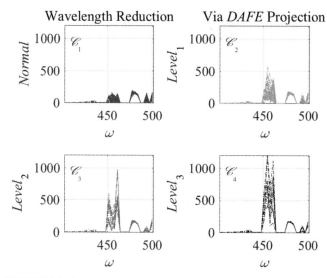

FIGURE 4.12   Variable reduction via *DAFE* projection weightings.

$\Pi_r = \Pi_3$. Using the scheme of dots ($\cdot$) for *Class 1* ($\mathscr{C}_1$), crosses ($\times$) for *Class 2* ($\mathscr{C}_2$), circles ($\bullet$) for *Class 3* ($\mathscr{C}_3$), and stars ($*$) for *Class 4* ($\mathscr{C}_4$), it appears from Figure 4.10 that three *DAFE* coordinates readily resolve these four classes.

The three rows of the projection matrix Π_3 computed for the aforementioned data determine which of the wavelengths are required for each *DAFE* coordinate. If a threshold (in this case, $\tau = 0.005$) is set so that when *all* rows of Π_3 are less than τ in absolute value, then the corresponding wavelengths are eliminated. Indeed the columns of Π_3 are the *DAFE* projection vectors. Therefore, if each *row* of Π_3 is less than the threshold, the weight at this coordinate can be set to zero and the corresponding wavelength excluded. Using this criterion, the wavelengths at indices 10, 11, 12, 17, 18, 19, 20, 25, 27, 30, and 32 can be eliminated. Thus, the rows of Π_r can be used as another systematic method to reduce unnecessary variables. As can be seen in Figure 4.11, the aforementioned wavelengths are less than the threshold τ *for all three DAFE projections*. The reduction of the data obtained by eliminating the wavelengths via *DAFE* weightings is illustrated in Figure 4.12. The following are MATLAB commands to determine these unnecessary wavelengths.

MATLAB Commands
(Determine Which Wavelengths Are Less Than the Threshold τ)

```
% Load in the data
Ddir = 'C:\PJC\Math_Biology\Chapter_4\Data';
load(fullfile(Ddir,'Multivariate_MultiClass_Data.mat'));
% Project the data classes into the DAFE coordinates comprising 99% of the information
[Xdafe,PIdafe] = dafe(D.X,0.99);
% Set the threshold
thres = 0.005;
```

```
% Determine which rows of the DAFE projection matrix are smaller than the threshold
find(all(abs(PIdafe) < thres,2)).' =

   10    11   12    17    18    19    20    25    27    30    32
```

The point of these projections and variable reduction methods is to establish a mathematical process for data classification. This will be the dominion of the remainder of the chapter.

EXERCISES

4.3 If the square matrix A is *positive*, then all eigenvalues of A are positive. Show that the eigenvalues of the matrix $B = \frac{1}{2}(A + A^T)$ are also *positive*.

4.4 If the square matrices A and B are *positive*, commutable, and symmetric, show that the matrix $C = A \cdot B$ is also positive. *Hint*: Use the Spectral Theorem to factor A as $V_A \Lambda_A V_A^T$ and write $A = A^{\frac{1}{2}} A^{\frac{1}{2}}$ where $A^{\frac{1}{2}} = V_A \sqrt{\Lambda_A} V_A^T$.

4.5 REGULARIZED DISCRIMINANT ANALYSIS AND CLASSIFICATION

To summarize the first four sections of this chapter, the following information is available.

(i) The data matrices $X_1, X_2,...,X_g$, $X_k \in \mathcal{Mat}_{n_k \times p}(\mathbb{R})$, representing the g data classes $\mathcal{C}_1, \mathcal{C}_2,..., \mathcal{C}_g$.

(ii) Methods to reduce measurement dimensionality via correlation, *principal component projection*, or *discriminant analysis feature extraction*.

(iii) The reduced data matrices $Y_1, Y_2,..., Y_g$ which have been projected into a lower $r < p$ dimensional space $Y_k \in \mathcal{Mat}_{n_k \times r}(\mathbb{R})$.

Suppose, in addition to (i)–(iii), a measurement $y \in \mathbb{R}^r$ is obtained. To which data class $\mathcal{C}_1, \mathcal{C}_2,..., \mathcal{C}_g$ does y belong? The answer is the class from which the Mahalanobis distance (4.15), taken with respect to the weighting matrix M, is smallest.

$$d_M^2(y, \bar{y}_k) = (y - \bar{y}_k) \cdot M^{-1} \cdot (y - \bar{y}_k)^T \qquad (4.15)$$

Here, as in Section 4.1, $\bar{y}_k$ is the (column-wise) mean vector of the (reduced dimension) data matrix Y_k and is constructed analogously to (4.1) and (4.2a). If $d_M^2(y, \bar{y}_\ell) = \min_{1 \le k \le g} d_M^2(y, \bar{y}_k)$, then $y \in \mathcal{C}_\ell$. This is the formal classification criterion. The class to which the measurement has the smallest Mahalanobis distance is the one to which the measurement is assigned. Therefore, the next matter to investigate is the weighting matrix M. What should it be?

One choice is the pooled within-class covariance C_{pool} of (4.11a). This selection $M = C_{pool}$ creates a hyperplane as a boundary in-between classes. Indeed, consider the simple case of two classes $\mathscr{C}_1$ and $\mathscr{C}_2$. For the (as yet) unidentified measurement y, compute the two Mahalanobis distances $d^2_{C_{pool}}(y, \bar{y}_1) = (y - \bar{y}_1) \cdot C^{-1}_{pool} \cdot (y - \bar{y}_1)^T$ and $d^2_{C_{pool}}(y, \bar{y}_2) = (y - \bar{y}_2) \cdot C^{-1}_{pool} \cdot (y - \bar{y}_2)^T$. Without loss of generality, suppose that $d^2_{C_{pool}}(y, \bar{y}_1) < d^2_{C_{pool}}(y, \bar{y}_2)$. Then

$$(y - \bar{y}_1) \cdot C^{-1}_{pool} \cdot (y - \bar{y}_1)^T < (y - \bar{y}_2) \cdot C^{-1}_{pool} \cdot (y - \bar{y}_2)^T. \tag{4.16a}$$

Recall that C_{pool} is a positive symmetric matrix. Hence, C^{-1}_{pool} is positive symmetric and $(\bar{y}_1 \cdot C^{-1}_{pool} \cdot y^T) = (y \cdot C^{-1}_{pool} \cdot \bar{y}_1^T)$. Expanding (4.16a) yields

$$y \cdot C^{-1}_{pool} \cdot y^T - 2\left(\bar{y}_1 \cdot C^{-1}_{pool} \cdot y^T\right) + \bar{y}_1 \cdot C^{-1}_{pool} \cdot \bar{y}_1^T$$
$$< y \cdot C^{-1}_{pool} \cdot y^T - 2\left(\bar{y}_2 \cdot C^{-1}_{pool} \cdot y^T\right) + \bar{y}_2 \cdot C^{-1}_{pool} \cdot \bar{y}_2^T.$$

Cancelling the quadratic terms (in y) and condensing the remaining terms produces the inequality

$$0 < 2\left((\bar{y}_1 - \bar{y}_2) \cdot C^{-1}_{pool} \cdot y^T\right) + \bar{y}_2 \cdot C^{-1}_{pool} \cdot \bar{y}_2^T - \bar{y}_1 \cdot C^{-1}_{pool} \cdot \bar{y}_1^T. \tag{4.16b}$$

By defining the vector $\alpha = 2(\bar{y}_1 - \bar{y}_2) \cdot C^{-1}_{pool}$ and the scalar $\beta = \bar{y}_2 \cdot C^{-1}_{pool} \cdot \bar{y}_2^T - \bar{y}_1 \cdot C^{-1}_{pool} \cdot \bar{y}_1^T$, (4.16b) becomes the *linear* inequality

$$0 < \alpha \bullet y^T + \beta. \tag{4.17}$$

The right-hand side of (4.17) defines a hyperplane which is the boundary in-between the classes $\mathscr{C}_1$ and $\mathscr{C}_2$. The choice of $M = C_{pool}$ and classification criterion $d^2_{C_{pool}}(y, \bar{y}_\ell) = \min_{1 \le k \le g} d^2_{C_{pool}}(y, \bar{y}_k) \Rightarrow y \in \mathscr{C}_\ell$ is called *linear discriminant analysis* (LDA).

Using the data generated in Section 4.1 and computing the corresponding pooled covariance matrix C_{pool}, it is seen that the minimum distance of the measurement $m = [0.8, 3.53]$ to the data classes $\mathscr{C}_1$ and $\mathscr{C}_2$ is to $\mathscr{C}_2$. Therefore, $m \in \mathscr{C}_2$.

Rather than selecting $M = C_{pool}$, choose $M = C_k$ for each $k = 1, 2, \ldots, g$. This results in a *quadratic* boundary separating the data classes. Therefore, $M = C_k$ and classification criterion $d^2_{C_k}(y, \bar{y}_\ell) = \min_{1 \le k \le g} d^2_{C_k}(y, \mathscr{C}_k) \Rightarrow y \in \mathscr{C}_\ell$ is referred to as *quadratic discriminant analysis* (QDA). It is left as an exercise to show that the two-class problem defines a quadratic boundary. As was seen in Section 4.1, the measurement $m = [0.8, 3.53]$ has a Mahalanobis distance of 5.3 to $Class_1$ when $M = C_1$ and $d_M = 5.5$ when $M = C_2$. Therefore, when using a quadratic boundary, $m \in \mathscr{C}_1$.

MATLAB Commands
(***Linear Discriminant Analysis***)

```
% Recover the data from the data directory
Ddir = 'C:\PJC\Math_Biology\Chapter_4\Data';
load(fullfile(Ddir,'Y1.mat'));
load(fullfile(Ddir,'Y2.mat'));
% Calculate the pooled within-class covariance from the (rotated) data
Y{1} = Y1.'; Y{2} = Y2.';
Cpool = pooledcov(Y);
Cpool =

    0.4820    -0.0266
   -0.0266     0.5548

% Compute the Mahalanobis distances for W = Cpool and m = [0.8, 3.53]
p1 = mahalanobis(Wo,mu1,Cpool); p2 = mahalanobis(Wo,mu2,Cpool);
[m,ii] = min([p1,p2])
% Minimum Mahalanobis distance ⇒ m ∈ C2
m =                                 ii =
    4.9153                           2
```

An illustration of the linear (dashed line $- \cdot -$) and quadratic (dotted curve $\cdots$) boundaries for the data depicted in Section 4.1 is presented in Figure 4.13. Observe that the *linear boundary* favors $Class_1$ over $Class_2$. Note that these data classes *are not* the same as in Figure 4.12. For the data illustrated in Figure 4.13, the line completely separates $Class_1$ from $Class_2$ but not vice versa. Consequently, if $Class_1$ represents the *positive* (i.e., positive for disease) class, then the linear discriminant method would provide a 100% probability of detection (p_d). As is evident from Figure 4.13, however, many $Class_2$ (*negative* for disease) points are to the left of the linear boundary. These points would be incorrectly assigned to the positive class and would contribute to a high probability of false alarm (p_{f_a}). If the consequence of *missing* disease (i.e., Ebola, HIV) is serious, then such a classification partition is acceptable. Conversely, if the cost of a false positive on a slow-growing cancer (i.e., prostate, cervical) or generally nonlethal disease (influenza) is high, then a quadratic boundary may be preferred. Again, as Figure 4.13 indicates, such a classification partition would still yield a high p_d but would lower the p_{f_a}. Plainly, the cost of treatment (i.e., surgery, chemotherapy, radiation) must be weighed against risk to patient health.

It is evident then that the choice of weighting matrix M will determine the shape of the class separation boundary as well as the ultimate classification.

While the matrices C_{pool} and C_k are readily understood choices for the weighting matrix M, Friedman [3] noticed that a weighted average of the *pooled* and

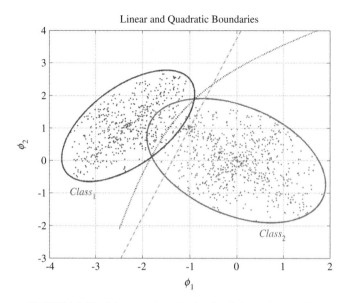

FIGURE 4.13   Linear and quadratic discriminant boundaries.

individual class covariance matrices produced a completely generalized discriminant technique. More specifically, the *Friedman matrix*

$$F_k(\gamma, \lambda) = (1 - \gamma)\left((1 - \lambda)C_k + \lambda C_{pool}\right) + \frac{\gamma}{r} \, trace\left((1 - \lambda)C_k + \lambda C_{pool}\right) \cdot I_{r \times r}$$

$$(4.18)$$

with associated *Friedman parameters* γ and λ, defines *regularized discriminant analysis* for r-dimensional measurements. *Regularized discriminant analysis (RDA)* has a decision boundary parameterized by a combination of quadratic ($\lambda = 0$, $\gamma = 0$) and linear ($\lambda = 1$, $\gamma = 0$) shapes. Thus, *RDA* is a smooth transition from linear to quadratic classification (e.g., see references [3, 7, 13]). Using the same example for *RDA* as for *LDA* and *QDA*, it is seen that the measurement $\boldsymbol{m} = [0.8, 3.53]$ and the choice of Friedman parameters $\gamma = \frac{1}{2} = \lambda$ results in Mahalanobis distances of (approximately) 5.5 and 5 to the $\mathscr{C}_1$ and $\mathscr{C}_2$ class centers, respectively. Thus, for these Friedman parameters, $\boldsymbol{m} \in \mathscr{C}_2$.

This approach to classification permits not only the "bending" of the class separation boundary (anywhere from linear to quadratic) but also helps temper the weighting matrix $W = F_k(\gamma, \lambda)$. If the data class matrices X_k contain fewer measurements than is required for a robust estimate of the class covariance matrices C_k, then values of (γ, λ) away from $\{0, 1\}$ can better "condition" the weighting matrix. For r-dimensional measurements, the data matrix $X_k \in \mathscr{Mat}_{n_k \times r}(\mathbb{R})$ and the covariance C_k as defined via (4.2b) requires $n_k \geq \frac{1}{2} r (r + 1)$ data points. If $n_k < \frac{1}{2} r (r + 1)$, then C_k will be ill-conditioned and possibly singular. This would make the Mahalanobis distance numerically unstable. Hence, the classification would be unreliable. As long as the Friedman parameter γ, however, is *not* zero, then the Friedman matrix $F_k(\gamma, \lambda)$ is a sum containing a multiple of the identity matrix. This will ensure the non-singular nature of $F_k(\gamma, \lambda)$ and the stability of the Mahalanobis distance.

Regardless of the choice of weighting matrix M, the classification criterion remains

$$d_M^2(y, \bar{y}_\ell) = \min_{1 \le k \le g} d_M^2(y, \bar{y}_k) \Rightarrow y \in \mathscr{C}_\ell \tag{4.19}$$

as first described in the beginning of this section. If the Friedman matrix $F_k(\gamma, \lambda)$ is chosen as the weighting matrix, then the Friedman parameters can be adjusted to shape the decision boundary. Are there additional techniques that will translate (i.e., move) the position of the boundary to balance probability of detection (p_d) and probability of false alarm (p_{f_a}) requirements? The answer, of course, is yes. The next section will offer a more general classification criterion along with other parameters whose adjustment can optimize classifier performance.

MATLAB Commands
(*Regularized Discriminant Analysis*)

```
% Calculate the Friedman matrices for γ = ¹/₂ = λ
F1 = fried(S1,Cpool,0.5,0.5); F2 = fried(S2,Cpool,0.5,0.5);
% Compute the Mahalanobis distances for W = Fₖ and m = [0.8, 3.53]
f1 = mahalanobis(Wo,mu1,F1); f2 = mahalanobis(Wo,mu2,F2);
[m,ii] = min([f1,f2]);
% Minimum Mahalanobis distance ⇒ m ∈ 𝒞₂
m =                              ii =
   4.9666                          2
```

The data for this computation are extracted via the previous MATLAB commands.

EXERCISE

4.5 Show, for the two-class problem on $\mathbb{R}^r$, that the choice of $W = C_k$ yields the quadratic inequality $0 < y \bullet C \bullet y^T + \alpha \bullet y^T + \beta$ with respect to the Mahalanobis comparison $d_{C_1}^2(y, \mathscr{C}_1) < d_{C_2}^2(y, \mathscr{C}_2)$. Hint: $C = C_2^{-1} - C_1^{-1}$, $\alpha = 2(\bar{y}_1 C_1^{-1} - \bar{y}_2 C_2^{-1})$, and $\beta = \bar{y}_2 C_2^{-1} \bar{y}_2^T - \bar{y}_1 C_1^{-1} \bar{y}_1^T$.

4.6   MINIMUM BAYES SCORE, MAXIMUM LIKELIHOOD, AND MINIMUM BAYES RISK

The previous section introduced the notion of classification by way of *discriminant analysis* and *minimum Mahalanobis distances* to data class centers. The methods discussed in Section 4.5 required only the class data matrices X_k and an unidentified measurement m to render a classification decision. Of the discriminant analysis (LDA, QDA, and RDA) methods described, only RDA has adjustable parameters that can determine the shape of the decision boundary and hence the outcome of the classification. These methods fail to use any additional *a priori* information available to aid in the class determination.

The probability that a measurement belongs to a specified data class is called the *prior probability* or *class prior*. Let r_k be the prior probability for class k: $r_k = \Pr(\mathscr{C}_k)$. How can an *a priori* probability influence classification? The answer lies in *Bayes Theorem*.

Bayes Theorem: The conditional probability of event A given event B is proportional to the prior probability of event B given event A. That is, $\Pr(A|B) = \Pr(B|A) \cdot \frac{\Pr(A)}{\Pr(B)}$.

Suppose, then, that $d_M^2(m, \bar{y}_k)$ is the Mahalanobis distance of a measurement $m \in \mathbb{R}^r$ to the center $\bar{y}_k$ of the data class $\mathscr{C}_k$ with respect to the weighting matrix M. The *Bayes score* $\beta_k(m; M, \lambda)$ is the natural logarithm of the posterior probability that m belongs to $\mathscr{C}_k$. This is defined with respect to the Friedman parameter λ.

$$\beta_k(m; M, \lambda) = d_M^2(m, \bar{y}_k) - 2\ln(r_k) + (1 - \lambda)\ln(|\det(M)|) \qquad (4.20)$$

The classification of a measurement m to *Class* ℓ occurs provided it is the index of the minimum Bayes score.

$$\min_{1 \le k \le g} \beta_k(m; M, \lambda) = \beta_\ell(m; M, \lambda) \Rightarrow m \in \mathscr{C}_\ell \qquad (4.21)$$

To see that β_k is derived from a probability, multiply by $-\frac{1}{2}$ and exponentiate both sides of (4.20) to obtain

$$\exp\left(-\frac{1}{2}\beta_k(m; M, \lambda)\right) = r_k \cdot \frac{1}{|\det(M)|^{\left(\frac{1-\lambda}{2}\right)}} \exp\left(-\frac{1}{2}d_M^2(m, \bar{y}_k)\right).$$

From the equation above, the *likelihood function* $\mathscr{L}_k(m; M, \lambda) \equiv \exp(-\frac{1}{2}\beta_k(m; M, \lambda))$ can be written as

$$\mathscr{L}_k(m; M, \lambda) = |\det(M)|^{\frac{\lambda}{2}} \cdot r_k \cdot \frac{1}{|\det(M)|^{\left(\frac{1}{2}\right)}} \exp\left(-\frac{1}{2}d_M^2(m, \bar{y}_k)\right). \qquad (4.22)$$

If A is the event of Class k ($\mathscr{C}_k$) and B is the event that the measurement is within Class k ($m \in \mathscr{C}_k$), then $\frac{1}{|\det(M)|^{\left(\frac{1}{2}\right)}} \exp(-\frac{1}{2}d_M^2(m, \bar{y}_k)) = \Pr(m \in \mathscr{C}_k \mid \mathscr{C}_k) = \Pr(B|A)$. Since $r_k = \Pr(\mathscr{C}_k) = \Pr(A)$, then the likelihood function (4.22) is an unnormalized version of Bayes Theorem: $\mathscr{L}_k(m; M, \lambda) = |\det(M)|^{\lambda/2} \bullet \Pr(A) \bullet \Pr(B|A)$.

Next, it will be seen that a minimum Bayes score is equivalent to a *maximum likelihood*. Again, consider a two-class case and assume that $\beta_1(m; M, \lambda) = \min_{1 \le k \le 2} \beta_k(m; M, \lambda)$. Then, by the classification criterion (4.21), $\beta_1(m; M, \lambda) < \beta_2(m; M, \lambda)$ or $d_M^2(m, \bar{y}_1) - 2\ln(r_1) + (1 - \lambda)\ln(|\det(M)|) < d_M^2(m, \bar{y}_2) - 2\ln(r_2) + (1 - \lambda)\ln(|\det(M)|)$. Multiplying through by $-\frac{1}{2}$ changes the direction of the inequality and produces

$$-\frac{1}{2}d_M^2(m, \bar{y}_1) + \ln(r_1) - \frac{(1-\lambda)}{2}|\det(M)| > $$
$$-\frac{1}{2}d_M^2(m, \bar{y}_2) + \ln(r_2) - \frac{(1-\lambda)}{2}|\det(M)| \Leftrightarrow$$

$$|\det(M)|^{\frac{\lambda}{2}} \cdot r_1 \cdot \frac{1}{|\det(M)|^{\left(\frac{1}{2}\right)}} \exp\left(-\frac{1}{2}d_M^2(m, \bar{y}_1)\right) >$$
$$|\det(M)|^{\frac{\lambda}{2}} \cdot r_2 \cdot \frac{1}{|\det(M)|^{\left(\frac{1}{2}\right)}} \exp\left(-\frac{1}{2}d_M^2(m, \bar{y}_2)\right)$$

$\Leftrightarrow \mathscr{L}_1(m; M, \lambda) > \mathscr{L}_2(m; M, \lambda) \Rightarrow \mathscr{L}_1(m; M, \lambda) = \max_{1 \le k \le 2}(\mathscr{L}_k(m; M, \lambda))$. The general case is left an as exercise.

The *probability* that a measurement m has been assigned to *Class k* is the normalized likelihood ratio.

$$\Pr(m \in C_\ell) = \frac{\mathscr{L}_\ell(m; M, \lambda)}{\sum_{k=1}^{g} \mathscr{L}_k(m; M, \lambda)} = \frac{r_\ell \cdot \exp\left(-\frac{1}{2}d_M^2(m; \bar{y}_\ell)\right)}{\sum_{k=1}^{g} r_k \cdot \exp\left(-\frac{1}{2}d_M^2(m; \bar{y}_k)\right)} \tag{4.23}$$

Why use a minimum Bayes score/maximum likelihood to classify a measurement? As was seen in Section 4.5, discriminant analysis methods permit the *shape* of the decision boundary to be "tuned" by the Friedman parameters (γ, λ). The Bayes score allows for a shift or translation of the boundary by varying the value of the prior probability r_k.

In theory, the priors r_k represent the *a priori* probabilities that a measurement belongs to a certain class. Reasonable estimates for priors are population proportions. For example, 75% of men in their 50s who are screened for prostate cancer are diagnosed as *negative* for the disease. Of the remaining 25% diagnosed with prostate cancer, 38% have indolent cancers (i.e., cancers which never escape the prostate capsule and cause no secondary disease). This means that prostate disease diagnoses can be divided into three categories: *Class 1* (normal tissue = no disease), *Class 2* (indolent cancers), and *Class 3* (invasive cancers). The population proportions are then $r_1 = 0.75$, $r_2 = 0.25 \times 0.38 = 0.095$, and $r_3 = 0.25 \times (1 - 0.38) = 0.155$.

To see how the priors affect the decision boundary, return to the two-class data of Figures 4.4 and 4.13. Employ the Friedman matrices F_1 and F_2 with equal parameter

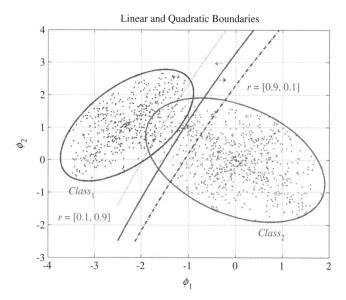

FIGURE 4.14   Minimum Bayes score and prior probabilities.

values (i.e., $\gamma = \lambda = \frac{1}{2}$). Furthermore, use equal priors ($r_1 = r_2$) so that the difference in Bayes scores is merely the difference in Mahalanobis distances (see Exercise 4.7). This gives way to the quadratic boundary illustrated by the solid curve in Figure 4.14. If, however, $r_1 > r_2$, then the boundary *shifts* to include more points from *Class* 1. This is depicted by the dashed curve (– –) in Figure 4.14. Conversely, the condition $r_2 > r_1$ results in a shift favoring *Class* 2 (dotted curve $\cdots$). Therefore, the priors can be used to "tune" the shape of the decision boundary to balance the probability of detection p_d versus the probability of false alarm p_{f_a} requirements.

Thus far, it has been discussed how the shape (Friedman parameters γ, λ) and translation (prior probabilities r_1, r_2,...) of the decision boundary can be altered to accommodate the (p_d, p_{f_a}) goals of the diagnostic system. This can be achieved via the use of the Friedman matrix in regularized discriminant analysis with Bayes score classification criterion (4.21). These are geometric considerations. What can be done to mitigate class-to-class errors? This is an algebraic issue.

More practically, it may be of little concern that a normal (non-diseased) sample is classified as low-grade disease. Classifying cancer as normal, however, can have deleterious consequences. That is, the *cost* of classifying cancer as normal is far greater than the cost of classifying normal as cancer. Such a misclassification is called an *inter-class error*. The manner in which such misclassifications are managed is via the *cost matrix*.

The (j, k)-element of the cost matrix $\mathcal{C} \in \mathcal{Mat}_{g \times g}(\mathbb{R})$ is the penalty or *cost* of classifying objects from *Class j* as being in *Class k*. Plainly, the (k, j)-element is the cost of classifying objects from *Class k* as *Class j*. It need not be the case that

$c_{j,k} = c_{k,j}$; indeed, generally, $c_{j,k} \neq c_{k,j}$. Since there is no penalty for classifying objects from *Class k* as *Class k*, then $c_{k,k} = 0$ for all k. The *cost matrix* thereby takes the form

$$
\mathcal{C} = \begin{bmatrix}
0 & c_{1,2} & \cdots & c_{1,g} \\
c_{2,1} & 0 & \vdots & c_{2,g} \\
\vdots & \vdots & \ddots & \vdots \\
c_{g,1} & c_{g,2} & \vdots & 0
\end{bmatrix}.
\tag{4.24}
$$

By recalling the definition of the Bayes likelihood of (4.22), a vector of length g can be formed which holds the likelihoods of a select measurement $\boldsymbol{m}$ to each of the g data centers with respect to the weighting matrix M and Friedman parameter λ.

$$
\mathcal{L}(\boldsymbol{m}; M, \lambda) \equiv [\mathcal{L}_1(\boldsymbol{m}; M, \lambda), \mathcal{L}_2(\boldsymbol{m}; M, \lambda), \ldots, \mathcal{L}_g(\boldsymbol{m}; M, \lambda)]^T
$$

The *Bayes risk* of a measurement $\boldsymbol{m}$ with respect to a weighting matrix M and Friedman parameter λ is then a linear combination of the Bayes likelihoods as weighted by the cost matrix

$$
\mathcal{R}(\boldsymbol{m}; M, \lambda) = \mathcal{C} \bullet \mathcal{L}(\boldsymbol{m}; M, \lambda).
\tag{4.25}
$$

With the reminder that the diagonal elements of the cost matrix $\mathcal{C}$ are all 0, $c_{k,k} = 0$, then the Bayes risk can be written as

$$
\left.
\begin{aligned}
\mathcal{R}(\boldsymbol{m}; M, \lambda) &= \left[\mathcal{R}_1(\boldsymbol{m}; M, \lambda), \mathcal{R}_2(\boldsymbol{m}; M, \lambda), \ldots, \mathcal{R}_g(\boldsymbol{m}; M, \lambda) \right]^T \\
\mathcal{R}_\ell(\boldsymbol{m}; M, \lambda) &= \ldots \sum_{j=1}^{g} c_{\ell,j} \cdot \mathcal{L}_j(\boldsymbol{m}; M, \lambda) \\
c_{\ell,\ell} &= 0 \text{ for every } \ell
\end{aligned}
\right\}.
\tag{4.26}
$$

The *Bayes risk decision rule* for the classification of the measurement $\boldsymbol{m} \in \mathbb{R}^r$ with respect to the weighting matrix M and Friedman parameter λ is the minimum Bayes risk as indicated in (4.27).

$$
\min_{1 \leq k \leq g} \mathcal{R}_k(\boldsymbol{m}; M, \lambda) = \mathcal{R}_\ell(\boldsymbol{m}; M, \lambda) \Rightarrow \boldsymbol{m} \in \mathcal{C}_\ell
\tag{4.27}
$$

Example Consider a three-group classification problem in which the likelihoods for a given measurement $\boldsymbol{m}$ are $\mathcal{L}_1 = 0.7$, $\mathcal{L}_2 = 0.6$, and $\mathcal{L}_3 = 0.5$. Then, using a maximum likelihood classification regiment, it is seen that the $\max_{1 \leq k \leq 3} \mathcal{L}_k = \mathcal{L}_1$ so that $\boldsymbol{m} \in \mathcal{C}_1$. Suppose that it is twice as expensive as the nominal cost to misclassify *Class 1* objects as *Class 2*; three times as expensive to misclassify *Class 1* objects as *Class 3*; and equally expensive to misclassify *Class 2* objects as *Class 3*, *Class 2* as *Class 1*, and *Class 3* as *Class 1*. Then the cost matrix is $\mathcal{C} = \begin{bmatrix} 0 & 2 & 3 \\ 1 & 0 & 1 \\ 1 & 1 & 0 \end{bmatrix}$. In this setting,

TABLE 4.3   Classification measures and decision criteria

Measure	Weighting matrix M	Parameters	Decision criterion
Mahalanobis distance d_M	C_k or C_{pool}	None	$\min\limits_{1 \le k \le g} d_M(\boldsymbol{m}, \bar{\boldsymbol{y}}_k) = d_M(\boldsymbol{m}, \bar{\boldsymbol{y}}_\ell) \Rightarrow \boldsymbol{m} \in \mathscr{C}_\ell$
	$F_k(\gamma, \lambda)$	(γ, λ)	
Bayes score $\beta(\boldsymbol{m}; M, \lambda)$	C_k or C_{pool}	$r_1, r_2, \dots, r_g, \lambda$	$\min\limits_{1 \le k \le g} \beta_k(\boldsymbol{m}; M, \lambda) = \beta_\ell(\boldsymbol{m}; M, \lambda) \Rightarrow \boldsymbol{m} \in \mathscr{C}_\ell$
	$F_k(\gamma, \lambda)$	$r_1, r_2, \dots, r_g,$ (γ, λ)	
Likelihood score $\mathscr{L}(\boldsymbol{m}; M, \lambda)$	C_k or C_{pool}	$r_1, r_2, \dots, r_g, \lambda$	$\max\limits_{1 \le k \le g} \mathscr{L}_k(\boldsymbol{m}; M, \lambda) = \mathscr{L}_\ell(\boldsymbol{m}; M, \lambda) \Rightarrow \boldsymbol{m} \in \mathscr{C}_\ell$
	$F_k(\gamma, \lambda)$	$r_1, r_2, \dots, r_g,$ (γ, λ)	
Bayes risk $\mathscr{R}(\boldsymbol{m}; M, \lambda)$	C_k or C_{pool}	$r_1, r_2, \dots, r_g,$ $\lambda, \mathscr{C}$	$\min\limits_{1 \le k \le g} \mathscr{R}_k(\boldsymbol{m}; M, \lambda) = \mathscr{R}_\ell(\boldsymbol{m}; M, \lambda) \Rightarrow \boldsymbol{m} \in \mathscr{C}_\ell$
	$F_k(\gamma, \lambda)$	$r_1, r_2, \dots, r_g,$ $(\gamma, \lambda), \mathscr{C}$	

the Bayes risk vector is $\mathscr{R} = \mathscr{C} \bullet \mathscr{L} = \begin{bmatrix} 0 & 2 & 3 \\ 1 & 0 & 1 \\ 1 & 1 & 0 \end{bmatrix} \cdot \begin{bmatrix} 0.7 \\ 0.6 \\ 0.5 \end{bmatrix} = \begin{bmatrix} 2.7 \\ 1.2 \\ 1.3 \end{bmatrix}$. The *minimum Bayes risk* classification measure determines that $\boldsymbol{m} \in \mathscr{C}_2$ since the minimum risk occurs in *Class* 2. Hence, the elements of the cost matrix can alter the shape of the decision boundary and influence the classification.

A summary of all of the classification measures and their associated decision rules are compiled in Table 4.3. For the "simplest" case of the Mahalanobis distance with a pooled within-group covariance weighting matrix, there are no parameters that can influence decision boundary shape or position. As the classification measure becomes more sophisticated (Bayes score, likelihood score, Bayes risk), additional parameters are added to the measure. These parameters can dramatically alter the decision boundary. As will be amplified in the next section, a set of *assessment metrics* will be discussed. These metrics permit the designer to gauge performance of the classifier.

EXERCISES

4.6 Prove that the classification of a minimum Bayes score is equivalent to a maximum likelihood. That is, if $\min\limits_{1 \le k \le g} \beta_k(\boldsymbol{m}; M, \lambda) = \beta_\ell(\boldsymbol{m}; M, \lambda) \Rightarrow \boldsymbol{m} \in \mathscr{C}_\ell$, then

$\max\limits_{1 \le k \le g} \left(\mathscr{L}_k(\boldsymbol{m}; M, \lambda) \right) = \mathscr{L}_\ell(\boldsymbol{m}; M, \lambda) \Rightarrow \boldsymbol{m} \in \mathscr{C}_\ell$.

4.7 Show that for a two-class problem, whenever $r_1 = r_2$, the Bayes score is equivalent to the Mahalanobis distance. *Hint*: Examine $\beta_1(\boldsymbol{m}; M, \lambda) - \beta_2(\boldsymbol{m}; M, \lambda)$.

4.7   THE CONFUSION MATRIX, RECEIVER–OPERATOR CHARACTERISTIC CURVES, AND ASSESSMENT METRICS

The description of classification, thus far, has focused exclusively on *how* a class assignment is made. Now the question of *how to gauge the success of the classifier* is addressed. The short answer is via assessment metrics. Before providing details about these measures, however, a bit more background development must be undertaken. In keeping with that charter, suppose there are two sets of data. The first set consists of data matrices $X_1, X_2,..., X_g$ whose disease state is *known*. That is, X_1 is [say] the $n_1 \times p$ matrix of known *normal* (i.e., negative for disease) samples, X_2 is the $n_2 \times p$ matrix of "known *disease level 1*" samples, X_3 is the $n_3 \times p$ matrix of "known *disease level 2*" samples, etc. These data matrices form the basis of any variable reduction transformation (e.g., *PCA, DAFE*) and the projected data matrices $Y_1, Y_2,..., Y_g$, where $Y_k \in \mathcal{M\!at}_{n_k \times r}$, $r < p$. The Y_k, in turn, give rise to the (projected) class mean vectors $\bar{y}_k \in \mathbb{R}^r$ and class covariance matrices $C_k \in \mathcal{M\!at}_{r \times r}$, $k = 1, 2,..., g$. The classifier is built from these projected data matrices Y_k, a choice of weighting matrix M, and the desired parameters as described in Table 4.3. The collection $\{Y_1, Y_2,..., Y_g\}$ is referred to as the *training data* of the classifier.

Next, suppose $\Psi_1, \Psi_2,..., \Psi_g$ are $m_k \times r$ matrices whose disease state is known and parallels the projected training data $Y_1, Y_2,..., Y_g$. Then the collection $\{\Psi_1, \Psi_2,..., \Psi_g\}$ is called the *test set*. Again, it is assumed that the measurements in Ψ_1 are *normal* (non-diseased), while measurements from $\Psi_2, \Psi_3,..., \Psi_g$ are *disease level 1, level 2,..., level g − 1*, respectively. Then the class to which any measurement $\psi_{k,j} \in \Psi_k$ is assigned determines the performance of the classifier. Here the notation $\psi_{k,j} \in \Psi_k$ means that $\psi_{k,j}$ is the *j*th row of the test matrix Ψ_k, $j = 1, 2,..., m_k$.

Consider the simplified case of $g = 2$. Specifically, *Class 1* is *normal* (no disease) and *Class 2* is disease. There are m_1 *normal* (i.e., negative) samples contained in the test matrix Ψ_1 and m_2 *disease* (positive) samples contained in Ψ_2. Suppose that $a_1 < m_1$ of the normal samples are classified as negative while $a_2 < m_2$ of disease samples are classified as positive. This means that $m_1 − a_1$ normal samples are *not* classified as negative. Rather, they are classified as *positive*. Similarly, $m_2 − a_2$ positive samples are classified as negative. All of this verbiage can be condensed into a *contingency table* as exhibited in Table 4.4a.

A direct computation shows *the probability that a [truly] negative sample is classified as negative* is a_1/m_1 and the *probability that a positive sample is classified as positive* is a_2/m_2. That is, $p_n = a_1/m_1 = \Pr(\psi$ is classified *negative* $\mid \psi \in \Psi_1)$. This is called the *specificity* of the classifier. Colloquially, it is a measure of how well a classifier detects *negative* (i.e., no disease) samples. Another way to examine

TABLE 4.4a   Contingency table for a two-class problem

		Normal	
		Negative	Positive
Disease	Negative	a_1	$m_2 - a_2$
	Positive	$m_1 - a_1$	a_2

TABLE 4.4b **Contingency table for a two-class problem**

		Normal	
		Negative	Positive
Disease	Negative	T_n	F_n
	Positive	F_p	T_p

Table 4.4a is to observe that a_1 is the *number of negatives correctly classified as negative*. Biostatisticians, medical researchers, and diagnostic experts refer to this quantity as the number of *true negatives*. Similarly, a_2 is the number of *true positives*, while $m_1 - a_1$ and $m_2 - a_2$ are the number of *false positives* and *false negatives*, respectively. Table 4.4a can now be rewritten as Table 4.4b.

The most popular metrics for assessing the performance of a classifier are the *specificity* and *sensitivity* as defined via equation (4.28a).

$$\left. \begin{aligned} Specificity &= \frac{T_n}{T_n + F_p} \cdot 100\% \\ Sensitivity &= \frac{T_p}{T_p + F_n} \cdot 100\% \end{aligned} \right\} \tag{4.28a}$$

Observe that the *specificity* and *sensitivity* are generally reported as percentages. They can be viewed as probabilities via (4.28b).

$$\left. \begin{aligned} specificity &= \frac{T_n}{T_n + F_p} = \Pr(\psi \in \mathscr{C}_{neg} | \psi \in \Psi_{neg}) \\ sensitivity &= \frac{T_p}{T_p + F_n} = \Pr(\psi \in \mathscr{C}_{pos} | \psi \in \Psi_{pos}) \end{aligned} \right\} \tag{4.28b}$$

The notation $\Pr(\psi \in \mathscr{C}_{neg} | \psi \in \Psi_{neg})$ = the probability that ψ is classified as *negative* given that ψ is selected from the *negative* test set. The meaning of $\Pr(\psi \in \mathscr{C}_{pos} | \psi \in \Psi_{pos})$ is comparable. The *accuracy* of a classifier is the ratio of the true positives and negatives to the total number of test samples: $acc = (T_p + T_n)/(T_p + F_n + T_n + F_p)$. This measure is rarely used in practice as it can give deceptive results when the number of negatives greatly overwhelms the number of positives. Adopt the notation $N_n = T_n + F_p$ as the number of negatives and $N_p = T_p + F_n$ as the number of positives. Suppose $N_n = 100 \cdot N_p$, the proportion of true negatives is high (say 95%), and the number of true positives is low (5%). In this case, the accuracy exceeds 94% (the details are left as Exercise 4.10).

Associated with the 2×2 contingency tables is the *confusion matrix*. Using the nomenclature of Table 4.4a, the corresponding confusion matrix is

$$C_f(2) = \begin{bmatrix} a_1 & m_2 - a_2 \\ m_1 - a_1 & a_2 \end{bmatrix}. \tag{4.29}$$

As stated at the beginning of this section, there are g distinct classes $\mathscr{C}_1, \mathscr{C}_2, ..., \mathscr{C}_g$. In the case in which $g > 2$, the contingency Tables 4.4a and 4.4b must be generalized. Making the assignments $\mathscr{C}_1 \mapsto$ *negative* (normal) and $\mathscr{C}_j \mapsto$ *disease level* $(j - 1)$ for

TABLE 4.5   **Contingency table for the g-class problem**

			Normal			
		Negative	Level$_1$	Level$_2$	$\cdots$	Level$_{g-1}$
Disease	Negative	$v_{1,1}$	$v_{1,2}$	$v_{1,3}$	$\cdots$	$v_{1,g}$
	Level$_1$	$v_{2,1}$	$v_{2,2}$	$v_{2,3}$	$\cdots$	$v_{2,g}$
	Level$_2$	$v_{3,1}$	$v_{3,2}$	$v_{3,3}$	$\cdots$	$v_{3,g}$
	$\vdots$	$\vdots$	$\vdots$	$\vdots$	$\ddots$	$\vdots$
	Level$_{g-1}$	$v_{g,1}$	$v_{g,2}$	$v_{g,3}$	$\cdots$	$v_{g,g}$

$j = 2, 3, \ldots, g$, the general contingency table is presented in Table 4.5. The notation $v_{j,k}$ means the number of *Class k* specimens which have been classified into *Class j*. If a classifier is perfect, then $v_{k,k} = m_k$ and $v_{j,k} = 0$ for all $j \neq k$.

There are no assessment metrics comparable to sensitivity and specificity for the general g-class problem. Typically, the diagnostic space is partitioned into disease levels that exceed a certain threshold versus those which do not. The generalized confusion matrix is given in (4.30).

$$C_f(g) = \begin{bmatrix} v_{1,1} & v_{1,2} & \cdots & v_{1,g} \\ v_{2,1} & v_{2,2} & \cdots & v_{2,g} \\ \vdots & \vdots & \ddots & \vdots \\ v_{g,1} & v_{g,2} & \cdots & v_{g,g} \end{bmatrix} \tag{4.30}$$

The assessment metrics presented in (4.28a) and (4.28b) along with the confusion matrix give a measure of how well a classifier performs with respect to a particular collection of tuning parameters (e.g., the Friedman parameters γ, λ; the prior probabilities $r = [r_1, r_2, \ldots, r_g]$). Rather than a haphazard combination of parameters, it is recommended that a *batch* of carefully mapped parameter values be applied to the classifier with respect to the *test* data. Then a systematic review of the *overall* classifier behavior can be determined from the ensemble of classifier assessment values.

What does this mean? It is known that the Friedman parameters (γ, λ) range over [0, 1]. Moreover, the prior probabilities have similar restrictions; $0 \leq r_j \leq 1$ for all j and $\sum_{j=1}^{g} r_j = 1$. Therefore, create a multidimensional lattice Λ of the form (4.31). Here, d_γ and d_λ are the number of values taken by the Friedman parameters over $[0, 1]$; $0 \leq \gamma_1 < \gamma_2 < \cdots < \gamma_{d_\gamma} \leq 1$ and $0 \leq \lambda_1 < \lambda_2 < \cdots < \lambda_{d_\lambda} \leq 1$. The vectors $r_{j,k}$ are the variations on the prior probabilities at each lattice point $\Lambda_{j,k}$. Each parameter collection $(\gamma_j, \lambda_k; r_{j,k})$ produces a classifier (with respect to a collection of *training data*).

$$\Lambda = \begin{bmatrix} (\gamma_1, \lambda_1; r_{1,1}) & (\gamma_1, \lambda_2; r_{1,2}) & \cdots & (\gamma_1, \lambda_{d_\lambda}; r_{1,d_\lambda}) \\ (\gamma_2, \lambda_1; r_{2,1}) & (\gamma_2, \lambda_2; r_{2,2}) & \cdots & (\gamma_2, \lambda_{d_\lambda}; r_{2,d_\lambda}) \\ \vdots & \vdots & \ddots & \vdots \\ (\gamma_{d_\gamma}, \lambda_1; r_{d_\gamma,1}) & (\gamma_{d_\gamma}, \lambda_2; r_{d_\gamma,2}) & \cdots & (\gamma_{d_\gamma}, \lambda_{d_\lambda}; r_{d_\gamma,d_\lambda}) \end{bmatrix} \tag{4.31}$$

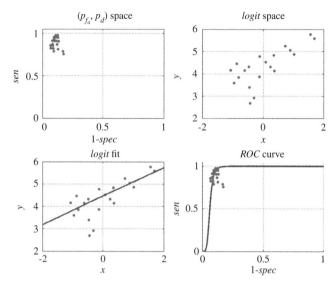

FIGURE 4.15   The *ROC* curve and transformation.

Therefore, one manner in which to gauge classifier performance is to compare the corresponding collection of (*sensitivity, specificity*) points. As an alternative, consider the following interpretation of these assessment metrics. Since *sensitivity* is the probability of detecting a (genuinely) positive sample, it can be thought of as the *probability of detection*; *sen* $= p_d$. Similarly, the complement of the *specificity* is the probability of identifying a *negative* specimen as positive. Namely, the probability of a *false alarm*; $1 - spec = p_{f_a}$. This coordinate system (p_{f_a}, p_d) is what radar engineers use to plot *receiver–operator characteristic* (*ROC*) curves as per references [6, 10, 11, 14]. This coordinate measure was first employed by signal processing engineers to analyze radar signals during the Second World War. Green and Swets [6] popularized *ROC* space for use in signal detection theory.

Consider Figure 4.15 in which the upper left-hand portion displays a series of $(1 - spec, sen) = (p_{f_a}, p_d)$ points. Each $(1 - spec_j, sen_j)$ pair represents a classifier with the associated parameter set $(\gamma_j, \lambda_{k(j)}; r_{k(j),j}); j = 1, 2, \ldots$. It is unclear what, if any, sensible curve can be fit to these data. Using the *logit* function as defined in (4.32), map the $(1 - spec, sen)$ data into a *linearized* coordinate space.

$$logit(z) = \ln\left(\frac{z}{1-z}\right) \quad \text{for } z \in (0, 1) \tag{4.32}$$

The linearization occurs via the *logit* transformation

$$\left.\begin{array}{l} x = logit(sen) + logit(1 - spec) = \ln\left(\frac{sen \cdot (1-spec)}{(1-sen) \cdot spec}\right) \\ y = logit(sen) - logit(1 - spec) = \ln\left(\frac{sen \cdot spec}{(1-sen) \cdot (1-spec)}\right) \end{array}\right\}. \tag{4.33}$$

This transformation is illustrated in the upper right-hand portion of Figure 4.15. Next, in the "*logit space*," calculate the line that has the "best fit" to these data. This is referred to as the *least squares* line $\mathcal{L}$: $y = mx + b$. The line $\mathcal{L}$ is the best linear fit of the *logit space data* as per Section 1 of the Appendix, equation (1.4). The mathematical details for constructing the formulae for the slope m and intercept b of the least squares line $\mathcal{L}$ are provided in Section 1 of the Appendix. The line $\mathcal{L}$ is depicted in the lower left-hand portion of Figure 4.15. Now, inverting the transformation (4.33) for the *sensitivity* as a function of the *specificity* produces the equation

$$ sen = \frac{1}{1 + \exp\left(\frac{b}{m-1}\right) \cdot \left(\frac{1-spec}{spec}\right)^{\frac{m+1}{m-1}}}. \tag{4.34a} $$

Set $u = sen$ and $v = 1 - spec$ and substitute into (4.34a) to obtain

$$ u = f_{ROC}(v) = \frac{1}{1 + \exp\left(\frac{b}{m-1}\right) \cdot \left(\frac{v}{1-v}\right)^{\frac{m+1}{m-1}}}. \tag{4.34b} $$

This is the transformation from *logit* space into *ROC* coordinates. The transformed *logit* space line $\mathcal{L}$ is the *ROC curve* as shown in the lower right-hand side of Figure 4.15. In theory, there exists a combination of parameters $(\gamma, \lambda; r)$ so that any point $(1 - spec, sen)$ on the *ROC curve*, $sen = f_{ROC}(1 - spec)$, can be attained.

The point $(1 - spec, sen)$ at which the parameter variation yields a desirable *sensitivity* and *specificity* for the classifier is called the *operating point*. In the next section, these concepts are combined to demonstrate how a classifier is tuned. The *Q-point* is the position on the *ROC curve* in which the *sensitivity* is equal to the *specificity*. Mathematically, the *Q-point* is a real-valued root of the polynomial

$$ p(z) = z \cdot (1 - z)^{m-1} + e^{\frac{b}{m-1}} \cdot z^{m+2} - (1 - z)^{m-1}. \tag{4.35} $$

Thus, the *Q-point* is the solution of $p(z) = 0$. For general slope m, this cannot be solved analytically and numerical methods must be utilized. The *Q-point* is an operator point equilibrium and expresses the theoretical "best" a classifier can achieve in *sensitivity* and *specificity* simultaneously.

EXERCISES

4.8 When the costs of inter-class misclassifications are all uniform, then the problem is *equally weighted* and the cost matrix takes the form $\mathcal{C} =$
$$ \begin{bmatrix} 0 & 1 & \cdots & 1 \\ 1 & 0 & \cdots & 1 \\ \vdots & \vdots & \ddots & \vdots \\ 1 & 1 & \cdots & 0 \end{bmatrix} $$. Show that, in the case of an equally weighted classification problem, the minimum Bayes risk and maximum likelihood are

equivalent. *Hint*: Assume, without loss of generality, that $\min\limits_{1 \le k \le g} \left(\mathscr{L}_k(\boldsymbol{m}; M, \lambda) \right)$
$= \mathscr{L}_1(\boldsymbol{m}; M, \lambda)$. Compute the Bayes risk from (4.25) and examine the kth row.

4.9   Derive equation (4.35). *Hint*: Simplify $z = f_{ROC}(z)$ as defined via (4.34b).

4.10   Show that if $N_n = 100 \cdot N_p$, $T_n = 0.95 \cdot N_n$, and $T_p = 0.05 \cdot N_p$, then $acc > 0.94$.

4.8   AN EXAMPLE

It is now time to use the ideas developed in Sections 4.4–4.7 on the data introduced in Section 4.1. To that end, the data exhibited in Figure 4.2 will be the *training data*. There are four data classes. The first is *normal* (or *negative* for disease) and the remaining three {*level*$_1$, *level*$_2$, *level*$_3$} are increasingly positive for disease, with *level*$_3$ being the most serious. The training data will be normalized via (4.5) and projected into (three) *DAFE* coordinates using (4.14b)–(4.14c). This projection is applied to the test data whose *DAFE* coordinates are depicted in Figure 4.16b.

Given the *training* and *test* data, designate the range of the Friedman parameters (γ, λ). Define a vector of prior probabilities $\boldsymbol{r}(\gamma, \lambda)$ associated with each Friedman parameter pair. These values are determined by the formulae in (4.36). The idea is to deflate the *negative* class probabilities while inflating the *positive* class probabilities. Let $\boldsymbol{r}_0 = [r_0(1), r_0(2),\ldots,r_0(g)]$ be the initial vector of prior probabilities (generally, population priors), $\boldsymbol{neg} = [1, 0, 0, 0]$ be the vector of *negative class* identifiers, $\boldsymbol{pos} = [0, 1, 1, 1]$ be the vector of *positive class* identifiers, and $W = \{w_{neg}, w_{pos}\}$ be a set of weights used on the population priors. Let $\boldsymbol{a} \odot \boldsymbol{b} \equiv \left[a_1 \cdot b_1, a_2 \cdot b_2,\ldots, a_m b_m \right]$ and $\Delta \boldsymbol{r}_j = \boldsymbol{r}_0 \odot \boldsymbol{neg} \bullet w_{neg}$ with a similar definition for $\Delta \boldsymbol{r}_k (= \boldsymbol{r}_0 \odot \boldsymbol{pos} \bullet w_{pos})$. Then the prior probabilities evolve as shown in equation (4.36).

$$\left. \begin{aligned} \gamma_j &= \gamma_0 + j \cdot \Delta\gamma \\ \lambda_k &= \lambda_0 + k \cdot \Delta\lambda \\ r_{j,k} &= \boldsymbol{r}_0 - \left(\Delta \boldsymbol{r}_j \right)^{(j-1)} + \left(\Delta \boldsymbol{r}_k \right)^{(k-1)} \end{aligned} \right\} \tag{4.36}$$

The precise values used in the tuning run for a minimum Bayes risk classifier applied to these data are listed in Table 4.6. Observe that two separate tuning experiments were conducted. In each case, the Friedman parameters and prior probabilities used the same values. For the first tuning run, however, the "equal cost" cost matrix was utilized. In the second run, the cost of misclassifying a *Class* 1 (i.e., *normal*) sample as *Class* 2 (i.e., *disease level*$_1$) is considered four times more "expensive" than the equal cost case. The resulting *ROC curves*, selected operating points, and *Q-points* are displayed in Figures 4.17a and 4.17b.

First, consider the result of the "equal cost" classifier tuning. The operating point is $(Sen, Spec) = (91\%, 94\%)$. The confusion matrix is $C_f(4) = \begin{bmatrix} 49 & 1 & 0 & 0 \\ 3 & 17 & 0 & 0 \\ 0 & 3 & 12 & 0 \\ 0 & 0 & 0 & 10 \end{bmatrix}$.

The population of the *test* data are 50 *Class* 1 (*normal*), 20 *Class* 2 (*disease level*$_1$),

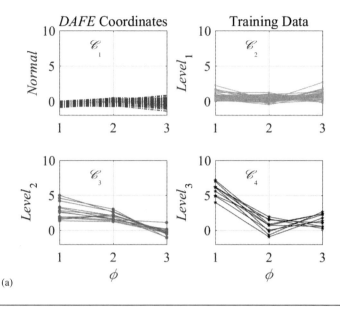

(a)

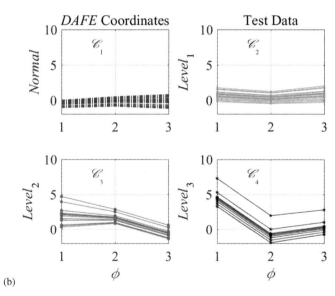

(b)

FIGURE 4.16   (a) *DAFE* coordinates (training). (b) *DAFE* coordinates (test).

15 *Class* 3 (*disease level$_2$*), and 10 *Class* 4 (*disease level$_3$*) samples. It is seen that the primary misclassification is in-between *Classes* 1 and 2.

Therefore, to improve the *sensitivity*, a penalty for misclassification is applied. Since the number of *false positives* (in this case, *disease level$_1$* classified as *normal*)

TABLE 4.6   Parameter settings for minimum Bayes risk classifier

Number of classes: $g = 4$		Bayes risk classifier		
Friedman parameters		Prior probabilities r_0	Weights W	Cost matrix $\mathcal{C}$
$\gamma_o = 0$ $\Delta\gamma = 0.2$ $\Leftrightarrow$ $\gamma \in \{0, 0.2, 0.4,$ $0.6, 0.8, 1\}$	$\lambda_o = 0$ $\Delta\lambda = 0.25$ $\Leftrightarrow$ $\lambda \in \{0, 0.25, 0.5,$ $0.75, 1\}$	$[0.78, 0.15, 0.05, 0.03]$	$w_{neg} = 0.1$ $w_{pos} = 0.2$	$\begin{bmatrix} 0 & 1 & 1 & 1 \\ 1 & 0 & 1 & 1 \\ 1 & 1 & 0 & 1 \\ 1 & 1 & 1 & 0 \end{bmatrix}$ $\begin{bmatrix} 0 & 1 & 1 & 1 \\ 4 & 0 & 1 & 1 \\ 1 & 1 & 0 & 1 \\ 1 & 1 & 1 & 0 \end{bmatrix}$

is small (namely, 3) and the clinical outcome of a false positive is generally benign (the patient is either retested or subject to more sophisticated tests), no penalty is applied in this instance. The converse (misclassifying a *disease level*$_1$ sample as *normal*), however, can have devastating consequences. The patient can go untreated and develop a life-threatening illness. Therefore, a penalty or *cost* for this misclassification is applied from *Class* 1 to *Class* 2. A factor of four was used as the penalty.

As the *ROC curves* (Figure 4.17) illustrate, the behavior of the classifier changes along with the cost matrix. The *sensitivity* increases (by 2.2%) as the cost of misclassification of the first positive class (*level*$_1$) increases. The *specificity*, however, decreases (by 6%). This tradeoff may be considered acceptable as the increase in the probability of detection ($p_d = sen$) compensates for the decrease in the probability of a false alarm ($p_{f_a} = 1 - spec$). The confusion matrix for the modified cost matrix

classifier is $C_{f,M}(4) = \begin{bmatrix} 46 & 4 & 0 & 0 \\ 2 & 18 & 0 & 0 \\ 0 & 3 & 12 & 0 \\ 0 & 0 & 0 & 10 \end{bmatrix}$. Thus, the number of false negatives has

decreased (from 3 to 2) while the number of false positives has increased (from 1 to 4). This tuning of the classifier allows for greater accuracy in disease detection with some degradation in false positives (or "overcalling"). The results of the two tunings are summarized in Table 4.7. If the disease in question is one which spreads rapidly and has serious consequences (fatality, permanent disability, lengthy recovery period), then a gain in sensitivity is more important than a loss in specificity.

Remark: The evolution of the prior probabilities as a function of Friedman parameter indices as described in equation (4.36) is entirely arbitrary. Other approaches can (and should) be utilized. Following the "update" equation $r_{j,k} = r_0 - (\Delta r_j)^{(j-1)} + (\Delta r_k)^{(k-1)}$, the prior probability vector is normalized so that its components sum to 1. That is, $r_{j,k} \mapsto \dfrac{1}{\sum_{\ell=1}^{g} r_{j,k}(\ell)} \cdot r_{j,k}.$

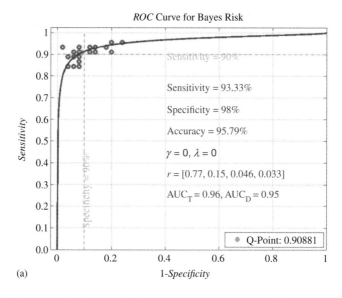

(a)

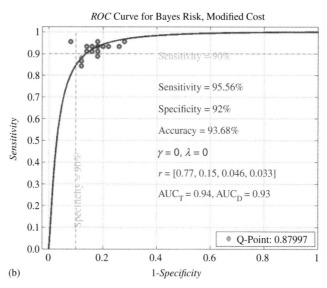

(b)

FIGURE 4.17   (a) *ROC* curve (equal cost). (b) *ROC* curve (cost penalty).

TABLE 4.7   Different operating points due to different cost matrices

Tuning	1	2
Cost matrix	Equal cost	Equal cost $+ \, \mathcal{C}_{2,1} = 4$
Friedman parameters	$\gamma = 0 = \lambda$	$\gamma = 0 = \lambda$
Prior probabilities	$r = [0.77, 0.15, 0.046, 0.033]$	$r = [0.77, 0.15, 0.046, 0.033]$
Sensitivity	93.33%	95.56%
Specificity	98%	92%

EXERCISE

4.11   Using the *training* and *test* data of Section 4.8, construct a new set of prior probabilities and determine what effect these have on the Bayes risk classifier performance. These data are contained in the Chapter 4 data directory as the MAT-files `Multivariate_MultiClass_Data.mat` and `Multivariate_MultiClass_Test.mat`.

4.9   NONLINEAR METHODS

Thus far the discussion of classification has been restricted to *linear* techniques. Even though *QDA* produces a quadratic boundary, the classification is made via matrix computations (Section 4.4). Hence, the process is *linear*. In this section, a *nonlinear* method, used in the two-class discrimination problem, is described. As earlier, let $X \in \mathcal{Mat}_{n \times p}(\mathbb{R})$ be the data matrix of the measurements $x_k = [x_{k,1}, x_{k,2}, \ldots, x_{k,p}] \in \mathbb{R}^p$ for $k = 1, 2, \ldots, n$. That is, $X = \begin{bmatrix} x_{1,1} & x_{1,2} & \cdots & x_{1,p} \\ x_{2,1} & x_{2,2} & \cdots & x_{2,p} \\ \vdots & \vdots & \ddots & \vdots \\ x_{n,1} & x_{n,2} & \cdots & x_{n,p} \end{bmatrix}$. Suppose further that $y_k \in \{-1, 1\}$ is an indicator of class. In particular, $y_j = -1$ indicates that measurement j is of the *negative* class, while $y_k = 1$ means that measurement k is from the *positive* class. The collection $\mathcal{S} = \{[x_1, y_1], [x_2, y_2], \ldots, [x_n, y_n]\}$ is the *data set* with respect to the classification problem.

As was seen in the case of *RDA* (Section 4.5), a classifier can be tuned and a boundary selected so that the "best achievable" operating point can be realized. The decision boundary, however, is selected by the needs of the designer rather than an objective criterion. If two data classes X_1 and X_2 can be separated by a line (plane for $p = 3$ dimensions or hyperplane for $p \geq 4$), can an optimal line be selected so that the classes are separated and the distance from each class to the boundary is maximized? The answer is *yes*, via *support vector machines*.

This discussion follows the presentations of Hastie et al. [7] and Burges [1].

The task at hand is to construct a weight vector w and an intercept b so that the equation $w^T x - b = 0$ separates the positive class vectors (associated with $y_k = 1$) and negative class vectors ($y_k = -1$) in an optimal manner. This plane $L_0: w^T x = b$ will be referred to as the *linear separator*. Moreover, the maximal *margin* that separates the two classes will help to form a "neutral zone" in which no classification can be rendered.

The construction of the linear separator L_0 is a matter of analytic geometry. To simplify matters, the case of $p = 2$ dimensions is examined. Here, the *linear sepa-rator* is a line. First, divide the data set $\mathbb{S}$ into two data matrices $X_\pi = \begin{bmatrix} -x_{j_1}- \\ -x_{j_2}- \\ \vdots \\ -x_{j_{n_\pi}}- \end{bmatrix}$,

so that $y_k = 1$ for $k = j_1, j_2, \ldots, j_{n_\pi}$, and $X_v = \begin{bmatrix} -x_{\ell_1}- \\ -x_{\ell_2}- \\ \vdots \\ -x_{\ell_{n_v}}- \end{bmatrix}$ with $y_j = -1$ for $j = \ell_1,$

$\ell_2, \ldots, \ell_{n_v}$. Let $\bar{x}_\pi = \frac{1}{n_\pi} \sum_{k=1}^{n_\pi} x_{j_k}$ and $\bar{x}_v = \frac{1}{n_v} \sum_{k=1}^{n_v} x_{\ell_k}$ be the mean vectors with respect to each data class. To construct L_0, first form the line L which connects the mean vectors $\bar{x}_\pi \equiv [\xi_\pi, \psi_\pi]^T$ and $\bar{x}_v \equiv [\xi_v, \psi_v]^T$. The connector line $L: y = mx + b$ has slope $m = \frac{\psi_\pi - \psi_v}{\xi_\pi - \xi_v}$ and intercept $b = \psi_\pi - m \xi_\pi$. The *separator* L_0 will be perpendicular to L and passing through the midpoint of the distinct class centers $x_m = \frac{1}{2}(\bar{x}_\pi + \bar{x}_v)$. Therefore, $L_0: y = m_0 x + b_0$ where $m_0 = -1/m$, and $b_0 = \frac{1}{2}(\psi_\pi + \psi_v) - m_0(\xi_\pi + \xi_v)$. These lines can be defined via the vector equations $w^T x - b = 0$ (for L) and $w_0^T x - b_0 = 0$ (for L_0) with $w^T = [m, 1]$, $w_0^T = [-1/m, 1]$, and $x = [x, y]^T$. These ideas are illustrated in Figure 4.18.

Linear *Support Vector Machine*

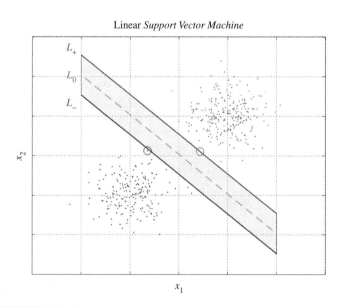

FIGURE 4.18   The *linear* support vector machine and indeterminate zone.

Next, the *margin* needs to be determined. To do so, compute the distances from each point in X_π to L_0 by projecting any $x \in X_\pi$ onto the *linear separator* L_0. This is accomplished as follows. Suppose $x = [x, y]^T$ is a point from the positive class. To project x onto L_0, the line passing through x and perpendicular to L_0 must be formed. Call this line L_π, which is of the form $y = m_\pi x + b_\pi$. Since $L_\pi \perp L_0$, then $m_\pi = -1/m_0$. Also, $x \in L_\pi$ means that $b_\pi = y - m_\pi x$. These two lines intersect when their representative equations are equal. In mathematical shorthand, $x \in L_0 \cap L_\pi \Rightarrow m_0 x - b_0 = y = m_\pi x - b_\pi \Rightarrow \hat{x} = [\hat{x}, \hat{y}] \in L_0 \cap L_\pi$, $\hat{x} = \frac{(b_0 - b_\pi)}{m_\pi - m_0}$, and $\hat{y} = m_0 \hat{x} - b_0$. Therefore, the distance from $x \in L_\pi$ to the *separator* is $d(x, L_0) = \sqrt{(x - \hat{x})^2 + (y - \hat{y})^2}$. The *smallest* such distance $M_+ = \min\limits_{x \in X_\pi} d(x, L_0)$ is the *positive margin*. In a similar fashion, $M_- = \min\limits_{x \in X_v} d(x, L_0)$ is the *negative margin*. Let $M = \min(M_+, M_-)$, $L_+: y = m_0 x + b_0 + M$, and $L_-: y = m_0 x + b_0 - M$. Then the strip defined by the area in-between L_+ and L_- is the *indeterminate zone*. In particular, measurements within this area cannot reliably be classified as either *positive* or *negative*. This zone is depicted as the parallelogram in Figure 4.18.

The problem can therefore be viewed as one of *optimization*. Specifically, what is the *maximum* margin M that forms the boundary in-between two distinct classes characterized by the data set $\mathbb{S}$? This can be stated as the optimization problem (4.37a)–(4.37b)

$$\max_{w_0, b_0, \|w_0\| = 1} M \tag{4.37a}$$

subject to

$$y_k \left(w_0^T x_k - b_0 \right) \geq M, \quad \text{for } k = 1, 2, \ldots, n. \tag{4.37b}$$

Observe that the weight vector w_0 is constrained to the unit sphere $\|w_0\| = 1$. To remove this constraint, replace (4.37b) with $\frac{1}{\|w_0\|} y_k (w_0^T x - b_0) \geq M$ or $y_k(w_0^T x - b_0) \geq M \|w_0\|$ for each k. Since scaling of w_0 and b_0 will not change the solution of (4.37a) and (4.37b), then setting $\|w_0\| = 1/M$ will permit a simpler statement of the problem

$$\max_{w_0, b_0} \|w_0\|^2 \tag{4.38a}$$

subject to

$$y_k \left(w_0^T x_k - b_0 \right) \geq 1, \quad \text{for } k = 1, 2, \ldots, n. \tag{4.38b}$$

This is a convex optimization problem to which Lagrange multipliers can be applied. Specifically, find the parameters α_k which minimize the Lagrangian

$$L_P = \tfrac{1}{2} \|w_0\|^2 - \sum_{k=1}^{n} \alpha_k \left(y_k \left(w_0^T x_k - b_0 \right) - 1 \right) \tag{4.39}$$

with respect to the normal vector w_0 and intercept b_0. Multivariate calculus theory notes that L_P is optimized when $\nabla L_P \equiv \left(\frac{\partial L_P}{\partial w_0}, \frac{\partial L_P}{\partial b_0} \right) = \mathbf{0}$. Since $\|w_0\|^2 = w_0^T w_0$, a straightforward calculation shows that $\frac{\partial L_P}{\partial w_0} = w_0^T - \sum_{k=1}^{n} \alpha_k y_k x_k$ and $\frac{\partial L_P}{\partial b_0} = \sum_{k=1}^{n} \alpha_k y_k$. If these partial derivatives are set to $\mathbf{0}$ and 0, respectively, then $w_0^T = \sum_{k=1}^{n} \alpha_k y_k x_k$ and $0 = \sum_{k=1}^{n} \alpha_k y_k$. Substituting these identities into (4.39) yields the (Wolfe) dual optimization problem

$$L_D = \sum_{k=1}^{n} \alpha_k - \frac{1}{2} \sum_{k=1}^{n} \sum_{j=1}^{n} \alpha_k \alpha_j y_k y_j x_k^T x_j \qquad (4.40a)$$

subject to

$$\alpha_k \geq 0, \quad k = 1, 2, \ldots, n. \qquad (4.40b)$$

The optimization problem is solved via the maximization of L_D on the positive "quadrant" of $\mathbb{R}^n$ in which all of the Lagrange multipliers α_k are non-negative. Observe that (4.40a) can be written in the more abstract form as $L_D[K] = \sum_{k=1}^{n} \alpha_k - \frac{1}{2} \sum_{k=1}^{n} \sum_{j=1}^{n} \alpha_k \alpha_j K(x_k, x_j)$, where K is the linear inner product kernel $K(x, y) = x^T y$. In the case of the linear kernel K, L_D is optimized by all vectors x_k for which $\alpha_k > 0$. This occurs when $y_k(w_0^T x_k - b_0) - 1 = 0$ or when $x_k \in L_0$ is on the boundary of the indeterminate zone parallelogram. When $y_k(w_0^T x_k - b_0) - 1 > 0$, then $x_k \notin L_0$ and $\alpha_k = 0$. The vectors $x_k \in L_0$ for which $\alpha_k \neq 0$ are the *support vectors* of the classifier.

Thus far, the discussion has centered around *linearly separable* data classes. In practicality, such data are rare. What should be done when the data are interwoven? The answer is to use a *nonlinear kernel*. For *nonlinear separator boundaries*, use any of the kernels listed hereafter.

$$\textit{Polynomial of degree d: } K_d(x, y) = \left(1 + x^T y \right)^d$$
$$\textit{Exponential radial: } K_\gamma(x, y) = \exp(-\gamma \, \|x - y\|^2)$$
$$\textit{Neural network: } K_\kappa(x, y) = \tanh(\kappa_1 x^T y + \kappa_2)$$

The nonlinear support vector machine is computed via the solution of the general quadratic optimization problem

$$L_D[K] = \sum_{k=1}^{n} \alpha_k - \frac{1}{2} \sum_{k=1}^{n} \sum_{j=1}^{n} \alpha_k \alpha_j y_k y_j K(x_k, x_j) \qquad (4.41a)$$

subject to

$$\alpha_k \geq 0, \quad k = 1, 2, \ldots, n. \qquad (4.41b)$$

If the further restriction of summation to unity on the Lagrange multipliers is enforced, $\sum_{k=1}^{n} \alpha_k = 1$, then the constraints (4.41b) are replaced by

$$0 \le \alpha_k \le 1 \quad \text{for } k = 1, 2, \ldots, n. \tag{4.41c}$$

This will be the environment in which the quadratic optimization problem is approached. To solve (4.41a)/(4.41c), the *QPC* package, developed by a team headed by Professor Brett Ninness of the University of Newcastle (New South Wales, Australia), is utilized [12]. In particular, the function gpip.m, which is part of the *QPC* package, is used as the quadratic program solver.

Once the Lagrange multipliers α_k are computed with respect to a kernel K, the functional $L_D[K]$, and the constraints (4.41c), the *SVM boundary function* $F_\partial(x)$ can be computed via (4.42a).

$$F_\partial(x) = \sum_{k \in N_{sv}} \alpha_k y_k K(x, x_k) + \beta_\partial \tag{4.42a}$$

Here N_{sv} is the set of indices of all *nonzero support vectors*. Practically, this means the set of all indices in which the Lagrange multipliers α_k are greater than a prescribed (positive) tolerance τ. The constant offset β_∂ is defined via the first index in N_{sv} corresponding to the first nonzero support vector x_0. More specifically, let $i_0 = \min\{N_{sv}\}$ be the minimum index of the nonzero support vectors, $x_0 = x_{i_0}$ be the i_0th row of the data matrix X, and $y_0 = y(i_0)$ be the identification of the corresponding support vector. Then,

$$\beta_\partial = sign(y_0) - \sum_{k \in N_{sv}} \alpha_k y_k K(x_k, x_0). \tag{4.42b}$$

The derivation of equation (4.42b) is left as an exercise (see 4.13). Figures 4.19a–4.19f illustrate examples of the *SVM boundaries* for six different kernels. In each case, the *positive class* data are represented via left cloud dots • while the *negative class* data are right cloud squares ■.

The data used to develop these support vector boundaries are the *DAFE* coordinates calculated from the MAT-file Multivariate_MultiClass_Data.mat contained in the Chapter 4 data directory.

Remarks:

1. The linear, cubic, and neural network kernels produce a single dichotomous boundary. That is, the positive (dots •) class is designated on the left of the boundary while the negative class (squares ■) is on the right side. A visual inspection suggests that the cubic kernel best partitions the positive/negative space.

2. The quadratic, radial, and exponential kernels appear to divide the classification space into *three* categories: positive (dots •), negative (squares ■), and

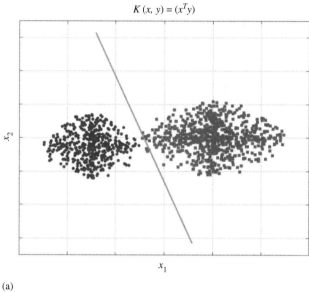

(a)

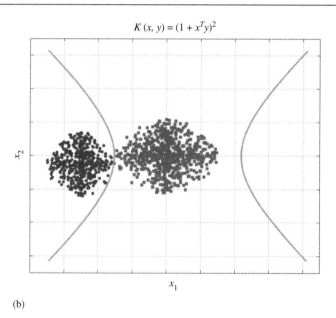

(b)

FIGURE 4.19   (a) Linear kernel. (b) Quadratic kernel. (c) Cubic kernel. (d) Radial kernel. (e) Exponential kernel. (f) Neural network kernel.

$$K(x, y) = (1 + x^T y)^3$$

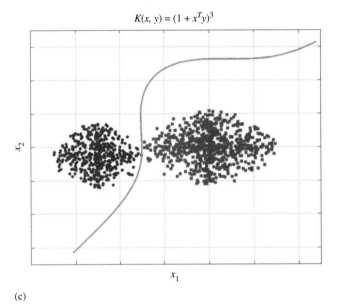

(c)

$$K(x, y) = \exp(-1/2 \cdot \|x - y\|^2)$$

(d)

FIGURE 4.19   (*Continued*)

$$K(x,y) = \exp(-1/2 \cdot \|x - y\|)$$

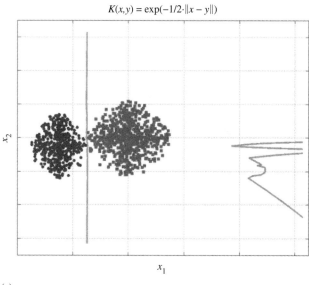

(e)

$$K(x,y) = \tanh(\kappa_1 \cdot (x^T y) + \kappa_2)$$

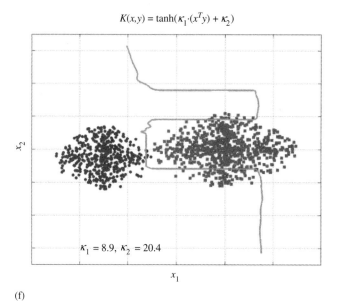

$\kappa_1 = 8.9, \; \kappa_2 = 20.4$

(f)

FIGURE 4.19 (*Continued*)

"other" (e.g., neither positive nor negative). This is readily seen in the case of the quadratic kernel that produces a hyperbola as a *SVM* boundary. The left portion of the hyperbola partitions the positive and negative classes. The right branch of the hyperbola can be viewed as a boundary in-between the negative class (to the left of the right branch) and "other" (to the right of the right branch).

3. A full *sensitivity/specificity* analysis as indicated in Section 4.8 should be undertaken on each *SVM* kernel. In particular, a *receiver–operator character* curve analysis of the $(1 - specificity, sensitivity)$ space should be conducted with respect to the parameters of the radial kernel $K_\gamma(x, y) = \exp(-\gamma \|x - y\|^2)$, the exponential kernel $K_\delta(x, y) = \exp(-\delta \|x - y\|)$, and the neural network kernel $K_\kappa(x, y) = \tanh(\kappa_1 \cdot x^T y + \kappa_2)$. For example, let $\gamma \in (0, 2)$ vary with respect to $K_\gamma(x, y)$. Calculate a sensitivity and specificity for each value of γ and plot the resulting $(1 - specificity, sensitivity)$ pairs on a *ROC* curve. Select the value of γ which returns the most desirable operating point. This is the process of *tuning* a *SVM* classifier.

4. Once a kernel has been selected and the resulting *SVM* boundary has been computed, an indeterminate band (as displayed in Figure 4.18) can be placed about the boundary. This would give rise to an indeterminate (i.e., neither positive nor negative) classification.

EXERCISES

4.12 Using the substitutions $w_0^T = \sum_{k=1}^n \alpha_k y_k x_k$ and $0 = \sum_{k=1}^n \alpha_k y_k$ in (4.39), show that L_P (the primal optimization functional) is equivalent to L_D (the dual functional) of equation (4.40a).

4.13 Derive formula (4.42b) by solving $1 = y_i F_\partial(x_i)$ for β_∂. Here the index i corresponds to any Lagrange multiplier so that $0 < \alpha_i < 1$. See Hastie et al. [7] for details.

REFERENCES

[1] Burges, C. J. C., "A tutorial on support vector machines for pattern recognition," *Data Mining and Knowledge Discovery*, vol. 2, pp. 121–167, 1998.

[2] Duda, R. O., P. E. Hart, and D. G. Stork, *Pattern Classification*, Second Edition, Wiley–Interscience, New York, 2001.

[3] Friedman, J. H., "Regularized discriminant analysis," *Journal of the American Statistical Association*, vol. 84, no. 405, pp. 165–175, March 1989.

[4] Fukunaga, K., *Statistical Pattern Recognition*, Second Edition, Morgan Kaufman (Academic Press), San Francisco, CA, 1990.

[5] Gnanadesikan, R., *Methods for Statistical Data Analysis of Multivariate Observations*, Second Edition, Wiley–Interscience, New York, 1997.

[6]   Green, D. M., and J. A. Swets, *Signal Detection Theory and Psychophysics*, John Wiley & Sons, Inc., New York, 1966.

[7]   Hastie, T., R. Tibshirani, and J. Friedman, *The Elements of Statistical Learning, Data Mining, Inference, and Prediction*, Second Edition, Springer, New York, 2009.

[8]   Johnson, R. A., and D.W. Wichern, *Applied Multivariate Statistical Analysis*, Fourth Edition, Prentice–Hall, Inc. Upper Saddle River, NJ, 1998.

[9]   Jolliffe, I. T., *Principal Component Analysis*, Second Edition, Springer, New York, 2004.

[10] Lee, C., and D. Landgrebe, "Feature extraction based on decision boundaries," *IEEE Pattern Analysis and Machine Intelligence Transactions*, vol. 15, no. 4, pp. 338–400, April 1993.

[11] Metz, C. E., "Basic Principles of ROC Analysis," *Seminars in Nuclear Medicine*, vol. III, no. 4, pp. 283–298, October 1978.

[12] Ninness, B., *QPC–Quadratic Programming in C*, University of Newcastle, New South Wales, Australia, http://sigpromu.org/quadprog/index.html

[13] Wu, W., Y. Mallet, B. Walczak, W. Pennickx, D. L. Massart, S. Heuerding, and F. Enri, "Comparison of regularized discriminant analysis, linear discriminant analysis, and quadratic analysis applied to NIR data," *Analytica Chimica Acta*, vol. 329, pp. 257–265, 1996.

[14] Zou, K. H., W. J. Hall, and D. E. Shapiro, "Smooth non–parametric receiver operator characteristic (ROC) curves for continuous diagnostic tests," *Statistics in Medicine*, vol. 16, pp. 2143–2156, 1997.

FURTHER READING

Demmel, J. W., *Applied Numerical Linear Algebra*, SIAM Books, Philadelphia, PA, 1997.

Feinstein, A. R., *Multivariate Analysis: An Introduction*, Yale University Press, New Haven, CT, 1996.

Krzanowski, W. J. and F. H. C. Marriott, *Multivariate Analysis, Part 2: Classification, Covariance Structures, and Repeated Measurements*, Arnold Press, London, 1995.

Landgrebe, D., "Information extraction principles and methods for multispectral and hyperspectral image data," in *Information Processing for Remote Sensing*, edited by C. H. Chen, World Scientific Publishing Company, River Edge, NJ, 1999.

5

BIOSTATISTICS AND HYPOTHESIS TESTING

The previous chapters have concerned mathematical models of biological processes and classification of patient status. These topics focus on deterministic, stochastic, and statistical methods to model biological behavior. An equally important aspect of mathematical biology is the notion of *biostatistics* and its crucial subtopic *hypothesis testing*. While models and classification methods can give *insight* into the behavior of a biological process, it is only through the analysis of measurements that *inferential* conclusions can be drawn. Put succinctly, models provide a notion of system evolution while statistical analysis permits an objective criterion to judge the efficacy of the model. This is accomplished via the statement and test of a *hypothesis*.

To that end, this chapter (and specifically Section 5.1) outlines the general principles of hypothesis testing. After this introduction, the mechanics of the hypothesis testing process are presented. This is done for the most popular parameters: Means, proportions, and variances. Non-parametric tests for both matched-pairs (*Wilcoxon signed-rank test*) and unpaired data (*Mann–Whitney U-test*) are also included. Furthermore, a previously unpublished and novel *exact* test for the coefficient of variation is developed in Section 5.1.

From Section 5.2 forward, this chapter resembles a statistical handbook. That is, the hypothesis test, its corresponding test statistic, associated confidence interval formulae, and sample size requirements are listed in a compact manner. Pertinent MATLAB functions are also inventoried with the hypothesis test.

This chapter then will be dedicated to *hypothesis testing* and the mathematical statistics required to properly test the "truth" of a hypothesis. The combination of

Applied Mathematics for the Analysis of Biomedical Data: Models, Methods, and MATLAB®, First Edition. Peter J. Costa.
© 2017 Peter J. Costa. Published 2017 by John Wiley & Sons, Inc.
Companion website: www.wiley.com/go/costa/appmaths_biomedical_data

statistics and hypothesis testing on biological data is commonly called *biostatistics*. All statistical models and analyses will be presented from a *frequentist* perspective. To learn about the *Bayesian* approach, see, for example, Link and Barker [19] and Bolstad [5].

5.1   HYPOTHESIS TESTING FRAMEWORK

The general idea behind *hypothesis testing* is to

    (i) measure something

   (ii) select a metric associated with the measurements

  (iii) devise a *hypothesis* about the behavior of the metric

  (iv) develop a mathematical model of the metric

   (v) apply the mathematical model to the measurements

  (vi) provide a criterion that enables a conclusion about the hypothesis

 (vii) compute estimates of the measurement in question and a range over which the estimate is most likely to occur.

This recipe is admittedly quite vague. Therefore, an example is in order. Suppose someone makes the casual remark, *basketball players are taller than hockey players*. How could such a statement be tested in a robust and scientifically acceptable manner?

First, a decision needs to be made as to what "taller" means. In aggregate? On average? Consider the data provided by the *National Hockey League* (NHL.com) and the *National Basketball Association* (NBA.com) as of 16 January 2015. By extracting and compiling the league-provided team roster data with respect to each player's position, height (in inches), weight (in pounds), and age (in years), a comprehensive set of measurements are obtained. Any player whose status was designated "injured reserve" is considered ineligible to play for the remainder of the season. Consequently, the player was *excluded* from the data collection.

This collection of data is called a *sample*. Plainly, the measurements are associated *only* with professional athletes who are active roster members as of mid-January 2015. It does not include players from previous years, players in other professional leagues, semi-professional players, college or high school players, or amateur enthusiasts. Therefore, only a subset of all hockey and basketball players are measured to test the assertion.

Each set of player data was entered into the files `Hockey_Data.xlsx` and `Basketball_Data.xlsx`, respectively. There are 684 eligible hockey players whose heights, weights, and ages are contained in the corresponding data file. Similarly, 438 basketball players were listed by the *NBA* during the same time frame.

Consequently, making a one-to-one comparison in-between hockey and basketball players is not directly possible. Therefore, an *average* height is considered.

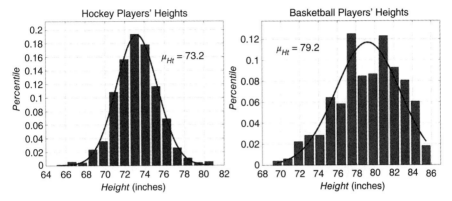

FIGURE 5.1   (a) Distribution of *NHL* heights. (b) Distribution of *NBA* heights.

Figures 5.1a and 5.1b present the sample distributions of player heights as *histograms*. Histograms display the *frequency* with which a selected measurement range occurs from a data sample. In this case, the histograms show the frequency of a given height with respect to a selected player sample. For Figure 5.1a, it can be seen that approximately 16% (0.16) of the *NHL* players are in the 72-inch height bin, while about 19% are at 73 inches. Basketball players appear to have two prominent peaks at 77 inches (12%) and 81 inches (12%). The well-known *bell-shaped curve* or *normal distribution* is superimposed on top of the histograms. As will be seen later in this section, the mathematical formulation of the *normal distribution* can be determined exclusively from the mean (μ) and variance (σ^2) of the random variable.

The average height of a year 2015 *NHL* player is 72.3 inches while the average *NBA* player is 79.2 inches. Is it scientifically valid to conclude that the average *NBA* player is taller than the average *NHL* player? The answer is to subject the data to a hypothesis test.

Using the hypothesis test framework outlined earlier, construct the list of steps.

1. *Measure something.* In this case, it is evident that the *average player height* is the measurement.
2. *Select a metric.* This is as much art as science. The average heights of the players from the various leagues are the starting point. Therefore, the *difference in averages* $\Delta\mu = \mu_1 - \mu_2$ seems a reasonable metric. Notice, however, that $\Delta\mu$ is measured in units of inches. To make the metric dimensionless (i.e., unitless), divide by a normalizing factor which is also in units of inches. The square root of the average variance (or *pooled standard deviation*) $\sqrt{\frac{1}{2}(\sigma_1^2 + \sigma_2^2)}$ is also in units of inches. Thus, the metric $\Delta\mu / \sqrt{\frac{1}{2}(\sigma_1^2 + \sigma_2^2)}$ is a dimensionless measurement. Since the actual means and standard deviations of the heights of professional hockey and basketball players cannot be known, the *sample means*

and *sample variances* defined in (5.1a) and (5.1b) will be used.

$$\bar{x} = \frac{1}{n} \sum_{j=1}^{n} x_j \tag{5.1a}$$

$$s^2 = \frac{1}{n-1} \sum_{j=1}^{n} (x_j - \bar{x})^2 \tag{5.1b}$$

As there are more hockey players (684) than basketball players (438), the sample variances are weighted inversely by the number of samples. Subsequently, the *test statistic* will be $T(X, Y) = \dfrac{\bar{x} - \bar{y}}{\sqrt{\frac{1}{n_x} s_x^2 + \frac{1}{n_y} s_y^2}}$. Here $X = \{x_1, x_2, \ldots, x_{n_1}\}$ and $Y = \{y_1, y_2, \ldots, y_{n_2}\}$ are samples of the professional *NHL* hockey (X) and *NBA* basketball (Y) players collected from the data posted by their respective leagues for the 2014–2015 season. In addition, n_1 and n_2 are the number of elements in each sample X and Y, respectively. In the case of the *NHL* players, $n_1 = 684$ while for the *NBA*, $n_2 = 438$.

3. *Devise a hypothesis.* The hypothesis to test is whether the average *NBA* height μ_B is greater than the average *NHL* height μ_H. The convention in hypothesis testing is to *reject the null hypothesis in favor of the alternate hypothesis*. Therefore, if the belief is that basketball players are taller than hockey players, the null and alternate hypotheses are as below.

$$H_0: \mu_B \leq \mu_H \qquad \qquad \textit{(Null hypothesis)}$$
$$H_1: \mu_B > \mu_H \qquad \qquad \textit{(Alternate hypothesis)}$$

As a point of emphasis, if the null hypothesis H_0 is *rejected in favor of the alternative hypothesis H_1*, then it can be stated that *the average height of professional basketball (NBA) players is statistically significantly greater than the average height of professional hockey (NHL) players.*

What does this word *significantly* mean? This refers to the *Type I error* or *significance level* of the hypothesis test. The *Type I error* occurs when the null hypothesis is rejected when, in fact, the null hypothesis is true. The probability of a Type I error or *significance level* is represented by the symbol $\alpha = $ Prob(Type I error). Typically, the significance level is set at 5% ($\alpha = 0.05$). The US government review agencies have established a 5% significance level as the baseline value. Thus, something is *statistically significant* at the prescribed *significance level.*

4. *Develop a mathematical model.* Since the *test statistic* is $T(X, Y) = \dfrac{\bar{x} - \bar{y}}{\sqrt{\frac{1}{n_x} s_x^2 + \frac{1}{n_y} s_y^2}}$, this can be modeled as the ratio of a normal random variable and a Chi-square random variable. More specifically, if $X_1, X_2, \ldots, X_n$ are independent and identically distributed normal random variables $X_j \sim \mathcal{N}(\mu, \sigma^2)$, then the *sample mean* $\overline{X} = \frac{1}{n_x} \sum_{j=1}^{n_x} X_j$ is a normal random variable

$\overline{X} \sim \mathcal{N}(\mu_x, \frac{1}{n_x}\sigma_x^2)$. Similarly, the sample mean from Y, $\overline{Y} = \frac{1}{n_y}\sum_{j=1}^{n_y} Y_j$ is normally distributed $\overline{Y} \sim \mathcal{N}(\mu_y, \frac{1}{n_y}\sigma_y^2)$. The difference of independent normally distributed random variables is also normal. Thus, $\overline{X} - \overline{Y} \sim \mathcal{N}(\mu_x - \mu_y, \frac{1}{n_x}\sigma_x^2 + \frac{1}{n_y}\sigma_y^2)$. The denominator of $T(X, Y)$, however, is the square root of the sum of squares of normally distributed random variables. This means that the measure $\frac{1}{n_x}s_x^2 + \frac{1}{n_y}s_y^2$ is modeled as the weighted sum of Chi-square random variables. Therefore, $T(X, Y)$ is now the ratio of a normally distributed random variable $\overline{x} - \overline{y}$ and the weighted sum of χ^2 random variables $\sqrt{\frac{1}{n_x}s_X^2 + \frac{1}{n_y}s_Y^2}$. The ratio $t = \dfrac{Z}{\sqrt{\frac{1}{df}W}}$, where $Z \sim \mathcal{N}(0, 1)$ and $W \sim \chi_{df}^2$ has a *student t-distribution*,

$t \sim \mathcal{T}_{df}$. The *probability density function (pdf)* $\phi_T(t; df)$ is defined via (5.2) while the *cumulative density function (cdf)* defines the probability of a $\mathcal{T}_{df}$-distributed random variable $\Phi_T(t; df) = \text{Prob}(T \leq t)$.

$$\phi_T(t; df) = \frac{\Gamma\left(\frac{1}{2}(df + 1)\right)}{\sqrt{\pi \cdot df} \cdot \Gamma\left(\frac{1}{2}df\right)} \cdot \frac{1}{\left(1 + \frac{1}{df}t^2\right)^{(df+1)/2}} \quad \text{for } -\infty < t < \infty \quad (5.2)$$

Here $\Gamma(t) = \int_0^\infty x^{t-1}e^{-x}\,dx$ is the *gamma function*; see Abramowitz and Stegun [1] or Olver et al. [29] for more detail. The *cdf* is merely the integral of the *pdf*, $\Phi_T(t; df) = \int_{-\infty}^t \phi_T(u; df)\,du$. This is a fundamental result of mathematical statistics, and the reader is referred to any of the references such as Hoel et al. [13], Hogg and Craig [14], and Bickel and Doksum [6].

5. To apply *test statistic* to the data: $T(X, Y) = 33.014$.

6. *Hypothesis decision criterion.* To determine whether the null hypothesis should be rejected, the *p*-value or *significance probability* of the test is calculated. The *p*-value is the *observed* or *sample significance level*. If $p < \alpha$ (the *a priori significance level*), then the null hypothesis is *rejected*. In this case, the *p*-value is computed via the cumulative density function (*cdf*) of the (right-tailed) test statistic: $p = \Phi_T(T; df)$, where Φ_T is the *cdf* of the student *t*-distribution with df degrees of freedom. In this instance, $df = 645$ and $p = \Phi_T(33.014; 645) < 10^{-4} < 0.05 = \alpha$. Therefore, the null hypothesis of the average height of a hockey player being larger than the average height of a basketball player is *rejected* (at the 5% significance level) in favor of the alternate hypothesis. Put succinctly, $p < 10^{-4} < \alpha \Rightarrow$ reject H_0: $\mu_B \leq \mu_H$ in favor of H_1: $\mu_B > \mu_H$. The hypothesis test is called *right-tailed* since the right most tail of the distribution determines the null hypothesis decision (i.e., to *reject* or *not to reject*).

7. *Compute estimates.* The average heights of the *NHL* and *NBA* players are, respectively, $\mu_H = 73.2$ inches and $\mu_B = 79.2$ inches. How precise are these averages?

As noted earlier, these estimates come from active players on team rosters in the 2014–2015 season. Since these data represent a *sample* of all such players from their respective sports, it is sage to present a *range of values* over which the probability that the measurement exists is relatively high.

This idea gives way to the notion of *confidence interval*. More specifically, the $(1 - \alpha)\cdot 100\%$ confidence interval for a parameter θ is a function of its estimate $\hat{\theta}$, the number of samples N_s used to calculate the estimate, the estimated *standard error*, and the *quantiles* of the distribution which models the parameter.

Suppose the ratio of difference of the actual to estimated parameter $\theta - \hat{\theta}$ to its standard error $SE(\hat{\theta})$ has a distribution with *cumulative density function* $\Phi_\theta(x)$. Then there are numbers a and b with $\mathrm{Prob}(a \le (\theta - \hat{\theta})/SE(\hat{\theta}) \le b) = 1 - \alpha$, where $\alpha \in (0, 1)$ is the designated *significance*. The numbers $a = \omega_\theta(\alpha/2)$ and $b = \omega_\theta(1 - \alpha/2)$ are referred to as the $(a/2)\cdot 100\%$ and $(1 - a/2)\cdot 100\%$ *quantiles* of the *probability density function* ϕ_θ. Following the algebra results in

$$
\begin{aligned}
1 - \alpha &= \mathrm{Prob}(\omega_\theta(\alpha/2) \le (\theta - \hat{\theta})/SE(\hat{\theta}) \le \omega_\theta(1 - \alpha/2)) \\
&= \mathrm{Prob}(\omega_\theta(\alpha/2)\cdot SE(\hat{\theta}) \le (\theta - \hat{\theta}) \le \omega_\theta(1 - \alpha/2)\cdot SE(\hat{\theta})) \\
&= \mathrm{Prob}(\hat{\theta} + \omega_\theta(\alpha/2)\cdot SE(\hat{\theta}) \le \theta \le \hat{\theta} + \omega_\theta(1 - \alpha/2)\cdot SE(\hat{\theta}))
\end{aligned}
$$

Equivalently, $CI(\theta) = [\hat{\theta} + \omega_\theta(\alpha/2)\cdot SE(\hat{\theta}), \hat{\theta} + \omega_\theta(1 - \alpha/2)\cdot SE(\hat{\theta})]$ is a $(1 - a)\cdot 100\%$ *confidence interval* for the parameter θ.

In the case of the difference of means $\Delta\mu$, the test statistic is modeled via a *t*-distribution with $df = 645$ degrees of freedom. Hence, $1 - \alpha = $

$$
\mathrm{Prob}\left(t_{df}(\alpha/2) \le (\Delta\mu - (\bar{x} - \bar{y}))/\sqrt{\frac{1}{n_x}s_x^2 + \frac{1}{n_y}s_y^2} \le t_{df}(1 - \alpha/2)\right) = \mathrm{Prob}\left((\bar{x} - \bar{y}) + \right.
$$

$$
t_{df}(\alpha/2)\cdot\sqrt{\frac{1}{n_x}s_x^2 + \frac{1}{n_y}s_y^2} \le \Delta\mu \le (\bar{x} - \bar{y}) + t_{df}(1 - \alpha/2)\cdot\sqrt{\frac{1}{n_x}s_x^2 + \frac{1}{n_y}s_y^2}\right) \text{ so that}
$$

$$
CI(\Delta\mu) = \left[(\bar{x} - \bar{y}) + t_{df}(\alpha/2)\cdot\sqrt{\frac{1}{n_x}s_x^2 + \frac{1}{n_y}s_y^2}, (\bar{x} - \bar{y}) + t_{df}(1 - \alpha/2)\cdot\sqrt{\frac{1}{n_x}s_x^2 + \frac{1}{n_y}s_y^2}\right]
$$

is a $(1 - \alpha)\cdot 100\%$ confidence interval for the difference in means $\Delta\mu = \mu_1 - \mu_2$. The quantiles $t_{df}(\alpha/2)$ and $t_{df}(1 - \alpha/2)$ are illustrated in Figure 5.2 (here $\alpha = 0.05$). For an individual population mean for the random variable X, the $(1 - \alpha)\cdot 100\%$ interval is

$$
CI(\mu) = \left[\bar{x} + t_{df}(\alpha/2)\cdot\frac{s_x}{\sqrt{n_x}}, \bar{x} + t_{df}(1 - \alpha/2)\cdot\frac{s_x}{\sqrt{n_x}}\right]
$$

A similar interval is constructed for the random variable Y. With respect to the *NHL/NBA* height data, Table 5.1 contains the estimates of the mean heights, the difference in means, the 95% confidence intervals for each mean height estimate, and the 95% confidence intervals in the difference of mean heights. Notice that the confidence level is 95% because the significance level is 5% (i.e., $\alpha = 0.05$).

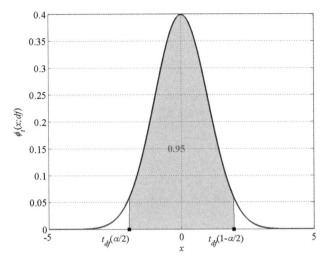

FIGURE 5.2 $\mathcal{T}$-Distribution with 2.5% and 97.5% quantiles.

This development of estimates, confidence intervals, and hypothesis testing is presented in an informal manner. Bickel and Doksum [6], Lehmann [16], and Pestman [33] give more rigorous and detailed explanations of *Neyman–Pearson* theory and hypothesis tests. The purpose of the presentation here is to expose the reader to the basic ideas of hypothesis testing. In what follows, a condensed summary of various hypothesis tests and the associated mathematical computations will be delineated.

More formally, the hypothesis test process is prescribed as below.

1.1. Measure a parameter θ. For example, the height of a professional athlete.

1.2. Develop a mathematical model that yields a measure of parameter behavior. Again, are the heights of basketball players greater than the heights of hockey players? This model is referred to as the *test statistic T*. Determine the *probability density function* $\phi_T(x)$ of the test statistic.

1.3. Establish a set of hypotheses that test the behavior in question. If the hypothesis is that the parameter from one population is larger than the same parameter from another population, the null hypothesis can be H_0: $\theta_1 \leq \theta_2$. Generally, to support a claim, it is desirable to *reject the null hypothesis H_0 in favor of the alternate hypothesis H_1*.

TABLE 5.1   **Estimates of mean player heights and differences**

Sample source	Mean estimate	95% confidence interval
NHL	$\hat{\mu}_H = 73.2$	$CI(\mu_H) = [73.06, 73.37]$
NBA	$\hat{\mu}_B = 79.2$	$CI(\mu_B) = [78.89, 79.53]$
	Difference	
NHL–NBA	$\hat{\mu}_H - \hat{\mu}_B = -6$	$CI(\mu_H - \mu_B) = [-6.32, -5.67]$

1.4. The phrase "test of equality" for the hypotheses H_0: $\theta_1 = \theta_2$ vs. H_1: $\theta_1 \neq \theta_2$ can be misunderstood. To *reject* H_0 permits the conclusion that there is a statistically significant difference in-between the two parameters θ_1 and θ_2. If the null H_0 *is not rejected* then *there is insufficient evidence of a statistically significant difference* in-between θ_1 and θ_2.

1.5. Compute the value of the test statistic at the measured data. Apply the mathematical model (i.e., the *pdf*) to calculate the *p*-value or significance probability. If the *p*-value is less than the significance level (α), then *reject the null hypothesis in favor of the alternate hypothesis*: $p < \alpha \Rightarrow$ *reject H_0 in favor of H_1*.

1.6. Using the mathematical model of the parameter distribution $\phi_\theta(x)$, provide a $(1 - \alpha) \cdot 100\%$ *confidence interval* for the estimate.

$$CI(\widehat{\theta}) = [\widehat{\theta} - \omega(1 - \alpha/2) \cdot SE(\widehat{\theta}), \widehat{\theta} + \omega(1 - \alpha/2) \cdot SE(\widehat{\theta})]$$

where $\omega(1 - \alpha/2)$ is the $(1 - \alpha/2) \cdot 100\%$ quantile of the distribution with *pdf* $\phi_\theta(x)$.

It should be emphasized that steps 1.1–1.6 are *guidelines*. They reflect the *spirit* of the Neyman–Pearson framework. As will be seen, the formulae associated with each distinct hypothesis test differ from those intimated in 1.1–1.6.

Moreover, the steps mentioned earlier are carried out only *after* data have been collected. The old adage of carpenters *measure twice, cut once* should be heeded before data collection is initiated. Indeed, Chapter 1 discusses the intricacies and subtleties involved in data collection. Such preparations come under the banner of *design of experiment*. In this milieu, the questions become: *What is being measured? How many samples are required so that the conclusions of the study are significant?* To answer these queries, some notation must be established.

NOMENCLATURE

X	A random variable
$\phi_X(x), x \in \Omega_X$	Probability density function (*pdf*) for X
Ω_X	Domain of definition for X
$\Phi_X(x)$	Cumulative density function (*cdf*) for X
$\phi(x), x \in \mathbb{R}$	*pdf* for the standard normal distribution $\mathcal{N}(0, 1)$
$\Phi(x)$	*cdf* for the standard normal distribution $\mathcal{N}(0, 1)$
z_γ	$\gamma \cdot 100\%$ quantile of the standard normal distribution $\mathcal{N}(0, 1)$
$\phi_F(x; \boldsymbol{df})$	*pdf* for an F-distributed random variable with $\boldsymbol{df} = [df_1, df_2]$ degrees of freedom
$\Phi_F(x; \boldsymbol{df})$	*cdf* for an F-distributed random variable with $\boldsymbol{df}$ degrees of freedom
$f_{df}(\gamma)$	$\gamma \cdot 100\%$ quantile of the F-distribution with $\boldsymbol{df}$ degrees of freedom

$\phi_T(x; df)$	*pdf* for a T-distributed random variable with *df* degrees of freedom
$\Phi_T(x; df)$	*cdf* for a T-distributed random variable with *df* degrees of freedom
$t_{df}(\gamma)$	$\gamma \bullet 100\%$ quantile of the T-distribution with *df* degrees of freedom
$\phi_{\chi^2}(x; df)$	*pdf* for a χ^2-distributed random variable with *df* degrees of freedom
$\Phi_{\chi^2}(x; df)$	*cdf* for a χ^2-distributed random variable with *df* degrees of freedom
$\chi^2_{df}(\gamma)$	$\gamma \bullet 100\%$ quantile of the χ^2-distribution with *df* degrees of freedom
$X = \{x_1, x_2, \ldots, x_n\}$	A sample of a random variable
n	The sample size
H_0, H_1	The null and alternate hypotheses
Type I error	To reject H_0 when H_0 is true
Type II error	To fail to reject H_0 when H_0 is false
$\alpha = Prob(\text{Type I error})$	The significance level of a hypothesis test
$\beta = Prob(\text{Type II error})$	The complement of power
$pwr = 1 - \beta$	The power of a hypothesis test
es	Effect size

The *probability density function* $\phi(x)$ and the associated *cumulative density function* $\Phi(x)$ for the *standard normal* distribution $\mathcal{N}(0, 1)$ are defined via the following formulae.

$$\phi(x) = \frac{1}{\sqrt{2\pi}} e^{-\frac{1}{2}x^2}, x \in \mathbb{R} \tag{5.3a}$$

$$\Phi(x) = \text{Prob}(Z \le x) = \int_{-\infty}^{x} \phi(t)\, dt, x \in \mathbb{R}, Z \sim \mathcal{N}(0, 1) \tag{5.3b}$$

Note that the *cdf* of an $\mathcal{N}(0, 1)$ random variable can be written in terms of the *error function erf*. That is, $\Phi(x) = \frac{1}{2}[1 + erf(x/\sqrt{2})]$, where $erf(x) = \frac{1}{\sqrt{\pi}} \int_{-x}^{x} \phi(t)\, dt$. See Abramowitz and Stegun [1] for details.

The *power (pwr)* of a hypothesis test is then the complement of the Type II error. That is, $pwr = 1 - \text{Prob}(\text{do not reject } H_0 | H_0 \text{ is false}) = \text{Prob}(\text{reject } H_0 | H_0 \text{ is false})$. Power, then, is the probability of correctly rejecting the null hypothesis in favor of the alternate hypothesis. The effect size es of a test is the "accuracy" by which a parameter can be measured. For example, if a manufacturer advertised a scale that reflects the weight of the person standing on it to within 2 ounces, then the *effect size* of the scale is ± 0.125 pounds. Given that adult humans weigh in-between 100 and 300 pounds, such an effect size is considered small. Such a scale reading 105.5 could be used to assume a true weight of 105.375 to 105.625 pounds. An effect size of 20 pounds, however, would render the same measurement in-between 85 and 125 pounds. Not a very accurate scale.

Thus, when deciding how many samples should be used to test a hypothesis, three items need to be balanced: Significance level or Type I error, power, and effect size. In the proceeding sections, the relationship in-between these competing factors will yield formulae for the required sample sizes as a function of the hypothesis test.

To get an idea of the interplay of significance α, power $1 - \beta$, and effect size es, consider once again the example of professional hockey and basketball player heights. Suppose now a test of *equal average height* is proposed. The hypotheses are

$$H_0: \mu_H = \mu_B \qquad \textit{(Null hypothesis)}$$
$$H_1: \mu_H \neq \mu_B \qquad \textit{(Alternate hypothesis)}$$

For simplicity's sake, assume that the variance of these populations (e.g., σ_B^2 basketball and σ_H^2 hockey players) is known. Further, suppose the two samples are of equal size and taken from normal populations. In this case, $X = \{x_1, x_2, ..., x_n\}$ and $Y = \{y_1, y_2, ..., y_n\}$ are independent and identically distributed; $x_j \sim \mathcal{N}(\mu_H, \sigma_H^2)$ and $y_k \sim \mathcal{N}(\mu_H, \sigma_B^2)$ for all j and k. Therefore, the means of the samples are also normally distributed. That is, $\overline{X} \sim \mathcal{N}(\mu_H, \frac{1}{n}\sigma_H^2)$ and $\overline{Y} \sim \mathcal{N}(\mu_B, \frac{1}{n}\sigma_B^2)$. The *test statistic* for these hypotheses is $T(X, Y) = \dfrac{\overline{x} - \overline{y}}{\sqrt{\frac{1}{n}(\sigma_H^2 + \sigma_B^2)}}$, where $\overline{x}$ and $\overline{y}$ are the means of the samples X and Y, respectively. This random variable is distributed as $\mathcal{N}(\mu_H - \mu_B, 1)$. Under the null hypothesis H_0, the difference in means is zero, $\mu_H - \mu_B = 0$. Thus, the test statistic can be modeled as a standard normal random variable, $T \sim \mathcal{N}(0, 1)$. The alternate hypothesis implies that the difference in the means is *not* zero. Without loss of generality, assume that this difference is a positive number, $\mu_H - \mu_B = \varepsilon > 0$. Then, the difference in sample means will also, for sufficiently large sample size n, be ε. Hence, the test statistic $T(X, Y)$ is distributed as a non-central normal random variable with mean $\mu(\varepsilon) = \dfrac{\varepsilon}{\sqrt{\frac{1}{n}(\sigma_H^2 + \sigma_B^2)}}$, $T \sim \mathcal{N}(\mu(\varepsilon), 1)$.

The null hypothesis is rejected if the value of the test statistic is *outside* of the $(1 - \alpha/2) \cdot 100\%$ quantiles as illustrated in Figure 5.3.

Therefore, the null hypothesis is *rejected* when the random variable T is in the far right or far left tails of the distribution; i.e., to the left of the $(\alpha/2) \cdot 100\%$ quantile $z_{\alpha/2} = -z_{1-\alpha/2}$ or to the right of the $(1 - \alpha/2) \cdot 100\%$ quantile $z_{1-\alpha/2}$. The *power* of a hypothesis is the probability of *rejecting* H_0 when H_1 is true. Mathematically, $pwr = 1 - \beta = \text{Prob}(|T| > z_{1-\alpha/2})$.

There are two cases for the power probability, namely $T > z_{1-\alpha/2}$ and $T < z_{\alpha/2}$. Under H_1, $T \sim \mathcal{N}(\mu(\varepsilon), 1)$ so that $Z = T - \mu(\varepsilon) \sim \mathcal{N}(0, 1)$. For $T > z_{1-\alpha/2}$, the power calculation is

$$
\begin{aligned}
1 - \beta &= \text{Prob}(T > z_{1-\alpha/2}) = \text{Prob}(T - \mu(\varepsilon) > z_{1-\alpha/2} - \mu(\varepsilon)) \\
&= \text{Prob}(Z > z_{1-\alpha/2} - \mu(\varepsilon)) = 1 - \text{Prob}(Z < z_{1-\alpha/2} - \mu(\varepsilon)) \\
&= 1 - \Phi(z_{1-\alpha/2} - \mu(\varepsilon))
\end{aligned}
$$

Standard Normal Quantiles

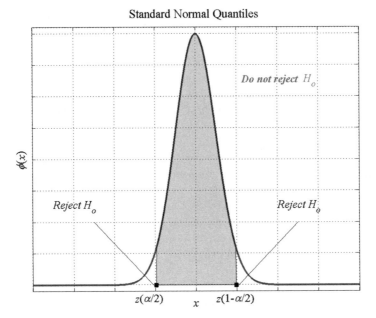

FIGURE 5.3   Regions to *reject* and *not reject* H_0.

Simplifying this equation produces $\beta = \Phi(z_{1-\alpha/2} - \mu(\varepsilon))$. Since $z_\beta = \Phi^{-1}(\beta)$, the equation reduces to $z_{1-\alpha/2} - z_\beta = \mu(\varepsilon)$. Recall that $-z_\beta = z_{1-\beta}$ and $\mu(\varepsilon) = \varepsilon/\sqrt{\frac{1}{n}(\sigma_H^2 + \sigma_B^2)}$. Subsequently,

$$z_{1-\alpha/2} + z_{1-\beta} = \frac{\varepsilon}{\sqrt{\frac{1}{n}\left(\sigma_H^2 + \sigma_B^2\right)}} \Leftrightarrow n^+ = \left[\!\left[\frac{\left(\sigma_H^2 + \sigma_B^2\right) \cdot (z_{1-\alpha/2} + z_{1-\beta})^2}{\varepsilon^2}\right]\!\right] \tag{5.4a}$$

where $[\![m]\!]$ is the *nearest integer function*. The sample size n is plainly a function of *effect size* ε, the significance level α, and the power $1 - \beta$.

For the second case of $T < z_{\alpha/2}$, the power equation $1 - \beta = \text{Prob}(T < z_{\alpha/2}) = \text{Prob}(Z < z_{\alpha/2} - \mu(\varepsilon)) = \Phi(z_{\alpha/2} - \mu(\varepsilon))$ gives way to (5.4b).

$$n^- = \left[\!\left[\frac{\left(\sigma_H^2 + \sigma_B^2\right) \cdot (z_{1-\alpha/2} + z_{1-\beta})^2}{\varepsilon^2}\right]\!\right] \tag{5.4b}$$

Since the full power equation is $1 - \beta = \text{Prob}(|T| > z_{1-\alpha/2}) = \text{Prob}(T > z_{1-\alpha/2}) + \text{Prob}(T < z_{\alpha/2})$, the total required sample size is

$$N = 2 \left[\!\left[\frac{\left(\sigma_H^2 + \sigma_B^2\right) \cdot (z_{1-\alpha/2} + z_{1-\beta})^2}{\varepsilon^2}\right]\!\right] \tag{5.5}$$

Since the sample populations are the same size, $n_1 = n_2 = n$, each sample population must sum to N. That is, $n_1 + n_2 = 2n = N$.

If, as in the case of the *NHL* and *NBA* players, the sample populations are *not* equal (e.g., $n_1 \neq n_2$), then the partition of the total sample size N becomes $n_1 = N/(1 + r_{12})$ and $n_2 = r_{12} \bullet N/(1 + r_{12})$, where $r_{12} = n_2/n_1$.

The sample size calculations for the one-sided hypothesis tests H_0: $\mu_H > \mu_B$ vs. H_1: $\mu_H \leq \mu_B$ and H_0: $\mu_H < \mu_B$ vs. H_1: $\mu_H \geq \mu_B$ follow directly from the cases above with $\alpha/2$ replaced by α.

Rather than testing the *equality* of two population means, however, it can be enough to determine whether the means are *equivalent*. To test whether two population means μ_1 and μ_2 are equal, the null hypothesis H_0: $\mu_1 = \mu_2$ is exercised. While failing to reject the null hypothesis is not *a priori* an error, the preferred Neyman–Pearson process is to *reject* the null hypothesis H_0 in favor of the alternative hypothesis H_1. Indeed, failing to reject H_0 means that there is insufficient evidence to *reject* H_0 *in favor of* H_1. Equivalence, conversely, indicates that the population means are "close enough" within an accepted margin.

What does *statistical equivalence* mean, and how does it differ from *statistical equality*? One way to view the statistical equality decision with respect to the null hypothesis H_0: $\mu_1 = \mu_2$ is as follows. If zero is contained in the confidence interval for the difference in means, the null hypothesis is *not* rejected: $0 \in CI(\Delta\mu) \Rightarrow do\ not$ *reject* H_0. If n is taken to be sufficiently large, however, the confidence interval will eventually be entirely on one side of zero. For example, see Figure 5.4.

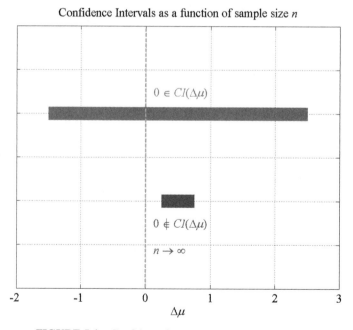

FIGURE 5.4   Confidence intervals and statistical equality.

In the case of statistical equality of means, the confidence interval of the difference $CI(\Delta\mu)$ is calculated from the samples $X = \{x_1, x_2, \ldots, x_{n_1}\}$, $Y = \{y_1, y_2, \ldots, y_{n_2}\}$. Without loss of generality, suppose $n_1 = n_2 = n$. Under this assumption, the confidence interval for the difference in means is $CI(\Delta\mu) = \left[\Delta\hat{\mu} + t_{df}(\alpha/2) \bullet \sqrt{\frac{1}{n} S^2_{pool}}, \right.$
$\left. \Delta\hat{\mu} + t_{df}(1 - \alpha/2) \bullet \sqrt{\frac{1}{n} S^2_{pool}} \right]$ so that the width of the interval is $2 \bullet t_{df}(1 - \alpha/2) \bullet$
$\sqrt{\frac{1}{n} S^2_{pool}}$, $S^2_{pool} = \frac{1}{2}(s_x^2 + s_x^2)$. Then, as n increases, the width of the confidence interval *decreases*. Indeed for well-behaved distributions, the sample population variances s_x^2 and s_y^2 will converge to their true variances σ_x^2 and σ_y^2, respectively. Thus, as $n \to \infty$, $S^2_{pool} \to \frac{1}{2}(\sigma_x^2 + \sigma_y^2)$ and hence, $CI(\Delta\mu) \to \Delta\hat{\mu}$.

For modest sample size n, as indicated by the appropriate formula earlier, it is easy to imagine a case in which the estimated difference in the means is [say] 0.5 and that the corresponding confidence interval ranges over $[-1.5, 2.5]$ (illustrated in Figure 5.4, top bar). For very large n, however, it may well be the case that the confidence interval *does not contain* 0 (lower bar). From the point of view of Altman et al. [3], this would imply that the null hypothesis H_0 *should be rejected*. This, in turn, means that *more samples* are undesirable for experimenters attempting to demonstrate statistical equality.

This is an untenable position. The *Central Limit Theorem* and the *Strong Law of Large Numbers* both confirm that *more* samples lead to the convergence of an estimate to the actual distribution mean. Thus, another approach is indicated: Statistical equivalence. Colloquially, this method requires a constant difference δ in means. Moreover, this difference (δ) is "small enough" to be clinically negligible. This constant is referred to as the clinical margin. It is the *judgment* of clinical practitioners as to what "close enough" means. A test of *equivalence of means* examines the hypotheses

$$H_0: |\mu_1 - \mu_2| \geq \delta \qquad \qquad \text{(Null hypothesis)}$$
$$H_1: |\mu_1 - \mu_2| < \delta \qquad \qquad \text{(Alternate hypothesis)}$$

The two means are *equivalent to within clinical margin δ* provided the null hypothesis H_0 *is rejected in favor of the alternate hypothesis H_1*.

Figure 5.5 illustrates the difference in-between statistical tests of equality and equivalence. The middle bar ($-$) depicts the region in which the confidence interval for the difference of means $CI(\Delta\mu)$ must be restricted to demonstrate statistically significant equality or equivalence, respectively. The interval bounded by w (on the left portion of the figure) contains the positive and negative extents of the confidence interval for the difference of means $CI(\Delta\mu) \subset [-w, w]$.

The determination of the clinical margin is a non-trivial matter. Depending on the parameter being estimated, it is the difference in which *clinical practice is not altered*. For example, if a diagnostic device reported a *sensitivity* of $90\% \pm 3\%$, then a reasonably clinical margin is anything less than 3%. There can be other criteria for establishing the clinical margin including historical data, outside studies, risk assessment, and medical judgment. The mathematics of *equivalence testing* alone will be presented in the subsequent sections.

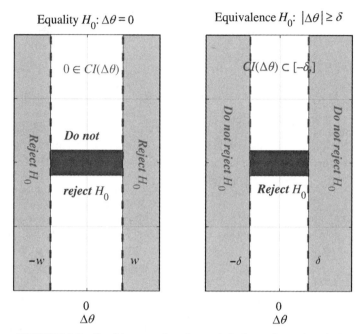

FIGURE 5.5   Confidence regions for statistical equality and equivalence.

In a comparable manner, the notions of *non-inferiority* and *superiority* can be established. Rather than examining population means, suppose that θ_1 and θ_2 are values of the same parameter (i.e., mean, proportion, variance, etc.) across two distinct populations. Further let θ_1 represent the established or *Control* population while θ_2 represents the *Test* population. Then θ_2 is (statistically significantly) at least as large as θ_1 provided $H_0: \theta_1 > \theta_2$ is *rejected in favor of* $H_1: \theta_1 \leq \theta_2$. This is known as a *left-tailed test.* Can the notion of *not worse than* or *substantially better than* be formalized? The answer is *yes* by considering the two tests below.

$$H_0(\delta): \theta_2 - \theta_1 \leq -\delta \qquad \text{(Null hypothesis)}$$
$$H_1(\delta): \theta_2 - \theta_1 > -\delta \qquad \text{(Alternate hypothesis)}$$

$$K_0(\delta): \theta_2 - \theta_1 \leq \delta \qquad \text{(Null hypothesis)}$$
$$K_1(\delta): \theta_2 - \theta_1 > \delta \qquad \text{(Alternate hypothesis)}$$

The hypotheses $H_0(\delta)$ vs. $H_1(\delta)$ is a *test of non-inferiority* whereas $K_0(\delta)$ vs. $K_1(\delta)$ is a *test of superiority.* To *reject* $H_0(\delta)$ *in favor of* $H_1(\delta)$ is to show that θ_2 is *no worse than* θ_1 by the margin δ. Hence θ_2, when supplemented by the margin δ, is at least as large as θ_1 and thereby *not inferior.* It is not difficult to see that the *non-inferiority* and *superiority* hypothesis tests become the *right-tailed tests* when $\delta = 0$.

TABLE 5.2   Hypothesis tests and null hypothesis criteria

Test	Hypotheses	Action	Confidence condition
Equality	$H_0: \Delta\theta = 0$ vs. $H_1: \Delta\theta \neq 0$	Do not reject H_0	$0 \in CI(\Delta\theta)$
Equivalence	$H_0: |\Delta\theta| \geq \delta$ vs. $H_1: |\Delta\theta| < \delta$	Reject H_0	$CI(\Delta\theta) \subset [-\delta, \delta]$
Left-tailed	$H_0: \Delta\theta \geq 0$ vs. $H_1: \Delta\theta < 0$	Reject H_0	$max[CI(\Delta\theta)] < 0$
Non-inferiority	$H_0: \Delta\theta \geq -\delta$ vs. $H_1: \Delta\theta > -\delta$	Reject H_0	$min[CI(\Delta\theta)] > -\delta$
Right-tailed	$H_0: \Delta\theta \leq 0$ vs. $H_1: \Delta\theta > 0$	Reject H_0	$min[CI(\Delta\theta)] > 0$
Superiority	$K_0: \Delta\theta \leq \delta$ vs. $K_1: \Delta\theta > \delta$	Reject K_0	$max[CI(\Delta\theta)] > \delta$

Table 5.2 summarizes these hypothesis tests using the notation $\Delta\theta = \theta_2 - \theta_1$ for the difference in parameters throughout. Figures 5.6 and 5.7 help illustrate the ideas of *right-tailed* and *left-tailed* tests along with *superiority* and *non-inferiority* tests. The conditions on the confidence intervals of the difference in parameters $CI(\Delta\theta)$ are listed as criteria for the required null hypothesis decision.

The remainder of this chapter will be delineated by specific hypothesis tests. The format will be to list the components of each formal test as follows:

Hypothesis test

Test statistic and distribution

p-Value formula

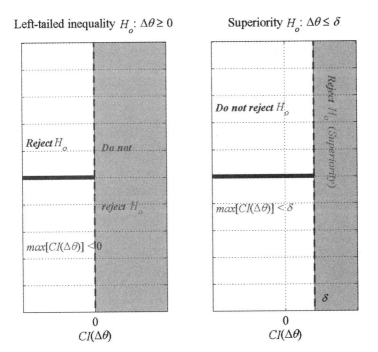

FIGURE 5.6   Left-tailed and superiority tests.

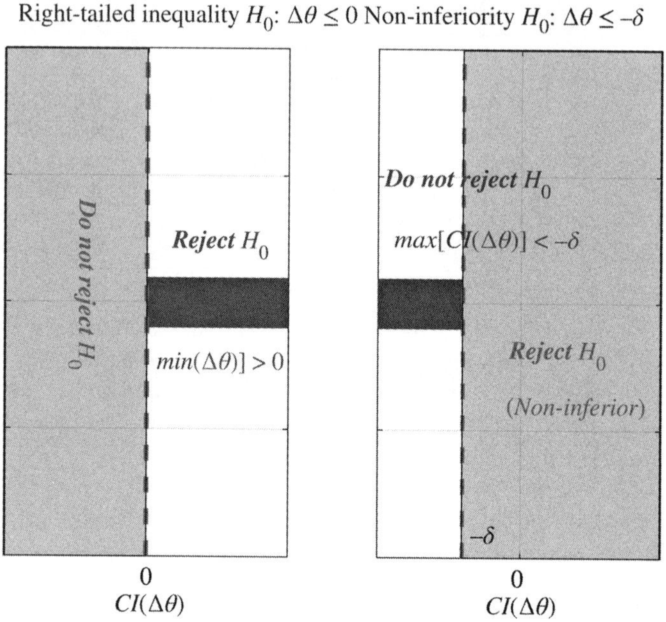

Right-tailed inequality H_0: $\Delta\theta \leq 0$ Non-inferiority H_0: $\Delta\theta \leq -\delta$

FIGURE 5.7   Right-tailed and non-inferiority tests.

Null hypothesis decision criterion
Confidence interval formulation
Achieved effect size
Required sample size
Associated MATLAB functions
Example using MATLAB commands

EXERCISES

5.1   Suppose $X_1, X_2, \ldots, X_n$ are independent and identically distributed (*i.i.d.*) random variables from a $\mathcal{N}(\mu, \sigma^2)$ distribution. Show that the mean $\overline{X} = \frac{1}{n}\sum_{j=1}^{n} X_j$ is a $\mathcal{N}(\mu, \frac{1}{n}\sigma^2)$ distributed random variable. *Hint*: $E[X_1 + X_2] = E[X_1] + E[X_2] = \mu + \mu$ and $\text{var}(aX) = a^2\,\text{var}(X)$.

5.2   Using the results from Exercise 5.1, show that if $X_1, X_2, \ldots, X_n$ are *i.i.d.* $\mathcal{N}(\mu, \sigma_1^2)$ and $Y_1, Y_2, \ldots, Y_n$ *i.i.d.* $\mathcal{N}(\mu, \sigma_2^2)$, then $Z = \dfrac{\overline{X} - \overline{Y}}{\sqrt{\frac{1}{n}(\sigma_1^2 + \sigma_2^2)}}$ is $\mathcal{N}(0, 1)$.

5.3   For the hypotheses H_0: $\mu_1 \geq \mu_2$ vs. H_1: $\mu_1 < \mu_2$ show that significance α, power $1 - \beta$, and effect size ε are related to sample size via $n = [\![\dfrac{(\sigma_1^2 + \sigma_2^2)\cdot(z_{1-\alpha} + z_{1-\beta})^2}{\varepsilon^2}]\!]$.

5.4 Show that, for *unequal* sample populations with ratio $r = n_2/n_1$, the total sample size requirement for the two-sided test H_0: $\mu_1 = \mu_2$ vs. H_1: $\mu_1 \neq \mu_2$ is $n_2 = $ $[\![\frac{(r\sigma_1^2+\sigma_2^2)(z_{1-\alpha/2}+z_{1-\beta})^2}{\varepsilon^2}]\!]$.

5.5 Show that the width of the confidence interval for the difference of means is $2 \bullet t_{df}(1 - \alpha/2) \bullet \sqrt{\frac{1}{n_1+n_2}S_{pool}^2}$. Hint: $t_{df}(\alpha/2) = -t_{df}(1 - \alpha/2)$.

5.6 Show that $S_{pool}^2 \to \frac{1}{2}(\sigma_x^2 + \sigma_y^2)$ provided $S_x^2 \to \sigma_x^2$ and $S_y^2 \to \sigma_y^2$.

MATLAB Commands
(One-Sided Test of Average Heights)

```
% Data Directory
Ddir = 'C:\PJC\Math_Biology\Chapter_5\Data';
% Load in the Hockey and Basketball data
load(fullfile(Ddir,'HockeyData.mat'))
load(fullfile(Ddir,'BasketballData.mat'))
% Collect the hockey (X) and basketball (Y) height data
Y = D.Ht; X = B.Ht;
% Set the significance level at 5%: α = 0.05
alpha = 0.05;
% Test the hypotheses H₀: μH ≤ μB vs. H₁: μB > μH
H = tmean(X,Y,alpha,2)
H =
              Ho: 'reject'
               T: 33.0137
              M1: 79.2123
              M2: 73.2164
             CIx: [78.8911 79.5336]
             CIy: [73.0609 73.3719]
           CIdm2: [5.6739 6.3180]
              dM: 5.9960
              CI: [5.6393   Inf]
              df: 645
          Pvalue: 5.1664e-141
       EqualVars: 'No'
            Test: 'one-sided, right tail, unequal variances'
```

5.2 TEST OF MEANS

This section will list the hypotheses that focus on the difference of means from two distinct populations. In each subsection, the *population samples* will be represented by the notation $X = \{x_1, x_2, ..., x_{n_1}\}$ and $Y = \{y_1, y_2, ..., y_{n_2}\}$. It will be assumed that these samples are independent, that X comes from the *Control* population, and Y represents the *Test* population. In select subsections, the development of the sample size formula is presented. For those sections that do not contain the derivation, it can be readily attained from another section.

5.2.1 Two-Sided Tests of Means

This section will present hypothesis tests that require the investigator to examine either side of the test statistic distribution to make a decision with respect to the null hypothesis. Figure 5.5 gives a geometric interpretation of the notion of a two-sided test. It also distinguishes between a test of equality and a test of equivalence.

5.2.1.1 Equal Variances   For this model, the assumption is that the samples are taken from two distinct populations with the same variance. That is, the random variables whence the samples are taken are distributed as $X \sim \mathcal{N}(\mu_1, \sigma^2)$ and $Y \sim \mathcal{N}(\mu_2, \sigma^2)$ with $\mu_1 \neq \mu_2$. The two-sided test of equal means, with equal variances, proceeds as follows.

Hypotheses

$$H_0: \mu_1 = \mu_2 \qquad\qquad \textit{(Null hypothesis)}$$
$$H_1: \mu_1 \neq \mu_2 \qquad\qquad \textit{(Alternate hypothesis)}$$

Test Statistic

$$T(X, Y) = \frac{\Delta\hat{\mu}}{SE}$$

$$\Delta\hat{\mu} = \bar{x} - \bar{y} \equiv \frac{1}{n_1}\sum_{j=1}^{n_1} x_j - \frac{1}{n_2}\sum_{k=1}^{n_2} y_k$$

$SE = \sqrt{\frac{1}{n_1}S_x^2 + \frac{1}{n_2}S_y^2} = \sqrt{(\frac{1}{n_1} + \frac{1}{n_2})S^2} = S\sqrt{(\frac{1}{n_1} + \frac{1}{n_2})}$ since, by assumption $S_x^2 = S_y^2 = S^2$,

$$S_x^2 = \frac{1}{n_1 - 1}\sum_{j=1}^{n_1}(x_j - \bar{x})^2, \, S_y^2 = \frac{1}{n_2 - 1}\sum_{k=1}^{n_2}(y_k - \bar{y})^2$$

Test Statistic Distribution
$T(X, Y) \sim \mathcal{T}_{df_e}$ = the *student t-distribution* with $df_e = n_1 + n_2 - 2$ degrees of freedom
$\phi_T(t; df_e)$ = *pdf* of the $\mathcal{T}$-distribution with df_e degrees of freedom as per (5.2) of Section 5.1.

p-Value

$$p = 2 \cdot \Phi_T(-|T(X, Y)|; df_e)$$

$\Phi_T(x; df_e)$ = *cumulative density function* of a $\mathcal{T}_{df_e}$ random variable

Null Hypothesis Decision

If the p-value is less than the significance level, *reject* the null hypothesis. More succinctly, $p < \alpha \Rightarrow$ *reject* H_0. Otherwise, *do not reject* H_0.

$(1 - \alpha)\bullet 100\%$ Confidence Interval for $\Delta\mu = \mu_1 - \mu_2$

$$CI(\Delta\mu) = [\Delta\hat{\mu} + t_{df_e}(\alpha/2)\bullet SE, \Delta\hat{\mu} + t_{df_e}(1 - \alpha/2)\bullet SE]$$

$(1 - \alpha)\bullet 100\%$ Confidence Intervals for μ_1 and μ_2

$$CI(\mu_1) = \left[\bar{x} + t_{df_e}(\alpha/2) \cdot \frac{1}{\sqrt{n_1}}S_x, \bar{x} + t_{df_e}(1 - \alpha/2) \cdot \frac{1}{\sqrt{n_1}}S_x\right]$$

$$CI(\mu_2) = \left[\bar{y} + t_{df_e}(\alpha/2) \cdot \frac{1}{\sqrt{n_2}}S_y, \bar{y} + t_{df_e}(1 - \alpha/2) \cdot \frac{1}{\sqrt{n_2}}S_y\right]$$

Required Sample Size

$$R = \frac{n_2}{n_1}$$

$$n_2 = \left[\!\!\left[\frac{(z_{1-\alpha/2} + z_{1-\beta})^2}{es^2} \cdot (R + 1) \cdot \sigma^2\right]\!\!\right]$$

$es = $ effect size ($es = |\Delta\mu| = |\mu_1 - \mu_2|$). When σ^2 is unknown, use S^2 in its place.

To demonstrate this formula, observe that the null hypothesis H_0 is *rejected* provided $|T(X, Y)| > z_{1-\alpha/2}$. This means that either $T(X,Y) > z_{1-\alpha/2}$ or $T(X,Y) < -z_{1-\alpha/2}$. In the first case, the *power* is approximated by the normal distribution as $\pi = \Phi(T(X,Y) - z_{1-\alpha/2})$. Since $\pi = 1 - \beta$ and a quantile is determined from the inverse *cdf*, $z_{1-\beta} = \Phi^{-1}(\pi) = T(X,Y) - z_{1-\alpha/2}$. Simplifying this equation produces

$$z_{1-\beta} - z_{1-\alpha/2} = \frac{\bar{x}-\bar{y}}{S\sqrt{\frac{1}{n_1}+\frac{1}{n_2}}} = \frac{(\bar{x}-\bar{y})\sqrt{n_2}}{S\sqrt{R+1}}, \text{ where } R = n_2/n_1.$$

This leads to $n_2 = (\frac{z_{1-\beta}+z_{1-\alpha/2}}{\bar{x}-\bar{y}})^2 \cdot S^2(R + 1)$. In a similar manner, $T(X, Y) < -z_{1-\alpha/2}$ produces the requirement $n_2 = (\frac{z_{1-\beta}+z_{1-\alpha/2}}{\bar{y}-\bar{x}})^2 \cdot S^2(R + 1)$. Replacing the difference in sample means by the effect size $es = |\mu_1 - \mu_2|$ and the sample variance with the actual variance σ^2 completes the sample size model.

MATLAB Function

```
tmean.m
```

TABLE 5.3   Hypothesis test summary for equal average ages for equal variances

Test	p-value	Reject H_0?	μ_1	$CI(\mu_1)$	μ_2	$CI(\mu_2)$
$H_0: \mu_1 = \mu_2$	0.0046	Yes ($p < \alpha$)	27.36	[27.03, 27.70]	26.6	[26.2, 27.01]
$CI(\mu_1 - \mu_2) = [0.24, 1.29]$						

Example: Test the hypothesis that the average age of a hockey player is the same as the average age of a basketball player. If A_H and A_B are the average hockey and basketball player ages for the data collected by the *NHL* and *NBA* for active players as of January 2015, then the hypotheses are

$$H_0: A_B = A_H \qquad \qquad \textit{(Null hypothesis)}$$
$$H_1: A_B \neq A_H \qquad \qquad \textit{(Alternate hypothesis)}$$

The mean ages are $A_B = 27.36$ and $A_H = 26.6$, respectively. The p-value of the test is $p = 0.0046 < \alpha = 0.05$. Therefore, at the 5% significance level, the null hypothesis of equal average ages is *rejected*. The MATLAB command tmean.m tests these hypotheses and determines via Levene's method (see Section 5.4.1.2) that the variances can be treated as equal. Table 5.3 summarizes the results.

MATLAB Commands
(***Test of Equal Means, Equal Variances***)

```
% Data Directory
Ddir = 'C:\PJC\Math_Biology\Chapter_5\Data';
% Load in the Hockey and Basketball data
load(fullfile(Ddir,'HockeyData.mat'))
load(fullfile(Ddir,'BasketballData.mat'))
% Collect the hockey (X) and basketball (Y) age data
X = D.Ages; Y = B.Ages;
% Set the significance level at 5%: α = 0.05
alpha = 0.05;

% Test the hypotheses H₀: μ_B = μ_H vs. H₁: μ_B ≠ μ_H
H = tmean(X,Y,alpha,2)
H =
            Ho: 'reject'
             T: 2.8408
            M1: 27.3611
            M2: 26.5982
           CIx: [27.0268 27.6955]
           CIy: [26.1900 27.0063]
         CIdm2: [0.2332 1.2927]
```

```
       dM: 0.7629
       CI: [0.2360 1.2899]
       df: 1120
    Pvalue: 0.0046
 EqualVars: 'Yes'
      Test: 'two-sided, equal variances'
```

5.2.1.2 *Unequal Variances*
In this case, the assumption is that the samples are taken from two distinct populations with the distinct variances: $X \sim \mathcal{N}(\mu_1, \sigma_1^2)$ and $Y \sim \mathcal{N}(\mu_2, \sigma_2^2)$, $\mu_1 \neq \mu_2$ and $\sigma_1^2 \neq \sigma_2^2$. The two-sided test of equal means parallels Section 5.2.1.1.

Hypotheses

$$H_0: \mu_1 = \mu_2 \qquad \qquad \text{(Null Hypothesis)}$$
$$H_1: \mu_1 \neq \mu_2 \qquad \qquad \text{(Alternate Hypothesis)}$$

Test Statistic

$$T(X, Y) = \frac{\Delta\hat{\mu}}{SE}$$

$$\Delta\hat{\mu} = \bar{x} - \bar{y} \equiv \frac{1}{n_1}\sum_{j=1}^{n_1} x_j - \frac{1}{n_2}\sum_{k=1}^{n_2} y_k$$

$$SE = \sqrt{\frac{1}{n_1}S_x^2 + \frac{1}{n_2}S_y^2}, \quad S_x^2 = \frac{1}{n_1-1}\sum_{j=1}^{n_1}(x_j - \bar{x})^2, \quad S_y^2 = \frac{1}{n_2-1}\sum_{j=1}^{n_2}(y_j - \bar{y})^2$$

Test Statistic Distribution

$T(X, Y) \sim \mathcal{T}_{df_e}$ = the *student t-distribution* with df_u degrees of freedom

$$df_u = \left[\!\left[\frac{\left(\frac{1}{n_1}S_x^2 + \frac{1}{n_2}S_y^2\right)^2}{\frac{1}{n_1}\left(\frac{1}{n_1}S_x^2\right)^2 + \frac{1}{n_2}\left(\frac{1}{n_2}S_y^2\right)^2} \right]\!\right], \; [\![n]\!] = \text{nearest integer function}$$

$\phi_{\mathcal{T}}(t; df_u)$ = pdf of the $\mathcal{T}$-distribution with df_u degrees of freedom as per (5.2) of Section 5.1.

p-Value

$$p = 2 \bullet \Phi_{\mathcal{T}}(-|T(X, Y)|; df_u)$$

$\Phi_{\mathcal{T}}(x; df_u)$ = *cumulative density function* of a $\mathcal{T}_{df_u}$ random variable

Null Hypothesis Decision
If the *p*-value is less than the significance level, *reject* the null hypothesis.
$p < \alpha \Rightarrow$ *reject H_0*. Otherwise, *do not reject H_0*.

$(1 - \alpha) \cdot 100\%$ Confidence Interval for $\Delta\mu = \mu_1 - \mu_2$

$$CI(\Delta\mu) = \left[\Delta\hat{\mu} + t_{df_u}(\alpha/2) \cdot \sqrt{\frac{1}{n_1 + n_2} S^2_{pool}}, \Delta\hat{\mu} + t_{df_u}(1 - \alpha/2) \cdot \sqrt{\frac{1}{n_1 + n_2} S^2_{pool}} \right]$$

$$S^2_{pool} = \frac{(n_1 - 1)S^2_x + (n_2 - 1)S^2_y}{(n_1 + n_2 - 2)}$$

$(1 - \alpha) \cdot 100\%$ Confidence Intervals for μ_1 and μ_2

$$CI(\mu_1) = \left[\bar{x} + t_{df_1}(\alpha/2) \cdot \frac{1}{\sqrt{n_1}} S_x, \bar{x} + t_{df_1}(1 - \alpha/2) \cdot \frac{1}{\sqrt{n_1}} S_x \right], df_1 = n_1 - 1$$

$$CI(\mu_2) = \left[\bar{y} + t_{df_2}(\alpha/2) \cdot \frac{1}{\sqrt{n_2}} S_y, \bar{y} + t_{df_2}(1 - \alpha/2) \cdot \frac{1}{\sqrt{n_2}} S_y \right], df_2 = n_2 - 1$$

Required Sample Size

$$R = \frac{n_2}{n_1}$$

$$n_2 = \left\lceil \left[\frac{(z_{1-\alpha/2} + z_{1-\beta})^2}{es^2} \cdot \left(R \cdot \sigma^2_x + \sigma^2_y \right) \right] \right\rceil$$

es = effect size ($es = |\Delta\mu| = |\mu_1 - \mu_2|$). When σ^2_x and σ^2_y are unknown, use S^2_x and S^2_y in their place, respectively.

This formula follows from the corresponding sample size requirement from Section 5.2.1.1.

MATLAB Function

`tmean.m`

Example: Test the hypothesis that the average weight of a hockey player is the same as the average weight of a baseball player. If W_H and W_B are the average hockey and basketball player weights, respectively, for the data collected by the *NHL* and *NBA* for active players as of January 2015, then the hypotheses are

$$H_0: W_B = W_H \qquad \text{(Null hypothesis)}$$
$$H_1: W_B \neq W_H \qquad \text{(Alternate hypothesis)}$$

The mean weights are $W_B = 222.36$ and $W_H = 202.1$, respectively. The p-value of the test is $p < 10^{-4} < \alpha = 0.05$. Moreover, the confidence interval of the difference is entirely negative. Therefore, at the 5% significance level, the null hypothesis of equal

TABLE 5.4   Hypothesis test summary for equal average weights for unequal variances

Test	p-value	Reject H_0?	μ_1	$CI(\mu_1)$	μ_2	$CI(\mu_2)$
$H_0: \mu_1 = \mu_2$	$<10^{-4}$	Yes ($p < \alpha$)	202.1	[200.97, 203.2]	222.36	[219.91, 224.82]
$CI(\mu_1 - \mu_2) = [-22.96, -17.56]$						

average weights is *rejected*. The MATLAB command `tmean.m` tested these hypotheses and determined via Levene's method (see Section 5.4.1.2) that the variances are *unequal*. Table 5.4 summarizes the results.

MATLAB Commands
(Test of Equal Means, Unequal Variances)

```
% Data Directory
Ddir = 'C:\PJC\Math_Biology\Chapter_5\Data';
% Load in the Hockey and Basketball data
load(fullfile(Ddir,'HockeyData.mat'))
load(fullfile(Ddir,'BasketballData.mat'))
% Collect the hockey (X) and basketball (Y) weight data
X = D.Wt; Y = B.Wt;
% Set the significance level at 5%: α = 0.05
alpha = 0.05;
```

```
% Test the hypotheses H0: μH = μB vs. H1: μH ≠ μB
H = tmean(X,Y,alpha,2)
H =
              Ho: 'reject'
               T: -14.7076
              M1: 202.1038
              M2: 222.3653
             CIx: [200.9658 203.2418]
             CIy: [219.9090 224.8216]
           CIdm2: [-22.6861 -17.8369]
              dM: -20.2615
              CI: [-22.9668 -17.5562]
              df: 628
          Pvalue: 2.7150e-42
       EqualVars: 'No'
            Test: 'two-sided, unequal variances'
```

5.2.1.3   *Test of Equivalent Means (Equal and Unequal Variances)*   As noted in Section 5.1, the notion of statistical equality of means can lead to contradictory results, as illustrated in Figure 5.4. Rather than equality, the test of statistical *equivalence* can be used in its stead. To execute such a test, the *clinical margin δ must* be produced as an input to the process. Ultimately, this is a value agreed upon by clinicians and reviewing agencies. For the purpose of this presentation, δ is assumed to be positive ($\delta > 0$). Now the description proceeds as in Sections 5.2.1.1 and 5.2.1.2. An illustration of the difference in hypothesis testing in *equality* and *equivalence* is illustrated in Figure 5.5. To provide an economic list of crucial formulae, the case of *equal* and *unequal* variances is listed in parallel.

Hypotheses

$$H_0: |\mu_2 - \mu_1| \geq \delta \qquad\qquad \text{(Null Hypothesis)}$$
$$H_1: |\mu_2 - \mu_1| < \delta \qquad\qquad \text{(Alternate Hypothesis)}$$

Test Statistics
Difference of Means

$$\Delta\hat{\mu} = \bar{x} - \bar{y} \equiv \frac{1}{n_1}\sum_{j=1}^{n_1} x_j - \frac{1}{n_2}\sum_{k=1}^{n_2} y_k$$

Equal Variances
There are two test statistics, $T_E^-(\delta) = \frac{\Delta\hat{\mu}-\delta}{SE_E}$ and $T_E^+(\delta) = \frac{\Delta\hat{\mu}+\delta}{SE_E}$, where

$$SE_E = S\sqrt{\frac{1}{n_1} + \frac{1}{n_2}}, \quad S_x^2 = S_y^2 = S^2 = \frac{1}{n_1-1}\sum_{j=1}^{n_1}(x_j - \bar{x})^2$$

Unequal Variances
As for the equal variances, the two test statistics are $T_U^-(\delta) = \frac{\Delta\hat{\mu}-\delta}{SE_U}$ and $T_U^+(\delta) = \frac{\Delta\hat{\mu}+\delta}{SE_U}$
with

$$SE_U = \sqrt{\frac{1}{n_1}S_x^2 + \frac{1}{n_2}S_y^2}, \quad S_x^2 = \frac{1}{n_1-1}\sum_{j=1}^{n_1}(x_j - \bar{x})^2, S_y^2 = \frac{1}{n_2-1}\sum_{j=1}^{n_2}(y_j - \bar{y})^2$$

Test Statistic Distribution
$T(X, Y) \sim \mathcal{T}_{df_e}$ = the *student t-distribution* with df_u degrees of freedom
$\phi_T(t; df_u)$ = *pdf* of the $\mathcal{T}$-distribution defined by equation (5.2) of Section 5.1 and with df degrees of freedom as per sections 5.2.1.1 (df_e) or 5.2.1.2 (df_u).

p-Values
There are two *p*-values for each variance condition. If $T_E^-(\delta) < t_{df}(\alpha)$ and $T_E^+(\delta) < t_{df}(1 - \alpha)$, then the null hypothesis H_0 is *rejected* in favor of H_1. Since the *cdf* for the $\mathcal{T}$-distribution is a monotone increasing function, the conditions aforementioned mean $p^- = \Phi_T(T_E^-(\delta); df) < \alpha$ and $\Phi_T(T_E^+(\delta); df) > 1 - \alpha$. This second inequality is equivalent to $\alpha > 1 - \Phi_T(T_E^+(\delta); df) = \Phi_T(-T_E^+(\delta); df) \equiv p^+$. Thus, if $p = \max\{p^-, p^+\} < \alpha$, then reject H_0 in favor of H_1. If $\pi^- \equiv \Phi_T(T_U^-(\delta); df)$ and

$\pi^+ \equiv \Phi_T(-T_U^+(\delta); df)$, then $\pi = \max\{\pi^-, \pi^+\}$ is the p-value for the case of unequal variances.

$$p = \begin{cases} \max\left\{\Phi_T\left(-T_E^+(\delta); df_e\right), \Phi_T\left(T_E^-(\delta); df_e\right)\right\} & \text{for equal variances} \\ \max\left\{\Phi_T\left(-T_U^+(\delta); df_e\right), \Phi_T\left(T_U^-(\delta); df_e\right)\right\} & \text{for unequal variances} \end{cases}$$

Null Hypothesis Decision
If the p-value is less than the significance level, *reject* the null hypothesis.
$p < \alpha \Rightarrow$ *reject* H_0 in favor of H_1. Otherwise, *do not reject* H_0. This criterion applies to *both* equal and unequal variances.

$(1 - \alpha)\bullet 100\%$ Confidence Interval for $\Delta\mu = \mu_1 - \mu_2$
Equal Variances

$$CI(\Delta\mu) = [\Delta\hat{\mu} + t_{df_e}(\alpha/2)\bullet SE_E, \Delta\hat{\mu} + t_{df_e}(1 - \alpha/2)\bullet SE_E]$$

Unequal Variances

$$CI(\Delta\mu) = [\Delta\hat{\mu} + t_{df_u}(\alpha/2)\bullet SE_U, \Delta\hat{\mu} + t_{df_u}(1 - \alpha/2)\bullet SE_U]$$

$(1 - \alpha)\bullet 100\%$ Confidence Intervals for μ_1 and μ_2

$$CI(\mu_1) = \left[\bar{x} + t_{df_1}(\alpha/2) \cdot \frac{1}{\sqrt{n_1}}S_x, \bar{x} + t_{df_1}(1 - \alpha/2) \cdot \frac{1}{\sqrt{n_1}}S_x\right], df_1 = n_1 - 1$$

$$CI(\mu_2) = \left[\bar{y} + t_{df_2}(\alpha/2) \cdot \frac{1}{\sqrt{n_2}}S_y, \bar{y} + t_{df_2}(1 - \alpha/2) \cdot \frac{1}{\sqrt{n_2}}S_y\right], df_2 = n_2 - 1$$

The individual confidence intervals are the same for *both* equal and unequal variances.

Required Sample Size

$$R = \frac{n_2}{n_1}$$

$$n_2 = \left[\!\left[\frac{(z_{1-\alpha/2} + z_{1-\beta})^2}{(\delta - es)^2} \cdot \left(R \cdot \sigma_x^2 + \sigma_y^2\right)\right]\!\right]$$

es = effect size ($es = |\Delta\mu| = |\mu_1 - \mu_2|$). When σ_x^2 and σ_y^2 are unknown, use S_x^2 and S_y^2 in their place, respectively.

See Chow et al. [8], equation (3.2.4).

MATLAB Function
```
nimean.m
```

TABLE 5.5   Hypothesis test summary for equivalent average heights

Test	p-value	Reject H_0?	μ_H	$CI(\mu_H)$	μ_B	$CI(\mu_B)$
H_0: $|\mu_B - \mu_H| > \delta$	1	No ($p > \alpha$)	73.22	[73.06, 73.37]	79.21	[78.89, 79.53]
$CI(\mu_B - \mu_H) = [5.67, 6.32]$						

Note: The MATLAB function `nimean.m` uses Levene's method to determine whether the variances of the two samples X and Y are equal or unequal. More details on Levene's test will be provided in Section 5.4 on tests of variance.

Example: Determine whether the average height of a hockey player is within 1 inch of the average height of a basketball player. Again, using the *NHL* and *NBA* posted data, the hypotheses are

$$H_0: |\mu_B - \mu_H| \geq \delta = 1 \qquad \textit{(Null hypothesis)}$$

$$H_1: |\mu_B - \mu_H| < \delta = 1 \qquad \textit{(Alternate hypothesis)}$$

where μ_B and μ_H are the average heights of *NBA* and *NHL* players, respectively. As can be seen in Table 5.5, the confidence interval for the difference between the mean basketball player (μ_B) and hockey player (μ_H) heights exceeds 5 (inches). Therefore, the mean heights are *not* within one inch and consequently are *not* equivalent to that margin.

MATLAB Commands
(*Test of Equivalent Means*)

```
% Data Directory
Ddir = 'C:\PJC\Math_Biology\Chapter_5\Data';

% Load in the Hockey and Basketball data
load(fullfile(Ddir,'HockeyData.mat'))
load(fullfile(Ddir,'BasketballData.mat'))

% Collect the hockey (X) and basketball (Y) height data
X = D.Ht; Y = B.Ht;

% Set the significance level at 5%: alpha = 0.05 and clinical margin delta = 1 (1 inch)
alpha = 0.05; delta = 1.0;

% Test the hypotheses H0: |muB - muH| >= delta vs. H1: |muB - muH| < delta
H = nimean(Y,X,alpha,delta)
```

```
H =
        Ho: 'do not reject'
         T: [27.5077 38.5197]
        Mx: 79.2123
        My: 73.2164
       CIx: [78.8911 79.5336]
       CIy: [73.0609 73.3719]
        dM: 5.9960
      CIdM: [5.6739 6.3180]
        df: 645
      Test: 'two-sided, equivalence, unequal variances'
    Pvalue: 1
```

EXERCISE

5.7 If being within 5 pounds is "close enough" from a clinical standpoint when comparing the weights of athletes, determine whether the average weight of a hockey player is heavier (i.e., *superior*) to that of the average weight of a basketball player. *Hint*: Set $\delta = 5$ and use nimean.m to test H_0: $\mu_H - \mu_B \leq \delta$ vs. H_1: $\mu_H - \mu_B > \delta$.

5.2.2   One-Sided Tests of Means

These hypothesis tests require the investigator to examine either the right or left tail of the test statistic's distribution to determine the disposition of the null hypothesis (i.e., whether to *reject H_0* or *not reject H_0*). Figure 5.7 was introduced to depict and compare *right-tailed* and *superiority tests*. This test (of *superiority*) is the one-sided associate of the equivalence test of Section 5.2.1.3. The details of *superiority* and *non-inferiority* tests will be provided in Sections 5.2.2.3 and 5.2.2.4. Parallel to right-tailed and superiority tests are *left-tailed and non-inferiority tests*. The geographic interpretation of these tests are provided in Figure 5.6 and detailed in the subsequent sections.

5.2.2.1   *Right-Tailed Test (Equal and Unequal Variances)*
Hypotheses

$$H_0: \mu_1 \leq \mu_2 \qquad \qquad \textit{(Null hypothesis)}$$
$$H_1: \mu_1 > \mu_2 \qquad \qquad \textit{(Alternate hypothesis)}$$

Test Statistics
Difference of Means

$$\Delta\hat{\mu} = \bar{x} - \bar{y} \equiv \frac{1}{n_1}\sum_{j=1}^{n_1} x_j - \frac{1}{n_2}\sum_{k=1}^{n_2} y_k$$

Equal Variances

$$T_E(\delta) = \frac{\Delta\hat{\mu}}{SE_E}$$

$$SE_E = S\sqrt{\frac{1}{n_1} + \frac{1}{n_2}}, \quad S_x^2 = S_y^2 = S^2 = \frac{1}{n_1-1}\sum_{j=1}^{n_1}(x_j - \bar{x})^2$$

Unequal Variances

$$T_U(\delta) = \frac{\Delta\hat{\mu}}{SE_U}$$

$$SE_U = \sqrt{\frac{1}{n_1}S_x^2 + \frac{1}{n_2}S_y^2}, \quad S_x^2 = \frac{1}{n_1-1}\sum_{j=1}^{n_1}(x_j - \bar{x})^2, S_y^2 = \frac{1}{n_2-1}\sum_{j=1}^{n_2}(y_j - \bar{y})^2$$

Test Statistic Distribution

$T(X, Y) \sim \mathcal{T}_{df_e}$ = the *student t-distribution* with df_u degrees of freedom

$\phi_T(t; df_u)$ = *pdf* of the $\mathcal{T}$-distribution defined by equation (5.2) of Section 5.1 and with *df* degrees of freedom as per sections 5.2.1.1 (df_e) or 5.2.1.2 (df_u).

p-Values

$$p = \begin{cases} \Phi_T(-T_E(\delta); df_e) & \text{for equal variances} \\ \Phi_T(-T_U(\delta); df_u) & \text{for unequal variances} \end{cases}$$

Null Hypothesis Decision

If the *p*-value is less than the significance level, *reject* the null hypothesis.

$p < \alpha \Rightarrow$ *reject* H_0 in favor of H_1. Otherwise, *do not reject* H_0. This criterion applies to *both* equal and unequal variances.

$(1 - \alpha)\cdot100\%$ Confidence Interval for $\Delta\mu = \mu_1 - \mu_2$

Equal Variances

$$CI(\Delta\mu) = [\Delta\hat{\mu} + t_{df_e}(\alpha)\cdot SE_E, +\infty)$$

Unequal Variances

$$CI(\Delta\mu) = [\Delta\hat{\mu} + t_{df_u}(\alpha)\cdot SE_U, +\infty)$$

The two-sided $(1 - \alpha)\cdot100\%$ *confidence intervals* for μ_1 and μ_2 are the same as noted in Section 5.2.1.1.

TABLE 5.6   Hypothesis test summary for larger average heights

Test	p-value	Reject H_0?	μ_H	$CI(\mu_H)$	μ_B	$CI(\mu_B)$
$H_0: \mu_B \leq \mu_H$	$<10^{-4}$	Yes $(p < \alpha)$	73.22	$[73.06, 73.37]$	79.21	$[78.89, 79.53]$
One-sided $CI(\mu_B - \mu_H) = [5.67, +\infty)$						

Required Sample Size

$$R = \frac{n_2}{n_1}$$

$$n_2 = \left\lceil \frac{(z_{1-\alpha} + z_{1-\beta})^2}{es^2} \cdot \left(R \cdot \sigma_x^2 + \sigma_y^2 \right) \right\rceil$$

es = effect size ($es = \Delta\mu = \mu_2 - \mu_1$). When σ_x^2 and σ_y^2 are unknown, use S_x^2 and S_y^2 in their place, respectively. See Chow et al. [8], equation (3.2.3), $\delta = 0$.

MATLAB Function

`tmean.m`

Example: Test whether the average height of a basketball player is larger than the average height of a hockey player. To show that the average basketball player's height is *larger* than the average hockey player's height, test the hypotheses

$$H_0: \mu_B \leq \mu_H \qquad\qquad \textit{(Null hypothesis)}$$
$$H_1: \mu_B > \mu_H \qquad\qquad \textit{(Alternate hypothesis)}$$

where μ_B and μ_H are the average heights of *NBA* and *NHL* players, respectively. If the null hypothesis is *rejected* in favor of the alternate hypothesis, then the assertion of taller basketball players can be made. As can be seen in Table 5.6, the one-sided confidence interval for the difference between the mean basketball player (μ_B) and hockey player (μ_H) heights exceeds 5 (inches). Therefore, the difference in mean heights is greater than 0 and consequently H_0 is *rejected in favor of H_1*. That is, $\mu_B > \mu_H$.

MATLAB Commands
(One-Sided Test of Larger Mean Heights)

```
% Data Directory
Ddir = 'C:\PJC\Math_Biology\Chapter_5\Data';
% Load in the Hockey and Basketball data
load(fullfile(Ddir, 'HockeyData.mat'))
load(fullfile(Ddir, 'BasketballData.mat'))
```

```
% Collect the hockey (X) and basketball (Y) height data
X = D.Ht; Y = B.Ht;
% Set the significance level at 5%: α = 0.05
alpha = 0.05;
% Test the hypotheses H₀: μ_B ≤ μ_H vs. H₁: μ_B > μ_H
H = tmean(X,Y,alpha,0)
H =

           Ho: 'reject'
            T: 33.0137
           M1: 79.2123
           M2: 73.2164
          CIx: [78.8911 79.5336]
          CIy: [73.0609 73.3719]
        CIdm2: [5.6739 6.3180]
           dM: 5.9960
           CI: [5.6968 Inf]
           df: 645
       Pvalue: 5.1664e-141
    EqualVars: 'No'
         Test: 'one-sided, right tail, unequal variances'
```

5.2.2.2 Left-Tailed Test (Equal and Unequal Variances)

Hypotheses

$$H_0: \mu_1 \geq \mu_2 \qquad \textit{(Null Hypothesis)}$$
$$H_1: \mu_1 < \mu_2 \qquad \textit{(Alternate hypothesis)}$$

Test Statistics
Difference of Means

$$\Delta\hat{\mu} = \bar{x} - \bar{y} \equiv \frac{1}{n_1}\sum_{j=1}^{n_1} x_j - \frac{1}{n_2}\sum_{k=1}^{n_2} y_k$$

Equal Variances

$$T_E(\delta) = \frac{\Delta\hat{\mu}}{SE_E}$$

$$SE_E = S\sqrt{\frac{1}{n_1} + \frac{1}{n_2}}, \quad S_x^2 = S_y^2 = S^2 = \frac{1}{n_1 - 1}\sum_{j=1}^{n_1}(x_j - \bar{x})^2$$

Unequal Variances

$$T_U(\delta) = \frac{\Delta\hat{\mu}}{SE_U}$$

$$SE_U = \sqrt{\frac{1}{n_1}S_x^2 + \frac{1}{n_2}S_y^2}, \quad S_x^2 = \frac{1}{n_1 - 1}\sum_{j=1}^{n_1}(x_j - \bar{x})^2, \quad S_y^2 = \frac{1}{n_2 - 1}\sum_{j=1}^{n_2}(y_j - \bar{y})^2$$

Test Statistic Distribution

$T(X, Y) \sim \mathcal{T}_{df_e}$ = the *student t-distribution* with df_u degrees of freedom

$\phi_\mathcal{T}(t; df_u)$ = *pdf* of the $\mathcal{T}$-distribution defined by equation (5.2) of Section 5.1 and with df degrees of freedom as per sections 5.2.1.1 (df_e) or 5.2.1.2 (df_u).

p-Values

$$p = \begin{cases} \Phi_\mathcal{T}(T_E(\delta); df_e) & \text{for equal variances} \\ \Phi_\mathcal{T}(T_U(\delta); df_u) & \text{for unequal variances} \end{cases}$$

Null Hypothesis Decision

If the *p*-value is less than the significance level, *reject* the null hypothesis.

$p < \alpha \Rightarrow$ *reject* H_0 in favor of H_1. Otherwise, *do not reject* H_0. This criterion applies to *both* equal and unequal variances.

$(1 - \alpha)\bullet 100\%$ Confidence Interval for $\Delta\mu = \mu_1 - \mu_2$

Equal Variances

$$CI(\Delta\mu) = (-\infty, \Delta\hat{\mu} + t_{df_e}(1 - \alpha)\bullet SE_E]$$

Unequal Variances

$$CI(\Delta\mu) = (-\infty, \Delta\hat{\mu} + t_{df_u}(1 - \alpha)\bullet SE_U]$$

The two-sided $(1 - \alpha)\bullet 100\%$ *confidence intervals* for μ_1 and μ_2 are the same as noted in Section 2.1.1.

Required Sample Size

$$R = \frac{n_2}{n_1}$$

$$n_2 = \left\lceil \frac{(z_{1-\alpha} + z_{1-\beta})^2}{es^2} \cdot \left(R \cdot \sigma_x^2 + \sigma_y^2\right) \right\rceil$$

es = effect size ($es = \Delta\mu = \mu_2 - \mu_1$). When σ_x^2 and σ_y^2 are unknown, use S_x^2 and S_y^2 in their place, respectively. See Chow et al. [8], equation (3.2.3), $\delta = 0$.

TABLE 5.7   Hypothesis test summary for smaller average ages

Test	p-value	Reject H_0?	μ_H	$CI(\mu_H)$	μ_B	$CI(\mu_B)$
$H_0: \mu_B \geq \mu_H$	0.0023	Yes ($p < \alpha$)	27.36	[27.03, 27.7]	26.6	[26.19, 27.01]
One-sided $CI(\mu_B - \mu_H) = (-\infty, -0.32]$						

MATLAB Function

```
tmean.m
```

Example: Test whether the average age of a basketball player is less than the average age of a hockey player. To show that the average basketball player's age is *less* than the average hockey player's age, test the hypotheses

$$H_0: \mu_B \geq \mu_H \qquad \qquad (Null\ hypothesis)$$
$$H_1: \mu_B < \mu_H \qquad \qquad (Alternate\ hypothesis)$$

where μ_B and μ_H are the average ages of *NBA* and *NHL* players, respectively. If the null hypothesis is *rejected* in favor of the alternate hypothesis, then the assertion of younger basketball players can be made. As can be seen in Table 5.7, the upper end of the one-sided confidence interval for the difference between the mean basketball player (μ_B) and hockey player (μ_H) ages is less than -0.32 (years). Therefore, the difference in mean age is less than 0 and consequently H_0 is rejected in favor of H_1. That is, $\mu_B < \mu_H$.

MATLAB Commands
(One-Sided Test of Smaller Mean Ages)

```
% Data Directory
Ddir = 'C:\PJC\Math_Biology\Chapter_5\Data';
% Load in the Hockey and Basketball data
load(fullfile(Ddir,'HockeyData.mat'))
load(fullfile(Ddir,'BasketballData.mat'))
% Collect the hockey (X) and basketball (Y) height data
Y = D.Ages; X = B.Ages;
% Set the significance level at 5%: alpha = 0.05
alpha = 0.05;
% Test the hypotheses Ho: muB >= muH vs. H1: muB < muH
H = tmean(X,Y,alpha,1)
H =
            Ho: 'reject'
             T: -2.8408
            M1: 26.5982
            M2: 27.3611
```

```
    CIx:  [26.1900  27.0063]
    CIy:  [27.0268  27.6955]
  CIdm2:  [-1.2927  -0.2332]
     dM:  -0.7629
     CI:  [-Inf  -0.3208]
     df:  1120
 Pvalue:  0.0023
EqualVars:  'Yes'
   Test:  'one-sided, left tail, equal variances'
```

5.2.2.3 Non-Inferiority Test (Equal and Unequal Variances)   As alluded to in the example from Section 5.2.2.2 on *left-tailed* tests and illustrated in Figure 5.7, *non-inferiority* tests determine whether the *test* mean (μ_2) exceeds the *control* mean (μ_1) when supplemented by a margin $\delta > 0$. Namely, $\mu_2 + \delta > \mu_1$ for some *a priori* selected margin δ. Equivalently, $\mu_2 > \mu_1 - \delta$. For example, it is plain *numerically* that $\mu_2 = 71.3 \not> 72 = \mu_1$. However, for $\delta = 2$, it may well be the case that $\mu_2 + \delta$ is *statistically significantly larger than* μ_1. Thus, within the margin δ, μ_2 is *not* considered to be *smaller* (i.e., *not inferior*) to μ_1. The mathematical particulars of the *non-inferiority test for means* are now provided here. Observe that unlike one-sided tests, the aim is to *reject the null hypothesis H_0 in favor of the alternate hypothesis H_1*.

Hypotheses

$$H_0: \mu_2 - \mu_1 \leq -\delta \qquad\qquad\qquad \text{(Null hypothesis)}$$
$$H_1: \mu_2 - \mu_1 > -\delta \qquad\qquad\qquad \text{(Alternate hypothesis)}$$

Clinical Margin

$$\delta > 0$$

Test Statistics
Difference of Means

$$\Delta\hat{\mu} = \bar{x} - \bar{y} \equiv \frac{1}{n_1} \sum_{j=1}^{n_1} x_j - \frac{1}{n_2} \sum_{k=1}^{n_2} y_k$$

Equal Variances

$$T_E(\delta) = \frac{\Delta\hat{\mu} + \delta}{SE_E}$$

$$SE_E = S\sqrt{\frac{1}{n_1} + \frac{1}{n_2}}, \quad S_x^2 = S_y^2 = S^2 = \frac{1}{n_1 - 1} \sum_{j=1}^{n_1} (x_j - \bar{x})^2$$

Unequal Variances

$$T_U(\delta) = \frac{\Delta\hat{\mu} + \delta}{SE_U}$$

$$SE_U = \sqrt{\frac{1}{n_1}S_x^2 + \frac{1}{n_2}S_y^2}, \quad S_x^2 = \frac{1}{n_1 - 1}\sum_{j=1}^{n_1}(x_j - \bar{x})^2, S_y^2 = \frac{1}{n_2 - 1}\sum_{j=1}^{n_2}(y_j - \bar{y})^2$$

Test Statistic Distribution

$T(X, Y) \sim \mathcal{T}_{df_e}$ = the *student t-distribution* with df_u degrees of freedom

$\phi_T(t; df_u)$ = *pdf* of the $\mathcal{T}$-distribution defined by equation (5.2) of Section 5.1 and with *df* degrees of freedom as per sections 5.2.1.1 (df_e) or 5.2.1.2 (df_u).

p-Values

$$p = \begin{cases} \Phi_T(-T_E(\delta); df_e) & \text{for equal variances} \\ \Phi_T(-T_U(\delta); df_u) & \text{for unequal variances} \end{cases}$$

Null Hypothesis Decision

If the *p*-value is less than the significance level, *reject* the null hypothesis.

$p < \alpha \Rightarrow$ *reject* H_0 in favor of H_1. Otherwise, *do not reject* H_0. This criterion applies to *both* equal and unequal variances.

$(1 - \alpha) \cdot 100\%$ Confidence Interval for $\Delta\mu = \mu_1 - \mu_2$
Equal Variances

$$CI(\Delta\mu) = [\Delta\hat{\mu} + t_{df_e}(\alpha) \cdot SE_E, +\infty)$$

Unequal Variances

$$CI(\Delta\mu) = [\Delta\hat{\mu} + t_{df_u}(\alpha) \cdot SE_U, +\infty)$$

The two-sided $(1 - \alpha) \cdot 100\%$ *confidence intervals* for μ_1 and μ_2 are the same as noted in Section 5.2.1.1.

Required Sample Size

$$R = \frac{n_2}{n_1}$$

$$n_2 = \left\lceil \frac{(z_{1-\alpha} + z_{1-\beta})^2}{(es - \delta)^2} \cdot \left(R \cdot \sigma_x^2 + \sigma_y^2\right) \right\rceil$$

es = effect size ($es = \Delta\mu = \mu_2 - \mu_1$). When σ_x^2 and σ_y^2 are unknown, use S_x^2 and S_y^2 in their place, respectively. In the special case that $\sigma_x^2 = \sigma_y^2$, the term $R \cdot \sigma_x^2 + \sigma_y^2$ can be replaced by $(R + 1) \cdot \sigma_x^2$. See Chow et al. [8], equation (3.2.3).

TABLE 5.8   Hypothesis test summary for non-inferior average weights

Test	p-value	Reject H_0?	μ_H	$CI(\mu_H)$	μ_B	$CI(\mu_B)$
H_0: $\mu_H - \mu_B \leq -\delta$	1	No ($p > \alpha$)	202.1	[200.97, 203.24]	222.36	[219.91, 224.82]
$CI(\mu_B - \mu_H) = [18.23, +\infty)$						

MATLAB Function

`nimean.m`

Example: To test whether the average weight of a hockey player μ_H when supplemented by $\delta = 10$ pounds is *at least* the weight of an average basketball player μ_B, exercise the hypotheses

$$H_0: \mu_H - \mu_B \leq -\delta = -10 \qquad \text{(Null hypothesis)}$$
$$H_1: \mu_H - \mu_B > -\delta = -10 \qquad \text{(Alternate hypothesis)}$$

In this setting, hockey players are the *Test* group while basketball players act as *Control*. If H_0 is *rejected*, then $\mu_H + \delta > \mu_B$. As can be seen in Table 5.8, however, the lower bound of the confidence interval for the difference between the mean hockey (μ_H) and basketball player (μ_B) weights exceeds 10 pounds. Indeed, it exceeds 18 pounds. Therefore, the mean weights are *not* within 10 pounds and subsequently the notion that the difference between the average basketball and hockey player weights exceeds 10 pounds (i.e., $\mu_B - \mu_H \geq 10$) *cannot be rejected*.

MATLAB Commands
(*Non-Inferiority Test*)

```
% Data Directory
Ddir = 'C:\PJC\Math_Biology\Chapter_5\Data';
% Load in the Hockey and Basketball data
load(fullfile(Ddir,'HockeyData.mat'))
load(fullfile(Ddir,'BasketballData.mat'))
% Collect the hockey (Y) and basketball (X) weight data
Y = D.Wt; X = B.Wt;
% Set the significance level at 5%: α = 0.05 and clinical margin δ = 10 (pounds)
alpha = 0.05; delta = 10;
% Test the hypotheses H₀: μH − μB ≤ −δ vs. H₁: μH − μB > −δ
H = nimean(X,Y,alpha,delta,0)
H =
         Ho: 'do not reject'
          T: -7.4487
         Mx: 222.3653
         My: 202.1038
        CIx: [219.9090 224.8216]
```

```
CIy:  [200.9658 203.2418]
 dM:  20.2615
CIdM:  [18.2276 Inf]
  df:  628
Test:  'one-sided, non-inferiority, unequal variances'
Pvalue:  1.0000
```

5.2.2.4 Superiority Test (Equal and Unequal Variances)   *Statistical superiority* of means implies that one mean is "larger than the other mean by a margin." For example, $\mu_2 = 71.3$ is numerically larger than $\mu_1 = 71$. But is the difference *statistically significant*? If the left endpoint of $CI(\mu_2 - \mu_1)$ exceeds a clinical meaningful margin δ, then it can be concluded that μ_1 is clinically larger (i.e., *superior*) to μ_2. Thus, $\mu_2' = \mu_1 + 3$ may well be significantly larger than μ_1. See Figure 5.6 for a geometric interpretation of *superiority*. The remainder of the section provides the mathematical detail.

Hypotheses

$$H_0: \mu_2 - \mu_1 \le \delta \qquad\qquad \textit{(Null Hypothesis)}$$
$$H_1: \mu_2 - \mu_1 > \delta \qquad\qquad \textit{(Alternate Hypothesis)}$$

Clinical Margin

$$\delta > 0$$

Test Statistics
Difference of Means

$$\Delta\hat{\mu} = \bar{x} - \bar{y} \equiv \frac{1}{n_1}\sum_{j=1}^{n_1} x_j - \frac{1}{n_2}\sum_{k=1}^{n_2} y_k$$

Equal Variances

$$T_E(\delta) = \frac{\Delta\hat{\mu} - \delta}{SE_E}$$

$$SE_E = S\sqrt{\frac{1}{n_1} + \frac{1}{n_2}}, \quad S_x^2 = S_y^2 = S^2 = \frac{1}{n_1 - 1}\sum_{j=1}^{n_1}(x_j - \bar{x})^2$$

Unequal Variances

$$T_U(\delta) = \frac{\Delta\hat{\mu} - \delta}{SE_U}$$

$$SE_U = \sqrt{\frac{1}{n_1}S_x^2 + \frac{1}{n_2}S_y^2}, \quad S_x^2 = \frac{1}{n_1 - 1}\sum_{j=1}^{n_1}(x_j - \bar{x})^2, S_y^2 = \frac{1}{n_2 - 1}\sum_{j=1}^{n_2}(y_j - \bar{y})^2$$

Test Statistic Distribution

$T(X, Y) \sim \mathcal{T}_{df_e}$ = the *student t-distribution* with df_u degrees of freedom

$\phi_T(t; df_u)$ = *pdf* of the $\mathcal{T}$-distribution defined by equation (5.2) of Section 5.1 and with *df* degrees of freedom as per sections 5.2.1.1 (df_e) or 5.2.1.2 (df_u).

p-Values

$$p = \begin{cases} \Phi_T(-T_E(\delta); df_e) & \text{for equal variances} \\ \Phi_T(-T_U(\delta); df_u) & \text{for unequal variances} \end{cases}$$

Null Hypothesis Decision

If the *p*-value is less than the significance level, *reject* the null hypothesis.

$p < \alpha \Rightarrow$ *reject H_0* in favor of H_1. Otherwise, *do not reject H_0*. This criterion applies to *both* equal and unequal variances.

$(1 - \alpha) \cdot 100\%$ Confidence Interval for $\Delta\mu = \mu_1 - \mu_2$

Equal Variances

$$CI(\Delta\mu) = (-\infty, \Delta\hat{\mu} + t_{df_e}(1 - \alpha) \cdot SE_E]$$

Unequal Variances

$$CI(\Delta\mu) = (-\infty, \Delta\hat{\mu} + t_{df_u}(1 - \alpha) \cdot SE_U]$$

The two-sided $(1 - \alpha) \cdot 100\%$ *confidence intervals* for μ_1 and μ_2 are the same as noted in Section 5.2.1.1.

Required Sample Size

$$R = \frac{n_2}{n_1}$$

$$n_2 = \left\lceil \left[\frac{(z_{1-\alpha} + z_{1-\beta})^2}{(es - \delta)^2} \cdot \left(R \cdot \sigma_x^2 + \sigma_y^2 \right) \right] \right\rceil$$

es = effect size ($es = \Delta\mu = \mu_2 - \mu_1$). When σ_x^2 and σ_y^2 are unknown, use S_x^2 and S_y^2 in their place, respectively. In the special case that $\sigma_x^2 = \sigma_y^2$, the term $R \cdot \sigma_x^2 + \sigma_y^2$ can be replaced by $(R + 1) \cdot \sigma_x^2$. See Chow et al. [8], equation (3.2.3).

MATLAB Function

`nimean.m`

Example: To test whether the average weight of a basketball player μ_B is at least $\delta = 10$ pounds heavier than the average weight of a hockey player μ_H, it must be

TABLE 5.9 Hypothesis test summary for superiority of average weights

Test	p-value	Reject H_0?	μ_H	$CI(\mu_H)$	μ_B	$CI(\mu_B)$
$H_0: \mu_B - \mu_H \leq \delta$	1	No $(p > \alpha)$	202.1	[200.97, 203.24]	222.36	[219.91, 224.82]
$CI(\mu_B - \mu_H) = (-\infty, 22.2954]$						

shown that $\mu_B \geq \mu_H + 10$. That is, $\mu_B - \mu_H \geq 10$. Let $\delta = 10$ so that the hypotheses are

$$H_0: \mu_B - \mu_H \leq \delta = 10 \qquad \qquad \textit{(Null hypothesis)}$$
$$H_1: \mu_B - \mu_H > \delta = 10 \qquad \qquad \textit{(Alternate hypothesis)}$$

In this setting, the hockey player weights are the *Control* measurements whereas the basketball player weights are the *Test*. Examining Table 5.9, it is seen that the upper bound of the confidence interval for the difference between the mean hockey player (μ_H) and basketball player (μ_B) weights is $\mu_H - \mu_B = -18$ pounds or $\mu_B - \mu_H = 18 > 10$ pounds $= \delta$. This fact combined with the *p*-value (greater than the significance level α) means that the null hypothesis is *rejected in favor of the alternate*. Hence, on average, basketball players are more than 10 pounds heavier than hockey players.

MATLAB Commands
(Test of Superiority)

```
% Data Directory
Ddir = 'C:\PJC\Math_Biology\Chapter_5\Data';
% Load in the Hockey and Basketball data
load(fullfile(Ddir,'HockeyData.mat'))
load(fullfile(Ddir,'BasketballData.mat'))
% Collect the hockey (X) and basketball (Y) weight data
Y = D.Wt; X = B.Wt;
% Set the significance level at 5%: α = 0.05 and clinical margin δ = 10 (pounds)
alpha = 0.05; delta = 10;
% Test the hypotheses H₀: μ_B − μ_H ≤ δ vs. H₁: μ_B − μ_H > δ
H = nimean(X,Y,alpha,delta,1)
H =
          Ho: 'do not reject'
           T: -21.9665
          Mx: 222.3653
          My: 202.1038
         CIx: [219.9090  224.8216]
         CIy: [200.9658  203.2418]
          dM: 20.2615
```

```
CIdM:  [-Inf 22.2954]
   df:  628
 Test:  'one-sided, superiority, unequal variances'
Pvalue:  1
```

5.3 TESTS OF PROPORTIONS

As with Section 5.2 on tests of means, this section will emphasize the hypothesis tests used to determine the difference of proportions from two distinct populations. In keeping with Section 5.2, the *population samples* will be represented by the notation $X = \{x_1, x_2,..., x_{n_1}\}$ and $Y = \{y_1, y_2,..., y_{n_2}\}$. As with tests of means, it will be assumed that X represents the *Control* population while Y reflects the *Test* population. A distinctive feature of this section is the relaxation of sample independence. That is, *correlated* proportions along with independent proportions will be modeled. Each population will be represented by binomially distributed random variables X and Y. Thus, the samples are from $X \sim B(p_1, n_1)$ and $Y \sim B(p_2, n_2)$.

Consider the case of n independent and identically distributed (*i.i.d.*) *Bernoulli* trials $B_1, B_2,..., B_n$ with probability p. That is, each B_j is a binary random variable with states 0 and 1. The probability of "success" is p while the probability of "failure" is $1 - p$. Succinctly put, $\text{Prob}(B_j = 1) = p$ and $\text{Prob}(B_j = 0) = 1 - p$. A binomial random variable U_n can be written as the sum of n independent Bernoulli trials. Thus, $U_n = B_1 + B_2 + \cdots + B_n = \sum_{j=1}^{n} B_j \sim B(p, n)$. Since $E[B_j] = \sum_{j=0}^{1} j \cdot f_{B_j}(j) = 0 \cdot \text{Prob}(B_j = 0) + 1 \cdot \text{Prob}(B_j = 1) = p$ for any j, $E[U_n] = \sum_{j=1}^{n} E[B_j] = np$. Moreover, as U_n is the sum of independent and identically distributed (*i.i.d.*) random variables, $\text{var}(U_n) = \text{var}(\sum_{j=1}^{n} E[B_j]) = \sum_{j=1}^{n} \text{var}(B_j) = n \cdot \text{var}(B_1) = n \cdot (E[B_1^2] - E[B_1]^2) = n \cdot (0 \cdot \text{Prob}(B_1^2 = 0) + 1 \cdot \text{Prob}(B_1^2 = 1) - p^2) = n \cdot (\text{Prob}(B_1 = 1) - p^2) = n \cdot (p - p^2) = np(1 - p)$.

In summary, $E[U_n] = np$ and $\text{Var}[U_n] = np(1 - p)$. Next one of the most important results in mathematical statistics is stated.

Central Limit Theorem: If $X_1, X_2,..., X_n$ are a collection of independent identically distributed random variables with mean μ and nonzero variance σ^2 and $S_n = \sum_{j=1}^{n} X_j$, then

$$\lim_{n \to \infty} \text{Prob}\left(\frac{S_n - n\mu}{\sigma \sqrt{n}} \leq x\right) = \Phi(x) \quad \text{for } -\infty \leq x \leq \infty$$

That is, for sufficiently large n, $\frac{S_n - n\mu}{\sigma \sqrt{n}}$ is approximately a standard normal random variable.

Reminder: As per the *nomenclature* developed in Section 5.1, $\Phi(x)$ is the cumulative density function for the standard normal distribution.

Returning to the example of the binomially distributed random variable U_n, it has been demonstrated that U_n is the sum of *i.i.d.* Bernoulli random variables with mean $\mu = p$ and variance $\sigma^2 = p(1 - p)$. Hence, $\frac{U_n - n\mu}{\sigma\sqrt{n}} = \frac{U_n - np}{\sqrt{p(1-p)}\sqrt{n}} = \frac{U_n - E[U_n]}{\sqrt{\text{var}(U_n)}} = \frac{\frac{1}{n}U_n - p}{\sqrt{\frac{1}{n}p(1-p)}} \approx \mathcal{N}(0, 1)$. Observe that $\frac{1}{n}U_n = \hat{p}$ is an estimate of the genuine proportion.

Thereby, $\frac{\hat{p} - p}{\sqrt{\frac{1}{n}p(1-p)}}$ is distributed as approximately $\mathcal{N}(0, 1)$.

Consequently, to examine the difference $\Delta p = p_1 - p_2$ of two independent proportions, the following test statistic is asymptotically normally distributed.

$$Z(X, Y) = \frac{p_1 - p_2}{\sqrt{\frac{1}{n_1}p_1(1 - p_1) + \frac{1}{n_2}p_2(1 - p_2)}} \equiv \frac{\Delta p}{SE} \sim \mathcal{N}(0, 1)$$

Here, the symbol SE represents the standard error. If the proportions are correlated with correlation coefficient ρ_{12}, then the test statistic becomes

$$Z_{cor}(X, Y) = \frac{p_1 - p_2}{\sqrt{\frac{1}{n_1}p_1(1 - p_1) + \frac{1}{n_2}p_2(1 - p_2) - 2\rho_{12}p_1(1 - p_1)p_2(1 - p_2)}}$$
$$\equiv \frac{\Delta p}{SE_{cor}} \sim \mathcal{N}(0, 1)$$

Now the list of hypothesis tests of proportion is presented in the further sections. Throughout Section 5.3, the following notations and comments are employed.

Notes:

(i) The samples X and Y are binary. That is, $x_j, y_\ell \in \mathbb{Z}_2 = \{0, 1\}$ for each $j = 1, 2,..., n_1$ and $\ell = 1, 2,..., n_2$. These integers represent a "success" (1) or "failure" (0) with respect to the binomial distribution. Therefore, $k_1 \equiv \sum_{j=1}^{n_1} x_j$ and $k_2 \equiv \sum_{\ell=1}^{n_2} y_\ell$ are the number of successes in the samples X and Y, respectively. Also X will represent a sample of the *Control* variable X while Y is taken from the *Test* population modeled by the random variable Y.

(ii) The difference in proportions will be designated by the symbol $\Delta p = p_1 - p_2$. The complementary difference is denoted by $\Delta p' = p_2 - p_1$.

(iii) Following Altman et al. [3], the null hypothesis decision will be determined via *both Newcombe score–method confidence interval NCI(Δp)* [26, 27] and the traditional p-value method. Specifically, if $0 \notin NCI(\Delta p)$, then the null hypothesis $H_0: p_1 = p_2$ is *rejected regardless of p-value*. In this sense, the confidence interval of the difference in proportions takes precedence over the p-value. Nevertheless, the p-values will be calculated for each test.

5.3.1   Two-Sided Tests of Proportions

This section will present hypothesis tests for both independent and correlated populations. Also the equivalence test, as illustrated in Section 5.1, Figure 5.5, will be outlined.

5.3.1.1   Test for Equality (Samples from Two Independent Populations)   Proceed as in Section 5.2.

Hypotheses

$$H_0: p_1 = p_2 \qquad \text{(Null hypothesis)}$$
$$H_1: p_1 \neq p_2 \qquad \text{(Alternate hypothesis)}$$

Test Statistic

$$Z(X, Y) = \frac{\Delta \hat{p}}{SE}$$

$$\Delta \hat{p} = \hat{p}_1 - \hat{p}_2 \equiv \frac{k_1}{n_1} - \frac{k_2}{n_2}, k_1 \equiv \sum_{j=1}^{n_1} x_j, k_2 \equiv \sum_{\ell=1}^{n_2} y_\ell$$

$$SE = \sqrt{\frac{1}{n_1}\hat{p}_1(1 - \hat{p}_1) + \frac{1}{n_2}\hat{p}_2(1 - \hat{p}_2)}$$

Test Statistic Distribution

$Z(X, Y) \sim \mathcal{N}(0, 1) =$ the *standard normal* distribution

$\phi(x) = \frac{1}{\sqrt{2\pi}}e^{-\frac{1}{2}x^2}$ for $-\infty < x < \infty =$ the *pdf* of the standard normal distribution

p-Value

$$\pi = 2 \cdot \Phi(-|Z(X, Y)|)$$

$\Phi(x) = $ *cumulative density function* of a standard normal random variable

Newcombe $(1 - \alpha) \cdot 100\%$ Score Method Confidence Interval for $\Delta p = p_1 - p_2$

$$NCI(\Delta p) = \left[\Delta \hat{p} - \sqrt{(\hat{p}_1 - \ell_1)^2 + (\hat{p}_2 - u_2)^2}, \Delta \hat{p} + \sqrt{(\hat{p}_2 - \ell_2)^2 + (\hat{p}_1 - u_1)^2}\right]$$

$$\left. \begin{array}{l} \ell_1 = \dfrac{Q_1 - z_{1-\alpha/2} S_1}{\Upsilon_1}, \quad \ell_2 = \dfrac{Q_2 - z_{1-\alpha/2} S_2}{\Upsilon_2}, \quad \Upsilon_j = 2\left(n_j + z_{1-\alpha/2}^2\right), j = 1, 2 \\[2mm] u_1 = \dfrac{Q_1 + z_{1-\alpha/2} S_1}{\Upsilon_1}, \quad u_2 = \dfrac{Q_2 + z_{1-\alpha/2} S_2}{\Upsilon_2} \\[2mm] Q_1 = 2k_1 + z_{1-\alpha/2}^2, \qquad Q_2 = 2k_2 + z_{1-\alpha/2}^2, \\[2mm] S_1 = \sqrt{z_{1-\alpha/2}^2 + 4k_1 \cdot (1 - \hat{p}_1)}, \quad S_2 = \sqrt{z_{1-\alpha/2}^2 + 4k_2 \cdot (1 - \hat{p}_2)} \end{array} \right\} \quad (5.6)$$

Null Hypothesis Decision

If $0 \notin NCI(\Delta p)$ or $\pi < \alpha$, then *reject* H_0. Otherwise, *do not reject* H_0.

Newcombe $(1 - \alpha) \cdot 100\%$ Score Method Confidence Intervals for p_1 and p_2

$$\left. \begin{array}{l} NCI(p_j) = \left[\dfrac{A_j - B_j}{C_j}, \dfrac{A_j + B_j}{C_j} \right] \\[3mm] A_j = 2k_j + z_{1-\alpha/2}^2 \\[2mm] B_j = z_{1-\alpha/2} \sqrt{z_{1-\alpha/2}^2 + 4k_j(1 - \hat{p}_j)} \\[2mm] C_j = 2\left(n_j + z_{1-\alpha/2}^2 \right) \end{array} \right\}, j = 1, 2 \qquad (5.7)$$

Required Sample Size

$$R = \frac{n_2}{n_1}$$

$$n_2 = \left[\left[\frac{(z_{1-\alpha/2} + z_{1-\beta})^2}{es^2} \cdot (R \cdot p_1(1 - p_1) + p_2(1 - p_2)) \right] \right]$$

es = effect size $(es = |\Delta p| = |p_1 - p_2|)$. See Chow et al. [8]; especially formula (4.2.2).

MATLAB Function

`smpropdiff.m`

Example: Consider the proportion of hockey players who are *strictly* taller than 6 feet versus the proportion of basketball players of the same height. Let p_H and p_B be the respective proportions. The hypotheses are

$$H_0: p_H = p_B \qquad \text{(Null hypothesis)}$$
$$H_1: p_H \neq p_B \qquad \text{(Alternate hypothesis)}$$

In Table 5.10, it is seen that the confidence interval for the difference between the proportion of hockey players (p_H) and basketball players (p_B) whose heights exceed 72 inches (= 6 feet) is *strictly negative*. This means that the difference $\Delta p = p_H - p_B$ *cannot* be 0 statistically. Hence, $0 \notin NCI(\Delta p)$ and the null hypothesis of equal proportions is *rejected*.

TABLE 5.10   Hypothesis test summary for test of equal independent proportions

Test	p-value	Reject H_0?	p_H	$NCI(p_H)$	p_B	$NCI(p_B)$
$H_0: p_H = p_B$	$<10^{-4}$	Yes $(p < \alpha)$	0.64	$[0.61, 0.68]$	0.96	$[0.94, 0.98]$
$NCI(p_H - p_B) = [-0.36, -0.28] \Rightarrow 0 \notin NCI(\Delta p) \Rightarrow$ reject H_0						

MATLAB Commands
(*Two-Sided Test of Equal Independent Proportions*)

```
% Data Directory
Ddir = 'C:\PJC\Math_Biology\Chapter_5\Data';
% Load in the Hockey and Basketball data
load(fullfile(Ddir,'HockeyData.mat'))
load(fullfile(Ddir,'BasketballData.mat'))
% Collect the hockey (X) and basketball (Y) height data
X = D.Ht; Y = B.Ht;
% Proportions of hockey and basketball players above 6 feet (72 inches) tall
Kh = sum(X > 72); Nh = numel(X);
Kb = sum(Y > 72); Nb = numel(Y);
% Set the significance level at 5%: α = 0.05
alpha = 0.05;
% Test the hypotheses H₀: pH = pB vs. H₁: pH ≠ pB
H = smpropdiff([Kh,Kb],[Nh,Nb],alpha,2)
H =
        Ho: 'reject'
        dp: -0.3202
         Z: -15.7019
        CI: [-0.3602 -0.2802]
       NCI: [-0.3594 -0.2789]
    Pvalue: 1.4681e-55
         p: [0.6433 0.9635]
       CI1: [0.6074 0.6792]
      NCI1: [0.6067 0.6783]
       CI2: [0.9459 0.9810]
      NCI2: [0.9415 0.9774]
       rho: 0
```

5.3.1.2 Test for Equality (Samples from Two Correlated Populations)   In this setting, the populations are *not* independent. It can be assumed that the samples X and Y with binary elements measure a "yes/no" quality as identified in the contingency Table 5.11. For illustrative purposes, it will be assumed that the proportions measure the "yes" rates as $p_X = p_1 = (x_{11} + x_{01})/n$ and $p_Y = p_2 = (x_{11} + x_{10})/n$ with $n = x_{11} + x_{10} + x_{01} + x_{00}$.

TABLE 5.11   Contingency for correlated dichotomous measurements

		X	
		Yes	No
Y	Yes	x_{11}	x_{10}
	No	x_{01}	x_{00}

Hypotheses

$$H_0: p_1 = p_2 \qquad \text{(Null hypothesis)}$$
$$H_1: p_1 \neq p_2 \qquad \text{(Alternate hypothesis)}$$

Test Statistic

$$Z(X, Y) = \frac{\Delta\hat{p}}{SE_{cor}}$$

$$\Delta\hat{p} = \hat{p}_1 - \hat{p}_2 \equiv \frac{k_1 - k_2}{n}, k_1 \equiv x_{11} + x_{01}, k_2 \equiv x_{11} + x_{10}$$

$$SE_{cor} = \frac{1}{z_{1-\alpha/2}}(L + U)$$

$$\left.\begin{array}{l}
L = \sqrt{(\hat{p}_1 - \ell_1)^2 + (u_2 - \hat{p}_2)^2 - 2\rho_{12}(\hat{p}_1 - \ell_1)(u_2 - \hat{p}_2)} \\[2mm]
U = \sqrt{(\hat{p}_2 - \ell_2)^2 + (u_1 - \hat{p}_1)^2 - 2\rho_{12}(\hat{p}_2 - \ell_2)(u_1 - \hat{p}_1)} \\[2mm]
\ell_j = \dfrac{2k_j + z_{1-\alpha/2}^2 - z_{1-\alpha/2}\sqrt{z_{1-\alpha/2}^2 + 4k_j(1 - \hat{p}_j)}}{2\left(n + z_{1-\alpha/2}^2\right)} \\[6mm]
u_j = \dfrac{2k_j + z_{1-\alpha/2}^2 + z_{1-\alpha/2}\sqrt{z_{1-\alpha/2}^2 + 4k_j(1 - \hat{p}_j)}}{2\left(n + z_{1-\alpha/2}^2\right)}
\end{array}\right\}, j = 1, 2 \qquad (5.8)$$

$$\rho_{12} = \text{correlation in-between the sample populations} = \begin{cases} 0 & \text{for } A = 0 \\ \dfrac{C}{\sqrt{A}} & \text{for } A \neq 0 \end{cases}$$

$$A = (x_{00} + x_{01})(x_{10} + x_{11})(x_{00} + x_{10})(x_{01} + x_{11})$$

$$B = x_{11} \bullet x_{00} - x_{01} \bullet x_{10}$$

$$C = \begin{cases} B - \dfrac{n}{2} & \text{for } B > \dfrac{n}{2} \\ 0 & \text{for } 0 \leq B \leq \dfrac{n}{2} \\ B & \text{for } B < 0 \end{cases}$$

Test Statistic Distribution

$Z(X, Y)$ is asymptotically *standard normal*

$$\phi(x) = \frac{1}{\sqrt{2\pi}}e^{-\frac{1}{2}x^2} \text{ for } -\infty < x < \infty = \text{the } pdf \text{ of the standard normal distribution}$$

p-Value

$$p = 2 \bullet \Phi(-|Z(X, Y)|)$$

$\Phi(x) = $ *cumulative density function* of a standard normal random variable

Newcombe $(1 - \alpha) \cdot 100\%$ Score Method Confidence Interval for $\Delta p = p_1 - p_2$

$$NCI(\Delta p) = [\Delta p - L, \Delta p + U]$$

L and U as for the *test statistic* in (5.8)

Null Hypothesis Decision
If $0 \notin NCI(\Delta p)$ or $p < \alpha$, then *reject H_0*. Otherwise, *do not reject H_0*.

Newcombe $(1 - \alpha) \cdot 100\%$ Score Method Confidence Intervals for p_1 and p_2
$NCI(p_j)$ is the same as formula (5.7).

Required Sample Size

$$n = \left\lceil \frac{(z_{1-\alpha/2} + z_{1-\beta})^2}{es^2} \cdot \left(p_1(1 - p_1) + p_2(1 - p_2) - 2\rho_{12}\sqrt{p_1(1 - p_1)p_2(1 - p_2)} \right) \right\rceil$$

es = effect size ($es = |\Delta p| = |p_1 - p_2|$).
This is derived directly from the case of independent proportions by noting that the samples are matched (i.e., $n_1 = n_2$) and by including the covariance term due to correlation in the variance.

MATLAB Function
smpropdiff.m

Example: Suppose a test of correlated proportions

$$H_0: p^+(C) = p^+(T) \text{ vs. } H_1: p^+(C) \neq p^+(T)$$

has the contingency matrix with the entries as following.

		Control	
		Yes	No
Test	Yes	125	2
	No	6	144

In this case, $p^+(C)$ and $p^+(T)$ are the proportions of the "yes" designation. The following MATLAB commands test the hypotheses stated earlier for the correlated populations of *Control* and *Test*. Table 5.12 summarizes the computations. Since $0 \in NCI(p^+(C) - p^+(T))$, the null H_0 *cannot be rejected*. Notice also that the correlation of the (*Control*, *Test*) pair is 93.5%; $\rho_{Control,Test} = 0.935$.

TABLE 5.12   Hypothesis test summary for test of equal correlated proportions

Test	p-value	Reject H_0?	$p^+(C)$	$NCI(p^+(C))$	$p^+(T)$	$NCI(p^+(T))$
$H_0: p^+(C) = p^+(T)$	0.18	No $(p > \alpha)$	0.47	[0.41, 0.53]	0.46	[0.40, 0.52]

$NCI(p_H - p_B) = [-0.007, 0.035] \Rightarrow 0 \in NCI(\Delta p^+) \Rightarrow$ *do not reject H_0*

MATLAB Commands
(Two-Sided Test of Equal Correlated Proportions)

```
% Contingency matrix
A = [125,2;6,144];
% Number of positives (i.e., "yes") for Control and Test
Kc = sum(A(:,1)); Kt = sum(A(1,:));
% Total number of samples
N = sum(A(:));
% Set the significance level at 5%: α = 0.05
alpha = 0.05;
% Test of hypotheses H₀: p⁺(Control) = p⁺(Test) vs. H₁: p⁺(Control) ≠ p⁺(Test)
H = smpropdiff([Kc,Kt],N*[1,1],alpha,2,A)
H =
         Ho: 'do not reject'
         dp: 0.0144
          Z: 1.3455
         CI: [-0.0066 0.0355]
        NCI: [-0.0066 0.0354]
     Pvalue: 0.1785
          p: [0.4729 0.4585]
        CI1: [0.4141 0.5317]
       NCI1: [0.4149 0.5317]
        CI2: [0.3998 0.5172]
       NCI2: [0.4008 0.5173]
        rho: 0.9351
```

5.3.1.3   Test for Equivalence (Samples from Two Independent Populations)   As with the tests for equivalent means in Section 5.2.1.3, the *margin δ* is considered an input established by clinical practitioners and experts. It will be considered a positive constant $\delta > 0$. To establish equivalence, the *null hypothesis is rejected in favor of the alternate hypothesis.*

Hypotheses

$$H_0: |p_1 - p_2| \geq \delta \qquad \text{(Null hypothesis)}$$
$$H_1: |p_1 - p_2| < \delta \qquad \text{(Alternate hypothesis)}$$

Test Statistic

$$Z_\ell(X, Y) = \frac{\Delta\hat{p} + \delta}{SE}, Z_u(X, Y) = \frac{\Delta\hat{p} - \delta}{SE}$$

$$\Delta\hat{p} = \hat{p}_1 - \hat{p}_2 \equiv \frac{k_1}{n_1} - \frac{k_2}{n_2}, k_1 \equiv \sum_{j=1}^{n_1} x_j, k_2 \equiv \sum_{\ell=1}^{n_2} y_\ell$$

$$SE = \sqrt{\frac{1}{n_1}\hat{p}_1(1 - \hat{p}_1) + \frac{1}{n_2}\hat{p}_2(1 - \hat{p}_2)}$$

Test Statistic Distribution

$Z_\ell(X, Y)$ and $Z_u(X, Y)$ are asymptotically *standard normal*

$$\phi(x) = \frac{1}{\sqrt{2\pi}}e^{-\frac{1}{2}x^2} \text{ for } -\infty < x < \infty = \text{the } pdf \text{ of the standard normal distribution}$$

p-Value

In this case, there are in essence two *p*-values. If $Z_\ell(X, Y) \geq z_{1-\alpha}$ and $Z_u(X, Y) \leq z_\alpha$, then H_0 is rejected in favor of H_1. As the standard normal *cdf* is a monotone increasing function, it can be applied to both sides of the first inequality to yield $\Phi(Z_\ell) \geq 1 - \alpha$ or $\alpha \geq 1 - \Phi(Z_\ell)$. Hence, $\alpha \geq \Phi(-Z_\ell) \equiv \pi_\ell$. Operating in the same manner on the second inequality produces $\alpha \geq \Phi(Z_u) \equiv \pi_u$. Therefore, if $a > \pi = \max(\pi_\ell, \pi_u)$, then reject H_0 in favor of H_1. The quantity π is the *p*-value.

$\Phi(x) = $ *cumulative density function* of a standard normal random variable

Newcombe $(1 - \alpha) \bullet 100\%$ Score Method Confidence Interval for $\Delta p = p_1 - p_2$

$NCI(\Delta p)$ is the same as formula (5.6) from Section 5.3.1.1.

Null Hypothesis Decision

If $NCI(\Delta p) \subset [-\delta, \delta]$ or $\pi < \alpha$, then *reject H_0*. Otherwise, *do not reject H_0*.

Newcombe $(1 - \alpha) \bullet 100\%$ Score Method Confidence Intervals for p_1 and p_2

$NCI(p_j)$ is the same as formula (5.7) from Section 5.3.1.1.

Required Sample Size

$$R = \frac{n_2}{n_1}$$

$$n_2 = \left\lceil \frac{(z_{1-\alpha} + z_{1-\beta/2})^2}{(\delta - es)^2} \cdot (R \cdot p_1(1 - p_1) + p_2(1 - p_2)) \right\rceil$$

$es = $ effect size $(es = |\Delta p| = |p_1 - p_2|)$. See Chow et al. [8]; especially formula (4.2.4).

MATLAB Function

```
nipropi.m
```

TABLE 5.13 **Hypothesis test summary for test of equivalence for independent proportions**

Test	p-value	Reject H_0?	p_H	$NCI(p_H)$	p_B	$NCI(p_B)$
$H_0: \|p_H - p_B\| \geq \delta$	1	No ($p > \alpha$)	0.47	[0.41, 0.53]	0.46	[0.40, 0.52]
$NCI(p_H - p_B) = [-0.285, -0.180] \Rightarrow NCI(\|p_H - p_B\|) = [0.180, 0.285] > \delta = 0.025$						

Example: In this example, a test of equivalence of the percentage of hockey and basketball players who weigh *at least* 200 pounds will be examined. If the percentages are within 2.5%, then they will be considered "equivalent." Hence, the hypotheses

$$H_0: |p(H_{wt} \geq 200) - p(B_{wt} \geq 200)| \geq \delta \text{ vs. } H_1: |p(H_{wt} \geq 200) - p(B_{wt} \geq 200)| < \delta$$

will be tested for $\delta = 0.025$. It is evident from Table 5.13 that the absolute value of the difference in proportions $|\Delta p| = |p(H_{wt} \geq 200) - p(B_{wt} \geq 200)| = |0.57 - 0.81| = 0.24 \gg \delta = 0.025$. Moreover, the low end of the Newcombe confidence interval for the difference $|\Delta p|$ is $0.18 > \delta$. Thus, $NCI(|p_H - p_B|) \not\subset [-\delta, \delta]$, the null hypothesis *cannot* be rejected, and the rates are not equivalent.

MATLAB Commands
(Two-Sided Test of Equivalent Proportions)

```
% Data Directory
Ddir = 'C:\PJC\Math_Biology\Chapter_5\Data';
% Load in the Hockey and Basketball data
load(fullfile(Ddir,'HockeyData.mat'))
load(fullfile(Ddir,'BasketballData.mat'))
% Collect the hockey (X) and basketball (Y) weight data
X = D.Wt; Y = B.Wt;
% Indices of hockey and basketball players at least 200 pounds in weight
Xh = (X >= 200); Yb = (Y >= 200);
% Set the significance level at 5%: alpha = 0.05 and clinical margin delta = 0.025
alpha = 0.05; delta = 0.025;
% Test the hypotheses
% Ho: |p(Hwt >= 200) - p(Bwt >= 200)| >= delta vs. H1: |p(Hwt >= 200) - p(Bwt >= 200)|
   < delta
H = nipropi(Xh,Yb,alpha,delta,2)
H =
        Pvalue: 1.0000
           Ho: 'do not reject'
           p1: 0.5716
          NCI1: [0.5343 0.6082]
           p2: 0.8059
          NCI2: [0.7663 0.8402]
           dp: -0.2343
```

```
      NCI:  [-0.2850 -0.1804]
  TestType:  'Equivalence'
    margin:  0.0250
```

5.3.1.4   Test for Equivalence (Samples from Two Correlated Populations)   This

section parallels Section 5.3.1.2 in that the populations are *not* independent. As stated
earlier, the model considers populations with binary outcomes (i.e., "yes" or "no").
This is identified via the contingency Table 5.11. Again, the assumption is that the
proportions measure the "yes" percentages as $p_1 = (x_{11} + x_{01})/n$ and $p_2 = (x_{11} + x_{10})/n$ with $n = x_{11} + x_{01} + x_{10} + x_{00}$. In this case, the difference between p_1 and p_2
is $p_1 - p_2 = (x_{11} + x_{01})/n - (x_{11} + x_{10})/n = (x_{01} - x_{10})/n$.

Before establishing the test statistic, several matters must be addressed as per the
work of Liu et al. [20]. Using Table 5.11 as a reference, let $x_{j,\bullet} = \sum_{k=0}^{1} x_{j,k}, j = 0, 1$;
$x_{\bullet,k} = \sum_{j=0}^{1} x_{j,k}, k = 0, 1$; and $n = \sum_{j=0}^{1} \sum_{k=0}^{1} x_{j,k}$. Liu et al. [20] note that the two-
sided equivalence hypothesis tests H_0: $|p_1 - p_2| \geq \delta$ vs. H_1: $|p_1 - p_2| < \delta$ can be
decomposed into two one-sided tests $H_{0\ell}$: $p_2 - p_1 \leq -\delta$ vs. $H_{1\ell}$: $p_2 - p_1 > -\delta$ and
H_{0u}: $p_2 - p_1 \geq \delta$ vs. H_{1u}: $p_2 - p_1 < \delta$ for $\delta > 0$. If both $H_{0\ell}$ and H_{0u} are *rejected*, then
H_0 is *rejected* and statistical equivalence to within the clinical margin δ is established.

This criteria of rejecting two one-sided tests requires the development of two test
statistics. In this section, the Wald asymptotic test and the *RMLE* (restricted maximum likelihood estimator) are presented. Both statistics are included in the M-file
`niprop.m` and are used to render a decision with respect to the null hypothesis H_0.

With these remarks in mind, the development now proceeds.

Hypotheses

$$H_0: |p_1 - p_2| \geq \delta \qquad\qquad\qquad \text{(Null hypothesis)}$$
$$H_1: |p_1 - p_2| < \delta \qquad\qquad\qquad \text{(Alternate hypothesis)}$$

Test Statistics

$$\widehat{\theta} = \hat{p}_{10} - \hat{p}_{01} \equiv \frac{1}{n}(x_{10} - x_{01}) = p_2 - p_1$$

$$\hat{p}_{j,k} = \frac{1}{n}x_{j,k}$$

Wald

$$Z_\ell(X, Y) = \frac{\widehat{\theta} + \delta}{\hat{\sigma}(\theta)} \equiv \frac{(x_{10} - x_{01}) + n\delta}{\sqrt{(x_{10} + x_{01}) - \frac{1}{n}(x_{10} - x_{01})^2}},$$

$$Z_u(X, Y) = \frac{\widehat{\theta} - \delta}{\hat{\sigma}(\theta)} \equiv \frac{(x_{10} - x_{01}) - n\delta}{\sqrt{(x_{10} + x_{01}) - \frac{1}{n}(x_{10} - x_{01})^2}},$$

$$\hat{\sigma}^2(\theta) = \frac{1}{n}(\hat{p}_{10} + \hat{p}_{01} - \hat{\theta}^2) = (x_{10} + x_{01}) - \frac{1}{n}(x_{10} - x_{01})^2$$

RMLE

$$Z_\ell^{RMLE} = \frac{\hat{\theta} + \delta}{\hat{\sigma}_\ell(\delta)}$$

$$\hat{\sigma}_\ell^2(\delta) = \frac{1}{n}(2\tilde{p}_{\ell,01} - \delta(1 + \delta)),$$

$$\tilde{p}_{\ell,01} = \frac{1}{4}\left(\tilde{a}_\ell + \sqrt{\tilde{a}_\ell^2 - 8\hat{p}_{01} \cdot \delta(1 + \delta)}\right), \tilde{a}_\ell = \hat{\theta} \cdot (1 - \delta) + 2(\hat{p}_{01} + \delta)$$

$$Z_u^{RMLE} = \frac{\hat{\theta} - \delta}{\hat{\sigma}_u(\delta)}$$

$$\hat{\sigma}_u^2(\delta) = \frac{1}{n}(2\tilde{p}_{u,01} + \delta(1 - \delta)),$$

$$\tilde{p}_{u,01} = \frac{1}{4}\left(\tilde{a}_u + \sqrt{\tilde{a}_u^2 + 8\hat{p}_{01} \cdot \delta(1 - \delta)}\right), \tilde{a}_u = \hat{\theta} \cdot (1 + \delta) + 2(\hat{p}_{01} - \delta)$$

Test Statistic Distribution

$Z_\ell(X, Y)$, $Z_u(X, Y)$, Z_ℓ^{RMLE}, and Z_u^{RMLE} all have an asymptotically *standard normal* distribution (see per Liu et al. [20])

$$\phi(x) = \frac{1}{\sqrt{2\pi}}e^{-\frac{1}{2}x^2} \text{ for } -\infty < x < \infty = \text{the } pdf \text{ of the standard normal distribution}$$

p-Value

In this case, there are in essence two p-values. For the case of the *Wald* statistics, $Z_\ell(X, Y) \geq z_{1-\alpha/2}$ and $Z_u(X, Y) \leq z_{\alpha/2}$ imply that H_0 is rejected in favor of H_1. As the standard normal *cdf* is a monotone increasing function, it can be applied to both sides of the first inequality to yield $\Phi(Z_\ell) \geq 1 - \alpha/2$ or $\alpha/2 \geq 1 - \Phi(Z_\ell)$. Hence, $\alpha \geq 2 \cdot \Phi(-Z_\ell) \equiv \pi_\ell$. Operating in the same manner on the second inequality produces $\alpha \geq 2 \cdot \Phi(Z_u) \equiv \pi_u$. Therefore, if $\alpha > \pi = \max(\pi_\ell, \pi_u)$, then reject H_0 in favor of H_1. The value π is the p-value. Notice that Liu et al. [20] do not require $Z_\ell(X, Y) \geq z_{1-\alpha/2}$ and $Z_u(X, Y) \leq z_{\alpha/2}$ despite the fact that the hypothesis is a two-sided test. In this treatment, however, the more conservative approach using the quantiles $z_{1-\alpha/2}$ and $z_{\alpha/2}$ is employed.

In a similar fashion, for the *RMLE* statistics, $\pi_\ell^{RMLE} = \Phi(-Z_\ell^{RMLE})$ and $\pi_u^{RMLE} = \Phi(Z_u^{RMLE})$ so that $\alpha > \pi^{RMLE} = \max(\pi_\ell^{RMLE}, \pi_u^{RMLE})$ means that the null hypothesis H_0 should be rejected. Thus, the p-value for the *RMLE* statistic is $p^{RMLE} = \pi^{RMLE}$. For the *RMLE* test, the recommendations of Liu et al. [20] are followed and the quantiles $z_{1-\alpha}$ and z_α are used to compute the p-values since the statistic is a maximum likelihood estimate.

If both tests of equivalence fail to reject the null, then both the conservative and general approach lead to the same conclusion supporting the decision.

$\Phi(x) = $ *cumulative density function* of a standard normal random variable

$(1 - \alpha) \cdot 100\%$ Score Method Confidence Interval for $\hat{\theta} = p_1 - p_2 = \hat{p}_{10} - \hat{p}_{01}$

Wald

$$CI(\hat{\theta}) = [\hat{\theta} + z_{\alpha/2} \cdot \hat{\sigma}(\theta), \hat{\theta} + z_{1-\alpha/2} \cdot \hat{\sigma}(\theta)]$$

RMLE

$$CI(\hat{\theta}) = [\hat{\theta}_\ell, \hat{\theta}_u]$$

$$\hat{\theta}_\ell = \inf_{\theta_o} \left| \frac{\hat{\theta} - \theta_o}{\hat{\sigma}_\ell(\theta_o)} - z_{1-\alpha/2} \right|, \hat{\theta}_u = \inf_{\theta_o} \left| \frac{\hat{\theta} - \theta_o}{\hat{\sigma}_u(\theta_o)} - z_{\alpha/2} \right|$$

Null Hypothesis Decision

If $\pi < \alpha$ or $\pi^{RMLE} < \alpha$, then *reject H_0*. Otherwise, *do not reject H_0*.

Required Sample Size

Wald

$$N_s = \left\lceil \left[(2p_{01}) \cdot \left(\frac{z_{1-\alpha/2} + z_{1-\beta}}{\delta} \right)^2 \right] \right\rceil$$

An *a priori* estimate of p_{01} is required.

RMLE

$$N_s = \left\lceil \left[(2p_{01}) \cdot \left(\frac{\frac{1}{\tilde{w}} \cdot z_{1-\alpha/2} + z_{1-\beta}}{\delta} \right)^2 \right] \right\rceil$$

$$\tilde{w} = \sqrt{\frac{2p_{01}}{2p_{\ell,01} - \delta(1 + \delta)}}$$

MATLAB Functions

`niprop.m, nisamplesize.m`

Example: In this example, the test of proportions of *positive* (or "yes") across the correlated *Control* and *Test* populations will be considered equivalent for a clinical margin of 2.5%, $\delta = 0.025$. The hypotheses are

$$H_0: |p^+(C) - p^+(T)| \geq \delta \text{ vs. } H_1: |p^+(C) - p^+(T)| < \delta$$

where $p^+(C)$ and $p^+(T)$ are the proportions of *Control* and *Test* samples that are designated as "yes."

TABLE 5.14 **Summary estimates of the proportions**

Method	$CI(p^+(C) - p^+(T))$	$CI(\Delta p^+) \subset [-\delta, \delta]$?
Wald	$[-0.0344, 0.0055]$	No
RMLE	$[-0.0309, 0.0042]$	No

If the associated contingency matrix is as below, then the p-values for *both* the Wald and RMLE methods are approximately 0.15 or larger which exceeds the significance level $a = 0.05$. Moreover, the left endpoint of both confidence intervals for the difference in proportions $\Delta p^+ = p^+(C) - p^+(T)$ is less than the negative of the margin. Thus, both test statistics indicate that the null H_0 *should not be rejected* for a clinical margin of 2.5% ($\delta = 0.025$). Table 5.14 summarizes the proportions and their Newcombe 95% confidence intervals.

		Control	
		Yes	No
Test	Yes	125	2
	No	6	144

MATLAB Commands
(*Two-Sided Test of Equivalent Correlated Proportions*)

```
% Contingency matrix
A = [125,2;6,144];
```
% Set the significance level at 5%: $\alpha = 0.05$ and clinical margin $\delta = 0.025$
```
alpha = 0.05; delta = 0.025;
```
% Test of hypotheses H_0: $|p^+(Control) - p^+(Test)| \geq \delta$ vs. H_1: $|p^+(Control) - p^+(Test)| < \delta$
```
H = niprop(A,alpha,delta,2)
H =
Wald
          Ho: 'do not reject'
       Tstat: [1.0379 -3.8766]
      Pvalue: 0.1497
    TestType: 'Equivalence'
CI
       Wald: [-0.0344 0.0055]
       RMLE: [-0.0309 0.0042]
RMLE
          Ho: 'do not reject'
       Tstat: [0.9026 -3.0005]
      Pvalue: 0.1834
    TestType: 'Equivalence'
```

5.3.2   One-Sided Tests of Proportions

This will parallel Section 5.2.1 and include *right-* and *left-tailed tests of proportions* along with *non-inferiority* and *superiority* tests. Again, special attention will be paid to samples from *independent* populations and samples with *correlated* proportions. Notes (i)–(iii) of Section 5.3 will apply throughout Section 5.3.2. Moreover for non-inferiority and superiority tests, the guidance of Lewis [18] and Liu et al. [20] will be adopted. Specifically, the $(1 - \alpha/2) \cdot 100\%$ quantile $z_{1-\alpha/2}$ will be used rather than the conventional $(1 - \alpha) \cdot 100\%$ quantile $z_{1-\alpha}$ for null hypothesis decisions. Indeed, the conventional quantile is considered anti-conservative as per Lewis [18]. The conventional $(1 - \alpha) \cdot 100\%$ quantile $z_{1-\alpha}$ will be retained for the *right-* and *left-tailed* tests. Similarly, $z_{1-\beta/2}$ and $z_{1-\beta}$ will be implemented on sample size calculations for {*right-* and *left-tailed* tests} and {*non-inferiority* and *superiority* tests}, respectively. This will become evident in the sample size formulae given in each subsection.

5.3.2.1   *Right-Tailed Test (Samples from Two Independent Populations)*

Hypotheses

$$H_0: p_1 \leq p_2 \qquad\qquad \textit{(Null hypothesis)}$$
$$H_1: p_1 > p_2 \qquad\qquad \textit{(Alternate hypothesis)}$$

Test Statistic

$$Z(X, Y) = \frac{\Delta\hat{p}}{SE}$$

$$\Delta\hat{p} = \hat{p}_1 - \hat{p}_2 \equiv \frac{k_1}{n_1} - \frac{k_2}{n_2}, k_1 \equiv \sum_{j=1}^{n_1} x_j, k_2 \equiv \sum_{\ell=1}^{n_2} y_\ell$$

$$SE = \sqrt{\frac{1}{n_1}\hat{p}_1(1 - \hat{p}_1) + \frac{1}{n_2}\hat{p}_2(1 - \hat{p}_2)}$$

Test Statistic Distribution
$Z(X, Y) \sim \mathcal{N}(0, 1) =$ the *standard normal* distribution

$\phi(x) = \frac{1}{\sqrt{2\pi}} e^{-\frac{1}{2}x^2}$ for $-\infty < x < \infty =$ the *pdf* of the standard normal distribution

p-Value
$\pi = \Phi(-Z(X, Y))$
$\Phi(x) = $ *cumulative density function* of a standard normal random variable

Newcombe $(1 - \alpha) \bullet 100\%$ Score Method Confidence Interval for $\Delta p = p_1 - p_2$

$NCI(\Delta p) = [\Delta \hat{p} - \sqrt{(\hat{p}_1 - \ell_1)^2 + (\hat{p}_2 - u_2)^2}, +\infty)$, ℓ_j and u_j as given by equation (5.6).

Null Hypothesis Decision

If $\min[NCI(\Delta p)] > 0$ or $\pi < \alpha$, then *reject H_0*. Otherwise, *do not reject H_0*.

Newcombe $(1 - \alpha) \bullet 100\%$ Score Method Confidence Intervals for p_1 and p_2

Same as equation (5.7).

Required Sample Size

$$R = \frac{n_2}{n_1}$$

$$n_2 = \left\lceil\left\lceil \frac{(z_{1-\alpha} + z_{1-\beta})^2}{es^2} \cdot (R \cdot p_1(1 - p_1) + p_2(1 - p_2)) \right\rceil\right\rceil$$

$es = $ effect size ($es = \Delta p = p_1 - p_2$).

See Chow et al. [8]; especially formula (4.2.3) with $\delta = 0$.

MATLAB Function

`smpropdiff.m`

Example: Examine the proportion of hockey players p_H who weigh *at least* 200 pounds in comparison against the proportion of basketball players p_B whose weight is *at least* 200 pounds. If a researcher wishes to demonstrate that there are more basketball players in this category, then the strategy is to test the hypotheses

$$H_0: p_B \le p_H \qquad \qquad \textit{(Null hypothesis)}$$

$$H_1: p_B > p_H \qquad \qquad \textit{(Alternate hypothesis)}$$

If the null is *rejected*, then it can be concluded that the proportion of basketball players in the 200+ pound category exceeds that of hockey players. As Table 5.15 indicates, the p-value is less than the significance level. Moreover, the left endpoint of the Newcombe confidence interval exceeds 0 so that $0 \notin NCI(p_B - p_H)$ and p_B is statistically significantly larger than p_H. Therefore, the null hypothesis H_0 is *rejected in favor of the alternate H_1*.

TABLE 5.15   Hypothesis test summary for one-sided test of independent proportions

Test	p-value	Reject H_0?	p_H	$NCI(p_H)$	p_B	$NCI(p_B)$
$H_0: p_B \le p_H$	$<10^{-4}$	Yes ($p < \alpha$)	0.57	[0.53, 0.61]	0.80	[0.76, 0.84]
$NCI(p_B - p_H) = [0.180, +\infty) \Rightarrow 0 \notin NCI(p_B - p_H) \Rightarrow$ reject H_0						

MATLAB Commands
(One-Sided, Right-Tailed Test of Independent Proportions)

```
% Data Directory
Ddir = 'C:\PJC\Math_Biology\Chapter_5\Data';
% Load in the Hockey and Basketball data
load(fullfile(Ddir,'HockeyData.mat'))
load(fullfile(Ddir,'BasketballData.mat'))
% Collect the hockey (Y) and basketball (X) weight data
Y = D.Wt; X = B.Wt;
% Proportions of hockey and basketball players who weigh at least 200 pounds
Kh = sum(X >= 200); Nh = numel(X);
Kb = sum(Y >= 200); Nb = numel(Y);
% Set the significance level at 5%: α = 0.05
alpha = 0.05;
% Right-tailed test the hypotheses H0: pH ≤ pB vs. H1: pH > pB
H = smpropdiff([Kh,Kb],[Nh,Nb],alpha,0)
H =
          Ho: 'reject'
          dp: 0.2343
           Z: 8.7618
          CI: [0.1819 Inf]
         NCI: [0.1804 Inf]
      Pvalue: 9.6088e-19
           p: [0.8059 0.5716]
         CI1: [0.7689 0.8430]
        NCI1: [0.7663 0.8402]
         CI2: [0.5346 0.6087]
        NCI2: [0.5343 0.6082]
         rho: 0
```

5.3.2.2   *Right-Tailed Test (Samples from Correlated Proportions)*

Hypotheses

$$H_0: p_1 \leq p_2 \qquad\qquad \textit{(Null hypothesis)}$$
$$H_1: p_1 > p_2 \qquad\qquad \textit{(Alternate hypothesis)}$$

Test Statistic

$$Z_{cor}(X, Y) = \frac{\Delta\hat{p}}{SE_{cor}}$$

$$\Delta\hat{p} = \hat{p}_1 - \hat{p}_2 \equiv \frac{k_1}{n_1} - \frac{k_2}{n_2}, k_1 \equiv \sum_{j=1}^{n_1} x_j, k_2 \equiv \sum_{\ell=1}^{n_2} y_\ell$$

$$SE_{cor} = \frac{1}{z_{1-\alpha}}(L + U) \text{ as defined by formula (5.8).}$$

Test Statistic Distribution

$Z_{cor}(X, Y) \sim \mathcal{N}(0, 1) =$ the *standard normal* distribution

$\phi(x) = \dfrac{1}{\sqrt{2\pi}} e^{-\frac{1}{2}x^2}$ for $-\infty < x < \infty =$ the *pdf* of the standard normal distribution

***p*-Value**

$$\pi = \Phi(-Z_{cor}(X, Y))$$

$\Phi(x) = $ *cumulative density function* of a standard normal random variable

Newcombe $(1 - \alpha) \bullet 100\%$ Score Method Confidence Interval for $\Delta p = p_1 - p_2$

$$\left. \begin{aligned} NCI(\Delta p) &= [\Delta\hat{p} - L, +\infty) \\ L &= \sqrt{(\hat{p}_1 - \ell_1)^2 + (\hat{p}_2 - u_2)^2 - 2\rho_{12}(\hat{p}_1 - \ell_1)(u_2 - \hat{p}_2)} \end{aligned} \right\} \quad (5.9)$$

ℓ_j and u_j defined as in equation (5.8) and ρ_{12} the sample correlation of Section 5.3.1.2.

Null Hypothesis Decision

If $\min[NCI(\Delta p)] > 0$ or $\pi < \alpha$, then *reject H_0*. Otherwise, *do not reject H_0*.

Newcombe $(1 - \alpha) \bullet 100\%$ Score Method Confidence Intervals for p_1 and p_2

$NCI(p_j)$ is the same as (5.7) from Section 5.3.1.1

Required Sample Size

$R = \dfrac{n_2}{n_1}$

$$n_2 = \left\lceil \frac{(z_{1-\alpha} + z_{1-\beta/2})^2}{es^2} \cdot \left(R \cdot p_1(1-p_1) + p_2(1-p_2) - 2\rho_{12}\sqrt{p_1(1-p_1)p_2(1-p_2)}\right) \right\rceil$$

$es = $ effect size ($es = \Delta p = p_1 - p_2$). This result is readily derived from the case for independent proportions.

MATLAB Function

`smpropdiff.m`

Example: In this example, the test of proportions of *positive* (or "yes") across the correlated *Control* and *Test* populations will be exercised against a *right-tailed test*.

TABLE 5.16   Hypothesis test summary for test of one-sided correlated proportions

Test	p-value	Reject H_0?	$p^+(C)$	$NCI(p^+(C))$	$p^+(T)$	$NCI(p^+(T))$
$H_0: p^+(C) \leq p^+(T)$	0.089	No ($p > \alpha$)	0.47	$[0.41, 0.53]$	0.46	$[0.40, 0.52]$

$NCI(p^+(C) - p^+(T)) = [-0.007, +\infty) \Rightarrow 0 \in NCI(p^+(C) - p^+(T)) \Rightarrow$ *do not reject* H_0

The contingency matrix and MATLAB commands are listed further with respect to the hypotheses

$$H_0: p^+(C) \leq p^+(T) \text{ vs. } H_1: p^+(C) > p^+(T)$$

		Control	
		Yes	No
Test	Yes	125	2
	No	6	144

Table 5.16 reveals that the p-value exceeds the significance level and the left end-point of the Newcombe confidence interval is less than 0. Therefore, it *cannot* be refuted that $p^+(C) \leq p^+(T)$ and so the null hypothesis is *not rejected*.

This example raises an important question. Even though the estimate for $p^+(T)$ is smaller than the estimate for $p^+(C)$, the null hypothesis that $p^+(C) \leq p^+(T)$ is *not rejected*. Why is this? The answer is that the difference $p^+(C) - p^+(T)$ is not *sufficiently* small to prevent 0 from being contained in the confidence interval. In the next section, it will be shown how the introduction of a clinical margin will make the difference "small enough."

MATLAB Commands
(One-Sided, Right-Tailed Test of Correlated Proportions)

```
% Contingency matrix
A = [125,2;6,144];
% Set the significance level at 5%: α = 0.05
alpha = 0.05;
% Control and Test "successes"
Kc = sum(A(:,1)); Kt = sum(A(1,:));
% Total number of samples.
N = sum(A(:));
% Test of hypotheses H0: p+(Control) ≤ p+(Test) vs. H1: p+(Control) > p+(Test)
H = smpropdiff([Kc,Kt],N*[1,1],alpha,0,A)
H =
         Ho: 'do not reject'
         dp: 0.0144
```

```
      Z: 1.3455
     CI: [-0.0066 Inf]
    NCI: [-0.0066 Inf]
 Pvalue: 0.0892
      p: [0.4729 0.4585]
    CI1: [0.4141 0.5317]
   NCI1: [0.4149 0.5317]
    CI2: [0.3998 0.5172]
   NCI2: [0.4008 0.5173]
    rho: 0.9351
```

5.3.2.3 Test of Non-Inferiority (Samples from Two Independent Populations)

As described in Section 5.1, a test of non-inferiority determines whether the proportion p_2 *is not worse than* an established proportion p_1 to within a *margin* $\delta > 0$. In this setting and indeed throughout this section, p_2 represents the response of the *Test* method while p_1 is the *Control*. Comparable to the test of equivalence, non-inferiority is established provided the *null hypothesis is rejected in favor of the alternate hypothesis*.

Hypotheses

$$H_0: p_2 - p_1 \leq -\delta \quad (\delta > 0) \qquad \qquad \text{(Null hypothesis)}$$
$$H_1: p_2 - p_1 > -\delta \quad (\delta > 0) \qquad \qquad \text{(Alternate hypothesis)}$$

Test Statistics

$$Z_\ell(X, Y) = \frac{\Delta\hat{p}' + \delta}{SE}$$

using the same definitions for $\Delta\hat{p}'$ and SE as in Sections 5.3 and 5.3.1.3.

Test Statistic Distribution
$Z_\ell(X, Y)$ is asymptotically *standard normal*

$\phi(x) = \dfrac{1}{\sqrt{2\pi}} e^{-\frac{1}{2}x^2}$ for $-\infty < x < \infty$ = the *pdf* of the standard normal distribution

p-Value

$$\pi = \Phi(-Z_\ell)$$

$\Phi(x)$ = *cumulative density function* of a standard normal random variable

Newcombe $(1 - \alpha) \bullet 100\%$ Score Method Confidence Interval for $\Delta p' = p_2 - p_1$
$NCI(\Delta p') = [\Delta\hat{p}' - \sqrt{(\hat{p}_1 - \ell_1)^2 + (\hat{p}_2 - u_2)^2}, +\infty)$, ℓ_j and u_j as given by equation (5.6).

TABLE 5.17 Summary statistics for non-inferiority test of independent proportions

Test	p-value	Reject H_0?	p_H	$NCI(p_H)$	p_B	$NCI(p_B)$
$H_0: p_B - p_H \leq -\delta$	$<10^{-4}$	Yes $(p < \alpha)$	0.57	[0.534, 0.608]	0.81	[0.766, 0.840]
$NCI(p_B - p_H) = [0.18, +\infty) \Rightarrow \min\{NCI(p_B - p_H)\} = 0.18 > -\delta = -0.025 \Rightarrow reject\ H_0$						

Null Hypothesis Decision

If $min[NCI(\Delta p')] > -\delta$ or $\pi < \alpha$, then *reject H_0*. Otherwise, *do not reject H_0*.

Newcombe $(1 - \alpha) \cdot 100\%$ Score Method Confidence Intervals for p_1 and p_2

$NCI(p_j)$ is the same as equation (5.7).

Required Sample Size

$$R = \frac{n_2}{n_1}$$

$$n_2 = \left\lceil \left[\frac{(z_{1-\alpha/2} + z_{1-\beta})^2}{(es - \delta)^2} \cdot (R \cdot p_1(1 - p_1) + p_2(1 - p_2)) \right] \right\rceil$$

es = effect size $(es = \Delta p' = p_2 - p_1)$.

See Chow et al. [8]; especially formula (4.2.3).

MATLAB Function

`nipropi.m`

Example: Here the hypothesis that the proportion of basketball players who weigh *at least* 200 pounds is *not* "worse" (i.e., smaller) than the corresponding proportion of hockey players to within a clinical margin of 2.5%. Specifically, the following hypotheses will be tested for $\delta = 0.025$.

$$H_0: p(B_{wt} \geq 200) - p(H_{wt} \geq 200) \leq -\delta \text{ vs. } H_1: p(B_{wt} \geq 200) - p(H_{wt} \geq 200) > -\delta$$

As can be seen from Table 5.17, the minimum of the confidence interval for the difference $\Delta p_{wt} = p(B_{wt} \geq 200) - p(H_{wt} \geq 200)$ in proportions exceeds the negative of the clinical margin. That is, $\min(NCI(\Delta p_{wt})) = 0.18 > -\delta = -0.025$. Thus, the null hypothesis is rejected.

MATLAB Commands
(Test of Non-Inferiority for Independent Proportions)

```
% Data Directory
Ddir = 'C:\PJC\Math_Biology\Chapter_5\Data';
% Load in the Hockey and Basketball data
load(fullfile(Ddir, 'HockeyData.mat'))
load(fullfile(Ddir, 'BasketballData.mat'))
```

```
% Collect the hockey (Y) and basketball (X) weight data
Y = D.Wt; X = B.Wt;
% Indices of basketball and hockey players at least 200 pounds in weight
Xh = (X >= 200); Yb = (Y >= 200);
% Set the significance level at 5%: α = 0.05 and clinical margin δ = 0.025
alpha = 0.05; delta = 0.025;
% Test the Non-inferiority hypotheses
% H₀: p(Bwt ≥ 200) − p(Hwt ≥ 200) ≤ −δ vs. H₀: p(Bwt ≥ 200) − p(Hwt ≥ 200) >
   −δ
H = nipropi(Xh,Yb,alpha,delta,0)
H =

        Pvalue: 1.5573e-22
            Ho: 'reject'
            p1: 0.8059
          NCI1: [0.7663 0.8402]
            p2: 0.5716
          NCI2: [0.5343 0.6082]
            dp: 0.2343
           NCI: [0.1804 Inf]
      TestType: 'Non-inferiority'
        margin: 0.0250
```

5.3.2.4 Test of Non-Inferiority (Samples from Two Correlated Populations) As in Section 5.3.1.2, the model considers correlated populations with binary outcomes (i.e., "yes" or "no") identified via the contingency Table 5.11. Again as noted in Section 5.3.2.3, p_2 represents the response of the *Test* method while p_1 is the *Control*. The assumption is that the proportions measure the "yes" percentages as $p_1 = (x_{11} + x_{01})/n$ and $p_2 = (x_{11} + x_{10})/n$.

Hypotheses

$$H_0: p_2 - p_1 \leq -\delta \quad (\delta > 0) \qquad\qquad (Null\ hypothesis)$$
$$H_1: p_2 - p_1 > -\delta \quad (\delta > 0) \qquad\qquad (Alternate\ hypothesis)$$

Test Statistics

$$\hat{\theta} = p_2 - p_1 = (x_{10} - x_{01})/n$$

Wald

$$Z_\ell(X, Y) = \frac{\hat{\theta} + \delta}{\hat{\sigma}(\theta)} \equiv \frac{(x_{10} - x_{01}) + n\delta}{\sqrt{(x_{10} + x_{01}) - \frac{1}{n}(x_{10} - x_{01})^2}}$$

RMLE

$$Z_\ell^{RMLE} = \frac{\hat{\theta} + \delta}{\hat{\sigma}_\ell(\delta)}$$

Remark: The symbols $\hat{\sigma}^2(\theta)$ and $\hat{\sigma}^2_{\ell}(\delta)$ are the same as in Section 5.3.1.4.

Test Statistic Distribution

$Z_{\ell}(X, Y)$ and Z_{ℓ}^{RMLE} have asymptotically *standard normal* distributions

$$\phi(x) = \frac{1}{\sqrt{2\pi}}e^{-\frac{1}{2}x^2} \text{ for } -\infty < x < \infty = \text{the } pdf \text{ of the standard normal distribution}$$

p-Values

$$\pi = \Phi(-Z_{\ell}), \pi^{RMLE} = \Phi\left(-Z_{\ell}^{RMLE}\right)$$

$\Phi(x) = $ *cumulative density function* of a standard normal random variable

$(1 - \alpha)\bullet 100\%$ Score Method Confidence Interval for $\hat{\theta} = p_2 - p_1 = \hat{p}_{10} - \hat{p}_{01}$

Wald

$$CI(\hat{\theta}) = [\hat{\theta} + z_{\alpha/2} \cdot \hat{\sigma}(\theta), 1]$$

RMLE

$$CI(\hat{\theta}) = [\hat{\theta}_{\ell}, 1]$$

$$\hat{\theta}_{\ell} = \inf_{\theta_o}\left|\frac{\hat{\theta} - \theta_0}{\hat{\sigma}_{\ell}(\theta_0)} - z_{1-\alpha/2}\right|$$

Null Hypothesis Decision

If $\pi < \alpha$ or $\pi^{RMLE} < \alpha$, then *reject H_0*. Otherwise, *do not reject H_0*.

Required Sample Size

Wald

$$N_s = \left\lceil\left[(2p_{01}) \cdot \left(\frac{z_{1-\alpha} + z_{1-\beta}}{\delta}\right)^2\right]\right\rceil$$

An *a priori* estimate of p_{01} is required.

RMLE

$$N_s = \left\lceil\left[(2p_{01}) \cdot \left(\frac{\frac{1}{\tilde{w}} \cdot z_{1-\alpha} + z_{1-\beta}}{\delta}\right)^2\right]\right\rceil$$

$$\tilde{w} = \sqrt{\frac{2p_{01}}{2p_{\ell,01} - \delta(1 + \delta)}}$$

The symbols $p_{\ell,01}$ and $\tilde{w}$ are defined in Section 5.3.1.4.

TABLE 5.18   Summary statistics for non-inferiority test of correlated proportions

Method	p-value	$p < \alpha$?	$CI(\Delta p^+)$	$CI(\Delta p^+) \subset [-\delta, \delta]$	Reject H_0?
Wald	0.006	Yes	[−0.03438, 1]	Yes	Yes
RMLE	0.032	Yes	[−0.03715, 1]	Yes	Yes

MATLAB Function

niprop.m

Example: Return to the example from Section 5.3.2.2. When subjected to a right-tailed test H_0: $p^+(C) \leq p^+(T)$, the result of the process was *not* to reject H_0 despite the point estimate indicating $p^+(C) \leq p^+(T)$. Therefore, a *non-inferiority test* with a clinical margin of 4% will be implemented via the hypotheses with contingency table.

$$H_0: p^+(C) - p^+(T) \leq -\delta \text{ vs. } H_1: p^+(C) - p^+(T) > -\delta$$

		Control	
		Yes	No
Test	Yes	125	2
	No	6	144

As a consequence of this test of correlated proportions, it is seen that for *either* the Wald or RMLE method $\min(CI(p^+(C) - p^+(T))) \geq -0.034 > -0.04$ and both p-values are less than the 5% significance level. Table 5.18 summarizes the results.

MATLAB Commands
(One-Sided, Non-Inferiority Test of Correlated Proportions)

```
% Contingency matrix
A = [125,2;6,144];
% Set the significance level at 5%: α = 0.05 and clinical margin δ = 0.04
alpha = 0.05; delta = 0.04;
% Test of hypotheses H₀: p⁺(Control) − p⁺(Test) ≤ −δ vs. H₁: p⁺(Control) −
  p⁺(Test) > −δ
H = niprop(A,alpha,delta,0)
H =
Wald
        Ho: 'reject'
     Tstat: 2.512
    Pvalue: 0.005998
  TestType: 'Non-inferiority'
```

```
CI
    Wald:  [-0.03438 1]
    RMLE:  [-0.03715 1]
RMLE
        Ho: 'reject'
     Tstat: 1.851
    Pvalue: 0.03209
  TestType: 'Non-inferiority'
```

5.3.2.5   Left-Tailed Test (Samples from Two Independent Populations)

Hypotheses

$$H_0: p_1 \geq p_2 \qquad \text{(Null hypothesis)}$$
$$H_1: p_1 < p_2 \qquad \text{(Alternate hypothesis)}$$

Test Statistic

$$Z(X, Y) = \frac{\Delta \hat{p}}{SE}$$

$$\Delta \hat{p} = \hat{p}_1 - \hat{p}_2 \equiv \frac{k_1}{n_1} - \frac{k_2}{n_2}, k_1 \equiv \sum_{j=1}^{n_1} x_j, k_2 \equiv \sum_{\ell=1}^{n_2} y_\ell$$

$$SE = \sqrt{\frac{1}{n_1}\hat{p}_1(1 - \hat{p}_1) + \frac{1}{n_2}\hat{p}_2(1 - \hat{p}_2)}$$

Test Statistic Distribution

$Z(X, Y) \sim \mathcal{N}(0, 1) =$ the *standard normal* distribution

$\phi(x) = \frac{1}{\sqrt{2\pi}} e^{-\frac{1}{2}x^2}$ for $-\infty < x < \infty =$ the *pdf* of the standard normal distribution

***p*-Value**

$$\pi = \Phi(Z(X, Y))$$

$\Phi(x) = $ *cumulative density function* of a standard normal random variable

Newcombe $(1 - \alpha) \bullet 100\%$ Score Method Confidence Interval for $\Delta p = p_1 - p_2$

$NCI(\Delta p) = (-\infty, \Delta \hat{p} + \sqrt{(\hat{p}_2 - \ell_2)^2 + (\hat{p}_1 - u_1)^2}]$, ℓ_j and u_j as given by equation (5.6).

Null Hypothesis Decision

If $\min[NCI(\Delta p)] < 0$ or $\pi < \alpha$, then *reject H_0*. Otherwise, *do not reject H_0*.

Newcombe $(1 - \alpha) \bullet 100\%$ Score Method Confidence Intervals for p_1 and p_2

Same as equation (5.7).

TABLE 5.19   Summary calculations for left-tailed test of independent proportions

H_0	p-value	Reject H_0	p_H	$NCI(p_H)$	p_B	$NCI(p_B)$
$p_H \geq p_B$	$<10^{-4}$	Yes ($p < \alpha$)	0.57	[0.534, 0.608]	0.80	[0.766, 0.84]

$NCI(p_H - p_B) = (-\infty, -0.18] \Rightarrow \max(NCI(p_H - p_B)) < 0 \Rightarrow$ reject H_0

Required Sample Size

$$R = \frac{n_2}{n_1}$$

$$n_2 = \left\lceil \frac{(z_{1-\alpha} + z_{1-\beta})^2}{es^2} \cdot (R \cdot p_1(1 - p_1) + p_2(1 - p_2)) \right\rceil$$

es = effect size ($es = \Delta p = p_1 - p_2$).
See Chow et al. [8]; especially formula (4.2.3) with $\delta = 0$.

MATLAB Function

`smpropdiff.m`

Example: To exercise the hypothesis that the proportion of professional basketball players weighing in excess of 200 pounds is larger than the comparable proportion of hockey players (see Table 5.19), test the hypotheses

$$H_0: p(H_{wt} \geq 200) \geq p(B_{wt} \geq 200) \text{ vs. } H_1: p(H_{wt} \geq 200) < p(B_{wt} \geq 200)$$

Use the shorthand $p_H = p(H_{wt} \geq 200)$ and $p_B = p(B_{wt} \geq 200)$

MATLAB Commands
(One-Sided, Left-Tailed Test of Independent Proportions)

```
% Data Directory
Ddir = 'C:\PJC\Math_Biology\Chapter_5\Data';
% Load in the Hockey and Basketball data
load(fullfile(Ddir,'HockeyData.mat'))
load(fullfile(Ddir,'BasketballData.mat'))
% Collect the hockey (X) and basketball (Y) weight data
X = D.Wt; Y = B.Wt;
% Proportions of hockey and basketball players who weigh at least 200 pounds
Kh = sum(X >= 200); Nh = numel(X);
Kb = sum(Y >= 200); Nb = numel(Y);
% Set the significance level at 5%: α = 0.05
alpha = 0.05;
```

```
% Left-tailed test H₀: pH ≥ pB vs. H₁: pH < pB
H = smpropdiff([Kh,Kb],[Nh,Nb],alpha,1)
H =
        Ho: 'reject'
        dp: -0.2343
         Z: -8.7618
        CI: [-Inf -0.1819]
       NCI: [-Inf -0.1804]
    Pvalue: 9.6088e-19
         p: [0.5716 0.8059]
       CI1: [0.5346 0.6087]
      NCI1: [0.5343 0.6082]
       CI2: [0.7689 0.8430]
      NCI2: [0.7663 0.8402]
       rho: 0
```

5.3.2.6   Left-Tailed Test (Samples from Correlated Proportions)

Hypotheses

$$H_0: p_1 \geq p_2 \qquad \text{(Null hypothesis)}$$
$$H_1: p_1 < p_2 \qquad \text{(Alternate hypothesis)}$$

Test Statistic

$$Z_{cor}(X, Y) = \frac{\Delta \hat{p}}{SE_{cor}}$$

$$\Delta \hat{p} = \hat{p}_1 - \hat{p}_2 \equiv \frac{k_1}{n_1} - \frac{k_2}{n_2}, k_1 \equiv \sum_{j=1}^{n_1} x_j, k_2 \equiv \sum_{\ell=1}^{n_2} y_\ell$$

$SE_{cor} = \frac{1}{z_{1-\alpha}}(L + U)$ as defined by formula (5.8).

Test Statistic Distribution

$Z_{cor}(X, Y) \sim \mathcal{N}(0, 1) =$ the *standard normal* distribution

$\phi(x) = \frac{1}{\sqrt{2\pi}} e^{-\frac{1}{2}x^2}$ for $-\infty < x < \infty =$ the *pdf* of the standard normal distribution

***p*-Value**

$$\pi = \Phi(Z_{cor}(X, Y))$$

$\Phi(x) = $ *cumulative density function* of a standard normal random variable

Newcombe $(1 - \alpha) \cdot 100\%$ Score Method Confidence Interval for $\Delta p = p_1 - p_2$

$$\left. \begin{array}{l} NCI(\Delta p) = (-\infty, \Delta \hat{p} + U] \\ U = \sqrt{(\hat{p}_2 - \ell_2)^2 + (\hat{p}_1 - u_1)^2 - 2\rho_{12}(\hat{p}_2 - \ell_2)(u_1 - \hat{p}_1)} \end{array} \right\} \quad (5.10)$$

ℓ_j and u_j defined as in equation (5.8) and ρ_{12} the sample correlation of Section 5.3.1.2.

Null Hypothesis Decision

If $\min[NCI(\Delta p)] < 0$ or $\pi < \alpha$, then *reject H_0*. Otherwise, *do not reject H_0*.

Newcombe $(1 - \alpha) \cdot 100\%$ Score Method Confidence Intervals for p_1 and p_2

$NCI(p_j)$ is the same as (5.7) from Section 5.3.1.1.

Required Sample Size

$$R = \frac{n_2}{n_1}$$

$$n_2 = \left[\left[\frac{(z_{1-\alpha} + z_{1-\beta})^2}{es^2} \cdot \left(R \cdot p_1(1 - p_1) + p_2(1 - p_2) - 2\rho_{12}\sqrt{p_1(1 - p_1)p_2(1 - p_2)} \right) \right] \right]$$

es = effect size ($es = \Delta p = p_1 - p_2$)

This can be derived directly from the sample size formula in Section 5.3.2.5.

MATLAB Function

smpropdiff.m

Example: Here the test of proportions of *positive* (or "yes") across the correlated *Control* and *Test* populations will be exercised against a *left-tailed test*. The contingency matrix used to test the hypotheses

$$H_0: p^+(C) \geq p^+(T) \text{ vs. } H_1: p^+(C) < p^+(T)$$

is discussed in Table 5.20. As can be seen from this table, the p-value exceeds the significance level $a = 0.05$. Moreover, the entire Newcombe confidence interval for the difference $\Delta p = p^+(C) - p^+(T)$ *does not* lie completely to the left of 0. Therefore, it *cannot* be concluded that $p^+(C) < p^+(T)$.

		Control	
		Yes	No
Test	Yes	125	2
	No	6	144

TABLE 5.20   Summary calculations for left-tailed test of independent proportions

H_0	p-value	Reject H_0	$p^+(C)$	$NCI(p^+(C))$	$p^+(T)$	$NCI(p^+(T))$
$p^+(C) \geq p^+(T)$	0.91	No $(p > \alpha)$	0.47	$[0.415, 0.532]$	0.459	$[0.401, 0.517]$

$NCI(p^+(C) - p^+(T)) = (-\infty, 0.035] \Rightarrow 0 \in NCI(p^+(C) - p^+(T)) \Rightarrow$ Do not reject H_0

MATLAB Commands
(One-Sided, Left-Tailed Test of Correlated Proportions)

```
% Contingency matrix
A = [125,2;6,144];
% Set the significance level at 5%: α = 0.05
alpha = 0.05;
% Control and Test "successes"
Kc = sum(A(:,1)); Kt = sum(A(1,:));
% Total number of samples.
N = sum(A(:));
% Test of hypotheses H0: p+(Control) ≥ p+(Test) vs. H1: p+(Control) < p+(Test)
H = smpropdiff([Kc,Kt],N*[1,1],alpha,1,A)
H =
        Ho: 'do not reject'
        dp: 0.0144
         Z: 1.3455
        CI: [-Inf 0.0355]
       NCI: [-Inf 0.0354]
    Pvalue: 0.9108
         p: [0.4729 0.4585]
       CI1: [0.4141 0.5317]
      NCI1: [0.4149 0.5317]
       CI2: [0.3998 0.5172]
      NCI2: [0.4008 0.5173]
       rho: 0.9351
```
% The proportions are $p^+(Control) = 0.473$ and $p^+(Test) = 0.458$ with p-value $>$
α. Therefore, the null hypothesis cannot be rejected.

5.3.2.7   Test of Superiority (Samples from Two Independent Populations)   Similar to a test of *non-inferiority*, *superiority* determines whether the proportion is p_1 *substantially larger than* an established proportion p_2 by a *margin* $\delta > 0$. As with non-inferiority, *superiority* is established by rejecting the *null hypothesis in favor of the alternate hypothesis*. It is emphasized as per Sections 5.3.2.3 and 5.3.2.4 that, p_2 represents the response of the *Test* method while p_1 is the *Control*.

Hypotheses

$$H_0: p_2 - p_1 \leq \delta \qquad \text{(Null hypothesis)}$$
$$H_1: p_2 - p_1 > \delta \qquad \text{(Alternate hypothesis)}$$

Test Statistic

$$Z_u(X, Y) = \frac{\Delta\hat{p}' - \delta}{SE}$$

using the same definitions for $\Delta\hat{p}'$ and SE as in Sections 5.3 and 5.3.1.3.

Test Statistic Distribution
$Z_u(X, Y)$ is asymptotically *standard normal*

$\phi(x) = \frac{1}{\sqrt{2\pi}}e^{-\frac{1}{2}x^2}$ for $-\infty < x < \infty$ = the *pdf* of the standard normal distribution

p-Value

$$\pi = \Phi(Z_u)$$

$\Phi(x)$ = *cumulative density function* of a standard normal random variable

Newcombe $(1 - \alpha)\bullet100\%$ Score Method Confidence Interval for $\Delta p' = p_2 - p_1$
$NCI(\Delta p') = (-\infty, \Delta\hat{p}' - \sqrt{(\hat{p}_2 - \ell_2)^2 + (\hat{p}_1 - u_1)^2}]$, ℓ_j and u_j as given by equation (5.6).

Null Hypothesis Decision
If $max[NCI(\Delta p)] < -\delta$ or $\pi < \alpha$, then *reject H_0*. Otherwise, *do not reject H_0*.

Newcombe $(1 - \alpha)\bullet100\%$ Score Method Confidence Intervals for p_1 and p_2
$NCI(p_j)$ is the same as equation (5.7).

Required Sample Size

$$R = \frac{n_2}{n_1}$$

$$n_2 = \left\lceil \left[\frac{(z_{1-\alpha} + z_{1-\beta/2})^2}{(\delta - es)^2} \cdot (R \cdot p_1(1 - p_1) + p_2(1 - p_2)) \right] \right\rceil$$

es = effect size $(es = |\Delta p| = |p_1 - p_2|)$
See Chow et al. [8], formula (4.2.4).

MATLAB Function
`nipropi.m`

Example: Here the hypothesis that the proportion of basketball players who weigh *at least* 200 pounds is *substantially better* (i.e., larger) than the corresponding proportion of hockey players within a clinical margin of 5%. That is, the hypotheses

$$H_0: p(B_{wt} \geq 200) - p(H_{wt} \geq 200) \leq \delta \text{ vs. } H_1: p(B_{wt} \geq 200) - p(H_{wt} \geq 200) > \delta$$

TABLE 5.21 Summary calculations for left-tailed test of independent proportions

H_0	p-value	Reject H_0	p_H	$NCI(p_H)$	p_B	$NCI(p_B)$
$p_B - p_H \leq \delta$	$<10^{-4}$	Yes $(p < \alpha)$	0.57	[0.534, 0.608]	0.806	[0.766, 0.84]
$NCI(p_B - p_H) = (-\infty, 0.285] \Rightarrow \max(NCI(p_B - p_H)) > \delta \Rightarrow$ reject H_0						

are tested for $\delta = 0.05$. Again, use the shorthand $p_H = p(H_{wt} \geq 200)$ and $p_B = p(B_{wt} \geq 200)$. Examining Table 5.21, it is seen that $\max(NCI(\Delta p)) > \delta$ so that $\Delta p > \delta$ and the null H_0 is *rejected*.

MATLAB Commands
(Test of Superiority for Independent Proportions)

```
% Data Directory
Ddir = 'C:\PJC\Math_Biology\Chapter_5\Data';
% Load in the Hockey and Basketball data
load(fullfile(Ddir,'HockeyData.mat'))
load(fullfile(Ddir,'BasketballData.mat'))
% Collect the hockey (Y) and basketball (X) weight data
Y = D.Wt; X = B.Wt;
% Indices of hockey and basketball players at least 200 pounds in weight
Xh = (X >= 200); Yb = (Y >= 200);
% Set the significance level at 5%: alpha = 0.05 and clinical margin delta = 0.05
alpha = 0.05; delta = 0.05;
% Test the Non-inferiority hypotheses
```

% $H_0: p(B_{wt} \geq 200) - p(H_{wt} \geq 200) \leq \delta$ vs. $H_1: p(B_{wt} \geq 200) - p(H_{wt} \geq 200) > \delta$

```
H = nipropi(Xh,Yb,alpha,delta,1)
H =
          Pvalue: 2.7507e-12
              Ho: 'reject'
              p1: 0.8059
            NCI1: [0.7663 0.8402]
              p2: 0.5716
            NCI2: [0.5343 0.6082]
              dp: 0.2343
             NCI: [-Inf 0.2850]
        TestType: 'Superiority'
          margin: 0.0500
```

5.3.2.8 Test of Superiority (Samples from Two Correlated Populations) Continue to use Table 5.11 for this model which considers correlated populations with dichotomous outcomes ("no" and "yes"). The assumption is that the proportions measure the "yes" percentages as $p_1 = (x_{11} + x_{10})/n$ and $p_2 = (x_{11} + x_{01})/n$ with

$n = x_{11} + x_{10} + x_{01} + x_{00}$. Once again as per Sections 5.3.2.3, 5.3.2.4, and 5.3.2.7, it is emphasized that p_2 represents the response of the *Test* method while p_1 is the *Control*.

Hypotheses

$$H_0: p_2 - p_1 \leq \delta \qquad \qquad \text{(Null hypothesis)}$$
$$H_1: p_2 - p_1 > \delta \qquad \qquad \text{(Alternate hypothesis)}$$

Test Statistics

$$\hat{\theta} = p_2 - p_1 \equiv \frac{1}{n}(x_{10} - x_{01})$$

Wald

$$Z_u(X, Y) = \frac{\hat{\theta} - \delta}{\hat{\sigma}(\theta)} \equiv \frac{(x_{10} - x_{01}) - n\delta}{\sqrt{(x_{10} + x_{01}) - \frac{1}{n}(x_{10} - x_{01})^2}},$$

RMLE

$$Z_u^{RMLE} = \frac{\hat{\theta} - \delta}{\hat{\sigma}_u(\delta)}$$

Remark: The symbols $\hat{\sigma}^2(\theta)$ and $\hat{\sigma}_u^2(\delta)$ are the same as in Section 5.3.1.4.

Test Statistic Distribution
$Z_u(X, Y)$ and Z_u^{RMLE} both have an asymptotically *standard normal* distribution (see Liu et al. [20]).

$\phi(x) = \frac{1}{\sqrt{2\pi}} e^{-\frac{1}{2}x^2}$ for $-\infty < x < \infty$ = the *pdf* of the standard normal distribution

p-Values
Wald

$$\pi_u = \Phi(Z_u(X, Y))$$

RMLE

$$\pi_u^{RMLE} = \Phi\left(Z_u^{RMLE}\right)$$

$\Phi(x)$ = *cumulative density function* of a standard normal random variable

$(1 - \alpha) \bullet 100\%$ Score Method Confidence Interval for $\hat{\theta} = p_2 - p_1 = \hat{p}_{10} - \hat{p}_{01}$
Wald

$$CI(\hat{\theta}) = [0, \hat{\theta} + z_{1-\alpha/2} \cdot \hat{\sigma}(\theta)]$$

RMLE

$$CI(\widehat{\theta}) = [0, \widehat{\theta}_u]$$

$$\widehat{\theta}_u = \inf_{\theta_o} \left| \frac{\widehat{\theta} - \theta_o}{\widehat{\sigma}_u(\theta_o)} - z_\alpha \right|$$

Null Hypothesis Decision

If $\pi < \alpha$ or $\pi_u^{RMLE} < \alpha$, then *reject H_0*. Otherwise, *do not reject H_0*.

Required Sample Size

Wald

$$N_s = \left\lceil \left\lceil (2p_{01}) \cdot \left(\frac{z_{1-\alpha} + z_{1-\beta}}{\delta} \right)^2 \right\rceil \right\rceil$$

An *a priori* estimate of p_{01} is required.

RMLE

$$N_s = \left\lceil \left\lceil (2p_{01}) \cdot \left(\frac{\frac{1}{\tilde{w}} \cdot z_{1-\alpha} + z_{1-\beta}}{\delta} \right)^2 \right\rceil \right\rceil, \quad \tilde{w} = \sqrt{\frac{2p_{01}}{2p_{\ell,01} - \delta(1 + \delta)}}$$

The symbols $p_{\ell,01}$ and $\tilde{w}$ are defined in Section 5.3.1.4.

MATLAB Function

`niprop.m`

Example: For the final example of Section 5.3, the test of proportions of *positive* (or "yes") across the correlated *Control* and *Test* populations will be examined as a *superiority test*. The contingency matrix and MATLAB commands presented here test the hypotheses

$$H_0: p^+(T) - p^+(C) \le \delta \text{ vs. } H_1: p^+(T) - p^+(C) > \delta$$

It is plain from the notation $\Delta p^+ = p^+(C) - p^+(T)$ and the contents of Table 5.22 that the right endpoint of the confidence interval for Δp^+ *does not* exceed the clinical margin δ. Therefore, the null *cannot be rejected*.

		Control	
		Yes	No
Test	Yes	125	2
	No	6	144

TABLE 5.22   Summary statistics for superiority test of correlated proportions

Method	p-value	$p < \alpha$?	$CI(\Delta p^+)$	$\max(CI(\Delta p^+)) > \delta$	Reject H_0?
Wald	0.9919	No	[0, 0.0055]	No	No
RMLE	0.9847	No	[0, 0.0042]	No	No

MATLAB Commands
(Test of Superiority for Correlated Proportions)

```
% Contingency matrix
A = [125,2;6,144];
% Set the significance level at 5%, α = 0.05 and a clinical margin of 1%, δ = 0.01
alpha = 0.05; delta = 0.01;
% Test of hypotheses H₀: p⁺(Test) − p⁺(Control) ≤ δ vs. H₁: p⁺(Test) −
  p⁺(Control) > δ
H = niprop(A,alpha,delta,1)
H =
Wald
       Ho: 'do not reject'
    Tstat: -2.4022
   Pvalue: 0.9919
 TestType: 'Superiority'
CI
     Wald:  [0 0.0055]
     RMLE:  [0 0.0042]
RMLE
       Ho: 'do not reject'
    Tstat: -2.1609
   Pvalue: 0.9847
 TestType: 'Superiority'
```

EXERCISE

5.8   Prove the Central Limit Theorem. *Hint*: This can be achieved by way of the *characteristic function* for the random variable $\tilde{S}_n = \frac{S_n - n\mu}{\sigma\sqrt{n}}$. The *characteristic function* of a random variable X is defined as $\chi_X(t) = E[e^{itX}] = \int_{\Omega_X} \phi_X(x)\, e^{itx}\, dx$. When $\Omega_X = \mathbb{R}$, then the characteristic function is the *inverse Fourier transform* of $\phi_X(x) = $ the *pdf* of the random variable X. That is, $\chi_X(t) = \int_{-\infty}^{\infty} \phi_X(x)\, e^{itx}\, dx = \mathcal{F}^{-1}[\phi_X(x)](t)$. For details, see Hoel et al. [13].

5.4   TESTS OF VARIANCES

Rather than considering the difference between means or proportions, the variation in the samples is often of interest, especially to review agencies. Indeed, for migration studies [15], the *ratio* of variances across two distinct samples is of primary concern. As in Section 5.3, let $X = \{x_1, x_2, \ldots, x_{n_x}\}$ and $Y = \{y_1, y_2, \ldots, y_{n_y}\}$ be samples of the independent random variables X and Y. The assumption will be that $X \sim \mathcal{N}(\mu_x, \sigma_x^2)$ and $Y \sim \mathcal{N}(\mu_y, \sigma_y^2)$ are independent with σ_x^2 and σ_y^2 unknown and not necessarily equal. Moreover, when such consideration is important (i.e., equivalence, non-inferiority, or superiority), X will represent a sample from the *Control* population while Y is taken from the *Test* group.

To test the relationship (i.e., statistically equal, greater than, etc.) between these variances, the ratio of their sample variances will be evaluated. Throughout this section, these definitions will be utilized.

$$S_X^2 \equiv \frac{1}{n_x - 1} \sum_{j=1}^{n_x} (x_j - \bar{x})^2, \bar{x} = \frac{1}{n_x} \sum_{j=1}^{n_x} x_j \tag{5.11a}$$

$$S_Y^2 \equiv \frac{1}{n_y - 1} \sum_{k=1}^{n_y} (y_k - \bar{y})^2, \bar{y} = \frac{1}{n_y} \sum_{k=1}^{n_y} y_k \tag{5.11b}$$

The test statistic will be

$$F(X, Y) = \frac{S_X^2}{S_Y^2} \tag{5.12}$$

Under the assumption of a sample from independent normal distributions, the sample variances are modeled as $\chi_{n_x}^2$ and $\chi_{n_y}^2$ random variables, respectively. Therefore, the statistic F is the ratio of χ^2 random variables and subsequently has an $\mathcal{F}$-distribution with $n_x - 1$ and $n_y - 1$ degrees of freedom. That is, $F \sim \mathcal{F}_{df}, df = (n_x - 1, n_y - 1)$. The probability density function for the $\mathcal{F}$-distribution with $df = (df_1, df_2)$ degrees of freedom is

$$\phi_\mathcal{F}(x; df) = \frac{\left(\frac{df_1}{df_2}\right)^{\frac{1}{2}df_1} x^{\frac{1}{2}(df_1 - 2)}}{B\left(\frac{1}{2}df_1, \frac{1}{2}df_2\right) \cdot \left(1 + \left(\frac{df_1}{df_2}\right) \cdot x\right)^{\frac{1}{2}(df_1 + df_2)}}, x > 0 \tag{5.13}$$

where $B(n, m) = \frac{\Gamma(n) \cdot \Gamma(m)}{\Gamma(n+m)}$ is the *beta function* and $\Gamma(n) = \int_0^\infty x^{n-1} e^{-x} dx$ is the *gamma function*. For more details about these *special functions*, see Olver et al. [29] or Abramowitz and Stegun [1]. Observe that an $\mathcal{F}$-distributed random variable is always *positive*. Hence, the *pdf* (5.13) is only defined for $x > 0$. The cumulative density function for the $\mathcal{F}$-distribution is denoted $\Phi_\mathcal{F}(x; df)$ and defined as the integral of $\phi_\mathcal{F}(x; df)$.

$$\Phi_\mathcal{F}(x; df) = \text{Prob}(F \leq x) = \int_0^x \phi_\mathcal{F}(u; df) \, du \tag{5.14}$$

The $\alpha \cdot 100\%$ quantile for the $\mathcal{F}$-distribution is expressed by the symbol $f_{df}(\alpha)$. As always, α will designate the test significance level. These definitions and assumptions will be used throughout Section 5.4. As for the tests of means and tests of proportions, this section will be subdivided into two-sided and one-sided tests.

To estimate a variance of a single sample, assume that $X = \{x_1, x_2, \ldots, x_{n_x}\}$ is composed of *i.i.d.* $\mathcal{N}(0, \sigma^2)$ elements. Then $\frac{1}{\sigma^2} \sum_{j=1}^{n_x} (x_j - \bar{x})^2$ is the sum of squares of $(n_x - 1)$ standard normal random variables. Note that there are only $(n_x - 1)$ independent random variables since $\bar{x} = \frac{1}{n_x} \sum_{j=1}^{n_x} x_j$. Therefore, once the variables $x_1, x_2, \ldots, x_{n_x-1}$ and $\bar{x}$ are specified, then $x_{n_x} = n_x \bar{x} - \sum_{j=1}^{n_x-1} x_j$. Thus, $\Xi = \frac{1}{\sigma^2} \sum_{j=1}^{n_x} (x_j - \bar{x})^2 \equiv \frac{(n_x-1)S_X^2}{\sigma^2}$ is a χ^2 random variable with $df = n_x - 1$ degrees of freedom; $\Xi \sim \chi^2_{n_x-1}$. In keeping with the nomenclature of Section 5.1, $\chi^2_{df}(\gamma)$ is a $\gamma \bullet 100\%$ quantile of a χ^2_{df}-distributed random variable. Subsequently, $1 - \alpha = \mathrm{Prob}\big(\chi^2_{df}(\alpha) \le \Xi \le \chi^2_{df}(1 - \alpha/2)\big) = \mathrm{Prob}\left(\chi^2_{df}(\alpha) \le \frac{(n_x-1)S_X^2}{\sigma^2} \le \chi^2_{df}(1 - \alpha/2)\right)$. This means that a $(1 - \alpha)\bullet 100\%$ confidence interval for σ^2 is

$$CI(\sigma^2) = \left[\frac{(df)S_X^2}{\chi^2_{df}(1 - \alpha/2)}, \frac{(df)S_X^2}{\chi^2_{df}(\alpha/2)}\right], df = n_x - 1 \qquad (5.15)$$

This computation follows Marques de Sá [21] and will be utilized throughout this chapter.

5.4.1   Two-Sided Tests of Variances

5.4.1.1   Test for Equality (Independent Samples from Normal Populations)

Hypotheses

$$H_0: \sigma_x^2 = \sigma_y^2 \qquad \text{(\textit{Null hypothesis})}$$
$$H_1: \sigma_x^2 \ne \sigma_y^2 \qquad \text{(\textit{Alternate hypothesis})}$$

Test Statistic
$F(X, Y)$ of equation (5.12)

Test Statistic Distribution
$F(X, Y)$ is distributed as $\mathcal{F}_{df}$ with $df = [n_x - 1, n_y - 1]$
$\phi_F(x; df)$ of equation (5.13)

p-Value

$$p_0 = \begin{cases} 2 \cdot (1 - \Phi_F(F(X, Y); df)) & \text{for} \quad F(X, Y) \ge 1 \\ 2 \cdot \Phi_F(F(X, Y); df) & \text{for} \quad F(X, Y) < 1 \end{cases}$$

$\Phi_{\mathcal{F}}(x; df) = $ *cumulative density function* of an $\mathcal{F}$-distributed random variable with df degrees of freedom of equation (5.14)

Null Hypothesis Decision
If $p < \alpha$, then *reject H_0* in favor of H_1. Otherwise, *do not reject H_0*.

$(1 - \alpha) \bullet 100\%$ Confidence Interval for σ_x^2 / σ_y^2

$$CI\left(\sigma_x^2 \big/ \sigma_y^2\right) = \left[\frac{F(X, Y)}{f_{(n_x-1, n_y-1)}(1 - \alpha/2)}, \frac{F(X, Y)}{f_{(n_x-1, n_y-1)}(\alpha/2)}\right]$$

$(1 - \alpha) \bullet 100\%$ Confidence Intervals for σ_x^2 and σ_y^2
$CI(\sigma_x^2)$ and $CI(\sigma_y^2)$ are formed from (5.15) with $df_x = n_x - 1$ and $df_y = n_y - 1$, respectively.

Required Sample Size

$$df = [df_1, df_2] \equiv [n_x - 1, n_y - 1]$$
$$\rho_{df} = \frac{df_1}{df_2} = \frac{n_x - 1}{n_y - 1}$$

If the samples are matched, then $\rho_{df} = 1$. Otherwise, set $n_x - 1 = n$ and $n_y - 1 = n/\rho_{df}$.

$$G(n) = \Phi_{\mathcal{F}}(es \cdot f_{[n, n/\rho_{df}]}(\alpha/2))$$
$$N_s = \arg\min_n\{|G(n) - (1 - \beta)|\} \Rightarrow n_x = N_s + 1 \text{ and } n_y = 1 + \left\lceil \frac{N_s}{\rho_{df}} \right\rceil.$$
$$power = 1 - \beta$$
$$es = \text{effect size } \left(es = \sigma_x^2 / \sigma_y^2\right)$$

MATLAB Function
ftest.m

Example: Test whether the variance in the weights of the professional hockey and basketball players is equal.

$$H_0: \sigma_H^2(wt) = \sigma_B^2(wt) \qquad \qquad \textit{(Null hypothesis)}$$
$$H_1: \sigma_H^2(wt) \neq \sigma_B^2(wt) \qquad \qquad \textit{(Alternate hypothesis)}$$

TABLE 5.23   Test of equal variances

H_0	p-value	Reject H_0	σ_H^2	$CI\left(\sigma_H^2\right)$	σ_B^2	$CI\left(\sigma_B^2\right)$
$\sigma_H^2 = \sigma_B^2$	$<10^{-4}$	Yes ($p < \alpha$)	229.7	[207.22, 256.22]	684.1	[601.76, 784.74]

$CI\left(\sigma_H^2/\sigma_B^2\right) = [0.2829, 0.3974] \Rightarrow 1 \notin CI\left(\sigma_H^2/\sigma_B^2\right) \Rightarrow$ reject H_0

The summary estimates are provided in Table 5.23. Since 1 is *not* contained within the confidence interval for the ratio $\sigma_H^2(wt)/\sigma_B^2(wt)$, the null hypothesis is rejected.

MATLAB Commands
(Two-Sided Test of Equal Variances)

```
% Data Directory
Ddir = 'C:\PJC\Math_Biology\Chapter_5\Data';
% Load in the Hockey and Basketball data
load(fullfile(Ddir,'HockeyData.mat'))
load(fullfile(Ddir,'BasketballData.mat'))
% Collect the hockey (X) and basketball (Y) weight data
X = D.Wt; Y = B.Wt;
% Set the significance level at 5%: α = 0.05
alpha = 0.05;
% Use the F-test method to test for equal variances H0: σH² = σB² vs. H1: σH² ≠ σB²
H = ftest(X,Y,alpha,2)
H =
          Ho: 'reject'
          Vx: 229.7623
        CIVx: [207.2159  256.2195]
        VarY: 684.1271
        CIVy: [601.7618  784.7357]
           F: 0.3358
          CI: [0.2829  0.3974]
          df: [683  437]
      Pvalue: 1.7503e-37
```

5.4.1.2   Test for Equality (Independent Samples from Non-Normal Populations)

It is not reasonable to assume *under all circumstances* that the samples X and Y are taken from normally distributed random variables. When this assumption becomes untenable, *Levene's Test* is applied. Here, the underlying assumption is that X and Y are samples from non-normal distributions.

Following Marques de Sá [21], let $\bar{x}$ and $\bar{y}$ be the sample means from X and Y, respectively, and define the *absolute mean deviations* as $D_{x,j} = |x_j - \bar{x}|, j = 1,$

$2, \ldots, n_x$ and $D_{y,k} = |y_k - \bar{y}|$, $k = 1, 2, \ldots, n_y$. The *expected values* and *variances* of the absolute mean deviations are computed as

$$
\left.
\begin{aligned}
E[D_x] \equiv \overline{D}_x &= \frac{1}{n_x} \sum_{j=1}^{n_x} D_{x,j} = \frac{1}{n_x} \sum_{j=1}^{n_x} |x_j - \bar{x}| \\
E[D_y] \equiv \overline{D}_y &= \frac{1}{n_y} \sum_{k=1}^{n_y} D_{y,k} = \frac{1}{n_y} \sum_{k=1}^{n_y} |y_k - \bar{y}|
\end{aligned}
\right\}
\tag{5.16}
$$

and

$$
\left.
\begin{aligned}
\text{var}[D_x] &= \frac{1}{n_x-1} \sum_{j=1}^{n_x} (D_{x,j} - \overline{D}_x)^2 \\
\text{var}[D_y] &= \frac{1}{n_y-1} \sum_{k=1}^{n_y} (D_{y,k} - \overline{D}_y)^2
\end{aligned}
\right\}
\tag{5.17}
$$

The *pooled variance* $V_{pool} = \frac{1}{n_x+n_y-2}\{(n_x - 1) \cdot \text{var}[D_x] + (n_y - 1) \cdot \text{var}[D_y]\}$ is the final ingredient utilized to define *Levene's test statistic*

$$
T(X, Y) = \frac{\overline{D}_x - \overline{D}_y}{\sqrt{\left(\dfrac{1}{n_x} + \dfrac{1}{n_y}\right) \cdot V_{pool}}}
\tag{5.18}
$$

The statistic T is distributed (approximately) as a $\mathcal{T}_{n_x+n_y-2}$ random variable.

Hypotheses

$$
\begin{aligned}
H_0&: \sigma_x^2 = \sigma_y^2 &&\text{(Null hypothesis)} \\
H_1&: \sigma_x^2 \neq \sigma_y^2 &&\text{(Alternate hypothesis)}
\end{aligned}
$$

Test Statistic
$T(X, Y)$ of equation (5.18)

Test Statistic Distribution
$T(X, Y)$ is distributed as $\mathcal{T}_{df}$ with $df = n_x + n_y - 2$
$\phi_T(x; df)$ of equation (5.2)

p-Value

$$
p = 2 \cdot \Phi_T(-|T|; df)
$$

$\Phi_T(x; df) = cdf$ of $\mathcal{T}$-distributed random variable with df degrees of freedom

Null Hypothesis Decision
If $p < \alpha$, then *reject H_0 in favor of H_1*. Otherwise, *do not reject H_0*.

$(1 - \alpha) \cdot 100\%$ Confidence Interval for $E(|X - \mu_x|) - E(|Y - \mu_y|)$
Observe that the numerator of the test statistic (5.18) is the sample difference of *first-order variations*. That is, $\overline{D}_x$ is the sample statistic for $E(|X - \mu_x|)$, where μ_x is the actual mean of the distribution from which the sample X is drawn. Similarly, $\overline{D}_y$ is the sample statistic for $E(|Y - \mu_y|)$. Therefore, the following confidence interval is for the difference in first-order variations. To obtain a confidence interval for the ratio of variances, use $CI(\sigma_x^2/\sigma_y^2)$ from Section 5.4.1.1.

$$CI(E(|X - \mu_x|) - E(|Y - \mu_y|)) = (\overline{D}_x - \overline{D}_y) + \left[t_{df}(\alpha/2) \cdot \sqrt{V_{pool} \left(\frac{1}{n_x} + \frac{1}{n_y} \right)}, \right.$$

$$\left. t_{df}(1 - \alpha/2) \cdot \sqrt{V_{pool} \left(\frac{1}{n_x} + \frac{1}{n_y} \right)} \right]$$

Required Sample Size
The sample size formula parallels that of the $\mathcal{F}$-distribution of Section 5.4.1.1. That is, set $df = n_x + n_y - 2$ to be the degrees of freedom of the test statistic $T(X, Y)$.

Let $n = df$ and $t_{df}(\alpha/2)$ be the $(\alpha/2) \cdot 100\%$ quantile of the $\mathcal{T}$-distribution with df degrees of freedom. For $G(n) = \Phi_T(es \cdot t_n(\alpha/2); n)$, the required number of samples is $N_s = \arg\min_n \{|G(n) - (1 - \beta)|\}$. Then select n_x and n_y so that n_x and $n - 2 = N_s$.

$es = $ effect size ($es = |\overline{D}_x - \overline{D}_y|$).

MATLAB Function
```
levene.m
```

Example: Determine whether the non-normal random variables distributed as uniform across the unit interval $X \sim \mathcal{U}[0, 1]$ and uniform over $[3, 4]$ $Y \sim \mathcal{U}[3, 4]$ have equal variances. That is, test $H_0: \sigma_1^2 = \sigma_2^2$ vs. $H_1: \sigma_1^2 \neq \sigma_2^2$. This is accomplished by taking a random sample X from $\mathcal{U}[0, 1]$ and another sample Y from $\mathcal{U}[3, 4]$. By applying Levene's test, it is seen *for this particular random sample* that the p-value $p = 0.8385 > \alpha = 0.05$. Therefore, H_0 *is not rejected*. This makes sense as the estimates of the variances of these samples are $\sigma_1^2 = 0.0837$ and $\sigma_2^2 = 0.0822$, respectively.

Note: Since X and Y are (pseudo) random samples, the results obtained earlier may not match the reader's repetition of this experiment. This example is provided to familiarize the reader with the M-file `levene.m`.

MATLAB Commands
(Two-Sided Test of Equal Variances, Non-Normal Populations)

```
% Generate two non-normal random variables, X ~ U[0, 1] and Y ~ U[a, b]
N = 1000; a = 3; b = 4;
X = rand(1,N); Y = a + (b-a)*rand(1,N);
% Set the significance level at 5%: α = 0.05
alpha = 0.05;
% Use Levene's test for equal variances H₀: σ₁² = σ₂² vs. H₁: σ₁² ≠ σ₂²
H = levene(X,Y,alpha,'plot')
H =
        Ho: 'do not reject'
         T: 0.2038
        CI: [-0.0402 0.0429]
    Pvalue: 0.8385
% Variances
[var(X), var(Y)] =
    0.0837    0.0822
```

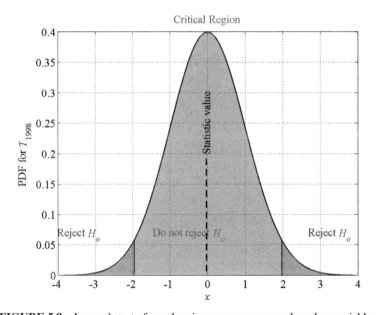

FIGURE 5.8   Levene's test of equal variances on non-normal random variables.

5.4.1.3 Test for Equivalence (Independent Samples from Normal Populations)

As was seen in Section 5.4.1.1, the hypothesis test for statistical equality $H_0: \sigma_x^2 = \sigma_y^2$ vs. $H_1: \sigma_x^2 \neq \sigma_y^2$ can be determined via an $\mathcal{F}$-distributed test statistic. Just as tests of statistical equality for proportions can be problematic, so too can tests of equality of variances. That is, statistical equivalence is a more stable approach to testing parameter behavior. This notion is discussed in Section 5.1.

For statistical equality, the ratio of the variances must be unity: $\sigma_x^2/\sigma_y^2 = 1$. For statistical equivalence, there is a *clinical margin* $\delta > 0$ so that the ratio is "close enough" to 1. That is, the clinical margin δ is in the open interval $(0, 1)$ and small enough that the presence of the ratio of variances within the inequality $1 - \delta < \sigma_x^2/\sigma_y^2 < 1 + \delta$ does *not* alter the clinical effectiveness of the medical treatment under review. Performing an algebraic simplification yields

$$-\delta < \sigma_x^2/\sigma_y^2 - 1 < \delta \text{ or } |\sigma_x^2/\sigma_y^2 - 1| < \delta$$

Thus, the hypotheses for equivalence of variances with clinical margin δ are as below.

Hypotheses

$$H_0: |\sigma_x^2/\sigma_y^2 - 1| \geq \delta \qquad\qquad (\textit{Null hypothesis})$$
$$H_1: |\sigma_x^2/\sigma_y^2 - 1| < \delta \qquad\qquad (\textit{Alternate hypothesis})$$

Test Statistic
Consider the null hypothesis as the two inequalities $\sigma_x^2/\sigma_y^2 - 1 \geq \delta$ and $\sigma_x^2/\sigma_y^2 - 1 \leq -\delta$. In the first case, $\sigma_x^2/\sigma_y^2 \geq 1 + \delta \Leftrightarrow \sigma_x^2 \geq (1 + \delta) \cdot \sigma_y^2$. Perform the substitution $\sigma_{z+}^2 = (1 + \delta) \cdot \sigma_y^2$. Then the first inequality becomes $\sigma_x^2 \geq \sigma_{z+}^2$. This is the left-tailed test for variance that has a p-value $p^+ = \Phi_F(F_{\delta+}(X, Y); df)$ where $df = [n_x - 1, n_y - 1]$ and $F_{\delta+}(X, Y) = \frac{S_X^2}{(1+\delta)\cdot S_Y^2}$.

Conversely, the second inequality $\sigma_x^2/\sigma_y^2 - 1 \leq -\delta$ is equivalent to $\sigma_x^2 \leq (1 - \delta) \cdot \sigma_y^2 \equiv \sigma_{z-}^2$; a right-tailed test. The p-value for a right-tailed test is $p^- = 1 - \Phi_F(F_{\delta-}(X, Y); df)$ where df is as above and $F_{\delta-}(X, Y) = \frac{S_X^2}{(1-\delta)\cdot S_Y^2}$.

Therefore, the overall p-value of *both* inequalities is the maximum of the two one-sided tests: $p = \max\{p^-, p^+\}$.

Test Statistic Distribution
$F_{\delta\pm}(X, Y) = \frac{F(X,Y)}{(1\pm\delta)}$ is a constant multiple of $F(X, Y) = \frac{S_X^2}{S_Y^2}$ which is distributed as $\mathcal{F}_{df}$ with $df = [n_x - 1, n_y - 1]$.

p-Value
$p = \max\{p^-, p^+\}$. See the computation for the test statistic.

Null Hypothesis Decision
If $p < \alpha$, then *reject H_0 in favor of H_1*. Otherwise, *do not reject H_0*.

$(1 - \alpha) \bullet 100\%$ Confidence Interval for σ_x^2 / σ_y^2

$$CI\left(\sigma_x^2 / \sigma_y^2\right) = \left[\frac{F(X, Y)}{f_{(n_x-1, n_y-1)}(1 - \alpha/2)}, \frac{F(X, Y)}{f_{(n_x-1, n_y-1)}(\alpha/2)}\right]$$

$F(X, Y)$ is the same statistic described in equation (5.12) of Section 5.4 and $f_{df}(\alpha)$ is the $\alpha \bullet 100\%$ quantile of the F_{df} distribution.

$(1 - \alpha) \bullet 100\%$ Confidence Intervals for σ_x^2 and σ_y^2

$CI(\sigma_x^2)$ and $CI(\sigma_y^2)$ are formed from (5.15) with $df_x = n_x - 1$ and $df_y = n_y - 1$, respectively.

Required Sample Size
The sample size formulation follows Section 5.4.1.1.

$$df = [df_1, df_2] \equiv [n_x - 1, n_y - 1]$$

$$\rho_{df} = \frac{df_1}{df_2} = \frac{n_x - 1}{n_y - 1}$$

If the samples are matched, then $\rho_{df} = 1$. Otherwise, set $n_x - 1 = n$ and $n_y - 1 = n/\rho_{df}$.

$$G(n) = \Phi_F(|es - \delta| \cdot f_{[n, n/\rho_{df}]}(\alpha/2))$$

$$N_s = \arg \min_n \{|G(n) - (1 - \beta)|\} \Rightarrow n_x = N_s + 1 \text{ and } n_y = 1 + \left\lceil \frac{N_s}{\rho_{df}} \right\rceil.$$

$$power = 1 - \beta$$

$$es = \text{effect size} \left(es = \sigma_x^2 / \sigma_y^2\right)$$

MATLAB Function
`niftest.m`

Example: Determine whether the variance for the weight of professional basketball players σ_B^2 is equivalent to the variance of the weight of hockey players σ_H^2 to within 10%. In this case, the hypotheses are

$$H_0: \left|\sigma_H^2 / \sigma_B^2 - 1\right| \geq \delta \text{ vs. } H_1: \left|\sigma_H^2 / \sigma_B^2 - 1\right| < \delta$$

where $\delta = 0.1$ is the clinical margin. As readily seen from Table 5.24, the left endpoint of the confidence interval for $\left|\sigma_H^2 / \sigma_B^2 - 1\right|$ *exceeds* the clinical margin, and hence, the estimate of $\left|\sigma_H^2 / \sigma_B^2 - 1\right|$ is significantly larger than δ. This information combined

TABLE 5.24   Summary results for test of equivalent variances

H_0	p-value	Reject H_0?	σ_H^2	$CI(\sigma_H^2)$	σ_B^2	$CI(\sigma_B^2)$
$\|\sigma_H^2/\sigma_B^2 - 1\| \geq \delta$	1	No ($p > \alpha$)	229.7	[207.2, 256.2]	684.1	[601.7, 784.7]
$CI(\sigma_H^2/\sigma_B^2) = [0.291, 0.397] \Rightarrow CI(\|\sigma_H^2/\sigma_B^2 - 1\|) = [0.6, 0.71] > \delta \Rightarrow$ do not reject H_0						

with the p-value (being greater than the significance level) indicates that the null hypothesis is *not rejected*.

MATLAB Commands
(Two-Sided Test of Equivalent Variances)

```
% Data Directory
Ddir = 'C:\PJC\Math_Biology\Chapter_5\Data';
% Load in the Hockey and Basketball data
load(fullfile(Ddir,'HockeyData.mat'))
load(fullfile(Ddir,'BasketballData.mat'))
% Collect the hockey (X) and basketball (Y) weight data
X = D.Wt; Y = B.Wt;
% Set the significance level at 5%: α = 0.05 and clinical margin δ = 0.1
alpha = 0.05; delta = 0.1;
% Test the hypotheses
% H₀: |Var(Hwt)/Var(Bwt) − 1| ≥ δ vs. H₁: |Var(Hwt)/Var(Bwt) − 1| < δ
H = niftest(X,Y,alpha,delta,2)
H =
         Pvalue: 1
             Ho: 'do not reject'
           VarX: 229.7623
           CIVx: [207.2159 256.2195]
           VarY: 684.1271
           CIVy: [601.7618 784.7357]
              F: 0.3358
             CI: [0.2908 0.3974]
       TestType: 'Equivalence'
         margin: 0.1000
```

5.4.2   One-Sided Tests of Variances

In this section, the usual one-sided tests (*right-* and *left-tailed* tests) are presented along with non-inferiority and superiority tests. Throughout the presentation, the assumption will be that the samples $X = \{x_1, x_2, ..., x_{n_x}\}$ and $Y = \{y_1, y_2, ..., y_{n_y}\}$

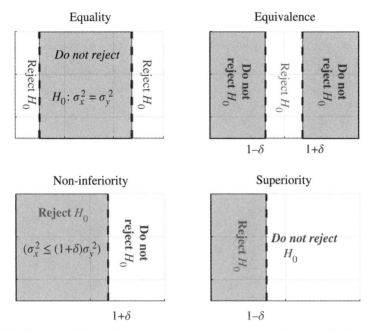

FIGURE 5.9   Statistical equality, equivalence, non-inferiority, and superiority of variances.

are taken from independent, normally distributed random variables X and Y, respectively. It is also assumed that X represents the *Control* population while Y is from the *Test* group. Parallel to the presentation of Figures 5.5–5.7, Figure 5.9 depicts the critical regions for statistical tests of equality, equivalence, non-inferiority, and superiority of variances. The shaded region in the upper left portion of the figure is the extent of the confidence interval for the estimated ratio of variances σ_x^2/σ_y^2 required to establish each scenario.

5.4.2.1   Right-Tailed Test (Independent Samples from Normal Populations)
Hypotheses

$$H_0: \sigma_x^2 \leq \sigma_y^2 \qquad \textit{(Null hypothesis)}$$
$$H_1: \sigma_x^2 > \sigma_y^2 \qquad \textit{(Alternate hypothesis)}$$

Test Statistic
$F(X, Y) = \dfrac{S_X^2}{S_Y^2}$ of equation (5.12)

Test Statistic Distribution
$F(X, Y)$ is distributed as $\mathcal{F}_{df}$ with $df = [n_x - 1, n_y - 1]$.

p-**Value**

$$p = 1 - \Phi_F(F(X, Y); df)$$

$\Phi_F(x; df) = $ *cumulative density function* of an F-distributed random variable with df degrees of freedom; see equation (5.14)

Null Hypothesis Decision
If $p < \alpha$, then *reject H_0* in favor of H_1. Otherwise, *do not reject H_0*.

$(1 - \alpha)\bullet100\%$ **Confidence Interval for** σ_x^2/σ_y^2

$$CI\left(\sigma_x^2/\sigma_y^2\right) = [\frac{F(X, Y)}{f_{(n_x-1, n_y-1)}(1 - \alpha)}, +\infty)$$

$(1 - \alpha)\bullet100\%$ **Confidence Intervals for** σ_x^2 **and** σ_y^2
$CI(\sigma_x^2)$ and $CI(\sigma_y^2)$ are formed from (5.15) with $df_x = n_x - 1$ and $df_y = n_y - 1$, respectively.

Required Sample Size
See Section 5.4.1.1.

$$df = [df_1, df_2] \equiv [n_x - 1, n_y - 1]$$
$$\rho_{df} = \frac{df_1}{df_2} = \frac{n_x - 1}{n_y - 1}$$

If the samples are matched, then $\rho_{df} = 1$. Otherwise, set $n_x - 1 = n$ and $n_y - 1 = n/\rho_{df}$.

$$G(n) = \Phi_F(es \cdot f_{[n, n/\rho_{df}]}(\alpha))$$
$$N_s = \arg\min_n\{|G(n) - (1 - \beta)|\} \Rightarrow n_x = N_s + 1 \text{ and } n_y = 1 + \left\lceil \frac{N_s}{\rho_{df}} \right\rceil$$
$$power = 1 - \beta$$
$$es = \text{effect size } (es = \sigma_x^2/\sigma_y^2)$$

MATLAB Function
ftest.m

Example: To test whether the variance of the weight of basketball players $\sigma_B^2(Wt)$ exceeds that of hockey players $\sigma_H^2(Wt)$, exercise the hypotheses

$$H_0: \sigma_B^2(Wt) \leq \sigma_H^2(Wt) \text{ vs. } H_1: \sigma_B^2(Wt) > \sigma_H^2(Wt)$$

TABLE 5.25   Summary results for right-tailed test of variances

H_0	p-value	Reject H_0?	σ_H^2	$CI(\sigma_H^2)$	σ_B^2	$CI(\sigma_B^2)$
$\sigma_B^2 \leq \sigma_H^2$	0	Yes ($p < \alpha$)	229.7	[207.2, 256.2]	684.1	[601.7, 784.7]

$CI\left(\sigma_H^2/\sigma_B^2\right) = [2.585, +\infty) \Rightarrow 1 \notin CI\left(\sigma_H^2/\sigma_B^2\right) \Rightarrow$ reject H_0

As can be seen from the summary computation in Table 5.25, both $p < \alpha$ and $1 \notin CI(\sigma_H^2/\sigma_B^2)$. Consequently, H_0 is rejected in favor of the alternate hypothesis H_1 that the variance σ_B^2 is larger than σ_H^2.

MATLAB Commands
(One-Sided Right-Tailed Test of Variances)

```
% Data Directory
Ddir = 'C:\PJC\Math_Biology\Chapter_5\Data';
% Load in the Hockey and Basketball data
load(fullfile(Ddir,'HockeyData.mat'))
load(fullfile(Ddir,'BasketballData.mat'))
% Collect the hockey (Y) and basketball (X) weight data
Y = D.Wt; X = B.Wt;
% Set the significance level at 5%: α = 0.05
alpha = 0.05;
% Use the F-test method to test for distinct variances
% H₀: σ²_B(Wt) ≤ σ²_H(Wt) vs. H₁: σ²_B(Wt) > σ²_H(Wt)
H = ftest(X,Y,alpha,0)
H =
        Ho: 'reject'
        Vx: 684.1271
      CIVx: [601.7618  784.7357]
        Vy: 229.7623
      CIVy: [207.2159  256.2195]
         F: 2.9775
        CI: [2.5851 Inf]
        df: [437 683]
    Pvalue: 0
```

5.4.2.2   *Left-Tailed Test (Independent Samples from Normal Populations)*

Hypotheses

$$H_0: \sigma_x^2 \geq \sigma_y^2 \qquad\qquad \textit{(Null hypothesis)}$$
$$H_1: \sigma_x^2 < \sigma_y^2 \qquad\qquad \textit{(Alternate hypothesis)}$$

Test Statistic
$F(X, Y) = \frac{S_X^2}{S_Y^2}$ of equation (5.12)

Test Statistic Distribution
$F(X, Y)$ is distributed as $\mathcal{F}_{df}$ with $df = [n_x - 1, n_y - 1]$.

p-Value

$$p = \Phi_F(F(X, Y); df)$$

$\Phi_F(x; df) = $ *cumulative density function* of an $\mathcal{F}$-distributed random variable with df degrees of freedom; see equation (5.14)

Null Hypothesis Decision
If $p < \alpha$, then *reject H_0* in favor of H_1. Otherwise, *do not reject H_0*.

$(1 - \alpha) \bullet 100\%$ Confidence Interval for σ_x^2 / σ_y^2

$$CI\left(\sigma_x^2 / \sigma_y^2\right) = (-\infty, \frac{F(X, Y)}{f_{(n_x-1, n_y-1)}(\alpha)}]$$

$(1 - \alpha) \bullet 100\%$ Confidence Intervals for σ_x^2 and σ_y^2
$CI(\sigma_x^2)$ and $CI(\sigma_y^2)$ are formed from (5.15) with $df_x = n_x - 1$ and $df_y = n_y - 1$, respectively.

Required Sample Size
See Section 5.4.1.1.

$$df = [df_1, df_2] \equiv [n_x - 1, n_y - 1]$$
$$\rho_{df} = \frac{df_1}{df_2} = \frac{n_x - 1}{n_y - 1}$$

If the samples are matched, then $\rho_{df} = 1$. Otherwise, set $n_x - 1 = n$ and $n_y - 1 = n/\rho_{df}$.

$$G(n) = \Phi_F(es \cdot f_{[n, n/\rho_{df}]}(\alpha))$$

$$N_s = \arg\min_n\{|G(n) - (1 - \beta)|\} \Rightarrow n_x = N_s + 1 \text{ and } n_y = 1 + \left\lceil \frac{N_s}{\rho_{df}} \right\rceil$$
$$power = 1 - \beta$$
$$es = \text{effect size } (es = \sigma_x^2 / \sigma_y^2)$$

MATLAB Function
ftest.m

TABLE 5.26   Summary results for left-tailed test of variances

H_0	p-value	Reject H_0?	σ_H^2	$CI(\sigma_H^2)$	σ_B^2	$CI(\sigma_B^2)$
$\sigma_B^2 \geq \sigma_H^2$	$<10^{-4}$	Yes ($p < \alpha$)	4.29	[3.87, 4.74]	11.7	[10.29, 13.42]
$CI\left(\sigma_H^2/\sigma_B^2\right) = [0, 0.4223] \Rightarrow 1 \notin CI\left(\sigma_H^2/\sigma_B^2\right)) \Rightarrow$ reject H_0						

Example: Test whether the variance in the heights of professional basketball players $\sigma_B^2(Ht)$ is greater than the variance in the heights of hockey players $\sigma_H^2(Ht)$. The hypotheses are

$$H_0: \sigma_H^2(Ht) \geq \sigma_B^2(Ht) \text{ vs. } H_1: \sigma_H^2(Ht) < \sigma_B^2(Ht)$$

The summary provided in Table 5.26 indicates that the p-value is less than the significance level and that 1 is *not* contained in the confidence interval for σ_H^2/σ_B^2. Thus, the null hypothesis is rejected in favor of the alternate hypothesis that $\sigma_H^2(Ht) < \sigma_B^2(Ht)$.

MATLAB Commands
(One-Sided Left-Tailed Test of Variances)

```
% Data Directory
Ddir = 'C:\PJC\Math_Biology\Chapter_5\Data';
% Load in the Hockey and Basketball data
load(fullfile(Ddir,'HockeyData.mat'))
load(fullfile(Ddir,'BasketballData.mat'))
% Collect the hockey (X) and basketball (Y) height data
X = D.Ht; Y = B.Ht;
% Set the significance level at 5%: α = 0.05
alpha = 0.05;
% Use the F-test method to test for distinct variances
% H₀: σ²H(Ht) ≥ σ²B(Ht) vs. H₁: σ²H(Ht) < σ²B(Ht)
H = ftest(X,Y,alpha,1)
H =
        Ho: 'reject'
        Vx: 4.2899
      CIVx: [3.8689 4.7838]
        Vy: 11.7008
      CIVy: [10.2921 13.4215]
         F: 0.3666
        CI: [0 0.4223]
        df: [683 437]
    Pvalue: 2.6756e-32
```

5.4.2.3   Non-Inferiority Test (Independent Samples from Normal Populations)

The *margin* $\delta \in (0, 1)$ should be viewed as a fraction over which one variance is *clinically* different from a second. For example, if a 25% change in variance is the largest to be clinically meaningful, then the margin is taken to be 0.25: $\delta = 0.25$. Since variances are positive, the inequality for non-inferiority $\sigma_x^2/\sigma_y^2 > 1 + \delta$ is equivalent to $\sigma_x^2 > (1 + \delta) \cdot \sigma_y^2$. This form will be used for the hypothesis test of *non-inferiority*. The remainder of the section now proceeds as throughout this chapter.

Hypotheses

$$H_0: \sigma_x^2 \geq (1 + \delta) \cdot \sigma_y^2 \qquad \text{(\textit{Null hypothesis})}$$
$$H_1: \sigma_x^2 < (1 + \delta) \cdot \sigma_y^2 \qquad \text{(\textit{Alternate hypothesis})}$$

Clinical Margin

$$\delta \in (0, 1)$$

Test Statistic

$F_\delta(X, Y) = \frac{1}{1+\delta} \bullet F(X, Y)$ where $F(X, Y) = \frac{S_X^2}{S_Y^2}$ of equation (5.12).

Test Statistic Distribution

$F(X, Y)$ has an F-distribution with $df = [n_x - 1, n_y - 1]$ degrees of freedom.

p-Value

$$p_\delta = \Phi_F(F_\delta(X, Y); df)$$

$\Phi_F(x; df) = $ *cumulative density function* of an F-distributed random variable with df degrees of freedom; see equation (5.14).

Null Hypothesis Decision

If $p_\delta < \alpha$, then *reject H_0* in favor of H_1. Otherwise, *do not reject H_0*.
The *two-sided* $(1 - \alpha) \bullet 100\%$ *confidence interval for* σ_x^2/σ_y^2 is presented in Section 5.4.1.1.

$(1 - \alpha) \bullet 100\%$ Confidence Intervals for σ_x^2 and σ_y^2

$CI(\sigma_x^2)$ and $CI(\sigma_y^2)$ are formed from (5.15) with $df_x = n_x - 1$ and $df_y = n_y - 1$, respectively.

Required Sample Size

The sample size formulation follows Section 5.4.1.1.

$$df = [df_1, df_2] \equiv [n_x - 1, n_y - 1]$$
$$\rho_{df} = \frac{df_1}{df_2} = \frac{n_x - 1}{n_y - 1}$$

TABLE 5.27   Summary results for non-inferiority test of variances

H_0	p-value	Reject H_0?	σ_H^2	$CI(\sigma_H^2)$	σ_B^2	$CI(\sigma_B^2)$
$\sigma_H^2 \geq (1+\delta) \cdot \sigma_B^2$	0.02	Yes $(p < \alpha)$	19.8	$[17.89, 22.12]$	18.89	$[16.62, 21.67]$

$CI(\sigma_H^2/\sigma_B^2) = [0.91, 1.24] \Rightarrow \max\{CI(\sigma_H^2/\sigma_B^2)\} < 1 + \delta \Rightarrow$ reject H_0

If the samples are matched, then $\rho_{df} = 1$. Otherwise, set $n_x - 1 = n$ and $n_y - 1 = n/\rho_{df}$.

$$G(n) = \Phi_F((es - \delta) \cdot f_{[n,n/\rho_{df}]}(\alpha))$$

$$N_s = \arg\min_n\{|G(n) - (1 - \beta)|\} \Rightarrow n_x = N_s + 1 \text{ and } n_y = 1 + \left\lceil \frac{N_s}{\rho_{df}} \right\rceil$$

$$power = 1 - \beta$$

$$es = \text{effect size } (es = \sigma_x^2/\sigma_y^2)$$

MATLAB Function

`niftest.m`

Example: Test whether the variance of the basketball player ages $\sigma_B^2(Ag)$ is not inferior to the variance in hockey player ages $\sigma_H^2(Ag)$ using a 25% margin. That is, test the hypotheses

$$H_0: \sigma_H^2(Ag) \geq (1 + \delta) \bullet \sigma_B^2(Ag) \text{ vs. } H_1: \sigma_H^2(Ag) < (1 + \delta) \bullet \sigma_B^2(Ag)$$

From Table 5.27 it is seen that $p < \alpha$ and $\max\{CI(\sigma_H^2/\sigma_B^2)\} < 1 + \delta$. Therefore, $\sigma_H^2(Ag) < (1 + \delta) \cdot \sigma_B^2(Ag)$ and the null H_0 is rejected in favor of the alternate H_1.

MATLAB Commands
(Non-Inferiority Test of Variances)

```
% Data Directory
Ddir = 'C:\PJC\Math_Biology\Chapter_3\Data';
% Load in the Hockey and Basketball data
load(fullfile(Ddir,'HockeyData.mat'))
load(fullfile(Ddir,'BasketballData.mat'))
% Collect the hockey (X) and basketball (Y) ages data
X = D.Ages; Y = B.Ages;
% Set the significance level at 5%: alpha = 0.05 and clinical margin delta = 0.25
alpha = 0.05; delta = 0.25;
% Test the hypotheses
% H_0: Var(H_age) >= (1 + delta) Var(B_age) vs. H_1: Var(H_age) < (1 + delta) Var(B_age)
```

```
H = niftest(X,Y,alpha,delta,0)
H =
      Pvalue: 0.0212
          Ho: 'reject'
        VarX: 19.8357
        CIVx: [17.8893 22.1198]
        VarY: 18.8908
        CIVy: [16.6164 21.6689]
           F: 1.0500
          CI: [0.9093 1.2426]
    TestType: 'Non-inferiority'
      margin: 0.2500
```

5.4.2.4   Superiority Test (Independent Samples from Normal Populations)
Hypotheses

$$H_0: \sigma_x^2 \geq (1 - \delta) \cdot \sigma_y^2 \qquad \text{(Null hypothesis)}$$
$$H_1: \sigma_x^2 < (1 - \delta) \cdot \sigma_y^2 \qquad \text{(Alternate hypothesis)}$$

Clinical Margin

$$\delta \in (0, 1)$$

Test Statistic

$F_\delta(X, Y) = \frac{1}{1-\delta} \cdot F(X, Y)$ where $F(X, Y) = \frac{S_X^2}{S_Y^2}$ of equation (5.12).

Test Statistic Distribution

$F(X, Y)$ has an $\mathcal{F}$-distribution with $df = [n_x - 1, n_y - 1]$ degrees of freedom.

p-Value

$$p_\delta = 1 - \Phi_F(F_\delta(X, Y); df)$$

$\Phi_F(x; df) = $ *cumulative density function* of an $\mathcal{F}$-distributed random variable with df degrees of freedom; see equation (5.14).

Null Hypothesis Decision

If $p_\delta < \alpha$, then *reject H_0* in favor of H_1. Otherwise, *do not reject H_0*.
The *two-sided $(1 - \alpha) \cdot 100\%$ confidence interval for* σ_x^2/σ_y^2 is presented in Section 5.4.1.1.

$(1 - \alpha) \cdot 100\%$ Confidence Intervals for σ_x^2 and σ_y^2

$CI(\sigma_x^2)$ and $CI(\sigma_y^2)$ are formed from (5.15) with $df_x = n_x - 1$ and $df_y = n_y - 1$, respectively.

TABLE 5.28   Summary results for superiority test of variances

H_0	p-value	Reject H_0?	σ_H^2	$CI(\sigma_H^2)$	σ_B^2	$CI(\sigma_B^2)$
$\sigma_H^2 \geq (1 - \delta) \cdot \sigma_B^2$	$<10^{-4}$	Yes ($p < \alpha$)	4.29	[3.87, 4.78]	11.7	[10.29, 13.42]
$CI(\sigma_H^2/\sigma_B^2) = [0.3175, 0.4339] \Rightarrow \max\{CI(\sigma_H^2/\sigma_B^2)\} < 1 - \delta \Rightarrow$ reject H_0						

Required Sample Size

The sample size formulation follows Section 5.4.1.1.

$$df = [df_1, df_2] \equiv [n_x - 1, n_y - 1]$$

$$\rho_{df} = \frac{df_1}{df_2} = \frac{n_x - 1}{n_y - 1}$$

If the samples are matched, then $\rho_{df} = 1$. Otherwise, set $n_x - 1 = n$ and $n_y - 1 = n/\rho_{df}$.

$$G(n) = \Phi_F\left((es - \delta) \cdot f_{[n,n/\rho_{df}]}(\alpha)\right)$$

$$N_s = \arg\min_n \{|G(n) - (1 - \beta)|\} \Rightarrow n_x = N_s + 1 \text{ and } n_y = 1 + \left\lceil \frac{N_s}{\rho_{df}} \right\rceil$$

$$power = 1 - \beta$$

$$es = \text{effect size } (es = \sigma_x^2/\sigma_y^2)$$

MATLAB Function

niftest.m

Example: Test whether the variance of the basketball player heights $\sigma_B^2(Ht)$ is *superior* to the variance in hockey player heights $\sigma_H^2(Ht)$ using a 10% margin. That is, even if the basketball variance is *deflated* by 10%, is it still larger than the hockey variance? The hypotheses are

$$H_0: \sigma_H^2(Ht) \geq (1 - \delta) \cdot \sigma_B^2(Ht) \text{ vs. } H_1: \sigma_H^2(Ht) < (1 - \delta) \cdot \sigma_B^2(Ht)$$

From Table 5.28 it is seen that $p < \alpha$ and $\max\{CI(\sigma_H^2/\sigma_B^2)\} < 1 - \delta$. Therefore, $\sigma_H^2(Ht) < (1 - \delta) \cdot \sigma_B^2(Ht)$ and the null H_0 is rejected in favor of the alternate H_1.

MATLAB Commands
(*Superiority Test of Variances*)

```
% Data Directory
Ddir = 'C:\PJC\Math_Biology\Chapter_5\Data';
% Load in the Hockey and Basketball data
load(fullfile(Ddir, 'HockeyData.mat'))
```

```
load(fullfile(Ddir,'BasketballData.mat'))
% Collect the hockey (X) and basketball (Y) height data
X = D.Ht; Y = B.Ht;
% Set the significance level at 5%: α = 0.05 and clinical margin δ = 0.1
alpha = 0.05; delta = 0.1;
% Test the hypotheses
% H₀: Var(H_ht) ≤ (1 − δ)·Var(B_ht) vs. H₁: Var(H_ht) < (1 − δ)·Var(B_ht)
H = niftest(X,Y,alpha,delta,1)
H =

        Pvalue: 2.8126e-26
            Ho: 'reject'
          VarX: 4.2899
          CIVx: [3.8689 4.7838]
          VarY: 11.7008
          CIVy: [10.2921 13.4215]
             F: 0.3666
            CI: [0.3175 0.4339]
      TestType: 'Superiority'
        margin: 0.1000
```

EXERCISE

5.9 Construct the test statistic and p-value for the equivalence test of variances for non-normally distributed random variables X and Y.

Hint: Define the absolute mean deviations $D_{x,j} = |x_j - \bar{x}|$ for $j = 1, 2,..., n_x$ and $D_{z,j} = (1 + \delta) \cdot D_{y,k} = (1 + \delta) \cdot |y_k - \bar{y}|$. Then $\overline{D}_z = \sqrt{(1 + \delta)} \cdot \overline{D}_y$, $\operatorname{var}[\overline{D}_z] = (1 + \delta) \cdot \operatorname{var}[\overline{D}_y]$, and $V_{pool}(\delta) = \frac{1}{n_x + n_y - 2}\{(n_x - 1) \cdot \operatorname{var}[D_x] + (n_y - 1) \cdot (1 + \delta) \cdot \operatorname{var}[D_y]\}$ so that the test statistic is $T_\delta(X, Y) = (\overline{D}_x - \sqrt{(1 + \delta)} \cdot \overline{D}_y) / \sqrt{(\frac{1}{n_x} + \frac{1}{n_y}) \cdot V_{pool}(\delta)}$. The p-value is $p_\delta = 2 \cdot \Phi_T(-|T_\delta|; df)$, $df = n_x + n_y - 2$.

5.5 OTHER HYPOTHESIS TESTS

The previous three sections have focused on the most prominent statistical tests: Means, proportions, and variances. There are other parameters, however, for which review agencies will request data analysis. A select few of these include *coefficient of variation (CV)*, *diagnostic likelihood ratios* (as the ratio of binomial variables), the *Mann–Whitney U test*, and the *Wilcoxon Signed-Rank test*. These measures, along with their formal hypothesis testing procedures, will be described in succeeding sections.

5.5.1 Coefficient of Variation

The *coefficient of variation* is the ratio of a random variable's standard deviation (σ) to its mean (μ). In this sense, it is a unitless measure of variability. More commonly in engineering, the ratio of the mean to standard deviation is referred to as the *signal-to-noise ratio*: $snr = \frac{\mu}{\sigma}$. Thus, the *coefficient of variation* is the *noise-to-signal ratio* $\frac{\sigma}{\mu}$ and is a measure of how "noisy" the system is under consideration. When comparing two distinct systems, statistically equal *CVs* will have comparable noise levels.

What is the hypothesis test formalism for the statistical equality of *CVs*? Consider the following scenario.

Let $X_1, X_2,..., X_{n_x}$ be independent and identically distributed (*i.i.d.*) normal random variables with mean μ_x and variance σ_x^2. Similarly, let $Y_1, Y_2,..., Y_{n_y}$ be i.i.d. $\mathcal{N}(\mu_y, \sigma_y^2)$. Then the coefficients of variance with respect to the different populations are $CV_x = \frac{\sigma_x}{\mu_x}$ and $CV_y = \frac{\sigma_y}{\mu_y}$. The focus of this section will be to determine the mathematical structure of the hypothesis test

$$H_0: \frac{\sigma_x}{\mu_x} = \frac{\sigma_y}{\mu_y} \qquad \text{(Null hypothesis)}$$

$$H_1: \frac{\sigma_x}{\mu_x} \neq \frac{\sigma_y}{\mu_y} \qquad \text{(Alternate hypothesis)}$$

A natural test statistic for H_0 is $T(X, Y) = \frac{S_x}{\bar{x}} - \frac{S_y}{\bar{y}}$ where $\bar{x}$ and $\bar{y}$ are the sample means, and S_x and S_y are the sample standard deviations. That is, $\bar{x} = \frac{1}{n_x} \sum_{j=1}^{n_x} x_j$ and $S_x = \sqrt{\frac{1}{n_x-1} \sum_{j=1}^{n_x} (x_j - \bar{x})^2}$ with parallel definitions for $\bar{y}$ and S_y. As mentioned in Section 5.1 (Exercise 5.1), the sample mean is distributed as a scaled normal random variable, $\bar{X} \sim \mathcal{N}(\mu_x, \frac{1}{n_x}\sigma_x^2)$. It is a well-established result that the sum of squares of normal random variables is a χ^2 variable; $\sum_{j=1}^{n_x} (X_j - \bar{X})^2 \sim \chi_{n_x-1}^2$. Consequently, the ratio of the sample mean to the sample standard deviation is distributed as a student-t variable with $n_x - 1$ degrees of freedom, $R_x = \frac{\bar{X}}{S_x} \equiv \frac{\bar{X}}{\sqrt{\frac{1}{n_x-1} \sum_{j=1}^{n_x} (X_j-\bar{X})^2}} \sim T_{n_x-1}$. To investigate the mathematical form of the distribution for CV_x, two important results from mathematical statistics need to be presented.

Change of Variable Theorem: Let X be a continuous random variable with *pdf* ϕ_x and domain of definition Ω_X. Further, let $g: \Omega_X \to \Omega_Y$ be a continuous one-to-one transformation with inverse g^{-1}. Then the random variable $Y = g(X)$ is continuous and has the *pdf* $\phi_y(y) = \phi_x(g^{-1}(y)) \cdot |\frac{dg^{-1}(y)}{dy}|$ with domain of definition Ω_Y.

The proof is a standard result from calculus and is left as an exercise.

The *change of variable theorem* can be extended to two dimensions as follows. Let $X = (X_1, X_2)$ be a continuous random vector with *joint probability density function* $\phi_X(x_1, x_2)$ and $x \in \Omega_X$. Moreover, let $U: \mathbb{R}^2 \to \mathbb{R}^2$ be a continuous, 1–1 transformation so that $(Y_1, Y_2) = Y = U(X) = \begin{bmatrix} u_1(x_1, x_2) \\ u_2(x_1, x_2) \end{bmatrix}$ and $\Omega_Y = U(\Omega_X)$. The inverse transform $X = U^{-1}(Y) = \begin{bmatrix} w_1(y_1, y_2) \\ w_2(y_1, y_2) \end{bmatrix}$ can be combined with the *Jacobian* of the transformation to compute the *joint pdf* $\phi_Y(y_1, y_2)$ of Y and $y \in \Omega_Y$. The *Jacobian J* of the mapping $X \mapsto U(X) = Y$ is defined as the determinant of the first-order partial derivatives.

$$J(x, y) = \det \left(\begin{bmatrix} \dfrac{\partial x_1}{\partial y_1} & \dfrac{\partial x_1}{\partial y_2} \\ \dfrac{\partial x_2}{\partial y_1} & \dfrac{\partial x_2}{\partial y_2} \end{bmatrix} \right) = \dfrac{\partial x_1}{\partial y_1} \cdot \dfrac{\partial x_2}{\partial y_2} - \dfrac{\partial x_2}{\partial y_1} \cdot \dfrac{\partial x_1}{\partial y_2}$$

Using this notation, the *joint pdf of the transformed random vector Y* is

$$\phi_Y(y_1, y_2) = \phi_X \left(w_1(y_1, y_2), w_2(y_1, y_2) \right) \cdot |J(x, y)|, y \in \Omega_Y \qquad (5.19)$$

The *pdf* of the elements of the random vector Y can be recovered from the joint *pdf* by simply "integrating out" the undesired element. The *pdf* of a single vector element is called the *marginal distribution*. For example, the *marginal pdfs* of Y_1 and Y_2 are given by (5.20a) and (5.20b).

$$\phi_{Y_1}(y_1) = \int_{\Omega_{Y_2}} \phi_Y(y_1, y_2) \, dy_2 \qquad (5.20a)$$

$$\phi_{Y_2}(y_2) = \int_{\Omega_{Y_1}} \phi_Y(y_1, y_2) \, dy_1 \qquad (5.20b)$$

Here, Ω_{Y_j} is the domain of definition of the random variable Y_j.

The domain of definition Ω or *support* of a function is the set of all points on which the function is not zero. This is denoted by $supp(\phi) = \{x: \phi(x) \neq 0\}$. The *pdf* of a continuous random variable can have finite support.

Example: The uniform distribution $\mathcal{U}([a, b])$. A random variable U has a uniform distribution over the interval $[a, b]$ provided $U = 1$ on $[a, b]$ and $U = 0$ on $\mathbb{R} \backslash [a, b] = \{x: x < a \text{ or } x > b\}$. The *pdf* for the uniform random variable on $[a, b]$ is $\phi_U(x) = \begin{cases} 1 & \text{for } a \leq x \leq b \\ 0 & \text{otherwise} \end{cases}$. Thus, the support for U is $supp(U) = [a, b]$.

The second result is based on the idea of *convolution*.

Definition: The *convolution* of two functions f and g, denoted as $f \circ g$ is

$$(f \circ g)(x) = \int_{-\infty}^{\infty} f(x - \xi) \cdot g(\xi) \, d\xi$$

If the support of f and g is, for example, the compact interval $[a, b]$, then the lower limit $(-\infty)$ of the integral in the definition is replaced by a. The following is a significant result of mathematical statistics. One way to prove the theorem requires the use of *characteristic functions* (see Exercise 5.8). A characteristic function of a random variable is the *inverse Fourier Transform* of its probability density function. A foray into this area, while immensely interesting, is beyond the present scope of this text. Andrews and Shivamoggi [4], Bracewell [7], Papoulis [30], Hogg and Craig [14], and other books on probability theory will discuss characteristic functions and Fourier transforms. Instead, the ideas extending the statement of the (one-dimensional) change of variables theorem will be utilized.

Convolution Theorem: If X_1 and X_2 are continuous random variables with *pdfs* ϕ_1 and ϕ_2, respectively, then the *pdf* of the random variable $Y = X_1 + X_2$ is $\phi_Y(y) = (\phi_1 \circ \phi_2)(y)$.

Proof: Suppose X_1 and X_2 are independent continuous random variables whose *pdfs* ϕ_1 and ϕ_2 both have supports equal to the real line $\mathbb{R}$. Then the *joint pdf* of the random vector X is $\phi_X(x_1, x_2) = \phi_1(x_1) \cdot \phi_2(x_2)$ for $x = (x_1, x_2) \in \mathbb{R} \times \mathbb{R} = \mathbb{R}^2$. Let $Y_1 = X_1 + X_2 \equiv u_1(X_1, X_2)$ and $Y_2 = X_1 - X_2 \equiv u_2(X_1, X_2)$. Then $U(X) = Y$, defined by the functions u_j, is continuous and invertible with $y_j \in \mathbb{R}$ for $j = 1, 2$. The inverse mapping is $U^{-1} = \begin{bmatrix} w_1(y_1, y_2) \\ w_2(y_1, y_2) \end{bmatrix} = \begin{bmatrix} \frac{1}{2}(Y_1 + Y_2) \\ \frac{1}{2}(Y_1 - Y_2) \end{bmatrix}$. The Jacobian of this transformation is

$$J(x, y) = \det\left(\begin{bmatrix} \frac{\partial x_1}{\partial y_1} & \frac{\partial x_1}{\partial y_2} \\ \frac{\partial x_2}{\partial y_1} & \frac{\partial x_2}{\partial y_2} \end{bmatrix}\right) = \det\left(\begin{bmatrix} \frac{1}{2} & \frac{1}{2} \\ \frac{1}{2} & -\frac{1}{2} \end{bmatrix}\right) = -\tfrac{1}{4} - \tfrac{1}{4} = -\tfrac{1}{2}.$$ Subsequently, by

formula (5.19), the *joint pdf* of the random vector Y, with $y \in \mathbb{R}^2$, is

$$\phi_Y(y_1, y_2) = \phi_1\left(w_1(y_1, y_2)\right) \cdot \phi_2\left(w_2(y_1, y_2)\right) \cdot |J(x,y)|$$

$$= \tfrac{1}{2}\phi_1\left(\tfrac{1}{2}(y_1 + y_2)\right) \cdot \phi_2\left(\tfrac{1}{2}(y_1 - y_2)\right)$$

Now to recover the *pdf* for $Y_1 = X_1 + X_2$, use (5.20a) to obtain

$$\phi_{Y_1}(y_1) = \int_{-\infty}^{\infty} \tfrac{1}{2}\phi_1\left(\tfrac{1}{2}(y_1 + y_2)\right) \cdot \phi_2\left(\tfrac{1}{2}(y_1 - y_2)\right) dy_2$$

Next, let $\xi = \frac{1}{2}(y_1 + y_2)$ so that $y_2 = 2\xi - y_1$, $\frac{1}{2} dy_2 = d\xi$, and the integral becomes

$$\phi_{Y_1}(y_1) = \int_{-\infty}^{\infty} \phi_1(\xi) \cdot \phi_2(\xi - y_2)\, d\xi = (\phi_1 \circ \phi_2)(\xi)$$

∎

With these two theorems in mind, the matter of the statistical equality of CVs can now be revisited. Recall that $T(X, Y) = \dfrac{S_x}{\bar{x}} - \dfrac{S_y}{\bar{y}}$ is a natural test statistic for the hypothesis H_0: $\sigma_x/\mu_x = \sigma_y/\mu_y$. To determine the *probability density function* of the test statistic T, the *pdfs* for $T_1 = \dfrac{S_x}{\bar{x}}$ and $T_2 = -\dfrac{S_y}{\bar{y}}$ must be constructed. Lehmann [16] notes that the statistic $Z_1 = \sqrt{n_x}/T_1 = \sqrt{n_x} \cdot \dfrac{\bar{x}}{S_x}$ has a *non-central T-distribution* with $df = n_x - 1$ degrees of freedom, non-centrality parameter $\lambda_x = \sqrt{n_x}\dfrac{\mu_x}{\sigma_x}$ and *pdf* $\phi_1(t) \equiv \phi_T(t; n_x - 1, \lambda_x)$ as per equation (5.21a).

$$\left.\begin{array}{c} \phi_T(t; n_x - 1, \lambda_x) = C(n_x) \displaystyle\int_0^{\infty} (\sqrt{\tau})^{(n_x - 2)} e^{-\frac{1}{2}\tau} e^{-\frac{1}{2}\left(t\sqrt{\frac{\tau}{n_x-1}} - \lambda_x\right)^2} d\tau \\[4mm] C(n_x) = \dfrac{1}{(\sqrt{2})^{(n_x-2)} \Gamma\left(\frac{1}{2}(n_x - 1)\right) \sqrt{\pi(n_x - 1)}} \end{array}\right\}, \quad \lambda_x = \sqrt{n_x}\dfrac{\mu_x}{\sigma_x}$$

$$(5.21a)$$

Notice that the change of variables $u = \dfrac{t\sqrt{\tau}}{\sqrt{n_x-1}}$ converts (5.21a) into (5.21b).

$$\left.\begin{array}{c} \phi_T(t; n_x - 1, \lambda_x) = C'(n_x) \displaystyle\int_0^{\infty} u^{(n_x-1)} e^{-\frac{1}{2}\left[u^2\left(\frac{n_x-1}{t^2}\right) + (u - \lambda_x)^2\right]} d\tau \\[4mm] C'(n_x) = \dfrac{2(n_x - 1)^{\frac{1}{2}n_x}}{(\sqrt{2})^{(n_x-2)} t^{n_x}\, \Gamma\left(\frac{1}{2}(n_x - 1)\right) \sqrt{\pi(n_x - 1)}} \end{array}\right\}, \quad \lambda_x = \sqrt{n_x}\dfrac{\mu_x}{\sigma_x}$$

$$(5.21b)$$

The support of Z_1 is the positive real line $\Omega_{Z_1} = \mathbb{R}^+ = [0, \infty)$. Consider the transformation $U(t; a) = a/t$. Since a *CV* is a noise-to-signal ratio, it will be assumed to be from a distribution with a non-zero mean. Therefore, $Z_1 = U(T_1; \sqrt{n_x}) = \sqrt{n_x}/T_1$ will have strictly positive support. Since the *pdf* for Z_1 is $\phi_T(t; n_x - 1, \lambda_x)$, as defined by equation (5.3a), and $U^{-1}(t; \sqrt{n_x}) = \sqrt{n_x}/T_1$ then $\dfrac{dU^{-1}(t; \sqrt{n_x})}{dt} = -\dfrac{\sqrt{n_x}}{t^2}$. Hence by the change of variable theorem, the *pdf* for Z_1 is $\phi_1(t) =$

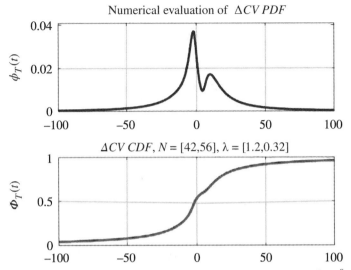

FIGURE 5.10   The *pdf* and *cdf* for the *CV* test statistic $T(X,Y) = \frac{S_x}{\bar{x}} - \frac{S_y}{\bar{y}}$

$\phi_T(\frac{\sqrt{n_x}}{t}; n_x - 1, \lambda_x) \cdot \frac{\sqrt{n_x}}{t^2}$. For $Z_2 = U(\tau; -\sqrt{n_y}) = -\sqrt{n_y}/\tau$ and $\tau = \frac{S_y}{\bar{y}}$ it is seen that $\phi_2(t) = \phi_T(-\frac{\sqrt{n_y}}{t}; n_y - 1, \lambda_y) \cdot \frac{\sqrt{n_y}}{t^2}$. Since the test statistic is $T(X, Y) = Z_1 + Z_2$, by the convolution theorem the *pdf* is a function of sample sizes n_x, n_y and noncentrality parameters λ_x, λ_y.

$$\phi_T(t; [n_x, n_y], [\lambda_x, \lambda_y]) = (\phi_1 \circ \phi_2)(t)$$
$$= \sqrt{n_x \cdot n_y} \int_{-\infty}^{\infty} \phi_T\left(\frac{\sqrt{n_x}}{(t-\tau)}; n_x - 1, \lambda_x\right) \cdot \phi_T\left(-\frac{\sqrt{n_y}}{\tau}; n_y - 1, \lambda_y\right) \cdot \frac{1}{(t-\tau)^2 \cdot \tau^2} d\tau$$

$$(5.21c)$$

The integral in (5.21c), however, resists reduction to either elementary or well-understood special functions. It can be evaluated *numerically* via the MATLAB function `integral.m`. The *cdf* for the test statistic is the integral of the *pdf* $\Phi_T(t; [n_x, n_y],$ $[\lambda_x, \lambda_y]) = \int_{-\infty}^{t} \phi_T(\tau; [n_x, n_y], [\lambda_x, \lambda_y]) \, d\tau$. Figure 5.10 illustrates the computation of (5.21b) for $n_x = 42$, $n_y = 56$, $\lambda_x = 1.2$, $\lambda_y = 0.32$, and $t \in [-100, 100]$ and the subsequent numerical construction of the corresponding *cdf*.

Hereafter, use the shorthand $\Phi_T(t)$ in place of $\Phi_T(t; [n_x, n_y], [\lambda_x, \lambda_y])$ for the remainder of this section.

The null hypothesis decision is then determined by the *p*-value, $p = 2 \cdot \min\{\Phi_T(T(X,Y)), 1 - \Phi_T(T(X,Y))\}$. This can be seen from the criteria that H_0 is *rejected* provided either $T(X,Y) > \zeta_T(1 - \alpha/2)$ or $T(X,Y) < \zeta_T(\alpha/2)$, where $\zeta_T(\gamma)$ is the $\gamma \cdot 100\%$ quantile associated with the distribution $\Phi_T(t)$ on $T(X,Y)$. If the *p*-value is less than the significance level, then the null hypothesis of equal CVs is rejected. That is, $p < \alpha \Rightarrow$ *reject* H_0.

To compute the confidence interval for $\Delta CV = CV_x - CV_y$, it is first necessary to calculate the confidence intervals for the individual coefficients of variance. This is achieved via the *closed form method* of Donner and Zou [9]. Let $t_{n-1}(\gamma;\lambda)$ be the $\gamma \bullet 100\%$ quantile of the non-central T-distribution with $df = n - 1$ degrees of freedom and non-centrality parameter λ. The $(1 - \alpha) \bullet 100\%$ confidence interval for a generic *coefficient of variance* is given by (5.22).

$$CI(CV) = \widehat{CV} \cdot \left[1 - \frac{1}{|t_{n-1}(1 - \alpha/2; \lambda)|}, 1 + \frac{1}{|t_{n-1}(\alpha/2; \lambda)|} \right] \equiv [CV_{low}, CV_{up}]$$

(5.22)

The corresponding confidence intervals for CV_x and CV_y utilize (5.22) with the pairs (n,λ) replaced by (n_x,λ_x) and (n_y,λ_y), respectively. Plainly, $\widehat{CV}_x = \frac{S_x}{\bar{x}} \equiv T_1$ and $\widehat{CV}_y = \frac{S_y}{\bar{y}} \equiv T_2$. Observe that the estimate of the difference in coefficients of variation is merely the test statistic T. That is, $\Delta\widehat{CV} = T(X,Y)$. Subsequently, the closed form method $(1 - \alpha) \bullet 100\%$ confidence interval for $\Delta CV = CV_X - CV_Y$ is

$$CI(\Delta CV) = T(X, Y) + \left[-\sqrt{(T_1 - CV_{x,low})^2 + (T_2 - CV_{y,up})^2}, \right.$$
$$\left. \sqrt{(T_1 - CV_{x,up})^2 + (T_2 - CV_{y,low})^2} \right]$$

(5.23)

The section can now be summarized as per Sections 5.5.2–5.5.4.

Hypothesis

$$H_0: \frac{\sigma_x}{\mu_x} = \frac{\sigma_y}{\mu_y} \qquad \text{(Null hypothesis)}$$

$$H_1: \frac{\sigma_x}{\mu_x} \neq \frac{\sigma_y}{\mu_y} \qquad \text{(Alternate hypothesis)}$$

Test Statistic

$$T(X, Y) = \frac{S_x}{\bar{x}} - \frac{S_y}{\bar{y}}$$

$$S_x^2 = \frac{1}{n_x-1} \sum_{j=1}^{n_x} (x_j - \bar{x})^2, \quad S_y^2 = \frac{1}{n_y-1} \sum_{j=1}^{n_y} (y_j - \bar{y})^2$$

Test Statistic Distribution

$\phi_T(t)$ of (5.21c) is the *pdf* with corresponding *cdf* $\Phi_T(t) = \int_{-\infty}^{t} \phi_T(\tau)\, d\tau$.

p-Value

$p = 2 \cdot \min\{\Phi_T(T(X,Y)), 1 - \Phi_T(T(X,Y))\}$

Null Hypothesis Decision
If $p < \alpha$, then *reject H_0* in favor of H_1. Otherwise, *do not reject H_0*.

$(1 - \alpha)\bullet 100\%$ Confidence Intervals for CV_x and CV_y

$$CI(CV_x) = \widehat{CV}_x \cdot \left[1 - \frac{1}{|t_{n_x-1}(1 - \alpha/2; \lambda_x)|}, 1 + \frac{1}{|t_{n_x-1}(\alpha/2; \lambda_x)|} \right]$$

with a similar formula for $CI(CV_y)$.

$(1 - \alpha)\bullet 100\%$ Confidence Interval for $\Delta CV = CV_x - CV_y$

$$CI(\Delta CV) = T(X, Y) + \left[-\sqrt{(T_1 - CV_{x,low})^2 + (T_2 - CV_{y,up})^2}, \right.$$

$$\left. \sqrt{(T_1 - CV_{x,up})^2 + (T_2 - CV_{y,low})^2} \right]$$

as per equation (5.23) above.

MATLAB Function
```
cvtest.m
```

Remark: There are several non-exact methods which have been developed to address the matter of the statistical equality of coefficients of variation. In particular, Pardo and Pardo [31] described the test statistics and null hypothesis rejection criteria for the methods of Bennett, Miller, and modified Bennett. All of these methods are included in the MATLAB function cvtest.m. As an example of these various methods, consider the independent and identically distributed samples $X = \{x_1, x_2, \ldots, x_{54}\} \sim \mathcal{N}(\varepsilon_x, 1)$ and $Y = \{y_1, y_2, \ldots, y_{72}\} \sim \mathcal{N}(\varepsilon_y, 1)$ with $\varepsilon_x, \varepsilon_y \ll 0.5$. These data are exhibited in Figure 5.11. Table 5.29 summarizes the analyses of these data by the exact method and three non-exact techniques discussed above. The test statistics for the Bennett and modified Bennett methods are asymptotically χ_1^2, while the method of Miller has an asymptotically normal test statistic. None of the aforementioned non-exact methods provide formulae for confidence intervals. These can be obtained via bootstrapping methods (see Efron [10]) which will be detailed in Chapter 6. Observe that the *exact* and *Bennett* methods agree on the null hypothesis decision, while the Miller method would *not* reject the null hypothesis of statistical equality between the CVs. Since the confidence interval for the difference of CVs *does not* contain 0, it is wise to reject H_0. Thus, the Miller method may only be suitable for very large sample sizes. One of the important consequences of this example is value of *exact* methods. The mathematics are more demanding but the results are more reliable.

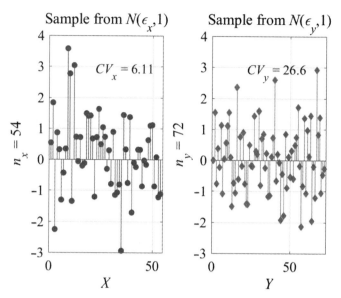

FIGURE 5.11   *I.I.D.* $\mathcal{N}(\varepsilon, 1)$ samples and *CV* estimates.

TABLE 5.29   CV analysis of the data samples *X* and *Y*

CV_x	$CI(CV_x)$	CV_y	$CI(CV_y)$	ΔCV	$CI(\Delta CV)$	*(Exact)*
6.11	[5.11, 7.5]	26.63	[22.8, 31.7]	−20.5	[−25.03, −16.02]	

Method	Test Statistic Distribution	*p*-value	H_0 decision
Exact	Equation (5.21b)	0.0170	*Reject*
Bennett	$\approx \chi_1^2$	0	*Reject*
Modified Bennett	$\approx \chi_1^2$	0	*Reject*
Miller	$\approx \mathcal{N}(0, 1)$	0.64	*Do not reject*

Example: To test the hypothesis of equal *coefficients of variation*

$$H_0: CV_x = CV_y \text{ vs. } H_1: CV_x \neq CV_y$$

the data illustrated in Figure 5.11 above are examined. Table 5.29 contains the results of several tests. It is seen that the null H_0 is *rejected in favor of H_1*.

MATLAB Commands
(Test of Equal CVs)

```
% Data Directory
Ddir = 'C:\PJC\Math_Biology\Chapter_5\Data';
% Load in i.i.d. samples from N(0, 1)
load(fullfile(Ddir,'Xcv.mat'))
load(fullfile(Ddir,'Ycv.mat'))
```

```
% Note size of each sample X and Y
Nx = numel(X); Ny = numel(Y);
% Set the significance level at 5%: α = 0.05
alpha = 0.05;
% Compute the CVs from each sample and test H₀: CVₓ = CVᵧ
H = cvtest(X,Y,alpha);

H =

CVx:   6.107
CVy:   26.63
CICVx:   [4.24 14.08]
CICVy:   [15.17 42.62]
Exact                    Bennett                    ModBen
      Ho:  reject           Ho:  reject              Ho:  reject
   Pvalue:  1.2411e-10   Pvalue:  0             Pvalue:  0
        T:  -20.53            T:  84.85                T:  84.65
       CI:  [-36.62 -6.56]
Miller
      Ho:  do not reject
   Pvalue:  0.6395
        T:  -0.3572
```

5.5.2   Wilcoxon Signed-Rank Test

Thus far all of the hypothesis tests have centered about the estimate of a particular parameter (mean, variance, proportion, coefficient of variation). This approach, called *parametric methods*, requires a signal measurement (i.e., parameter) to gauge the comparison between two distinct populations. Suppose, instead, that the investigator wanted to determine how two populations interact in aggregate. For example, imagine two teams within the same sport played a series of games against one another. Further, suppose that when team A is victorious over team B, the margin is 2 points. Conversely, when team B defeats team A, the margin of victory is 9 points. If these teams play n games with $m \ll n$ ties and an equal number of wins for each team, can the observer determine whether one team is better than the other? The answer is *yes* via the *Wilcoxon signed-rank test*.

To develop this statistic, there is one major restriction that must be imposed on the measurements. That is, the measurements must consist of *matched pairs*. In the vernacular of a sporting event, this means that each team must play the same number of games. To that end, let $X = \{X_1, X_2, ..., X_n\}$ and $Y = \{Y_1, Y_2, ..., Y_n\}$ be the scores for each game played by team A against team B. For example, the score for game k is $A = X_k$ and $B = Y_k$. The matched pair then is the score double (X_k, Y_k). Let $Z_k = X_k - Y_k$ be the difference in scores. The *signum* and *indicator* functions are defined as

$$sign(Z) = \begin{cases} 1 & \text{for } Z > 0 \\ 0 & \text{for } Z = 0 \\ -1 & \text{for } Z < 0 \end{cases} \qquad (5.24)$$

$$\mathscr{I}(Z) = \begin{cases} 1 & \text{if } Z > 0 \\ 0 & \text{otherwise} \end{cases}. \tag{5.25}$$

Observe $|Z_k|$ measures the *magnitude* of the difference in scores. Arrange these numbers via the indices $i_1, i_2, \ldots, i_n$ so that they are in ascending order. That is, $|Z_{i_1}| \le |Z_{i_2}| \le \cdots \le |Z_{i_n}|$. In this case, $|Z_{i_n}|$ is the largest magnitude and consequently has a *rank* of 1; $|Z_{i_{n-1}}|$ is of next highest magnitude with *rank* 2; etc. Rather than examining the ordered absolute magnitudes, consider the *signed ranks* of the $|Z_k|$. That is, $R_k = sign(Z_k) \bullet rank(Z_k)$.

The *Wilcoxon signed-rank test* (*WSRT*) is defined as

$$W = \sum_{j=1}^{n} \mathscr{I}(Z_j) \cdot rank(|Z_j|) \tag{5.26a}$$

Notice that the *WSRT* can be written as $W = W^+ + W^-$, where

$$W^+ = \sum_{j=1}^{n} \mathscr{I}(Z_j) \cdot rank(|Z_j|) \tag{5.26b}$$

$$W^- = \sum_{j=1}^{n} \mathscr{I}(-Z_j) \cdot rank(|Z_j|) \tag{5.26c}$$

are the *positive-* and *negative-signed ranks*. Since $W^+ - W^- = \sum_{j=1}^{n} rank(|Z_j|) = \frac{1}{2} n_z(n_z + 1)$, it is sufficient to use either W^+ or W^-. Here n_z is the number of *non-zero differences* Z_j. Both test statistics will be used in this section.

If $J_p = \{j_1, j_2, \ldots, j_p\}$ is a subset of the full set of indices $\{1, 2, \ldots, n\}$ so that $Z_{j_k} > 0$ for each $k = 1, 2, \ldots, p$ and $|Z_{j_1}| \le |Z_{j_2}| \le \cdots \le |Z_{j_p}|$, then (5.26a) can be written as $W^+ = \sum_{k=1}^{p} rank(|Z_{j_k}|)$. Similarly, let $I_t = \{i_1, i_2, \ldots, i_t\} \subset \{1, 2, \ldots, n\}$ be a subset of indices so that $Z_i < 0$ for each $i \in I_t$ and $|Z_{i_1}| \le |Z_{i_2}| \le \cdots \le |Z_{i_t}|$. Then the number of nonzero differences Z_j is $n_z = p + t$.

Note that (5.26b) is referred to as the *one-sample Wilcoxon signed-rank test*. This statistic measures the superiority of the *Treatment* or *Test* (in this case the random variable Y) to the *Control* (here, X). To determine whether the *Treatment* is inferior to the *Control*, the statistic W^- is utilized.

If m is the number of differences equal to 0 (i.e., the number of ties), then $n = n_z + m$. Further, it can be shown that

$$\mu_W \equiv E[W] = \frac{1}{4} n_z(n_z + 1) \tag{5.27a}$$

and

$$\sigma_W^2 \equiv var(W) = \frac{1}{24} n_z(n_z + 1)(2n_z + 1) \tag{5.27b}$$

Moreover, as $-W^- = \frac{1}{2} n_z(n_z + 1) - W^+$, a direct computation shows that $E[-W^-] = \frac{1}{2} n_z(n_z + 1) - \frac{1}{4} n_z(n_z + 1) = \frac{1}{4} n_z(n_z + 1) \equiv \mu_{W^+}$ and $var(-W^-) = \frac{1}{24} n_z(n_z + 1)(2n_z + 1) \equiv \sigma_{W^+}^2$. Hereafter, the mean and variance for either W^+ or

$-W^-$ will simply be identified as μ_W and σ_W^2, respectively. See Bickel and Doksum [6] or Marques de Sá [21] for details. Invoking the Central Limit Theorem it is seen that $T^+(X, Y) = \dfrac{W^+ - \mu_W}{\sigma_W^2}$ and $T^-(X, Y) = \dfrac{-W^- - \mu_W}{\sigma_W^2}$ are approximately distributed as a standard normal random variable.

Warning: Bickel and Doksum recommend that n_z be *greater than 16* to ensure that the approximation to the standard normal distribution $\Phi(x)$ is accurate.

Given this situation, what does the *Wilcoxon signed-rank statistic* actually test? Since this is a non-parametric test (i.e., the test does not concern the value of a particular statistic such as mean or variance), the Wilcoxon signed-rank tests whether one random variable is stochastically equal (larger or smaller) than another. Two random variables are stochastically equal provided their distributions are the same. Thus, $X = Y$ *stochastically* $\Leftrightarrow \Phi_X(x) = \Phi_Y(x)$.

Returning to the origin of this discussion, how can the Wilcoxon signed-rank test be used to determine whether *Team A* is better than *Team B*? For a test of stochastic equality, this means computing a *p*-value and comparing it against a preset significance level α. Again, for $n_z > 16$, $T^+(X, Y) \sim \mathcal{N}(0, 1)$. Hence, the *p*-value for a two-sided test is $p = 2 \cdot \Phi(-|T(X, Y)|)$. Assume that when either team wins, it is via "shutout." Since *Team A* wins by 2 points, an *A-victory* occurs via a score of $A = 2$ and $B = 0$. Similarly, *Team B* wins by 9 to 0. Assume all ties have the scores 1 to 1. Finally, suppose each team wins $n_w = 25$ games and ties $m = 4$ games for a total of $25 + 25 + 4 = 54$ games played. Then, using the data illustrated in Figure 5.12, the Wilcoxon signed-rank statistic has the value $T(X, Y) = 3.1124$ and subsequent

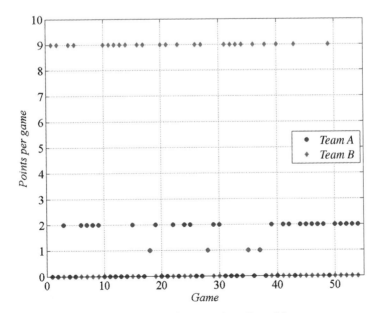

FIGURE 5.12   Scores for *Team A* vs. *Team B* by game.

p-value $p = 2 \cdot \Phi(-|T(X, Y)|) = 0.0019 < 0.05 = \alpha$. Thus, the null hypothesis H_0: $\Phi_X(x) = \Phi_Y(x)$ is *rejected*.

To address the issue of one-sided tests (i.e., the left-tailed test H_0: $\Phi_X(x) \geq \Phi_Y(x) \Rightarrow X \geq Y$ *stochastically*), the p-value becomes $p = \Phi(T^-(X, Y))$ and $p = \Phi(-T^-(X, Y))$ for the right-tailed test H_0: $\Phi_X(x) \leq \Phi_Y(x)$. As will be seen in succeeding sections, a correction factor is added to the right- and left-tailed test statistics to enhance accuracy of the normal approximation.

The last matter to address is the relationship between significance (α), power $(1-\beta)$, and sample size presented in the work of Shieh et al. [34]. The test of stochastic equality H_0: $\Phi_X(x) = \Phi_Y(x)$ is equivalent to stating that the distributions are translations with a 0 offset. That is, $\Phi_Y(x) = \Phi_X(x + \theta)$. Under the null hypothesis H_0: $\Phi_X(x) = \Phi_Y(x)$, $\theta = 0$. This parameter θ then acts as the effect size in a sample size calculation. The hypothesis test can now be written as

$$H_0: \Phi_X(x + \theta) = \Phi_Y(x), \theta = 0 \qquad \text{(Null hypothesis)}$$
$$H_1: \Phi_X(x + \theta) = \Phi_Y(x), \theta \neq 0. \qquad \text{(Alternate hypothesis)}$$

Under the null hypothesis, the mean and variance of the Wilcoxon signed-rank test are

$$\mu_0(n) = \frac{1}{4}n(n + 1) \text{ and } \sigma_0^2(n) = \frac{1}{24}n(n + 1)(2n + 1)$$

whereas, under the alternative hypothesis, these statistics are (approximated by)

$$\mu_\theta(n) = n \cdot \Phi(\theta) + \frac{1}{2}n(n + 1) \cdot \Phi(\sqrt{2}\theta)$$

and

$$\sigma_\theta^2(n) = n \cdot \Phi(\theta)(1 - \Phi(\theta)) + \frac{1}{2}n(n - 1)\Phi(\sqrt{2}\theta)\left(1 - \Phi(\sqrt{2}\theta)\right)$$
$$+ 2n(n - 1)\left(\frac{1}{2}\Phi^2(\theta) + \frac{1}{2}\Phi(\sqrt{2}\theta) - \Phi(\theta) \cdot \Phi(\sqrt{2}\theta)\right)$$
$$+ n(n - 1)(n - 2)\left(\int_{-\infty}^{\infty} \Phi^2(2\theta + x)\phi(x) \, dx - \Phi^2(\sqrt{2}\theta)\right).$$

Recall, as per the nomenclature of Section 5.1, ϕ and Φ are the *pdf* and *cdf* of the standard normal distribution, respectively. The general form for the alternate hypothesis mean $\mu_\theta(n)$ and variance $\sigma_\theta^2(n)$ replaces the normal *cdf* by the general cumulative distribution Φ_X of the random variable X. Since the normal *cdf* Φ is used, the effect size θ must be normalized to $\mathcal{N}(0, 1)$. Thus, $\theta = \dfrac{\Delta\mu}{\sqrt{\sigma_{pool}^2}}$ where $\Delta\mu$ is the difference in the means of X and Y and σ_{pool}^2 is the pooled variance. Finally, let

$$G(n; \theta) = (\mu_\theta(n) - \mu_0(n)) - \left(z_{1-\alpha/2}\sqrt{\sigma_0^2(n)} + z_{1-\beta}\sqrt{\sigma_\theta^2(n)}\right). \text{ Then the sample size}$$

required to obtain an effect size θ with respect to the significance level α and power $1 - \beta$ is the value $n = N_s$ which minimizes $G(n; \theta)$. That is,

$$G(N_s; \theta) = \arg\min_n \{G(n; \theta)\}$$

$$= \arg\min_n \left\{ \mu_\theta(n) - \mu_0(n) - \left[z_{1-\alpha/2}\sqrt{\sigma_0^2(n)} + z_{1-\beta}\sqrt{\sigma_\theta^2(n)} \right] \right\}. \quad (5.28)$$

For a one-sided test, merely replace the $(1 - \alpha/2) \cdot 100\%$ quantile in (5.28) with the $(1 - \alpha) \cdot 100\%$ quantile; $z_{1-\alpha/2} \to z_{1-\alpha}$.

5.5.2.1 Wilcoxon Signed-Rank Test for Stochastic Equality

As noted at the end of Section 5.5.2, two random variables X and Y are stochastically equal provided their distributions (*cumulative density functions*) are equal over their common domain of definition Ω. The hypothesis testing methodology now proceeds as per Section 5.4.

Hypothesis

$$H_0: \Phi_X(x) = \Phi_Y(x), x \in \Omega \qquad \text{(Null hypothesis)}$$
$$H_1: \Phi_X(x) \neq \Phi_Y(x), x \in \Omega \qquad \text{(Alternate hypothesis)}$$

Test Statistic

$$T(X, Y) = \frac{W^+ - \mu_W}{\sigma_W^2}$$

$$W^+ = \sum_{j=1}^{n} \vartheta(Z_j) \cdot rank(|Z_j|), Z_k = X_k - Y_k$$

$$\mu_W = \frac{1}{4}n_z(n_z + 1), \sigma_W^2 = \frac{1}{24}n_z(n_z + 1)(2n_z + 1)$$

n_z = total number of nonzero differences Z_k.

Test Statistic Distribution

$$T(X, Y) \sim \mathcal{N}(0, 1) \text{ for } n_z > 16$$

p-Value

$$p = 2 \cdot \Phi(-|T(X, Y)|)$$

$(1 - \alpha) \bullet 100\%$ Critical Region for W

$$CR(W^+) = [\min(k), \max(k)]$$

$$k = \left(\left\lfloor \mu_W + z_{\alpha/2} \cdot \sqrt{\sigma_W^2} \right\rfloor, \left\lceil \mu_W + z_{1-\alpha/2} \cdot \sqrt{\sigma_W^2} \right\rceil \right)$$

Here $\lfloor x \rfloor = floor(x)$ and $\lceil x \rceil = ceil(x)$ are the floor and ceiling functions, respectively.

Null Hypothesis Decision

If $p < \alpha$, then *reject H_0* in favor of H_1. Otherwise, *do not reject H_0*. Equivalently, if $W^+ \notin CR(W^+)$, then reject H_0.

Required Sample Size

Find the value $n = N_s$ which minimizes (5.28).

θ = effect size, which is a fraction of what can practically be expected to be the minimum difference in the samples. As a rule of thumb, $\theta \leq \dfrac{\min_j(|Z_j|)}{\sqrt{\sum_{j=1}^{n_z} |Z_j|^2}}$.

MATLAB Function

wilcoxonsr.m

Example: To determine whether *Teams A* and *B* are *stochastically equal* (i.e., are "the same" since they won equal numbers of games), test

$$H_0: \Phi_X(x) = \Phi_Y(x) \text{ against } H_1: \Phi_X(x) \neq \Phi_Y(x)$$

via the *WSRT*. It is seen that the *p*-value is less than the significance level $p = 0.0019 < \alpha = 0.05$ and that the test statistic *does not* reside in the critical region $W^+ = 950 \notin [434, 841] = CR(W^+)$. Hence, the *null H_0* is *rejected in favor of H_1*. To decide whether A is a better team than B, one-sided Wilcoxon tests must be performed. These are the focus of the next two sections.

MATLAB Commands
(Two-Sided Wilcoxon Test)

```
% Data Directory
Ddir = 'C:\PJC\Math_Biology\Chapter_5\Data';
% Load in the competing scores from teams A and B
load(fullfile(Ddir,'WilcoxonA.mat'))
load(fullfile(Ddir,'WilcoxonB.mat'))
% Set the significance level at 5%, alpha = 0.05
alpha = 0.05;
% Test the hypothesis H_0: Phi_X(x) = Phi_Y(x)
H = wilcoxonsr(A,B,alpha,2)
```

```
H =
          Ho: 'reject'
           W: 950
          CR: [434 841]
      Pvalue: 0.0019
       normT: -3.1124
           n: 50
       Nties: 4
    TestType: 'Two-tailed'
```

5.5.2.2   Wilcoxon Signed-Rank Test for One-Sided, Left-Tailed Test   Here the question of whether X is stochastically smaller than Y over their common domain of definition Ω is approached by *rejecting* the null hypothesis of X having a larger distribution function than Y.

Hypothesis

$$H_0: \Phi_X(x) \geq \Phi_Y(x), x \in \Omega \qquad \text{(Null hypothesis)}$$

$$H_1: \Phi_X(x) < \Phi_Y(x), x \in \Omega \qquad \text{(Alternate hypothesis)}$$

Test Statistic

$$T^-(X, Y) = \frac{-W^- - \mu_W + \frac{1}{2}}{\sigma_W^2}$$

Note that the numerator is enhanced by $1/2$. This is a correction factor that makes the normal approximation more accurate. See Gibbons and Chakraborti [12] for details.

$$W^- = \sum_{j=1}^{n} \vartheta(-Z_j) \cdot rank(|Z_j|), Z_k = X_k - Y_k$$

$$\mu_W = \tfrac{1}{4}n_z(n_z + 1), \sigma_W^2 = \tfrac{1}{24}n_z(n_z + 1)(2n_z + 1)$$

$$n_z = \text{total number of nonzero differences } Z_k.$$

Test Statistic Distribution

$$T^-(X, Y) \sim \mathcal{N}(0, 1) \text{ for } n_z > 16$$

p-Value

$$p = \Phi(T^-(X, Y))$$

$(1 - \alpha) \bullet 100\%$ Critical Region for W

$$CR(W^-) = (-\infty, \max(k)]$$

$$k = \left(\left\lfloor \mu_W + z_\alpha \cdot \sqrt{\sigma_W^2} \right\rfloor, \left\lceil \mu_W + z_{1-\alpha} \cdot \sqrt{\sigma_W^2} \right\rceil \right)$$

Here $\lfloor x \rfloor = floor(x)$ and $\lceil x \rceil = ceil(x)$ are the floor and ceiling functions, respectively.

Null Hypothesis Decision

If $p < \alpha$, then *reject H_0* in favor of H_1. Otherwise, *do not reject H_0*. Equivalently, if $W^- \notin CR(W^-)$, then reject H_0.

Required Sample Size

Find the value $n = N_s$ which minimizes (5.28) with the exception that $z_{1-\alpha/2}$ is replaced by $z_{1-\alpha}$.

θ = effect size, which is a fraction of what can practically be expected to be the minimum difference in the samples. As a rule of thumb, $\theta < \dfrac{\min_j(|Z_j|)}{\sqrt{\sum_{j=1}^{n_z} |Z_j|^2}}$.

MATLAB Function

`wilcoxonsr.m`

Example: Recall the example in Section 5.5.2 of *Team A* vs. *Team B* as illustrated in Figure 5.12. To determine whether *Team B* is "better" (i.e., stochastically larger) than *Team A*, test the hypotheses

$$H_0: \Phi_A(x) \geq \Phi_B(x) \text{ vs. } H_1: \Phi_A(x) < \Phi_B(x).$$

Using the Wilcoxon signed-rank test, it is seen that the *p*-value $p = 9 \times 10^{-4} < \alpha$. Also, $W^- = 950 \notin (-\infty, 808] = CR(W^-)$ so that H_0 *is rejected in favor of H_1*. Consequently, it can be concluded that *Team B* is better than *Team A* since its margin of victory is substantially greater than its margin of defeat.

MATLAB Commands
(Left-Tailed Wilcoxon Test)

```
% Data Directory
Ddir = 'C:\PJC\Math_Biology\Chapter_5\Data';
% Load in the competing scores from teams A and B
load(fullfile(Ddir, 'WilcoxonA.mat'))
load(fullfile(Ddir, 'WilcoxonB.mat'))
% Set the significance level at 5%, α = 0.05
alpha = 0.05;
```

```
% Test the hypothesis H₀: ΦA(x) ≥ ΦB(x)
H = wilcoxonsr(A,B,alpha,1)
H =
         Ho: 'reject'
          W: 950
         CR: [-Inf 808]
     Pvalue: 9.4370e-04
      normT: -3.1074
          n: 50
      Nties: 4
   TestType: 'Left-tailed'
```

5.5.2.3   Wilcoxon Signed-Rank Test for One-Sided Right-Tailed Test   Finally, it is determined whether X is stochastically larger than Y over their common domain of definition Ω, by *rejecting* the null hypothesis of X having a smaller distribution function than Y.

Hypothesis

$$H_0: \Phi_X(x) \leq \Phi_Y(x), x \in \Omega \qquad \text{(Null hypothesis)}$$
$$H_1: \Phi_X(x) > \Phi_Y(x), x \in \Omega \qquad \text{(Alternate hypothesis)}$$

Test Statistic

$$T^-(X, Y) = \frac{-W^- - \mu_W - \frac{1}{2}}{\sigma_W^2}$$

Note that the numerator is degraded by $\frac{1}{2}$. This is a correction factor that makes the normal approximation more accurate. See Gibbons and Chakraborti [12] for details.

$$W^- = \sum_{j=1}^{n} \mathcal{I}(-Z_j) \cdot rank(|Z_j|), Z_k = X_k - Y_k$$

$$\mu_W = \tfrac{1}{4}n_z\left(n_z + 1\right), \sigma_W^2 = \frac{1}{24}n_z\left(n_z + 1\right)\left(2n_z + 1\right)$$

$n_z = $ total number of nonzero differences Z_k.

Test Statistic Distribution

$$T^-(X, Y) \sim \mathcal{N}(0, 1) \text{ for } n_z > 16$$

p-Value

$$p = \Phi\left(-T^-(X, Y)\right)$$

$(1 - \alpha)\bullet 100\%$ Critical Region for W^-

$$CR(W^-) = [\min(k), +\infty)$$

$$k = \left(\left\lfloor \mu_W + z_\alpha \cdot \sqrt{\sigma_W^2} \right\rfloor, \left\lceil \mu_W + z_{1-\alpha} \cdot \sqrt{\sigma_W^2} \right\rceil \right)$$

Here $\lfloor x \rfloor = floor(x)$ and $\lceil x \rceil = ceil(x)$ are the floor and ceiling functions, respectively.

Null Hypothesis Decision

If $p < \alpha$, then *reject H_0* in favor of H_1. Otherwise, *do not reject H_0*. Equivalently, if $W^- \notin CR(W^-)$ then *reject H_0*.

Required Sample Size

Find the value $n = N_s$ that minimizes (5.28) with the exception that $z_{1-\alpha/2}$ is replaced by $z_{1-\alpha}$.

θ = effect size, which is a fraction of what can practically be expected to be the minimum difference in the samples. As a rule of thumb, $\theta < \dfrac{\min_j(|Z_j|)}{\sqrt{\sum_{j=1}^{n_z} |Z_j|^2}}$.

MATLAB Function

`wilcoxonsr.m`

Example: The example of *Team A* vs. *Team B* as illustrated in Figure 5.12 and considered in Section 5.5.2.2 is tested to determine whether *Team A* is "better" (i.e., stochastically larger) than *Team B*. Test the hypotheses

$$H_0: \Phi_A(x) \leq \Phi_B(x) \text{ vs. } H_1: \Phi_A(x) > \Phi_B(x)$$

For these data, the Wilcoxon signed-rank test produces a p-value of $p = 0.999 > \alpha = 0.05$. Moreover, the test statistic *is* within the critical region $W^- = 950 \in [467, \infty) = CR(W)$. Hence, H_0 *is not rejected*. Therefore, the margin of victory is the determining factor. The left-tailed test of 5.2.2.2 can be used to show that *Team B* is "better" than *Team A*.

MATLAB Commands
(*Right-Tailed Wilcoxon Test*)

```
% Data Directory
Ddir = 'C:\PJC\Math_Biology\Chapter_5\Data';
% Load in the competing scores from teams A and B
load(fullfile(Ddir,'WilcoxonA.mat'))
load(fullfile(Ddir,'WilcoxonB.mat'))
% Set the significance level at 5%, α = 0.05
alpha = 0.05;
```

```
% Test the hypothesis H₀: Φₓ(x) ≤ Φₓ(x)
H = wilcoxonsr(A,B,alpha,0)
H =
          Ho: 'do not reject'
           W: 950
          CR: [467 Inf]
      Pvalue: 0.9991
       normT: -3.1174
           n: 50
       Nties: 4
    TestType: 'Right-tailed'
```

5.5.3   The Mann–Whitney U-Test

The Wilcoxon signed-rank test of Section 5.5.2 is a non-parametric test of equal distributions for *matched pairs*. That is, the samples $X = \{x_1, x_2,\ldots, x_{n_x}\}$ and $Y = \{y_1, y_2,\ldots, y_{n_y}\}$ are the same size; $n_x = n_y$. When the sample sizes are not the same, however, the need to test the equality of distributions remains. Therefore, a comparable test is desired. This is accomplished via the *Mann–Whitney* U-test. Just as the Wilcoxon signed-rank test is non-parametric, so too is the Mann–Whitney U-test.

To establish the U-test, let $X = \{x_1, x_2,\ldots, x_{n_x}\}$ and $Y = \{y_1, y_2,\ldots, y_{n_y}\}$ be unmatched samples so that $n_x \neq n_y$. Without loss of generality, assume $n_x < n_y$ (if this is not the case, simply reverse X and Y). Now set $W = \{X, Y\} = \{x_1, x_2,\ldots, x_{n_x}, y_1, y_2,\ldots, y_{n_y}\}$ and rank the elements of W from smallest to largest, $R_1 < R_2 < \cdots < R_n$, where $n = n_x + n_y$. If N_t is the number of tied ranks, set the variance correction term to be $V_c = \frac{1}{2}N_t(N_t^2 - 1)$. The mean and variance of the U-test are $E[U] = \frac{1}{2} n_x \bullet n_y$ and $\text{var}[U] = \frac{1}{12}n_x \cdot n_y \cdot (n + 1) - V_c = \frac{1}{12}n_x \cdot n_y \cdot (n_x + n_y + 1) - \frac{1}{2}N_t(N_t^2 - 1)$. Then $U_x = \sum_{j=1}^{n_x} R_j - \frac{1}{2} n_x \bullet (n_x + 1)$ and $U_y = \sum_{j=1+n_x}^{n_y+n_x} R_j - \frac{1}{2} n_y \bullet (n_y + 1)$ are the corrected sum of the ranks. The statistic $U = \min\{U_x, U_y\}$ is the *Mann–Whitney* U-test for the hypothesis of stochastic equality.

$$H_0: \Phi_X(x) = \Phi_Y(x) \qquad \text{(Null hypothesis)}$$

$$H_1: \Phi_X(x) \neq \Phi_Y(x) \qquad \text{(Alternate hypothesis)}$$

The test statistic is the normalized U-test; $Z = \dfrac{U - E[U]}{\sqrt{\text{var}[U]}} =$

$$\dfrac{U - \frac{1}{2}n_x \cdot n_y}{\sqrt{\frac{1}{12}n_x \cdot n_y(n_x + n_y + 1) - \frac{1}{2}N_t(N_t^2 - 1)}}.$$

The sample size calculation for the U-test is more complicated than the *Wilcoxon signed-rank test*. One approach, offered by Chow et al. [8], is to use the ratio of sample sizes and pilot study data. More specifically, $n_x = \kappa \bullet n_y$ and

$$n_y = \frac{\left(z_{1-\alpha/2}\sqrt{\frac{1}{12}\kappa(\kappa+1)} + z_{1-\beta}\sqrt{\kappa^2(p_2 - p_1^2) + \kappa(p_3 - p_1^2)}\right)^2}{\kappa\left(\frac{1}{2} - p_1\right)^2} \tag{5.29a}$$

$$\left.\begin{aligned}
\hat{p}_1 &= \frac{1}{n_x n_y} \sum_{k=1}^{n_y} \sum_{j=1}^{n_x} \mathcal{I}(\tilde{y}_k \geq \tilde{x}_j) \\
\hat{p}_2 &= \frac{1}{n_x n_y(n_x - 1)} \sum_{k=1}^{n_y} \sum_{\substack{j_1, j_2 = 1 \\ j_1 \neq j_2}}^{n_x} \mathcal{I}(\tilde{y}_k \geq \tilde{x}_{j_1} \& \tilde{y}_k \geq \tilde{x}_{j_2}) \\
\hat{p}_3 &= \frac{1}{n_x n_y(n_y - 1)} \sum_{\substack{k_1, k_2 = 1 \\ k_1 \neq k_2}}^{n_y} \sum_{j=1}^{n_x} \mathcal{I}(\tilde{y}_{k_1} \geq \tilde{x}_j \& \tilde{y}_{k_2} \geq \tilde{x}_j)
\end{aligned}\right\} \tag{5.29b}$$

Here the $\hat{p}_\ell$ are computed via the indicator function $\mathcal{I}$ from (5.25) with respect to the *pilot study* data $\tilde{X} = \{\tilde{x}_1, \tilde{x}_2, \ldots, \tilde{x}_{n_x}\}$, $\tilde{Y} = \{\tilde{y}_1, \tilde{y}_2, \ldots, \tilde{y}_{n_y}\}$. It is noted that a pilot study is a "proof of concept" analysis conducted on samples that are typically of smaller sample size than is required by the formal Neyman–Pearson method (of significance α, power $1 - \beta$, and effect size *es*). In the absence of a pilot sample, the ratio $\kappa = n_x/n_y$ is coupled along with the Wilcoxon signed-rank sample size estimate (5.28).

The remainder of the section is dedicated to the, by now, usual delineation of hypothesis test and corresponding companion formulae.

5.5.3.1 Mann–Whitney U-Test for Stochastic Equality
This test determines whether two unmatched samples of unequal size X and Y are taken from the same distribution.

Hypothesis

$$H_0: \Phi_X(x) = \Phi_Y(x) \qquad\qquad \text{(Null hypothesis)}$$
$$H_1: \Phi_X(x) \neq \Phi_Y(x) \qquad\qquad \text{(Alternate hypothesis)}$$

Test Statistic

$$T(X, Y) = \frac{U - E[U]}{\sqrt{\operatorname{var}[U]}}$$

$$U = \min\{U_x, U_y\}, \; U_x = \sum_{j=1}^{n_x} R_j - \tfrac{1}{2} n_x \bullet (n_x + 1) \; \text{ and}$$

$$U_y = \sum_{j=1+n_x}^{n_y + n_x} R_j - \tfrac{1}{2} n_y \bullet (n_y + 1)$$

$R_1 \leq R_2 \leq \cdots \leq R_n$, $n = n_x + n_y$ are the ranks of $Z = \{X, Y\}$.

$$E[U] = \tfrac{1}{2} n_x \bullet n_y$$
$$\mathrm{var}[U] = \tfrac{1}{12} n_x \cdot n_y \cdot (n_x + n_y + 1) - \tfrac{1}{2} N_t \left(N_t^2 - 1\right)$$
$$N_t = \text{total number of tied ranks.}$$

Test Statistic Distribution

$$T(X, Y) \sim \mathcal{N}(0, 1) \text{ for } n_x, n_y > 8$$

p-Value

$$p = 2 \bullet \Phi\left(-|T(X, Y)|\right)$$

For n_x, $n_y \leq 8$, there are exact tables for p-values incorporated into `mwutest.m`.

$(1 - \alpha) \bullet 100\%$ Critical Region for U

$$CR(U) = \left[E[U] + z_{\alpha/2} \cdot \sqrt{\mathrm{var}[U]}, E[U] + z_{1-\alpha/2} \cdot \sqrt{\mathrm{var}[U]}\right]$$

Null Hypothesis Decision

If $p < \alpha$, then *reject H_0* in favor of H_1. Otherwise, *do not reject H_0*. Equivalently, if $U \notin CR(U)$, then *reject H_0*.

Required Sample Size

Either conduct a pilot study and use equations (5.29a) and (5.29b) or use the Wilcoxon signed-rank method. That is, find the value $n = n_y$ that minimizes (5.28) and set $n_x = \kappa \bullet n_y$ where κ is the ratio of the X samples to Y samples.

θ = effect size, which is a fraction of what can practically be expected to be the *normalized* minimum difference in the sample distribution means.

Remark   Suppose that $X_m = \min\{X\}$ and $X_M = \max\{X\}$ are the minimum and maximum values of the sample X with similar definitions for Y_m and Y_M. Then θ can be selected as the minimum or mean of the set of values $\{|X_m - Y_m|/\sigma, |X_M - Y_M|/\sigma, |X_m - Y_M|/\sigma, |X_M - Y_m|/\sigma\}$, σ is the sample standard deviation of the sample

$$W = \{X, Y\}, \sigma = \sqrt{\sum_{j=1}^{n} (W_j - \overline{w})^2}, \text{ and } \overline{w} = \tfrac{1}{n} \left(\sum_{j=1}^{n_x} X_j + \sum_{j=1}^{n_y} Y_j\right).$$

MATLAB Function

`mwutest.m`

Example: Determine whether the heights of hockey and basketball players are *stochastically equal* by testing the hypotheses

$$H_0: \Phi_X(x) = \Phi_Y(x) \text{ vs. } H_1: \Phi_X(x) \neq \Phi_Y(x)$$

The p-value for this test is less than the significance level $p < 10^{-4} < \alpha = 0.05$ and the value of the test statistic *is not* contained within the critical region $U = 275426 \notin [140608, 158983] = CR(U)$. Hence, H_0 is *rejected in favor of* H_1.

MATLAB Commands
(*Two-Sided Mann–Whitney U-Test*)

```
% Data Directory
Ddir = 'C:\PJC\Math_Biology\Chapter_5\Data';
% Load in the height data
load(fullfile(Ddir,'HockeyData.mat'))
load(fullfile(Ddir,'BasketballData.mat'))
% Hockey and Basketball height data
X = D.Ht; Y = B.Ht;
% Set the significance level at 5%, α = 0.05
alpha = 0.05;
% Test the hypothesis H₀: Φ_X(x) = Φ_Y(x)
H = mwutest(X,Y,alpha,2)
H =

          Ho: 'reject'
           U: 2.4166e+04
          CR: [1.4061e+05 1.5898e+05]
      Pvalue: 3.2246e-158
```

5.5.3.2 Mann–Whitney U-Test for One-Sided, Left-Tailed Test As with the Wilcoxon signed-rank test, this test for unmatched samples determines whether X is stochastically smaller than Y and is approached by *rejecting* the null hypothesis of X having a larger distribution function than Y.

Hypothesis

$$H_0: \Phi_X(x) \geq \Phi_Y(x) \qquad \text{(Null hypothesis)}$$
$$H_1: \Phi_X(x) < \Phi_Y(x) \qquad \text{(Alternate hypothesis)}$$

Test Statistic

$$T(X, Y) = \frac{U - E[U]}{\sqrt{\text{var}[U]}}$$

$$U = \max\left\{U_x, U_y\right\}, U_x = \sum_{j=1}^{n_x} R_j - \frac{1}{2} n_x \bullet (n_x + 1) \text{ and}$$

$$U_y = \sum_{j=1+n_x}^{n_y+n_x} R_j - \frac{1}{2} n_y \bullet (n_y + 1)$$

$R_1 \leq R_2 \leq \cdots \leq R_n, n = n_x + n_y$ are the ranks of $Z = \{X, Y\}$.

$$E[U] = \frac{1}{2} n_x \bullet n_y$$

$$\text{var}[U] = \frac{1}{12} n_x \cdot n_y \cdot (n_x + n_y + 1) - \frac{1}{2} N_t \left(N_t^2 - 1\right)$$
$$N_t = \text{total number of tied ranks.}$$

Test Statistic Distribution

$$T(X, Y) \sim \mathcal{N}(0, 1) \text{ for } n_x, n_y > 8$$

p-Value

$$p = \Phi\left(-T(X, Y)\right)$$

For $n_x, n_y \leq 8$, there are exact tables for p-values incorporated into `mwutest.m`.

$(1 - \alpha)\bullet 100\%$ Critical Region for U

$$CR(U) = (-\infty, E[U] + z_{1-\alpha} \cdot \sqrt{\text{var}[U]}]$$

Null Hypothesis Decision
If $p < \alpha$, then *reject H_0* in favor of H_1. Otherwise, *do not reject H_0*. Equivalently, if $U \notin CR(U)$ then *reject H_0*.

Required Sample Size
Either conduct a pilot study and use equations (5.29a) and (5.29b) or use the Wilcoxon signed-rank method. That is, find the value $n = n_y$ that minimizes (5.28) and set $n_x = \kappa \bullet n_y$ where κ is the ratio of the X samples to Y samples. Instead of the quantile $z_{1-\alpha/2}$, however, use $z_{1-\alpha}$.
$\theta =$ effect size which is a fraction of what can practically be expected to be the minimum difference in the sample distribution means. See the comment just prior to Section 5.5.2.1 for guidance as to the value of θ.

MATLAB Function
`mwutest.m`

Example: Consider the following example from Hogg and Craig [14]. To test whether the sample $Y = \{5.5, 7.9, 6.8, 9.0, 5.6, 6.3, 8.5, 4.6, 7.1\}$ of the random variable Y is stochastically larger than the sample $X = \{4.3, 5.9, 4.9, 3.1, 5.3, 6.4, 6.2, 3.8, 7.5, 5.8\}$ of X, use the *Mann–Whitney U-test* on the hypotheses

$$H_0: \Phi_X(x) \geq \Phi_Y(x) \text{ vs. } H_1: \Phi_X(x) < \Phi_Y(x)$$

The p-value is less than the significance level $p = 0.025 < \alpha = 0.05$ and $U = 69 \notin (-\infty, 24.85] = CR$. Therefore, *reject H_0* in favor of the alternate hypothesis H_1 that Y is stochastically larger than X.

MATLAB Commands
(*Left-Tailed Mann–Whitney U-Test*)

```
% Data from two distinct samples X and Y
X = [4.3, 5.9, 4.9, 3.1, 5.3, 6.4, 6.2, 3.8, 7.5, 5.8];
Y = [5.5, 7.9, 6.8, 9.0, 5.6, 6.3, 8.5, 4.6, 7.1];
% Set the significance level at 5%, α = 0.05
alpha = 0.05;
% Test the hypothesis H₀: Φx(x) ≥ ΦY(x)
H = mwutest(X,Y,alpha,1)
H =

        Ho: 'reject'
         U: 69
        CR: [-Inf 24.8547]
    Pvalue: 0.0250
```

5.5.3.3   Mann–Whitney U-Test for One-Sided, Right-Tailed Test

This test determines whether X is stochastically larger than Y by *rejecting* the null hypothesis of X having a smaller distribution function than Y for unmatched samples.

Hypothesis

$$H_0: \Phi_X(x) \leq \Phi_Y(x) \qquad \qquad \text{(Null hypothesis)}$$
$$H_1: \Phi_X(x) > \Phi_Y(x) \qquad \qquad \text{(Alternate hypothesis)}$$

Test Statistic

$$T(X, Y) = \frac{U - E[U]}{\sqrt{\text{var}[U]}}$$

$$U = \max\{U_x, U_y\}, U_x = \sum_{j=1}^{n_x} R_j - {}^{1}\!/_{2}\, n_x \bullet (n_x + 1) \text{ and}$$

$$U_y = \sum_{j=1+n_x}^{n_y+n_x} R_j - \tfrac{1}{2} n_y \bullet (n_y + 1)$$

$R_1 \leq R_2 \leq \cdots \leq R_n, n = n_x + n_y$ are the ranks of $Z = \{X, Y\}$.

$E[U] = \tfrac{1}{2} n_x \bullet n_y$

$\text{var}[U] = \tfrac{1}{12} n_x \cdot n_y \cdot (n_x + n_y + 1) - \tfrac{1}{2} N_t (N_t^2 - 1)$

N_t = total number of tied ranks.

Test Statistic Distribution

$$T(X, Y) \sim \mathcal{N}(0, 1) \text{ for } n_x, n_y > 8$$

p-Value

$$p = \Phi(T(X, Y))$$

For $n_x, n_y \leq 8$, there are exact tables for p-values incorporated into mwutest.m.

Null Hypothesis Decision
If $p < \alpha$, then *reject H_0* in favor of H_1. Otherwise, *do not reject H_0*.

$(1 - \alpha) \bullet 100\%$ Critical Region for U

$$CR(U) = [E[U] + z_\alpha \cdot \sqrt{\text{var}[U]}, +\infty)$$

Required Sample Size
Either conduct a pilot study and use equations (5.29a) and (5.29b) or use the Wilcoxon signed-rank method. Find the value $n = n_y$ that minimizes (5.28) and set $n_x = \kappa \bullet n_y$, where κ is the ratio of the X samples to Y samples. Instead of the quantile $z_{1-\alpha/2}$, however, use $z_{1-\alpha}$.

θ = effect size which is a fraction of what can practically be expected to be the minimum difference in the sample distribution means. See the comment just prior to Section 5.5.2.1 for guidance as to the value of θ.

MATLAB Function
mwutest.m

Example: To test the example (Hogg and Craig [14]) from Section 5.5.3.2 that the sample $Y = \{5.5, 7.9, 6.8, 9.0, 5.6, 6.3, 8.5, 4.6, 7.1\}$ of the random variable Y is stochastically smaller than the sample $X = \{4.3, 5.9, 4.9, 3.1, 5.3, 6.4, 6.2, 3.8, 7.5, 5.8\}$ of X, test the hypotheses

$$H_0: \Phi_X(x) \leq \Phi_Y(x) \text{ vs. } H_1: \Phi_X(x) > \Phi_Y(x)$$

The p-value is greater than the significance level $p = 0.975 > \alpha = 0.05$ and $U = 69 \in [65.14, +\infty) = CR$. Therefore, *do not reject H_0*.

MATLAB Commands
(***Right-Tailed Mann–Whitney U-Test***)

```
% Data from two distinct samples X and Y
X = [4.3, 5.9, 4.9, 3.1, 5.3, 6.4, 6.2, 3.8, 7.5, 5.8];
Y = [5.5, 7.9, 6.8, 9.0, 5.6, 6.3, 8.5, 4.6, 7.1];
% Set the significance level at 5%, α = 0.05
alpha = 0.05;
% Test the hypothesis H₀: Φ_X(x) ≤ Φ_Y(x)
H = mwutest(X,Y,alpha,0)
H =
        Ho: 'do not reject'
         U: 69
        CR: [65.1453 Inf]
    Pvalue: 0.9750
```

5.5.4 Diagnostic Likelihood Ratios

As was first discussed in Chapter 3, one measure of how well a diagnostic method/ device determines the presence of disease is the *sensitivity*. In particular, suppose a *Control* process can determine the actual disease state of a patient. Then a diagnosis of *positive* by the *Control* is considered a correct determination of disease, while a reading of *negative* echoes the genuine absence of disease. If a new or *Test* device measures the same samples as does the *Control*, then the 2×2 contingency Table 5.30 arises.

The *sensitivity* and *specificity* are then defined via (5.30a) and (5.30b).

$$sen = \frac{n_{2,2}}{n_{2,2} + n_{1,2}} = \frac{T_p}{T_p + F_n} \tag{5.30a}$$

$$spec = \frac{n_{1,1}}{n_{1,1} + n_{2,1}} = \frac{T_n}{T_n + F_p} \tag{5.30b}$$

It is not difficult to imagine two contingency tables parallel to Table 5.30: First a table for $Test_1$ and a second table for $Test_2$. These can be combined into a single

TABLE 5.30   Contingency for *sensitivity* and *specificity*

		Control			
		Negative (−)		Positive (+)	
Test	Negative (−)	$n_{1,1}$	(T_n)	$n_{1,2}$	(F_n)
	Positive (+)	$n_{2,1}$	(F_p)	$n_{2,2}$	(T_p)

TABLE 5.31   **Contingency for competing diagnostic tests against gold standard verification**

Gold standard diagnosis	$T_1 = 1$		$T_1 = 0$		
	$T_2 = 1$	$T_2 = 0$	$T_2 = 1$	$T_2 = 0$	Totals
$D_x = 1$	s_{11}	s_{10}	s_{01}	s_{00}	$s = \sum_{k=0}^{1}\sum_{j=0}^{1} s_{jk}$
$D_x = 0$	r_{11}	r_{10}	r_{01}	r_{00}	$r = \sum_{k=0}^{1}\sum_{j=0}^{1} r_{jk}$
Totals	n_{11}	n_{10}	n_{01}	n_{00}	$n = s + r$

contingency illustrated in Table 5.31. With respect to this mnemonic, the expression $T_1 = 1$ indicates that $Test_1$ determined the sample has a *positive* diagnosis (i.e., disease is present), whereas $T_2 = 0$ means that $Test_2$ yields a *negative* finding (i.e., absence of disease). The symbols $D_x = 1$ and $D_x = 0$ mean that the *Control* or "gold standard" diagnosis is *positive* (1) and *negative* (0), respectively.

For example, the sensitivity with respect to $Test_1$ is $sen_1 = (s_{11} + s_{10})/s$ while the $Test_2$ sensitivity is $sen_2 = (s_{11} + s_{01})/s$. Similarly, $spec_1 = (r_{01} + r_{00})/r$ and $spec_2 = (r_{10} + r_{00})/r$. In this manner, a hypothesis test of equal or equivalent *sensitivities* can be established by treating each sensitivity as a correlated proportion. These ideas are detailed in Sections 5.3.1.2 and 5.3.1.4.

For a variety of reasons, however, comparing sensitivities across two distinct systems/devices may not be sensible. Suppose, for example, that $Test_1$ represents a device that is presently used by clinicians and received approval for use by the *US Food and Drug Administration* in 2002. Further, suppose the device diagnoses a particular disease and that the *standard of care* among diagnosticians is that there are three levels of the disease: *Normal* (i.e., no disease), *level 1* (an early disease state), and *level 2* (full-blown disease). Now posit that in 2015 a review board of leading diagnosticians determines that this disease is best managed via a five-level triage: *Normal, level 1* (a pre-disease state), *level 2* (early disease state), *level 3* (mid-disease state), and *level 4* (full disease state). What does a sensitivity of [say] 75% in 2002 mean vs. a sensitivity of 70% in 2015? It is not sensible to link the two measurements as they are derived from a different "ground truth." Therefore, two different assessment metrics are analyzed: The *positive diagnostic likelihood ratio*, defined as the ratio of the *sensitivity* to the *complement of the specificity* and the *negative diagnostic likelihood ratio* which is the *complement of the sensitivity* divided by the *specificity*. These definitions are made explicit in (5.31a) and (5.31b).

$$DLR^+ = \frac{sen}{1 - spec} \tag{5.31a}$$

$$DLR^- = \frac{1 - sen}{spec} \tag{5.31b}$$

More precisely, the *positive diagnostic ratio* is the ratio of the probability that a diagnostic test yields a positive (for disease) result given that disease is truly present to the probability that the diagnostic test is positive given the absence of disease. That is,

$$DLR^+ = \frac{\text{Prob}\,(Test = +|Disease)}{\text{Prob}\,(Test = +|No\ Disease)} \qquad (5.32a)$$

If $DLR^+ > 1$, then the outcome of the diagnostic test is more likely to be positive for a patient who has disease than it would be for a patient without disease. Similarly,

$$DLR^- = \frac{\text{Prob}\,(Test = -|Disease)}{\text{Prob}\,(Test = -|No\ Disease)} \qquad (5.32b)$$

so that $DLR^- \leq 1$ implies that the diagnostic test is more likely to be negative for a patient without disease than it would be for a patient with disease. For more details on the clinical meaning of this metric, see Pepe [32] and Zhou et al. [35].

Now return to Table 5.31 and obtain the two sets of sensitivities (sen_1 and sen_2) and specificities ($spec_1$ and $spec_2$). This, of course, produces two sets of diagnostic likelihood ratios (DLR_1^+, DLR_2^+) and (DLR_1^-, DLR_2^-). Exercising the hypothesis tests of equal diagnostic likelihood ratios

$$H_0^+: DLR_1^+ = DLR_2^+ \qquad \text{(Null hypothesis)}$$
$$H_1^+: DLR_1^+ \neq DLR_2^+ \qquad \text{(Alternate hypothesis)}$$

and

$$H_0^-: DLR_1^- = DLR_2^- \qquad \text{(Null hypothesis)}$$
$$H_1^-: DLR_1^- \neq DLR_2^- \qquad \text{(Alternate hypothesis)}$$

is not a direct application of the previous sections. Indeed, diagnostic likelihood ratios can be thought of as the ratios of binomial proportions. See, for example, Koopman [17], Gart and Nam [11], and Nam [23–25]. Rather than extending these methods, the work of Nofuentes and Luna del Castillo [28] will be detailed.

To that end, make the following representations.

$$\rho^+ = \frac{DLR_1^+}{DLR_2^+}, \rho^- = \frac{DLR_1^-}{DLR_2^-} \qquad (5.33a)$$

$$\omega^+ = \ln(\rho^+), \omega^- = \ln(\rho^-) \qquad (5.33b)$$

Rather than testing $H_0^+: DLR_1^+ = DLR_2^+$ against $H_1^+: DLR_1^+ \neq DLR_2^+$, it is equivalent to examine the hypotheses $H_0^+: \omega^+ = 0$ vs. $H_1^+: \omega^+ \neq 0$. This can be readily seen, by noting for nonzero DLR, the null hypothesis $H_0^+: DLR_1^+ = DLR_2^+$ implies $DLR_1^+/DLR_2^+ = 1$ and thereby $\omega^+ = \ln(DLR_1^+/DLR_2^+) = 0$. Mimicking this computation for the negative diagnostic likelihood ratio leads to $H_0^-: \omega^- = 0$ vs. $H_1^-: \omega^- \neq 0$.

How then to proceed with the hypothesis test? The answer is quite complicated and the details are provided in the paper of Nofuentes and Luna del Castillo [28]. The basic outline of the process is as follows.

Define the conditional $\theta_{jk} = \text{Prob}(D_x = 1|T_1 = j, T_2 = k)$ and unconditional $\eta_{jk} = \text{Prob}(T_1 = j, T_2 = k)$ probabilities for $j, k = 0, 1$. The diagnostic likelihood ratios can be written as

$$DLR_1^+ = \frac{1 - \sum_{k=0}^{1}\sum_{j=0}^{1}\theta_{jk}\eta_{jk}}{\sum_{k=0}^{1}\sum_{j=0}^{1}\theta_{jk}\eta_{jk}} \cdot \frac{\sum_{k=0}^{1}\theta_{1k}\eta_{1k}}{\sum_{k=0}^{1}(1-\theta_{1k})\eta_{1k}}, DLR_2^+ = \frac{1 - \sum_{k=0}^{1}\sum_{j=0}^{1}\theta_{jk}\eta_{jk}}{\sum_{k=0}^{1}\sum_{j=0}^{1}\theta_{jk}\eta_{jk}} \cdot \frac{\sum_{j=0}^{1}\theta_{j1}\eta_{j1}}{\sum_{j=0}^{1}(1-\theta_{j1})\eta_{j1}}$$

$$DLR_1^- = \frac{1 - \sum_{k=0}^{1}\sum_{j=0}^{1}\theta_{jk}\eta_{jk}}{\sum_{k=0}^{1}\sum_{j=0}^{1}\theta_{jk}\eta_{jk}} \cdot \frac{\sum_{k=0}^{1}\theta_{0k}\eta_{0k}}{\sum_{k=0}^{1}(1-\theta_{0k})\eta_{0k}}, \text{ and } DLR_2^- = \frac{1 - \sum_{k=0}^{1}\sum_{j=0}^{1}\theta_{jk}\eta_{jk}}{\sum_{k=0}^{1}\sum_{j=0}^{1}\theta_{jk}\eta_{jk}} \cdot \frac{\sum_{j=0}^{1}\theta_{j0}\eta_{j0}}{\sum_{j=0}^{1}(1-\theta_{j0})\eta_{j0}}$$

It is seen that $\omega^+ = \ln(DLR_1^+/DLR_2^+)$ is a nonlinear function of $\theta = [\theta_{00}, \theta_{01}, \theta_{10}, \theta_{11}]$ and $\eta = [\eta_{00}, \eta_{01}, \eta_{10}]$. Observe that η_{11} is excluded from the vector η since $\eta_{11} = 1 - \eta_{00} - \eta_{01} - \eta_{10}$ is not an independent variable. Also, use the notation $s = [s_{00}, s_{01}, s_{10}, s_{11}]$, $r = [r_{00}, r_{01}, r_{10}, r_{11}]$, and $n = [n_{00}, n_{01}, n_{10}, n_{11}]$ where s_{jk}, r_{jk}, and n_{jk} are defined in Table 5.31. Hence, a candidate test statistic for the null hypothesis $H_0^+: \omega^+ = 0$ is $T(\omega^+) = \omega^+ / \sqrt{\text{var}(\omega^+)}$. Nofuentes and Luna del Castillo [28] note that $T(\omega^+)$ is asymptotically normal $\mathcal{N}(0, 1)$. The same logic applies to ω^-. To compute the variance of ω (either ω^+ or ω^-), the delta method of Agresti [2] will be utilized. In this technique

$$\text{var}(\omega) = \left(\frac{\partial\omega}{\partial\theta}\right)I^{-1}(\theta)\left(\frac{\partial\omega}{\partial\theta}\right)^T + \left(\frac{\partial\omega}{\partial\eta}\right)I^{-1}(\eta)\left(\frac{\partial\omega}{\partial\eta}\right)^T \tag{5.34}$$

where $\frac{\partial\omega}{\partial\theta}$ and $\frac{\partial\omega}{\partial\eta}$ are the Jacobians of ω with respect to θ and η, and $I^{-1}(\theta)$ and $I^{-1}(\eta)$ are the *Fisher information matrices*. The formulae for these components are now listed sequentially.

$$\frac{\partial\omega^+}{\partial\theta} = \left[\frac{\partial\omega^+}{\partial\theta_{00}}, \frac{\partial\omega^+}{\partial\theta_{01}}, \frac{\partial\omega^+}{\partial\theta_{10}}, \frac{\partial\omega^+}{\partial\theta_{11}}\right]$$

$$\frac{\partial\omega^+}{\partial\theta_{00}} = 0, \frac{\partial\omega^+}{\partial\theta_{01}} = \frac{-\eta_{01}(\eta_{01}+\eta_{11})}{\sum_{j=0}^{1}(1-\theta_{j1})\eta_{j1}\sum_{j=0}^{1}\theta_{j1}\eta_{j1}}, \frac{\partial\omega^+}{\partial\theta_{10}} = \frac{-\eta_{10}(\eta_{10}+\eta_{11})}{\sum_{k=0}^{1}(1-\theta_{1k})\eta_{1k}\sum_{k=0}^{1}\theta_{1k}\eta_{1k}},$$

$$\frac{\partial\omega^+}{\partial\theta_{11}} = \frac{(1-\theta_{01})\eta_{01}}{(1-\theta_{11})\sum_{j=0}^{1}(1-\theta_{j1})\eta_{j1}} - \frac{(1-\theta_{10})\eta_{10}}{(1-\theta_{11})\sum_{k=0}^{1}(1-\theta_{1k})\eta_{1k}} + \frac{\theta_{01}\eta_{01}}{\theta_{11}\sum_{j=0}^{1}\theta_{j1}\eta_{j1}}$$

$$- \frac{\theta_{10}\eta_{10}}{\theta_{11}\sum_{k=0}^{1}\theta_{1k}\eta_{1k}}$$

$$\tag{5.35a}$$

The corresponding *Fisher information* matrix is

$$
I^{-1}(\boldsymbol{\theta}) =
\begin{bmatrix}
\dfrac{\theta_{00}^2(1+\theta_{00})^2}{s_{00}(1+\theta_{00})^2+r_{00}\theta_{00}^2} & 0 & 0 & 0 \\[3mm]
0 & \dfrac{\theta_{01}^2(1+\theta_{01})^2}{s_{01}(1+\theta_{01})^2+r_{01}\theta_{01}^2} & 0 & 0 \\[3mm]
0 & 0 & \dfrac{\theta_{10}^2(1+\theta_{10})^2}{s_{10}(1+\theta_{10})^2+r_{10}\theta_{10}^2} & 0 \\[3mm]
0 & 0 & 0 & \dfrac{\theta_{11}^2(1+\theta_{11})^2}{s_{11}(1+\theta_{11})^2+r_{11}\theta_{11}^2}
\end{bmatrix}
\tag{5.35b}
$$

$$
\frac{\partial \omega^+}{\partial \boldsymbol{\eta}} = \left[\frac{\partial \omega^+}{\partial \eta_{00}}, \frac{\partial \omega^+}{\partial \eta_{01}}, \frac{\partial \omega^+}{\partial \eta_{10}} \right]
$$

$$
\left.
\begin{aligned}
\frac{\partial \omega^+}{\partial \eta_{00}} &= \frac{1-\theta_{11}}{\sum\limits_{k=0}^{1}(1-\theta_{1k})\eta_{1k}} - \frac{1-\theta_{11}}{\sum\limits_{j=0}^{1}(1-\theta_{j1})\eta_{j1}} - \frac{\theta_{11}}{\sum\limits_{k=0}^{1}\theta_{1k}\eta_{1k}} + \frac{\theta_{11}}{\sum\limits_{j=0}^{1}\theta_{j1}\eta_{j1}} \\[3mm]
\frac{\partial \omega^+}{\partial \eta_{01}} &= \frac{\theta_{11}}{\sum\limits_{k=0}^{1}\theta_{1k}\eta_{1k}} + \frac{1-\theta_{11}}{\sum\limits_{k=0}^{1}(1-\theta_{1k})\eta_{1k}} + \frac{\theta_{11}-\theta_{01}}{\sum\limits_{j=0}^{1}\theta_{j1}\eta_{j1}} + \frac{\theta_{11}-\theta_{01}}{\sum\limits_{j=0}^{1}(1-\theta_{j1})\eta_{j1}} \\[3mm]
\frac{\partial \omega^+}{\partial \eta_{10}} &= \frac{\theta_{11}}{\sum\limits_{j=0}^{1}\theta_{j1}\eta_{j1}} - \frac{1-\theta_{11}}{\sum\limits_{j=0}^{1}(1-\theta_{j1})\eta_{j1}} - \frac{\theta_{11}-\theta_{01}}{\sum\limits_{k=0}^{1}\theta_{1k}\eta_{1k}} - \frac{\theta_{11}-\theta_{01}}{\sum\limits_{k=0}^{1}(1-\theta_{1k})\eta_{1k}}
\end{aligned}
\right\}
\tag{5.35c}
$$

The *Fisher information* matrix associated with $\dfrac{\partial \omega^+}{\partial \boldsymbol{\eta}}$ is

$$
I^{-1}(\boldsymbol{\eta}) =
\begin{bmatrix}
\dfrac{\eta_{00}^2}{n_{00}} & 0 & 0 \\[3mm]
0 & \dfrac{\eta_{01}^2}{n_{01}} & 0 \\[3mm]
0 & 0 & \dfrac{\eta_{10}^2}{n_{10}}
\end{bmatrix} - M(\boldsymbol{\eta}, \boldsymbol{n})
\tag{5.35d}
$$

where $M(\boldsymbol{\eta}, \boldsymbol{n}) = \dfrac{1}{\sum_{j=0}^{1}\sum_{k=0}^{1}\dfrac{\eta_{jk}^2}{n_{jk}}}
\begin{bmatrix}
\left(\dfrac{\eta_{00}^2}{n_{00}}\right)^2 & \dfrac{\eta_{00}^2}{n_{00}}\dfrac{\eta_{01}^2}{n_{01}} & \dfrac{\eta_{00}^2}{n_{00}}\dfrac{\eta_{10}^2}{n_{10}} \\[3mm]
\dfrac{\eta_{00}^2}{n_{00}}\dfrac{\eta_{01}^2}{n_{01}} & \left(\dfrac{\eta_{01}^2}{n_{01}}\right)^2 & \dfrac{\eta_{10}^2}{n_{10}}\dfrac{\eta_{01}^2}{n_{01}} \\[3mm]
\dfrac{\eta_{00}^2}{n_{00}}\dfrac{\eta_{10}^2}{n_{10}} & \dfrac{\eta_{10}^2}{n_{10}}\dfrac{\eta_{01}^2}{n_{01}} & \left(\dfrac{\eta_{10}^2}{n_{10}}\right)^2
\end{bmatrix}$. Consequently, the vari-

ance of ω^+, var(ω^+), is the product of the 1×4 vector $\dfrac{\partial \omega^+}{\partial \boldsymbol{\theta}}$, the 4×4 matrix $I^{-1}(\boldsymbol{\theta})$, and the 4×1 vector $\left(\dfrac{\partial \omega^+}{\partial \boldsymbol{\theta}}\right)^T$ along with the product of the 1×3 vector

$\frac{\partial \omega^+}{\partial \eta}$, the 3×3 matrix $I^{-1}(\eta)$, and the 3×1 vector $\left(\frac{\partial \omega^+}{\partial \eta}\right)^T$. In a similar manner,
$\text{var}(\omega^-) = \left(\frac{\partial \omega^-}{\partial \theta}\right) I^{-1}(\theta) \left(\frac{\partial \omega^-}{\partial \theta}\right)^T + \left(\frac{\partial \omega^-}{\partial \eta}\right) I^{-1}(\eta) \left(\frac{\partial \omega^-}{\partial \eta}\right)^T$ is computed via the following formulae.

$$\frac{\partial \omega^-}{\partial \theta} = \left[\frac{\partial \omega^-}{\partial \theta_{00}}, \frac{\partial \omega^-}{\partial \theta_{01}}, \frac{\partial \omega^-}{\partial \theta_{10}}, \frac{\partial \omega^-}{\partial \theta_{11}}\right]$$

$$\left.\begin{array}{l} \frac{\partial \omega^-}{\partial \theta_{00}} = \eta_{00} \left(\frac{1}{\sum\limits_{k=0}^{1}(1-\theta_{0k})\eta_{0k}} + \frac{1}{\sum\limits_{k=0}^{1}\theta_{0k}\eta_{0k}} - \frac{1}{\sum\limits_{j=0}^{1}(1-\theta_{j1})\eta_{j1}} - \frac{1}{\sum\limits_{j=0}^{1}\theta_{j1}\eta_{j1}} \right) \\[1em] \frac{\partial \omega^-}{\partial \theta_{01}} = \frac{-\eta_{01}(\eta_{00}+\eta_{01})}{\sum\limits_{k=0}^{1}(1-\theta_{0k})\eta_{0k}\sum\limits_{k=0}^{1}\theta_{0k}\eta_{0k}}, \quad \frac{\partial \omega^-}{\partial \theta_{10}} = \frac{-\eta_{01}(\eta_{00}+\eta_{01})}{\sum\limits_{j=0}^{1}(1-\theta_{j1})\eta_{j1}\sum\limits_{j=0}^{1}\theta_{j1}\eta_{j1}}, \quad \frac{\partial \omega^-}{\partial \theta_{11}} = 0 \end{array}\right\}$$

$$\text{(5.35e)}$$

$$\frac{\partial \omega^-}{\partial \eta} = \left[\frac{\partial \omega^-}{\partial \eta_{00}}, \frac{\partial \omega^-}{\partial \eta_{01}}, \frac{\partial \omega^-}{\partial \eta_{10}}\right]$$

$$\left.\begin{array}{l} \frac{\partial \omega^-}{\partial \eta_{00}} = \frac{(\theta_{00}-\theta_{01})\eta_{01}}{\sum\limits_{k=0}^{1}(1-\theta_{0k})\eta_{0k}\sum\limits_{k=0}^{1}\theta_{0k}\eta_{0k}} + \frac{1-\theta_{00}}{\sum\limits_{j=0}^{1}(1-\theta_{j0})\eta_{j0}} - \frac{\theta_{00}}{\sum\limits_{j=0}^{1}\theta_{j0}\eta_{j0}} \\[1em] \frac{\partial \omega^-}{\partial \eta_{01}} = \frac{(\theta_{01}-\theta_{00})\eta_{00}}{\sum\limits_{k=0}^{1}(1-\theta_{0k})\eta_{0k}\sum\limits_{k=0}^{1}\theta_{0k}\eta_{0k}}, \quad \frac{\partial \omega^-}{\partial \eta_{10}} = \frac{(\theta_{00}-\theta_{01})\eta_{00}}{\sum\limits_{j=0}^{1}(1-\theta_{j0})\eta_{j0}\sum\limits_{j=0}^{1}\theta_{j0}\eta_{j0}} \end{array}\right\}$$

$$\text{(5.35f)}$$

The variance $\sigma^2(\omega^+) \equiv \text{var}(\omega^+)$ is modeled after (5.34) and constructed via (5.35a), (5.35b), (5.35c), and (5.35d). Similarly, $\sigma^2(\omega^-) \equiv \text{var}(\omega^-)$ is determined via (5.35e), (5.35b), (5.35f), and (5.35d). The $(1-\alpha) \cdot 100\%$ confidence interval for $\omega^+ = \ln(\rho^+)$, $\rho^+ \equiv \frac{DLR_1^+}{DLR_2^+}$ is

$$CI(\omega^+) = \left[\omega^+ + z_{\alpha/2}\sqrt{\sigma^2(\omega^+)}, \omega^+ + z_{1-\alpha/2}\sqrt{\sigma^2(\omega^+)}\right]$$
$$= \left[\omega^+ + z_{\alpha/2}\sigma(\omega^+), \omega^+ + z_{1-\alpha/2}\sigma(\omega^+)\right] \qquad \text{(5.36a)}$$

and, after applying the exponential function,

$$CI(\rho^+) = \left[\rho^+ \cdot e^{z_{\alpha/2}\cdot\sigma(\omega^+)}, \rho^+ \cdot e^{z_{1-\alpha/2}\cdot\sigma(\omega^+)}\right] \qquad \text{(5.36b)}$$

The formulae for the $(1 - \alpha) \cdot 100\%$ confidence interval for $\omega^- = \ln(\rho^-)$, $\rho^- \equiv \frac{DLR_1^-}{DLR_2^-}$ are analogous.

$$
CI(\omega^-) = \left[\omega^- + z_{\alpha/2} \sqrt{\sigma^2(\omega^-)}, \omega^- + z_{1-\alpha/2} \sqrt{\sigma^2(\omega^-)} \right]
$$
$$
= \left[\omega^- + z_{\alpha/2} \sigma(\omega^-), \omega^- + z_{1-\alpha/2} \sigma(\omega^-) \right] \tag{5.37a}
$$

$$
CI(\rho^-) = \left[\rho^- \cdot e^{z_{\alpha/2} \cdot \sigma(\omega^-)}, \rho^- \cdot e^{z_{1-\alpha/2} \cdot \sigma(\omega^-)} \right] \tag{5.37b}
$$

The formatting of Sections 5.2–5.4 can now be implemented.

Hypotheses

$$H_0^+: DLR_1^+ = DLR_2^+ \qquad\qquad \textit{(Null hypothesis)}$$
$$H_1^+: DLR_1^+ \neq DLR_2^+ \qquad\qquad \textit{(Alternate hypothesis)}$$

$$H_0^-: DLR_1^- = DLR_2^- \qquad\qquad \textit{(Null hypothesis)}$$
$$H_1^-: DLR_1^- \neq DLR_2^- \qquad\qquad \textit{(Alternate hypothesis)}$$

Test Statistics
For H_0^+ vs. H_1^+

$$
T^+(X, Y) = \frac{\omega^+}{\sqrt{\operatorname{var}(\omega^+)}}, \omega^+ = \ln(\rho^+), \rho^+ \equiv \frac{DLR_1^+}{DLR_2^+}
$$

$\operatorname{var}(\omega^+)$ computed via equations (5.34) and (5.35a)–(5.35d), the estimates $\hat{\theta}_{jk} = \frac{s_{jk}}{n_{jk}}$, $\hat{\eta}_{jk} = \frac{n_{jk}}{n}$, s_{jk}, n_{jk}, and n from Table 5.31.
For H_0^- vs. H_1^-

$$
T^-(X, Y) = \frac{\omega-}{\sqrt{\operatorname{var}(\omega^-)}}, \omega^- = \ln(\rho^-), \rho^- \equiv \frac{DLR_1^-}{DLR_2^-}
$$

$\operatorname{var}(\omega^-)$ computed via equations (5.34), (5.35b), (5.35d), (5.35e), (5.35f), the estimates $\hat{\theta}_{jk} = \frac{s_{jk}}{n_{jk}}$, $\hat{\eta}_{jk} = \frac{n_{jk}}{n}$, s_{jk}, n_{jk}, and n from Table 5.31.

Test Statistic Distribution

$$
T^+(X, Y), T^-(X, Y) \sim \mathcal{N}(0, 1)
$$

p-Values

$$p^+ = 2 \cdot \Phi(-|T^+(X, Y)|)$$

$$p^- = 2 \cdot \Phi(-|T^-(X, Y)|)$$

Null Hypothesis Decision

If $p^+ < \alpha$, then *reject* H_0^+. Otherwise, *do not reject* H_0^+.

If $p^- < \alpha$, then *reject* H_0^-. Otherwise, *do not reject* H_0^-.

$(1 - \alpha) \cdot 100\%$ Confidence Intervals for $\rho^+ = \dfrac{DLR_1^+}{DLR_2^+}$ and $\rho^- = \dfrac{DLR_1^-}{DLR_2^-}$

$$CI(\rho^+) = [\rho^+ \cdot e^{z_{\alpha/2} \cdot \sigma(\omega^+)}, \rho^+ \cdot e^{z_{1-\alpha/2} \cdot \sigma(\omega^+)}]$$

$$CI(\rho^-) = [\rho^- \cdot e^{z_{\alpha/2} \cdot \sigma(\omega^-)}, \rho^- \cdot e^{z_{1-\alpha/2} \cdot \sigma(\omega^-)}]$$

Required Sample Size

For H_0^+ vs. H_1^+

$$n^+ = \left(\frac{z_{1-\alpha/2} + z_\beta}{\delta^+} \right)^2 \cdot \text{var}(\hat{\omega}^+)$$

$$\delta^+ = \ln \left(\frac{\widehat{DLR}_1^+}{\widehat{DLR}_2^+} \right) = \text{the (positive) effect size}$$

$\text{var}(\hat{\omega}^+)$ can either be estimated from a pilot study by way of Table 5.31 and equation (5.34), or from disease prevalence and the conditional probabilities $\xi_{jk} = \text{Prob}(T_1 = j, T_2 - k|D_x = 1)$, $\psi_{jk} = \text{Prob}(T_1 = j, T_2 - k|D_x = 0)$ as per Appendix B of Nofuentes and Luna del Castillo [28].

For H_0^- vs. H_1^-

$$n^- = \left(\frac{z_{1-\alpha/2} + z_\beta}{\delta^-} \right)^2 \cdot \text{var}(\hat{\omega}^-)$$

$$\delta^- = \ln \left(\frac{\widehat{DLR}_1^-}{\widehat{DLR}_2^-} \right) = \text{the (negative) effect size}$$

$\text{var}(\hat{\omega}^-)$ can either be estimated from a pilot study by way of Table 5.31 and equation (5.34), or from disease prevalence and the conditional probabilities $\xi_{jk} = \text{Prob}(T_1 = j, T_2 - k|D_x = 1)$, $\psi_{jk} = \text{Prob}(T_1 = j, T_2 - k|D_x = 0)$ as per Appendix B of Nofuentes and Luna del Castillo [28].

MATLAB Function

```
dlratios.m
```

Example: The example takes the following data table from Nofuentes and Luna del Castillo [28].

TABLE 5.32   Input matrix for `dlratios.m`

| Gold standard diagnosis | $T_1 = 1$ | | $T_1 = 0$ | | |
	$T_2 = 1$	$T_2 = 0$	$T_2 = 1$	$T_2 = 0$	Totals
$D_x = 1$	786	29	183	25	1023
$D_x = 0$	69	46	176	151	442
Totals	855	75	359	176	1465

This table will be used to test the two sets of hypothesis tests $H_0^+: DLR_1^+ = DLR_2^+$ vs. $H_1^+: DLR_1^+ \neq DLR_2^+$ and $H_0^-: DLR_1^- = DLR_2^-$ vs. $H_1^-: DLR_1^- \neq DLR_2^-$. Results are given for H_0^+ (HoPlus) and H_0^- (HoMinus) along with estimates of $\rho = DLR_1/DLR_2$ and $\omega = \ln(DLR_1/DLR_2)$.

MATLAB Commands
(*Two-Sided Diagnostic Likelihood Ratios*)

% Data Table as presented in Nofuentes and Luna del Castillo [28]

```
A = [786, 29, 183, 25;69, 46, 176, 151];
```
% Set the significance level at 5%, $\alpha = 0.05$
```
alpha = 0.05;
```
% Test the hypotheses $H_0^+: DLR_1^+ = DLR_2^+$ and $H_0^-: DLR_1^- = DLR_2^-$
```
S = dlratios(A,alpha);
S =
T1                              T2
    DLRplus                         DLRplus
        3.062                           1.709
    DLRminus                        DLRminus
        0.2748                          0.1184
    CI                              CI
        DLRplus                         DLRplus
            [2.609 3.594]                   [1.57 1.86]
        DLRminus                        DLRminus
            [0.2405 0.314]                  [0.08954 0.1566]
    HoPlus                          HoMinus
        Reject                          Reject
    Pplus                           Pminus
        9.142e-08                       1.756e-08
```

```
Ratios
    Std                          LnCI
        Wplus                        Wplus
            0.1092                       [0.3693 0.7972]
        Wminus                       Wminus
            0.1494                       [0.549 1.135]
    Wplus
        0.5832
    Wminus
        0.8418
    RHOplus
        1.792
    RHOminus
        2.321
    CI
        RHOplus
            [1.447 2.219]
        RHOminus
            [1.731 3.11]
```

EXERCISES

5.10 Prove the change of variable theorem. *Hint:* Let $F(x) = \int f(x)dx$ be the *anti-derivative* of the function f. Then, by the Fundamental Theorem of Calculus, $\frac{dF}{dx}(x) = F'(x) = f(x)$. If $g: [a, b] \rightarrow \mathbb{R}$ is a continuous and differentiable function, then $\frac{d}{dx}F(g(x)) = F'(g(x)) \cdot g'(x) = f(g(x)) \cdot g'(x)$ by the chain rule. Therefore, $\int_a^b f(g(x)) \cdot g'(x)\,dx = \int_a^b \frac{d}{dt}F(g(x))\,dx = F(g(b)) - F(g(a)) = \int_{g(a)}^{g(b)} f(x)\,dx$ by the Fundamental Theorem of Calculus. Now recall that a *pdf* is a positive function which integrates to 1 over its domain of definition.

5.11 Show via the change of variables $u = t\sqrt{\tau}/\sqrt{n_x - 1}$ that (5.21a) becomes (5.21b).

5.12 Test the null hypothesis of equal *CV*s for the hockey and basketball height data mentioned in Section 5.1. Note, in this case, the *exact method* and the method of *Miller do not reject* the null hypothesis while the *Bennett* and *modified Bennett* methods *do* reject the null. The following table gives the results for the *exact* method. This is another indication that the exact method gives the most accurate result.

CV_H	$CI(CV_H)$	CV_B	$CI(CV_B)$	ΔCV	$CI(\Delta CV)$	*(Exact)*
0.028	[0.007, 1.78]	0.043	[0.011, 3.4]	−0.015	[−4.53, 4.5]	

Project

Following the development for *diagnostic likelihood ratios* in Section 5.5.4, develop the *test statistics, confidence intervals*, and *sample size* formulae for the one-sided, *left-tailed* and *right-tailed* hypothesis tests listed below.

$$H_0^+: DLR_1^+ \leq DLR_2^+ \text{ vs. } H_1^+: DLR_1^+ > DLR_2^+ \qquad \textit{(Right-tailed test)}$$
$$H_0^-: DLR_1^- \leq DLR_2^- \text{ vs. } H_1^-: DLR_1^- > DLR_2^- \qquad \textit{(Right-tailed test)}$$

$$K_0^+: DLR_1^+ \geq DLR_2^+ \text{ vs. } K_1^+: DLR_1^+ < DLR_2^+ \qquad \textit{(Left-tailed test)}$$
$$K_0^-: DLR_1^- \geq DLR_2^- \text{ vs. } K_1^-: DLR_1^- < DLR_2^- \qquad \textit{(Left-tailed test)}$$

Hint: Follow the example of proportions from Sections 5.3.2.1 and 5.3.2.5. The test statistics for the one-sided tests will be the same as for the two-sided tests but the *p*-values will change. Also, the confidence intervals will all be one-sided.

REFERENCES

[1] Abramowitz, M., and I. A. Stegun, *Handbook of Mathematical Functions: With Formulas, Graphs, and Mathematical Tables*, Dover Books, New York, 2002.

[2] Agresti, A., *Categorical Data Analysis*, Second Edition, Wiley–Interscience Books, Hoboken, NJ, 2002.

[3] Altman, D. G., D. Machin, T. N. Bryant, and M. J. Gardner, *Statistics with Confidence*, Second Edition, British Medical Journal Books, London, 2000.

[4] Andrews, L. C., and B. K. Shivamoggi, *Integral Transforms for Engineers and Applied Mathematicians*, Macmillan Publishing Company, New York, 1988.

[5] Bolstad, W. M., *Introduction to Bayesian Statistics*, Wiley–Interscience, Hoboken, NJ, 2004.

[6] Bickel, P. J., and K. A. Doksum, *Mathematical Statistics: Basic Ideas and Selected Topics*, Holden–Day, Incorporated, 1977, San Francisco, CA.

[7] Bracewell, R., *The Fourier Transform and Its Applications*, McGraw–Hill, New York, 1965.

[8] Chow, S–C, J. Shao, and H. Wang, *Sample Size Calculations in Clinical Research*, Second Edition, Chapman & Hall/CRC Press, Boca Raton, FL, 2008.

[9] Donner, A., and G. Y. Zou, "Closed–form confidence intervals for functions of the normal mean and standard deviation," *Statistical Methods in Medical Research*, vol. 21, no. 4, pp. 347–359, 2010.

[10] Efron, B., *The Jackknife, the Bootstrap and Other Resampling Plans*, SIAM Books, Philadelphia, PA, 1982.

[11] Gart, J. J., and J–m Nam, "Approximate interval estimation of the ratio of binomial parameters: a review and corrections for skewness," *Biometrics*, vol. 44, pp. 323–338, June 1998.

[12] Gibbons, J. D., and S. Chakraborti, *Nonparametric Statistical Inference*, Fifth Edition, Chapman & Hall/CRC Press, Taylor & Francis Group, Boca Raton, FL, 2011.

[13] Hoel, P. G., S. C. Port, and C. J. Stone, *Introduction to Probability Theory*, Houghton Mifflin Company, Boston, MA, 1971.

[14] Hogg, R. V., and A. T. Craig, *Introduction to Mathematical Statistics*, Fourth Edition, Macmillan, New York, 1978.

[15] Hovjat, S., and M. V. Kondratovich, Assay Migration Studies for In Vitro Diagnostic Devices: Guidance for Industry and FDA Staff, 25 April 2013, U.S. Department of Health and Human Services, Food and Drug Administration. http://www.fda.gov/downloads/MedicalDevices/DeviceRegulationandGuidance/GuidanceDocuments/UCM092752.pdf

[16] Lehmann, E. L., *Testing Statistical Hypotheses*, Second Edition, Wiley–Interscience, Hoboken, NJ, 1986.

[17] Koopman, P. A. R., "Confidence intervals for the ratio of two binomial proportions," *Biometrics*, vol. 40, pp. 513–517, June 1984.

[18] Lewis, J. A., "Statistical principles for clinical trials (ICH E9): An introductory note to an international guidance," *Statistics in Medicine*, vol. 18, pp. 1903–1942, 1999.

[19] Link, W. A., and R. J. Barker, *Bayesian Inference with Ecological Applications*, Academic Press, London, 2010.

[20] Liu, J–p, H–m Hsueh, E. Hsieh, and J. J. Chen, "Tests for equivalence or non–inferiority for paired binary data," *Statistics in Medicine*, vol. 21, pp. 231–245, 2002.

[21] Marques de Sá, J. P., *Applied Statistics, Using SPSS, STATISTICA, MATLAB, and R*, Second Edition, Springer, Berlin, 2007.

[22] Mahmoudvand, R., and H. Hassani, "Two new confidence intervals for the coefficient of variation in a normal distribution," *Journal of Applied Statistics*, vol. 36, no. 4, pp. 429–442, 2009.

[23] Nam, J–m, "Confidence limits for the ratio of two binomial proportions based on likelihood scores non–iterative method," *Biometrical Journal*, vol. 37, no. 3, pp. 375–379, 1995.

[24] Nam, J–m, "Establishing equivalence of two treatments and sample size requirements in matched–pairs design," *Biometrics*, vol. 53, pp. 1422–1430, December 1997.

[25] Nam, J–m, "Power and sample size requirements for non–inferiority in studies comparing two matched proportions where the events are correlated," *Computational Statistics and Data Analysis*, vol. 55, pp. 2880–2887, 2011.

[26] Newcombe, R. G., "Interval estimation for the difference between independent proportions: comparison of eleven methods," *Statistics in Medicine*, vol. 17, pp. 873–890, 1998.

[27] Newcombe, R. G., "Improved confidence intervals for the difference between binomial proportions based on paired data," *Statistics in Medicine*, vol. 17, pp. 2635–2650, 1998.

[28] Nofuentes, J., and J. Dios Luna del Castillo, "Comparison of the likelihood ratios of two binary diagnostic tests in paired designs," *Statistics in Medicine*, vol. 26, pp. 4179–4201, 2007.

[29] Olver, F. W. J., D. W. Lozier, R. F. Boisvert, and C. W. Clark, *NIST Handbook of Mathematical Functions*, Cambridge University Press, 2010.

[30] Papoulis, A., *Probability, Random Variables, and Stochastic Processes*, Third Edition, WCB/McGraw–Hill, Boston, MA, 1991.

[31] Pardo, M. C. and J. A. Pardo, "Use of Renyi's divergence to test for the equality of the coefficients of variation," *Journal of Computational and Applied Mathematics*, vol. 116, pp. 93–104, 2000.

[32] Pepe, M. S., *The Statistical Evaluation of Medical Tests for Classification and Prediction*, Oxford Statistical Science Series, vol. 31, Oxford University Press, 2003.

[33] Pestman, W. R., *Mathematical Statistics*, Walter de Gruyter, Berlin, 1998.

[34] Shieh, G., S. L. Jan, and R. H. Randles, "Power and sample size determination for the Wilcoxon signed–rank test," *Journal of Statistical Computation and Simulation*, vol. 77, no. 8, pp. 717–724, August 2007.

[35] Zhou, X–H, N. A. Obuchowski, and D. K. McClish, *Statistical Methods in Diagnostic Medicine*, Second Edition, John Wiley & Sons, Inc., Hoboken, NJ, 2011.

FURTHER READING

Cohen, J., *Statistical Power Analysis for the Behavioral Sciences*, Second Edition, Lawrence Erlbaum Associates, 1988.

Cohen, Y. and Cohen J., *Statistics and Data with R: An Applied Approach Through Examples*, Wiley, West Sussex, UK, 2008.

Daniel, W. W., *Biostatistics*, Ninth Edition, Wiley, Hoboken, NJ, 2009.

Hogg, R. V. and Tanis E. A., *Probability and Statistical Inference*, Macmillan, New York, 1977.

6

CLUSTERED DATA AND ANALYSIS OF VARIANCE

Many elements of diagnostic medicine are *interpretive*. That is, a diagnostician (e.g., a pathologist or radiologist) will review a patient specimen and pronounce a disease state based on his/her reading of a tissue sample, a slide containing cells, or an image. Figure 6.1 provides an example of a magnetic resonance image (MRI) scan of a human brain.

The impression one radiologist garners from these images may well differ from a second radiologist. Therefore, to determine the ultimate state of the patient, a *panel* of radiologists can review these images. Each radiologist will grade an image for disease state. For example, a score of 0 could indicate the lack of disease (no tumor = normal patient), whereas scores of $\ell = 1, 2,..., L$ indicate a disease state of level ℓ. How can these various scores from a group of diagnosticians lead to an assessment of the efficacy of the device? This collection of measurements is commonly referred to as multi-reader, multi-category (MRMC) or *clustered data*. Unlike previous examples of statistical analyses (see Chapter 5), the data obtained from clustered data studies are multi-indexed. In particular, the data are referenced by reader and sample. The first two sections of this chapter will examine the mathematical techniques needed to assess the efficacy/assessment metrics of a medical device from clustered data.

This chapter will focus on the mathematical methods required to analyze clustered and multi-indexed data. Since these data structures are more complicated than the one-dimensional measurements presented in Chapter 5, the corresponding mathematical models are more complex. The chapter begins with the notion of non-inferiority

Applied Mathematics for the Analysis of Biomedical Data: Models, Methods, and MATLAB®, First Edition. Peter J. Costa.
© 2017 Peter J. Costa. Published 2017 by John Wiley & Sons, Inc.
Companion website: www.wiley.com/go/costa/appmaths_biomedical_data

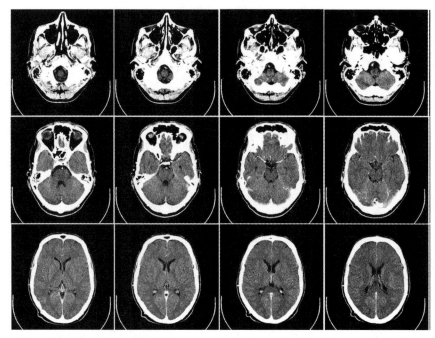

FIGURE 6.1 Example of an MRI brain scan. (From http://braincancerpictures.com/large/20/Brain-Cancer-MRIs-3.jpg)

for matched-pair clustered data. The associated assessment metrics and their corresponding mathematical forms are then detailed. The analysis of variance (ANOVA) for clustered data is also described along with several examples. The final section of the chapter contains a description of the highly regarded bootstrapping method of Efron [3] and its use in computing a confidence interval for clustered data.

As a point of emphasis, *assessment metrics* are measures of diagnostic performance. The most common assessment metrics, and those referred to in this chapter exclusively, are *sensitivity, specificity, positive/negative diagnostic likelihood ratios* (DLR^+ or DLR^-), and *positive/negative relative diagnostic likelihood ratios* ($rDLR^+$ or $rDLR^-$). If human interpretation of data is a topic of investigation, so too must be the evaluation of biological information processed on diagnostic devices. Plainly, in the manufacture of a medical instrument (such as the aforementioned MRI), it is critical that the variation from device to device be small. No one would conclude that a new diagnostic technology is efficacious if the variation in output across machines alters the diagnosis of a patient. Consequently, review agencies require that the diagnostic output of a set of devices be equivalent over a sequence of days, runs, and sample replicates. Thus, a second topic within this chapter, also concerning multi-indexed data, is the *analysis of variance*. These are variations inherent in data collected over a series of days, runs, and replicates, or alternately, over a set of tests, readers, and specimens.

Finally, clustered data will be combined with multi-indexed data to produce a section on the analysis of variance for clustered data. The exposition will require the following notation.

K = the total number of data clusters (total number of readers)

T_j = the diagnostic value of test j

$T_j = 0 \Rightarrow$ test j does not indicate disease, $T_j = 1 \Rightarrow$ test j indicates the presence of disease

D_x = the true (gold standard) diagnostic state

$D_x = 0 \Rightarrow$ the sample is disease free, $D_x = 1 \Rightarrow$ the sample is positive for disease

N_s = the total number of specimens/samples

6.1 CLUSTERED MATCHED-PAIR DATA AND NON-INFERIORITY

In Chapter 5, correlated data by way of contingency tables are investigated. In particular, statistical tests of equivalence, non-inferiority, and superiority with respect to correlated proportions are detailed in Sections 5.3.1.3, 5.3.2.4, and 5.3.2.8. Suppose that K readers are asked to interpret a set of images taken on an established *Control* device and a new *Test* device. Further, suppose that a "yes" response by the reader corresponds to the detection of disease on the image (*yes* = *positive* for disease) while "no" is equivalent to *negative*. Rather than Table 5.11 of Chapter 5, Table 6.1 is the new paradigm for each of the $k = 1, 2, \ldots, K$ readers.

Here $n_k = \sum_{i=0}^{1} \sum_{j=0}^{1} x_{ij,k}$ is the total number of images/samples read by reader k. The subscripted symbol • is used to indicate a sum across the corresponding subscript. Thus, $x_{i\bullet,k} = \sum_{j=0}^{1} x_{ij,k}$, $x_{\bullet j,k} = \sum_{i=0}^{1} x_{ij,k}$, $x_{i\bullet,\bullet} = \sum_{j=0}^{1} \sum_{k=1}^{K} x_{ij,k}$, $x_{\bullet j,\bullet} = \sum_{i=0}^{1} \sum_{k=1}^{K} x_{ij,k}$, and $n_\bullet = \sum_{k=1}^{K} n_k$ = total number of samples examined. The measurement $p_{1,k} = (x_{11,k} + x_{01,k})/n_k$ is the proportion of *Control* samples diagnosed as positive (i.e., yes) while $p_{2,k} = (x_{11,k} + x_{10,k})/n_k$ is the proportion of *Test* samples diagnosed as positive. The difference in proportions is $p_{2,k} - p_{1,k} = (x_{11,k} + x_{10,k})/n_k - (x_{11,k} + x_{01,k})/n_k = (x_{10,k} - x_{01,k})/n_k \equiv p_{10,k} - p_{01,k}$. Notice that $p_{i,k} = x_{i\bullet,k}/n_k$ for $i = 1, 2$.

To determine whether the clustered data yield comparable non-inferior positive proportions, one approach would be to test the K hypotheses $H_0(k)$: $p_{2,k} - p_{1,k} = \delta_k$

TABLE 6.1 **Contingency table for each of the**
$k = 1, 2, \ldots, K$ clustered data sets

		Control		
		Yes	No	Totals
Test	Yes	$x_{11,k}$	$x_{10,k}$	$x_{1\bullet,k}$
	No	$x_{01,k}$	$x_{00,k}$	$x_{0\bullet,k}$
	Totals	$x_{\bullet 1,k}$	$x_{\bullet 0,k}$	n_k

for $k = 1, 2, \ldots, K$. By the comment stated earlier, this is the same as testing the collection of hypotheses $H_0(k)$: $p_{10,k} - p_{01,k} = \delta_k$. Obuchowski [9] along with Nam and Kwon [6], however, suggest a single non-inferiority test for a *negative clinical margin* $\delta < 0$. By setting $p_{test}(+) = p_{2,\bullet} = \sum_{k=1}^{K} p_{2,k} = \sum_{k=1}^{K} \frac{1}{n_k}(x_{11,k} + x_{10,k})$, $p_{control}(+) = p_{1,\bullet} = \sum_{k=1}^{K} p_{1,k} = \sum_{k=1}^{K} \frac{1}{n_k}(x_{11,k} + x_{01,k})$, and $\delta = \sum_{k=1}^{K} \frac{1}{n_k}\delta_k$, then the set of hypotheses $H_0(k)$, $k = 1, 2, \ldots, K$ can be written as the single ensemble (non-inferiority) hypothesis test

$$H_{0\ell}: p_{2,\bullet} - p_{1,\bullet} = \delta \, (< 0) \qquad \text{(Null hypothesis)}$$

$$H_{1\ell}: p_{2,\bullet} - p_{1,\bullet} > \delta \, (< 0). \qquad \text{(Alternate hypothesis)}$$

As in the case of the hypotheses for each reader, the aggregate hypothesis test is equivalent to the form below.

Hypothesis Test

$$H_{0\ell}: p_{10,\bullet} - p_{01,\bullet} = \delta (< 0) \qquad \text{(Null hypothesis)}$$

$$H_{1\ell}: p_{10,\bullet} - p_{01,\bullet} > \delta (< 0) \qquad \text{(Alternate hypothesis)}$$

Here, of course, $p_{10,\bullet} = \sum_{k=1}^{K} p_{10,k}$ and $p_{01,\bullet} = \sum_{k=1}^{K} p_{01,k}$. In keeping with the format of Chapter 5, the mathematical structure of the hypothesis test will now be laid out sequentially.

Clinical Margin

$$\delta < 0$$

Test Statistic
This hypothesis test uses the modified Obuchowski [9] statistic Z_O as presented by Nam and Kwon [6].

$$Z_O = \frac{\hat{\theta}_{\bullet\bullet} - \delta}{\sqrt{\text{var}(\hat{\theta}_{\bullet\bullet} - \delta)}}$$

$$\hat{\theta}_{\bullet\bullet} = \hat{p}_{1\bullet,\bullet} - \hat{p}_{\bullet1,\bullet} \equiv \hat{p}_{2,\bullet} - \hat{p}_{1,\bullet}$$

$$\hat{p}_{i\bullet,\bullet} = \frac{\sum_{j=0}^{1}\sum_{k=1}^{K} x_{ij,k}}{\sum_{k=1}^{K} n_k} = \frac{x_{i\bullet,\bullet}}{n_\bullet}, \hat{p}_{\bullet j,\bullet} = \frac{\sum_{i=0}^{1}\sum_{k=1}^{K} x_{ij,k}}{\sum_{k=1}^{K} n_k} = \frac{x_{\bullet j,\bullet}}{n_\bullet} \quad \text{for } i, j = 0, 1.$$

Variance

$$\mathrm{var}(\hat{\theta}_{\bullet\bullet} - \delta) = \mathrm{var}(\hat{p}_{1\bullet\bullet} - \hat{p}_{\bullet1\bullet} - \delta) = \mathrm{var}(\hat{p}_{1\bullet\bullet}) + \mathrm{var}(\hat{p}_{\bullet1\bullet}) - 2\,\mathrm{cov}(\hat{p}_{1\bullet\bullet},\hat{p}_{\bullet1\bullet})$$

$$\mathrm{var}(\hat{p}_{1\bullet\bullet}) = \frac{K}{(K-1)\cdot(n_\bullet)^2}\sum_{k=1}^{K}(x_{1\bullet k} - n_k\bar{p}_{1\bullet\bullet})^2,\ \ \bar{p}_{1\bullet\bullet} = \frac{1}{2}\left(\frac{x_{1\bullet\bullet} + x_{\bullet1\bullet}}{n_\bullet} + \delta\right)$$

$$\mathrm{var}(\hat{p}_{\bullet1\bullet}) = \frac{K}{(K-1)\cdot(n_\bullet)^2}\sum_{k=1}^{K}(x_{\bullet1 k} - n_k\bar{p}_{\bullet1\bullet})^2,\ \ \bar{p}_{\bullet1\bullet} = \frac{1}{2}\left(\frac{x_{1\bullet\bullet} + x_{\bullet1\bullet}}{n_\bullet} - \delta\right)$$

$$\mathrm{cov}(\hat{p}_{1\bullet\bullet},\hat{p}_{\bullet1\bullet}) = \frac{K}{(K-1)\cdot(n_\bullet)^2}\sum_{k=1}^{K}(x_{1\bullet k} - n_k\bar{p}_{1\bullet\bullet})(x_{\bullet1 k} - n_k\bar{p}_{\bullet1\bullet}).$$

Note: Since $\mathrm{var}(\hat{\theta}_{\bullet\bullet} - \delta) = \mathrm{var}(\hat{\theta}_{\bullet\bullet}) \equiv \sigma^2(\hat{\theta}_{\bullet\bullet})$, the symbols $\mathrm{var}(\hat{\theta}_{\bullet\bullet} - \delta)$ and $\sigma^2(\hat{\theta}_{\bullet\bullet})$ will be used interchangeably.

Test Statistic Distribution
Z_O is the ratio of the asymptotically standard normal distributed numerator and the square root of the χ^2 variable $\mathrm{var}(\hat{\theta}_{\bullet\bullet} - \delta)$ with $df = K - 1$ degrees of freedom. Therefore, $Z_O \sim T_{df}$ as per (5.2) of Chapter 5.

Remark: It is left as an exercise (6.1) to show that $E[\theta_{\bullet\bullet} - \delta] = 0$. This calculation shows why each proportion's mean has the margin offset δ to ensure that $\theta_{\bullet\bullet} - \delta$ is a zero-mean random variable.

p-Value and Null Hypothesis Decision
As per Nam and Kwon [6], the null rejection criterion is $Z_O \geq z_{1-\alpha}$. It should be noted, however, that $\hat{\theta}_{\bullet\bullet} - \delta$ can be viewed as $\mathcal{N}(0, \sigma^2(\theta_{\bullet\bullet}))$. Further, $\sigma^2(\theta_{\bullet\bullet})$ can be treated as a χ^2-random variable with $df = K - 1$ degrees of freedom. Therefore, $Z_O = \dfrac{\theta_{\bullet\bullet} - \delta}{\sqrt{\sigma^2(\theta_{\bullet\bullet})}}$ can be viewed as the ratio of a normal and the square root of a χ^2-random variable. That is, $Z_O \sim T_{K-1}$. Following the work of Liu et al. [5], the null hypothesis will be rejected when $Z_O \geq t_{df}(1 - \alpha/2) \Leftrightarrow \alpha \geq 2\Phi_T(-Z_O; df)$, $df = K - 1$. This is more conservative than the criterion of Nam and Kwon [6] which requires a rejection of the null provided $Z_O \geq z(1 - \alpha) \Leftrightarrow \alpha \geq \Phi(-Z_O)$. In this more conservative approach, the *p-value* is $p = 2\Phi_T(-Z_O; df)$.

$(1 - a) \cdot 100\%$ Confidence interval for $\theta_{\bullet\bullet} = p_{2\bullet} - p_{1\bullet} = p_{test}(+) - p_{control}(+)$

$$CI(\theta_{\bullet\bullet}) = [\hat{\theta}_{\bullet\bullet} + t_{K-1}(\alpha/2)\cdot\sigma(\hat{\theta}_{\bullet\bullet}), \hat{\theta}_{\bullet\bullet} + t_{K-1}(1 - \alpha/2)\cdot\sigma(\hat{\theta}_{\bullet\bullet})].$$

Required Sample Size
The cluster size K calculation, for the one-sided non-inferiority test, depends upon the clinical margin δ, the in-between-cluster variance σ_b^2, and the within-cluster variance

σ_w^2. Here $\delta_1 > \delta$ is a margin that makes the hypotheses $H_{0\ell}: p_{10,\bullet} - p_{01,\bullet} = \delta \ (< 0)$ vs. $H_{1\ell}: p_{10,\bullet} - p_{01,\bullet} > \delta$ equivalent to $H_{0\ell}: p_{10,\bullet} - p_{01,\bullet} = \delta \ (< 0)$ vs. $H_{1\ell}: p_{10,\bullet} - p_{01,\bullet} = \delta_1 > \delta$. As a practical matter, δ_1 can be set to 0 (conservative estimate) or some fraction of δ (say, $\delta_1 = \delta/2$). See Nam [7] for details.

$$K = \left[\!\left[\left(\frac{z_{1-\alpha}\sqrt{\sigma_b^2(\delta) + \frac{1}{n_0}\sigma_w^2} + z_{1-\beta}\sqrt{\sigma_b^2(\delta_1) + \frac{1}{n_0}\sigma_w^2}}{\delta} \right)^2 \right]\!\right] \tag{6.1a}$$

$$\left. \begin{aligned} \sigma_b^2(\vartheta) &= \frac{1}{(K-1)} \sum_{k=1}^{K} (p_{10,k} - p_{01,k} - \vartheta)^2 \\ \sigma_w^2 &= \frac{1}{(K-1)} \sum_{k=1}^{K} (p_{10,k} + p_{01,k} - (p_{10,k} - p_{01,k})^2) \end{aligned} \right\} \tag{6.1b}$$

The divisor n_0 that controls the contribution of the within-cluster variance to the cluster size calculation in (6.1a) can be taken as either the median or mean of the cluster unit sizes $n_1, n_2, \ldots, n_K$. That is, $n_0 = \text{median}\{n_1, n_2, \ldots, n_K\}$ or $n_0 = \frac{1}{K}\sum_{k=1}^{K} n_k$.

In general, sample size formulae have one or more parameters that must be estimated from preliminary experiments or pilot study data. Since the sample size formula for the Obuchowski test statistic involves an entire set of clustered data, it will usually be based on a pilot study. Such a pilot study, by its nature, will be smaller than the planned study. Thus, either the *Control* or the *Test* proportion estimate may turn out to be substantially larger than the other simply due to sampling error. To obtain the most conservative (i.e., largest) sample size estimate, the *Control* for this calculation should be the method that yields the smallest discordance proportion $p_{10,k}$. Therefore, it is recommended that the in-between-cluster variance $\sigma_b^2(\vartheta)$ operates on the absolute value in-between the discordance proportions. That is, replace (6.1b) by (6.1c). This is more conservative as it leads to larger sample sizes.

$$\left. \begin{aligned} \sigma_{b'}^2(\vartheta) &= \frac{1}{(K-1)} \sum_{k=1}^{K} (|p_{10,k} - p_{01,k}| - \vartheta)^2 \\ \sigma_w^2 &= \frac{1}{(K-1)} \sum_{k=1}^{K} (p_{10,k} + p_{01,k} - (p_{10,k} - p_{01,k})^2) \end{aligned} \right\} \tag{6.1c}$$

MATLAB Functions

`clustmatchedpair.m, modobuchowski.m`

Example: Consider the data garnered from three readers as presented in the contingency tables below.

To test the hypothesis that all of the *Control* and *Test* positive proportions are within a 3% clinical margin, compare $H_0: p_{2,\bullet} - p_{1,\bullet} = -0.03$ vs. $H_1: p_{2,\bullet} - p_{1,\bullet}$

> -0.03 where $p_{2,\bullet}$ represents the *positive* proportion developed via the *Test* method and $p_{1,\bullet}$ is the *positive* proportion determined via the *Control* method. It is seen that the p-value is greater than the significance level, $p = 0.3338 > \alpha = 0.05$. Moreover, note that the confidence interval for the difference in proportions $\Delta p = p_{2,\bullet} - p_{1,\bullet} = -0.0014$ includes zero, $0 \in CI(Dp) = [-0.0988, 0.096]$. Therefore, the null hypothesis *is not rejected*. Finally, it is seen that $p_{2,\bullet} = 0.4259$ and $p_{1,\bullet} = 0.4273$.

	Reader₁				Reader₂				Reader₃	
		Control				*Control*				*Control*	
		Yes	No			Yes	No			Yes	No
Test	Yes	379	30	*Test*	Yes	348	57	*Test*	Yes	363	33
	No	36	500		No	40	509		No	48	498

MATLAB Commands
(*Clustered Data Modified Obuchowski Method*)

```
% Set the data directory ...
Ddir = 'C:\PJC\Math_Biology\Chapter_6\Data';
% ... and recover the K × 2 × 2 data cube
D = load(fullfile(Ddir,'DataCube.mat'));
% Set the significance level at 5% (α = 0.05) and a (negative) clinical margin of 3% (δ = –0.03)
alpha = 0.05; delta = -0.03;
% Test the hypothesis H₀: p_test(+) – p_control(+) = δ (< 0)
H = clustmatchedpair(D.Cube,delta,alpha)
H =
         Zo: 1.2632
         Ho: 'do not reject'
         Dp: -0.0014
         CI: [-0.0988 0.0960]
     Pvalue: 0.3338
      PTest: 0.4259
   PControl: 0.4273
```

EXERCISES

6.1 Show that $E[\theta_{\bullet\bullet} - \delta] = 0$.

   Hint: Observe that $\bar{p}_{1\bullet,\bullet} = \frac{1}{2}\left(\dfrac{x_{1\bullet,\bullet} + x_{\bullet 1,\bullet}}{n_\bullet} + \delta\right)$, $\bar{p}_{\bullet 1,\bullet} = \frac{1}{2}\left(\dfrac{x_{1\bullet,\bullet} + x_{\bullet 1,\bullet}}{n_\bullet} - \delta\right)$,
   and $\theta_{\bullet\bullet} = p_{1\bullet,\bullet} - p_{\bullet 1,\bullet}$. Then $E[\theta_{\bullet\bullet}] = \bar{p}_{1\bullet,\bullet} - \bar{p}_{\bullet 1,\bullet}$.

6.2 If $CI(\theta_{\bullet\bullet}) \subset [\delta, -\delta]$ for $\delta < 0$, what conclusions can be drawn about the null hypothesis H_0? That is, does $CI(\theta_{\bullet\bullet}) \subset [\delta, -\delta]$ imply a rejection of H_0? If so, why?

6.2   CLUSTERED DATA, ASSESSMENT METRICS, AND DIAGNOSTIC LIKELIHOOD RATIOS

In Chapter 5 (Section 5.5.4), the assessment metrics and *diagnostic likelihood ratios* (*DLRs*) are detailed. As noted previously, the traditional metrics of sensitivity and specificity can prove less than robust in the presence of changing diagnostic standards. Moreover, in certain instances, sensitivities from competing methods can be computed as statistically significantly distinct but have little clinical or practical difference. An example exhibiting this phenomenon is provided at the end of this section. Competing binary diagnostic tests evaluated by K readers can be compared via *DLRs*. In this setting, Table 6.2 provides the paradigm. The symbols T_ℓ indicate the diagnostic test under consideration ($\ell = 1, 2$), D_x is the "gold standard" diagnostic result. The integers $k = 1, 2, \ldots, K$ refer to the reader index. Each reader examines n_k patient samples.

From these definitions and Table 6.2, the usual assessment metrics and corresponding *DLRs* can be formulated as indexed by reader k.

$$\left.\begin{array}{cc} Sen_k(T_1) = \dfrac{s_{11}^k + s_{10}^k}{s_{\bullet\bullet}^k}, & Sen_k(T_2) = \dfrac{s_{11}^k + s_{01}^k}{s_{\bullet\bullet}^k} \\[3mm] Spec_k(T_1) = \dfrac{r_{01}^k + r_{00}^k}{r_{\bullet\bullet}^k}, & Spec_k(T_2) = \dfrac{r_{10}^k + r_{00}^k}{r_{\bullet\bullet}^k} \end{array}\right\} \tag{6.2}$$

$$\left.\begin{array}{l} DLR_k^+(T_\ell) = \dfrac{Sen^{(k)}(T_\ell)}{1 - Spec^{(k)}(T_\ell)} \\[3mm] DLR_k^-(T_\ell) = \dfrac{1 - Sen^{(k)}(T_\ell)}{Spec^{(k)}(T_\ell)} \end{array}\right\}, \ell = 1, 2 \tag{6.3a}$$

If the detection of disease is considered the focus of the diagnostic test, then *sensitivity* is the probability of detection (p_d) while the complement of *specificity* is the

TABLE 6.2   Multiple reader contingency for clustered data on two binary diagnostic tests

Gold standard diagnosis	$T_1 = 1$		$T_1 = 0$		Totals
	$T_2 = 1$	$T_2 = 0$	$T_2 = 1$	$T_2 = 0$	
$D_x = 1$	s_{11}^k	s_{10}^k	s_{01}^k	s_{00}^k	$s_{\bullet\bullet}^k = \sum\limits_{i=0}^{1}\sum\limits_{j=0}^{1} s_{ij}^k$
$D_x = 0$	r_{11}^k	r_{10}^k	r_{01}^k	r_{00}^k	$r_{\bullet\bullet}^k = \sum\limits_{i=0}^{1}\sum\limits_{j=0}^{1} r_{ij}^k$
Totals	n_{11}^k	n_{10}^k	n_{01}^k	n_{00}^k	$n_k = \sum\limits_{i=0}^{1}\sum\limits_{j=0}^{1} n_{ij}^k$

probability of false alarm (p_{f_a}). That is, *sensitivity* $= p_d$ and $1 -$ *specificity* $= p_{f_a}$. Hence, the *positive diagnostic likelihood* is the ratio of the probability of detecting a true positive to the probability of detecting a false positive $DLR^+ = p_d/p_{f_a} =$ $\text{Prob}(T_p)/\text{Prob}(F_p)$. In a similar manner, the *negative diagnostic likelihood* is the ratio of the probability of detecting a false negative to the probability of detecting a true negative $DLR^- = \text{Prob}(F_n)/\text{Prob}(T_n)$. Thus, $DLR^+ > 1$ implies that the diagnostic test is more likely to detect disease correctly. If $DLR^- < 1$, then the diagnostic test is more likely to *correctly* determine that the patient is healthy (namely, detect the *absence* of disease).

Notice that if the identifications $s_0^k(1) = s_{00}^k + s_{01}^k, s_0^k(2) = s_{00}^k + s_{10}^k, s_1^k(1) = s_{11}^k + s_{10}^k, s_1^k(2) = s_{11}^k + s_{01}^k, r_0^k(1) = r_{00}^k + r_{01}^k, r_0^k(2) = r_{00}^k + r_{10}^k, r_1^k(1) = r_{11}^k + r_{10}^k, r_1^k(2) = r_{11}^k + r_{01}^k$ are made, then equation (6.3b) can be written directly from Table 6.2.

$$\left.\begin{aligned}
DLR_k^+(T_\ell) &= \frac{s_1^k(\ell)}{r^{(k)} - r_0^k} \cdot \frac{r_{\cdot\cdot}^k}{s_{\cdot\cdot}^k} \\[2ex]
DLR_k^-(T_\ell) &= \frac{s^{(k)} - s_1^k(\ell)}{r_0^k(\ell)} \cdot \frac{r_{\cdot\cdot}^k}{s_{\cdot\cdot}^k}
\end{aligned}\right\}, \ell = 1, 2 \qquad (6.3b)$$

Analogous to the notation adapted prior to equation (6.3b), set $n_{ij}^k = s_{ij}^k + r_{ij}^k, n(1) = n_{1\bullet}^\bullet + n_{0\bullet}^\bullet = \sum_{k=1}^K (n_{11}^k + n_{10}^k + n_{01}^k + n_{00}^k) = \sum_{k=1}^K n_k = n_{\bullet 1}^\bullet + n_{\bullet 0}^\bullet = n(2)$. With this notation in mind, the *overall sensitivity, specificity*, and *diagnostic likelihood ratios* are defined via (6.4a) and (6.4b), (6.5a) and (6.5b).

$$Sen(T_\ell) = \frac{1}{n(\ell)} \sum_{k=1}^K n_k \, Sen_k(T_\ell) = \frac{1}{n(\ell)} \sum_{k=1}^K n_k \frac{s_1^k(\ell)}{s_{\cdot\cdot}^k} \qquad (6.4a)$$

$$Spec(T_\ell) = \frac{1}{n(\ell)} \sum_{k=1}^K n_k \, Spec_k(T_\ell) = \frac{1}{n(\ell)} \sum_{k=1}^K n_k \frac{r_1^k(\ell)}{r_{\cdot\cdot}^k} \qquad (6.4b)$$

$$DLR^+(T_\ell) = \frac{Sen(T_\ell)}{1 - Spec(T_\ell)} \qquad (6.5a)$$

$$DLR^-(T_\ell) = \frac{1 - Sen(T_\ell)}{Spec(T_\ell)} \qquad (6.5b)$$

To test for statistical equality of the overall sensitivities, the ideas of test statistic and *p*-value, as detailed in Chapter 5, must be replaced by the behavior of the respective confidence intervals. Following Zhou et al. [15], the $(1 - \alpha)\bullet 100\%$ confidence intervals for the overall sensitivity take the form

$$CI(Sen(T_\ell)) = Sen(T_\ell) + [z_{\alpha/2}, z_{1-\alpha/2}] \cdot \sqrt{\text{var}(Sen(T_\ell))}, \qquad (6.6a)$$

where z_γ is the $\gamma \bullet 100\%$ quantile of the standard normal distribution. In a similar manner, the $(1 - \alpha) \bullet 100\%$ confidence interval for the *overall specificity* is

$$CI(Spec(T_\ell)) = Spec(T_\ell) + [z_{\alpha/2}, z_{1-\alpha/2}] \cdot \sqrt{\text{var}(Spec(T_\ell))}. \qquad (6.6b)$$

The formulae for the variances of the overall sensitivity and specificity are provided in Zhou et al. [15] and reported here. The symbol $\bar{n}(\ell) = \frac{1}{K} n^\bullet_{\bullet\ell} = \frac{1}{K} \sum_{k=1}^{K} \sum_{i=0}^{1} n^k_{ij}$ is the mean cluster size.

$$\text{var}(Sen(T_\ell)) = \frac{1}{\bar{n}(\ell)^2 \cdot K \cdot (K-1)} \sum_{k=1}^{K} (n_k)^2 \cdot (Sen_k(T_\ell) - Sen(T_\ell))^2 \qquad (6.7a)$$

$$\text{var}(Spec(T_\ell)) = \frac{1}{\bar{n}(\ell)^2 \cdot K \cdot (K-1)} \sum_{k=1}^{K} (n_k)^2 \cdot (Spec_k(T_\ell) - Spec(T_\ell))^2 \qquad (6.7b)$$

To determine the variance of the difference in sensitivities, observe that

$$\text{var}(\Delta Sen) = \text{var}(Sen(T_1)) + \text{var}(Sen(T_2)) - 2 \bullet \text{cov}(Sen(T_1), Sen(T_2)) \qquad (6.8)$$

for $\Delta Sen = Sen(T_1) - Sen(T_2)$. A direct computation from (6.4a) and (6.4b) shows

$$\text{cov}(Sen(T_1), Sen(T_2)) = \frac{K}{\bar{n}(1) \cdot \bar{n}(2) \cdot (K-1)}$$
$$\bullet \sum_{k=1}^{K} n_k (Sen_k(T_1) - Sen(T_1)) n_k (Sen_k(T_2) - Sen(T_2)).$$

The tests of statistical equality for overall sensitivity and specificity can now be summarized as per the format of Chapter 5 and in keeping with the clarification provided in hypothesis process step 1.4 at the end of Section 5.1.

Hypothesis Test (Equal Overall Sensitivity)

$H_0: Sen(T_1) = Sen(T_2)$ (*Null hypothesis*)

$H_1: Sen(T_1) \neq Sen(T_2)$ (*Alternate hypothesis*)

Test Statistic

$$Z_s = \frac{\Delta Sen}{\sqrt{\text{var}(\Delta Sen)}} \equiv \frac{Sen(T_1) - Sen(T_2)}{\sqrt{\text{var}(Sen(T_1)) + \text{var}(Sen(T_2)) - 2 \cdot \text{cov}(Sen(T_1), Sen(T_2))}}$$

p-Value

$$p = 2 \bullet \Phi(-|Z_s|)$$

$\Phi(x) = $ cumulative density function for the standard normal distribution

Null Hypothesis Decision
If $0 \notin CI(\Delta Sen)$ or $p < \alpha$ (the significance level), then *reject H_0 in favor of H_1*. Otherwise, *do not reject H_0*.

$(1 - \alpha)$•100% Confidence Interval

$$CI(\Delta Sen) = \Delta Sen + [z_{\alpha/2}, z_{1-\alpha/2}] \cdot \sqrt{\text{var}(\Delta Sen)}$$

Hypothesis Test (Equal Overall Specificity)

$K_0: Spec(T_1) = Spec(T_2)$ *(Null hypothesis)*
$K_1: Spec(T_1) \neq Spec(T_2)$ *(Alternate hypothesis)*

Test Statistic

$$Z_p = \frac{\Delta Spec}{\sqrt{\text{var}(\Delta Spec)}} \equiv \frac{Spec(T_1) - Spec(T_2)}{\sqrt{\text{var}(Spec(T_1)) + \text{var}(Spec(T_2)) - 2 \cdot \text{cov}(Spec(T_1), Spec(T_2))}}$$

p-Value

$$p = 2 \cdot \Phi(-|Z_p|)$$

$\Phi(x)$ = cumulative density function for the standard normal distribution

Null Hypothesis Decision
If $0 \notin CI(\Delta Spec)$ or $p < \alpha$ (the significance level), then *reject K_0 in favor of K_1*. Otherwise, *do not reject K_0*.

$(1 - \alpha)$•100% Confidence Interval

$$CI(\Delta Spec) = \Delta Spec + [z_{\alpha/2}, z_{1-\alpha/2}] \cdot \sqrt{\text{var}(\Delta Spec)}$$

The approach to diagnostic likelihood ratios, while similar, requires an additional idea. From (6.5a) it follows that $\ln(DLR^+(T_\ell)) = \ln\left(\dfrac{Sen(T_\ell)}{1 - Spec(T_\ell)}\right) = \ln(Sen(T_\ell)) - \ln(1 - Spec(T_\ell))$. By setting $p_1 = Sen(T_\ell)$ and $p_2 = 1 - Spec(T_\ell)$, it is seen that $DLR^+(T_\ell) = p_1/p_2$ is the ratio of two independent proportions. Consequently, $\ln(DLR^+(T_\ell)) = \ln\left(\dfrac{p_1}{p_2}\right) = \ln(p_1) - \ln(p_2)$ and, by the independence of the proportions, $\text{var}(\ln(DLR^+(T_\ell))) = \text{var}(\ln(p_1)) + \text{var}(\ln(p_2))$. The *delta method* (see, e.g., Agresti [1]) for estimating the variances of nonlinear functions of a random variable notes that $\text{var}(f(x)) \approx \dfrac{1}{n}\left(\dfrac{df}{dx}(x)\right)^2 \text{var}(x)$. Since $f(x) = \ln(x)$ then $\dfrac{df}{dx}(x) = 1/x$. Thus, $\text{var}(\ln(p_j)) = \dfrac{1}{n_j}\left(\dfrac{1}{p_j}\right)^2 p_j(1 - p_j) = \dfrac{1 - p_j}{n_j p_j}, j = 1, 2$. Recall the definitions of $s_0^k(\ell)$, $s_1^k(\ell)$, $r_0^k(\ell)$ and $r_1^k(\ell)$ from the comments just above equation (6.3b) and set $s_i(\ell) = \sum_{k=1}^{K} s_i^k(\ell)$, $r_i(\ell) = \sum_{k=1}^{K} r_i^k(\ell)$ for $i = 0, 1$. Then the delta method yields the formulae

$$\text{var}(\ln(DLR^+(T_\ell))) = \frac{1}{s_1(\ell)}(1 - Sen(T_\ell)) + \frac{1}{r_1(\ell)}Spec(T_\ell) \qquad (6.9a)$$

$$\text{var}(\ln(DLR^-(T_\ell))) = \frac{1}{s_0(\ell)}Sen(T_\ell) + \frac{1}{r_0(\ell)}(1 - Spec(T_\ell)). \qquad (6.9b)$$

Again, under the assumption of normality for $\ln(DLR^+(T_\ell))$ and $\ln(DLR^-(T_\ell))$, then the $(1 - \alpha) \cdot 100\%$ confidence interval for the natural logarithm of a diagnostic likelihood ratio is

$$CI\,(\ln(DLR)) = \ln(DLR) + [z_{\alpha/2}, z_{1-\alpha/2}] \cdot \sqrt{\text{var}\,(\ln(DLR))}.$$

Now applying the monotonic increasing and continuous function "exp" to this confidence interval produces

$$CI(DLR) = DLR \cdot \exp([z_{\alpha/2}, z_{1-\alpha/2}] \cdot \sqrt{\text{var}(\ln(DLR))}).$$

Combining these results with (6.9a) and (6.9b) provides the confidence intervals for $DLR^+(T_\ell)$ and $DLR^-(T_\ell)$ as below.

$$CI(DLR^+(T_\ell)) = DLR^+(T_\ell)$$
$$\cdot \exp\left([z_{\alpha/2}, z_{1-\alpha/2}] \cdot \sqrt{\frac{1}{s_1(\ell)}(1 - Sen(T_\ell)) + \frac{1}{r_1(\ell)}Spec(T_\ell)}\right)$$
$$(6.10a)$$

$$CI(DLR^-(T_\ell)) = DLR^-(T_\ell)$$
$$\cdot \exp\left([z_{\alpha/2}, z_{1-\alpha/2}] \cdot \sqrt{\frac{1}{s_0(\ell)}Sen(T_\ell) + \frac{1}{r_0(\ell)}(1 - Spec(T_\ell))}\right)$$
$$(6.10b)$$

While formulae (6.10a) and (6.10b) provide confidence intervals for the positive and negative diagnostic likelihood ratios, it is premature to develop a hypothesis test of equal *DLR*s. Instead, consider the following example of equal *overall sensitivities* and *overall specificities* from a multilevel diagnostic state, clustered data set.

Example: A $K \times 2 \times 4$ data cube of measurements is analyzed. These data compare *Control* (T_1) and *Test* (T_2) methods against a gold standard diagnosis. In this example, a set of matched samples, prepared via *Control* and *Test* methods, were read by $K = 3$ reviewers. The reviewers could select eight distinct diagnoses ranging from indeterminate (diagnostic grade 0), normal (or no evidence of disease = grade 1), and six distinct disease categories (grades 2–7, respectively). The higher the diagnostic grade, the more serious is the disease level. These data are contained in the file `DataCube4.mat` and are illustrated below.

Gold standard diagnosis	$T_1 = 1$		$T_1 = 0$		
	$T_2 = 1$	$T_2 = 0$	$T_2 = 1$	$T_2 = 0$	$Reader_1$
$D_x = 1$	355	9	12	38	
$D_x = 0$	23	27	18	462	
Gold standard diagnosis	$T_1 = 1$		$T_1 = 0$		
	$T_2 = 1$	$T_2 = 0$	$T_2 = 1$	$T_2 = 0$	$Reader_2$
$D_x = 1$	323	27	33	34	
$D_x = 0$	23	13	24	475	
Gold standard diagnosis	$T_1 = 1$		$T_1 = 0$		
	$T_2 = 1$	$T_2 = 0$	$T_2 = 1$	$T_2 = 0$	$Reader_3$
$D_x = 1$	335	21	13	41	
$D_x = 0$	28	27	20	454	

The *overall sensitivities* and *specificities* for the *Control* and *Test* methods are compared for samples judged *positive* provided the diagnostic grade exceeds or equals level 1. Table 6.3 summarizes the results of the clustered data analysis.

TABLE 6.3   Summary of *overall sensitivity/specificity* analysis for clustered data

System	Control (T_1)	Test (T_2)
Sensitivity	0.8799	0.8872
CI(Sen)	[0.8715, 0.8884]	[0.8787, 0.8957]
$H_0: Sen(T_1) = Sen(T_2)$	Reject	
p-Value	0	
$\Delta Sen = Sen(T_1) - Sen(T_2)$	−0.00725	
$CI(\Delta Sen)$	[−0.0073, −0.0071]	
Specificity	0.9028	0.9197
CI(Spec)	[0.8955, 0.9101]	[0.9123, 0.9272]
$K_0: Spec(T_1) = Spec(T_2)$	Reject	
p-Value	0	
$\Delta Spec = Spec(T_1) - Spec(T_2)$	−0.0169	
$CI(\Delta Spec)$	[−0.0171, −0.0168]	

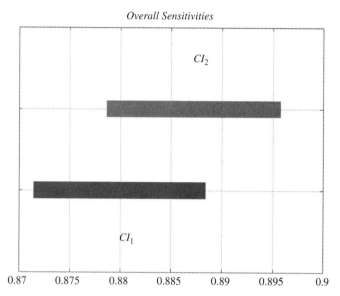

FIGURE 6.2   95% confidence intervals for the overall sensitivities.

Remark: The number of readers $K = 3$ is not an uncommon set of arbiters for review agency validation.

Observe that the *Control* sensitivity is 88% while the *Test* sensitivity is 88.7%. Even though the 95% confidence interval for each estimate overlaps (see Figure 6.2), the confidence interval for the difference in sensitivities does not contain 0 so that the estimates are statistically significantly different. But is there truly a *clinical* difference in-between the *Control* and *Test* systems? The answer depends upon which disease is being diagnosed. How deadly is the disease? How quickly must treatment be administered in order to be effective? How many people are exposed and susceptible to the disease? These questions can help public health professionals determine a clinical margin of error. For the example stated earlier with a positive diagnosis occurring at Level 1 of the disease, it is unlikely that this difference is clinically significant; especially if disease Level 1 is an indicator of a slowly progressing malady. What assessment metric can give a less sensitive measure of diagnostic performance? As will be detailed in the next section, the answer is *relative diagnostic likelihood ratios*.

EXERCISE

6.3   Write down the formula for $\mathrm{cov}(Spec(T_1), Spec(T_2))$ analogous to (6.8).

MATLAB Commands
(Overall Sensitivity and Specificity for Clustered Data)

```
% Data directory containing the K–by–2–by–4 data cube as per the table on
page 283
Ddir = 'C:\PJC\Math_Biology\Chapter_6\Data';
% The data cube
E = load(fullfile(Ddir,'DataCube4.mat'));
% Set the significance level at 5%: α = 0.05
alpha = 0.05;
% Compute the overall sensitivities, specificities, and DLRs for the Control and
Test measurements
H = dlrtest(E.Cube4,alpha);
```

```
%% Hypothesis test results
% H0: Sen(T1) = Sen(T2)
Sen
    Ho
        reject
    Pvalue
        0
    dSen
        -0.007252
    CIdSen
        [-0.007322  -0.007183]

% H0: Spec(T1) = Spec(T2)
Spec
    Ho
        reject
    Pvalue
        0
    dSpec
        -0.01693
    CIdSpec
        [-0.01706  -0.01679]
```

```
% Estimates for Sen(T1), Spec(T1)
Sen
    Value
        0.8799
    CI
        [0.8715 0.8884]
Spec
    Value
        0.9028
    CI
        [0.8955 0.9101]
% Estimates for Sen(T2), Spec(T2)
Sen
    Value
        0.8872
    CI
        [0.8787 0.8957]
Spec
    Value
        0.9197
    CI
        [0.9123 0.9272]
```

Remark: The contents of the MATLAB Commands table above returns only a *portion* of the output of `dlrtest.m`. The remaining output, concerning diagnostic likelihood ratios, is given at the end of Section 6.3.

6.3   RELATIVE DIAGNOSTIC LIKELIHOOD RATIOS

The diagnostic likelihood ratios DLR^+ and DLR^- discussed in Section 6.2 are assessment metrics used to determine how a particular diagnostic test performs. When comparing two distinct diagnostic tests (say T_1 and T_2), however, the relative diagnostic likelihood ratios $\rho^+ = rDLR^+ = DLR^+(T_1)/DLR^+(T_2)$ and $\rho^- = rDLR^- = DLR^-(T_1)/DLR^-(T_2)$ are comparative measures of diagnostic value. That is, $\rho^+ > 1$ implies T_1 is better than T_2 in diagnosing disease, while $\rho^- < 1$ means T_2 is a more accurate test for healthy (i.e., non-diseased) patients. Therefore, if ρ^+ is statistically significantly larger than 1, then T_1 is a significantly better test for the detection of disease. How can this be formalized via a hypothesis test? Certainly the ratio

$$\rho^+ = \frac{DLR^+(T_1)}{DLR^+(T_2)} = \frac{Sen(T_1)}{1 - Spec(T_1)} \cdot \frac{1 - Spec(T_2)}{Sen(T_2)}$$ does not readily conform to a known

distribution even under the assumption of normality for the independent variables $Sen(T_\ell)$ and $Spec(T_\ell)$. Instead, to the ratios ρ^+ and ρ^-, apply the natural logarithm and obtain

$$\omega^+ = \ln(\rho^+) = \ln(Sen(T_1)) + \ln(1 - Spec(T_2)) - \ln(Sen(T_2)) - \ln(1 - Spec(T_1))$$
$$\tag{6.11a}$$

$$\omega^- = \ln(\rho^-) = \ln(1 - Sen(T_1)) + \ln(Spec(T_2)) - \ln(1 - Sen(T_2)) - \ln(Spec(T_1)).$$
$$\tag{6.11b}$$

Set $X_\ell = \ln(Sen(T_\ell))$ and $Y_\ell = \ln(1 - Spec(T_\ell))$ for $\ell = 1, 2$. Then using the formula $var(\sum_{j=1}^{N} \Xi_j) = \sum_{j=1}^{n} var(\Xi_j) + 2 \sum_{j<k} cov(\Xi_j, \Xi_k)$ with the identifications $\Xi_1 = X_1, \Xi_2 = Y_2, \Xi_3 = -X_2$, and $\Xi_4 = -Y_1$ results in

$$var(\omega^+) = var(X_1) + var(X_2) + var(Y_1) + var(Y_2) + 2 cov(X_1, Y_2) - 2 cov(X_1, Y_1)$$
$$- 2 cov(X_1, X_2) - 2 cov(Y_1, Y_2) - 2 cov(X_2, Y_2) + 2 cov(X_2, Y_1).$$

Now, by assumption, the pairs (X_1, Y_1) and (X_2, Y_2) are independent random variables (i.e., X_1 and Y_1 are independent as are X_2 and Y_2). Hence, $cov(X_j, Y_j) = 0$ for $j = 1, 2$. Thus, the formula above becomes

$$var(\omega^+) = var(X_1) + var(X_2) + var(Y_1) + var(Y_2) + 2 cov(X_1, Y_2)$$
$$- 2 cov(X_1, X_2) - 2 cov(Y_1, Y_2) + 2 cov(X_2, Y_1).$$
$$\tag{6.12a}$$

Proceeding in a parallel fashion with $Z_\ell = \ln(1 - Sen(T_\ell))$ and $W_\ell = \ln(Spec(T_\ell))$ for $\ell = 1, 2$ results in

$$var(\omega^-) = var(Z_1) + var(Z_2) + var(W_1) + var(W_2) - 2 cov(Z_1, Z_2)$$
$$+ 2 cov(Z_1, W_2) + 2 cov(Z_2, W_1) - 2 cov(W_1, W_2).$$
$$\tag{6.12b}$$

Now, by the independence of X_j and Y_j, it is seen that

$$\text{var}(\ln(DLR^+(T_j))) = \text{var}(X_j - Y_j) = \text{var}(X_j) + \text{var}(Y_j) - 2\,\text{cov}(X_j, Y_j)$$
$$= \text{var}(X_j) + \text{var}(Y_j)$$

for $j = 1, 2$. But by equation (6.9a) of Section 6.2,

$$\text{var}(\ln(DLR^+(T_j))) = \frac{1}{s_1(j)}(1 - Sen(T_j)) + \frac{1}{r_1(j)}Spec(T_j)$$

so that $\text{var}(X_j) = \frac{1}{s_1(j)}(1 - Sen(T_j))$ and $\text{var}(Y_j)\,\frac{1}{r_1(j)}Spec(T_j)$. Subsequently equation (6.12a) takes the form

$$\text{var}(\omega^+) = \frac{1}{s_1(1)}(1 - Sen(T_1)) + \frac{1}{r_1(1)}Spec(T_1) + \frac{1}{s_1(2)}(1 - Sen(T_2)) + \frac{1}{r_1(2)}Spec(T_2)$$
$$+ 2\,\text{cov}(X_1, Y_2) - 2\,\text{cov}(X_1, X_2) - 2\,\text{cov}(Y_1, Y_2) + 2\,\text{cov}(X_2, Y_1). \quad (6.13a)$$

To complete the formulation of $\text{var}(\omega^+)$, it remains to calculate $\text{cov}(X_1, X_2)$, $\text{cov}(Y_1, Y_2)$, and $\text{cov}(X_j, Y_k)$ for j, $k = 1$, 2. These quantities are now computed directly.

$$\left.\begin{aligned}
\text{cov}(X_1, X_2) &= \frac{K}{(K-1)}\sum_{k=1}^{K}\frac{n_{1\bullet}^k}{n_{1\bullet}^\bullet}(X_{1,k} - \bar{X}_1)\cdot\frac{n_{\bullet1}^k}{n_{\bullet1}^\bullet}(X_{2,k} - \bar{X}_2) \\
\text{cov}(Y_1, Y_2) &= \frac{K}{(K-1)}\sum_{k=1}^{K}\frac{n_{0\bullet}^k}{n_{0\bullet}^\bullet}(Y_{1,k} - \bar{Y}_1)\cdot\frac{n_{\bullet0}^k}{n_{\bullet0}^\bullet}(Y_{2,k} - \bar{Y}_2) \\
\text{cov}(X_1, Y_2) &= \frac{K}{(K-1)}\sum_{k=1}^{K}\frac{n_{1\bullet}^k}{n_{1\bullet}^\bullet}(X_{1,k} - \bar{X}_1)\cdot\frac{n_{\bullet0}^k}{n_{\bullet0}^\bullet}(Y_{2,k} - \bar{Y}_2) \\
\text{cov}(Y_1, X_2) &= \frac{K}{(K-1)}\sum_{k=1}^{K}\frac{n_{0\bullet}^k}{n_{0\bullet}^\bullet}(Y_{1,k} - \bar{Y}_1)\cdot\frac{n_{\bullet1}^k}{n_{\bullet1}^\bullet}(X_{2,k} - \bar{X}_2)
\end{aligned}\right\} \quad (6.13b)$$

The notation $n_{\bullet j}^k = \sum_{i=0}^{1}n_{ij}^k$ and $n_{i\bullet}^k = \sum_{j=0}^{1}n_{ij}^k$ is derived from Table 6.2 and detailed following (6.3b) of Section 6.2. The remaining variables are $X_{\ell,k} = \ln(Sen_k(T_\ell))$, $\bar{X}_\ell = \frac{1}{K}\sum_{k=1}^{K}X_{\ell,k}$, $Y_{\ell,k} = \ln(1 - Spec_k(T_\ell))$, and $\bar{Y}_\ell = \frac{1}{K}\sum_{k=1}^{K}Y_{\ell,k}$.

Employ the notation $\rho^+ = \dfrac{DLR^+(T_1)}{DLR^+(T_2)}$ so that $\omega^+ = \ln(\rho^+) = \ln(DLR^+(T_1)) - \ln(DLR^+(T_2))$. Observe that the hypothesis test H_0: $DLR^+(T_1) = DLR^+(T_2)$ vs. H_1: $DLR^+(T_1) \neq DLR^+(T_2)$ is equivalent to H_0: $\rho^+ = 1$ vs. H_1: $\rho^+ \neq 1$. Then, as in the case for the overall sensitivity, the hypothesis tests of equal DLR^+s can proceed directly.

Hypothesis Test (Equal Overall Positive Diagnostic Likelihood Ratio)

H_0: $\rho^+ = 1$ (*Null hypothesis*)

H_1: $\rho^+ \neq 1$ (*Alternate hypothesis*)

Test Statistic

$$Z_+ = \frac{\omega^+}{\sqrt{\text{var}(\omega^+)}}$$

$\text{var}(\omega^+)$ defined via (6.12a)

p-Value

$$p = 2 \cdot \Phi(-|Z_+|)$$

$\Phi(x)$ = cumulative density function for the standard normal distribution

Null Hypothesis Decision

If $p < \alpha$ (the significance level) or $1 \notin CI(\rho^+)$, then *reject H_0* in favor of H_1. Otherwise, *do not reject H_0*.

$(1 - \alpha) \cdot 100\%$ Confidence Interval for $\rho^+ = DLR^+(T_1)/DLR^+(T_2)$

$$CI(\rho^+) = \left[\rho^+ \cdot \exp\left(z_{\alpha/2} \sqrt{\text{var}(\omega^+)} \right), \rho^+ \cdot \exp\left(z_{1-\alpha/2} \sqrt{\text{var}(\omega^+)} \right) \right]$$

Using a parallel notation, $\rho^- = \dfrac{DLR^-(T_1)}{DLR^-(T_2)}$ and $\omega^- = \ln(\rho^-) = \ln(DLR^-(T_1)) - \ln(DLR^-(T_2))$, the statistical equality of negative diagnostic likelihood ratios proceeds as above. The form of the variance for the difference $\text{var}(\omega^-)$ parallels (6.13a) and is left as an exercise.

Hypothesis Test (Equal Overall Negative Diagnostic Likelihood Ratio)

K_0: $\rho^- = 1$ (*Null hypothesis*)

K_1: $\rho^- \neq 1$ (*Alternate hypothesis*)

Test Statistic

$$Z_- = \frac{\omega^-}{\sqrt{\text{var}(\omega^-)}}$$

p-Value

$$p = 2 \cdot \Phi(-|Z_-|)$$

$\Phi(x)$ = cumulative density function for the standard normal distribution

Null Hypothesis Decision

If $p < \alpha$ (the significance level) or $1 \notin CI(\rho^-)$, then *reject K_0* in favor of K_1. Otherwise, *do not reject K_0*.

$(1 - \alpha) \cdot 100\%$ Confidence Interval for $\rho^- = DLR^-(T_1)/DLR^-(T_2)$

$$CI(\rho^-) = \left[\rho^- \cdot \exp\left(z_{\alpha/2} \sqrt{\operatorname{var}(\omega^-)} \right), \rho^- \cdot \exp\left(z_{1-\alpha/2} \sqrt{\operatorname{var}(\omega^-)} \right) \right]$$

Example: (*Section 6.2, Continued*):   As is described in the example from Section 6.2, the *Control* and *Test* overall sensitivities and overall specificities are statistically significantly distinct. The differences, however, are 0.7% (ΔSen) and 1.7% ($\Delta Spec$), respectively. Are these differences clinically significant? As can be seen from Table 6.4, the null hypotheses of statistically equal diagnostic likelihood ratios cannot be rejected. Equivalently, the hypotheses that the relatively diagnostic likelihood ratios are significantly different from 1 *cannot* be rejected.

Indeed for both the *Control* and *Test* methodologies, the positive likelihood ratios are significantly larger than 1; $DLR^+(T_j) \gg 1$ for $j = 1, 2$. Thus both tests are considerably more likely to be positive for patients who in fact have disease. Similarly, $DLR^-(T_j) \ll 1$ for $j = 1, 2$. Consequently, both tests will produce negative results for patients who are disease-free. Combine this with the results from the *relative diagnostic likelihood ratios* not being significantly different from 1 and the reviewer can conclude that these tests have equivalent diagnostic functionality. Table 6.4 summarizes these results.

TABLE 6.4   Summary of *sensitivity/specificity* analysis for clustered data

System	Control (T_1)	Test (T_2)
$DLR^+(T_1)$	9.06	11.06
$CI(DLR^+(T_1))$	[6.94, 11.81]	[8.23, 14.86]
$H_0: DLR^+(T_1) = DLR^+(T_2)$	Do not reject	
p-Value	0.2981	
$\rho^+ = DLR^+(T_1)/DLR^+(T_2)$	0.819	
$CI(\rho^+)$	[0.562, 1.193]	
$DLR^-(T_1)$	0.133	0.123
$CI(DLR^-(T_1))$	[0.102, 0.173]	[0.094, 0.161]
$K_0: DLR^-(T_1) = DLR^-(T_2)$	Do not reject	
p-Value	0.2706	
$\rho^- = DLR^-(T_1)/DLR^-(T_2)$	1.08	
$CI(\rho^-)$	[0.94, 1.25]	

These findings contrast the conclusions reached when comparing *sensitivity* and *specificity*. Therefore, the comparison of relative diagnostic ratios can be a better gauge of *clinical* performance than the traditional assessment metrics *sensitivity* and *specificity*.

MATLAB Commands
(*Relative Positive and Negative Diagnostic Likelihood Ratios for Clustered Data*)

```
% Data directory containing the K–by–2–by–4 data cube as per the table on
page 283
Ddir = 'C:\PJC\Math_Biology\Chapter_6\Data';
% The data cube
E = load(fullfile(Ddir,'DataCube4.mat'));
% Set the significance level at 5%: α = 0.05
alpha = 0.05;
% Compute the overall sensitivities, specificities, and DLRs for the Control and
  Test measurements
H = dlrtest(E.Cube4,alpha);
```

```
%% Hypothesis test results

% H0: DLR+(T1) = DLR+(T2)
rDLRp
    Ho
        do not reject
    Pvalue
        0.2981
    rDLRp
        0.8191
    CIrDLRp
        [0.5624 1.1929]
% H0: DLR-(T1) = DLR-(T2)
rDLRm
    Ho
        do not reject
    Pvalue
        0.2706
    rDLRm
        1.084
    CIrDLRm
        [0.9389 1.2521]
```

```
% Estimates for DLR+(T1) and DLR-(T1)
DLRplus
    Value
        9.055
    CI
        [6.942 11.81]
DLRminus
    Value
        0.133
    CI
        [0.1024 0.1727]
% Estimates for DLR+(T2) and DLR-(T2)
DLRplus
    Value
        11.06
    CI
        [8.226 14.86]
DLRminus
    Value
        0.1226
    CI
        [0.09359 0.1607]
```

EXERCISES

6.4 From equation (6.5a) of Section 6.2 and the identifications $X_\ell = \ln(Sen(T_\ell))$ and $Y_\ell = \ln(1 - Spec(T_\ell))$, show that $\ln(DLR^+(T_\ell)) = X_\ell - Y_\ell$.

6.5 From the definition $\rho^- = \dfrac{DLR^-(T_1)}{DLR^-(T_2)} = \dfrac{1 - Sen(T_1)}{Spec(T_1)} \cdot \dfrac{Spec(T_2)}{1 - Sen(T_2)}$, derive equation (6.11b).

6.6 Derive equation (6.12b). *Hint*: Follow the formulation of (6.12a) using the identifications $Z_\ell = \ln(1 - Sen(T_\ell))$, $W_\ell = \ln(Spec(T_\ell))$ for $\ell = 1, 2$ and $\Xi_1 = Z_1, \Xi_2 = -Z_2, \Xi_3 = -W_1$, and $\Xi_4 = W_2$.

6.7 Produce formulae comparable to (6.13a) for the variance of the negative relative diagnostic ratios var(ω^-).

6.4 ANALYSIS OF VARIANCE FOR CLUSTERED DATA

Before launching into the analysis of variance for clustered data, a brief review of analysis of variance (ANOVA) methods is provided. Assessment metrics such as *sensitivity*, *specificity*, or *diagnostic likelihood ratios* are used to determine whether a *new* (*Test*) method is equivalent to an established (*Control*) method. These metrics, however, do not measure the *variation* in method application. Generally, to determine how a particular variable affects the variation in a procedure, a (linear) model is established.

For example, consider the following table of data in which two competing tests are interpreted by a single reader (i.e., a diagnostician) over a set of specimens. The elements $Y_{i,j}$ of Table 6.5 represent a measured signal of sample i via the [sole] reader using test j.

These measurements can be represented by a *two-way model*

$$Y_{ik} = \mu + \tau_i + C_k + (\tau C)_{ik} + \varepsilon_{ik}, \tag{6.14}$$

where μ is the overall mean value of the measurements, τ_i is the effect of the ith test, C_k is the effect of the kth specimen, $(\tau C)_{ik}$ is the interaction of test i and specimen k,

TABLE 6.5 Diagnostic classifications over two diagnostic tests for two-way model

Specimen	$Test_1$ Reader	$Test_2$ Reader
C_1	$Y_{1,1}$	$Y_{1,2}$
C_2	$Y_{2,1}$	$Y_{2,2}$
$\vdots$	$\vdots$	$\vdots$
C_K	$Y_{K,1}$	$Y_{K,2}$

TABLE 6.6   ANOVA for the two-way layout

Factor	Sum of squares	df	MS	F-statistic
Test	$SS_\tau = K \cdot \sum\limits_{i=1}^{I} (\bar{Y}_{i\bullet} - \bar{Y}_{\bullet\bullet})^2$	$df_\tau = I - 1$	$MS_\tau = \dfrac{SS_\tau}{df_\tau}$	$F_\tau = \dfrac{MS_\tau}{MS_W}$
Specimen	$SS_C = I \cdot \sum\limits_{k=1}^{K} (\bar{Y}_{\bullet k} - \bar{Y}_{\bullet\bullet})^2$	$df_C = K - 1$	$MS_C = \dfrac{SS_C}{df_C}$	$F_C = \dfrac{MS_C}{MS_W}$
Residual	$SS_W = \sum\limits_{i=1}^{I} \sum\limits_{k=1}^{K} (Y_{ik} - \bar{Y}_{i\bullet} - \bar{Y}_{\bullet k} + \bar{Y}_{\bullet\bullet})^2$	$df_W = (I-1) \cdot (K-1)$	$MS_W = \dfrac{SS_W}{df_W}$	
Total	$SS_T = \sum\limits_{i=1}^{I} \sum\limits_{k=1}^{K} (Y_{ik} - \bar{Y}_{\bullet\bullet})^2$	$df_T = IK - 1$		

and ε_{ik} is the random error within the ith classification for the k observations. In this setting, K is the number of specimens and I is the number of tests so that $i = 1, 2,..., I$ and $k = 1, 2,..., K$. It is assumed that the tests τ_i and specimens C_k are normally distributed and that the errors ε_{ik} are *i.i.d.* $\mathcal{N}(0, 1)$.

How can the data contained in Table 6.5 be combined with the two-way model to determine the variation induced by each effect (in this case, *test* and *specimen*)? The answer is the *analysis of variance*. To establish the mathematical formulation of ANOVA, the following notation is employed. Borrowing the "•" subscript from Sections 6.2 and 6.3, let a "dot – bar" combination indicate a mean over the indicated index. That is,

$$\bar{Y}_{i\bullet} = \frac{1}{K} \sum_{k=1}^{K} Y_{ik}, \quad \bar{Y}_{\bullet k} = \frac{1}{I} \sum_{i=1}^{I} Y_{ik}, \quad \bar{Y}_{\bullet\bullet} = \frac{1}{I \cdot K} \sum_{i=1}^{I} \sum_{k=1}^{K} Y_{ik}. \qquad (6.15)$$

Using these definitions and the ideas of error sums of squares (SS), degrees of freedom (df), and mean squared error (MS), Table 6.6 provides the foundations of ANOVA for the two-way model (6.14).

To test the hypotheses of *equality among all tests* (that is, the choice of test τ_i has no effect on the measurement)

$H_0: \tau_1 = \tau_2 = \cdots = \tau_I$ *(Null hypothesis)*

$H_1: \tau_i \neq \tau_\ell$ for some $i \neq \ell$ *(Alternate hypothesis)*

it is sufficient to determine whether the p-value $p_\tau = 1 - \Phi_F(F_\tau; df_\tau, df_W)$ is less than the significance level. That is, $p_\tau < \alpha \Rightarrow$ reject H_0. Here, $\Phi_F(x; df_1, df_2)$ is the F_{df}-cumulative density function (see equations (5.13) and (5.14) of Chapter 5) with $df = (df_1, df_2)$ degrees of freedom. Observe that from Table 6.6, $df_\tau = I - 1$ while $df_W = (I - 1) \bullet (K - 1)$.

TABLE 6.7 Diagnostic classifications over two diagnostic tests for three-way model

Specimen	$Test_1$				$Test_2$			
	$Reader_1$	$Reader_2$	...	$Reader_J$	$Reader_1$	$Reader_2$	...	$Reader_J$
C_1	$Y_{1,1,1}$	$Y_{1,2,1}$	...	$Y_{1,J,1}$	$Y_{2,1,1}$	$Y_{2,2,1}$	...	$Y_{2,J,1}$
C_2	$Y_{1,1,2}$	$Y_{1,2,2}$	...	$Y_{1,J,2}$	$Y_{2,1,2}$	$Y_{2,2,2}$	...	$Y_{2,J,2}$
$\vdots$	$\vdots$	$\vdots$	$\ddots$	$\vdots$	$\vdots$	$\vdots$	$\ddots$	$\vdots$
C_K	$Y_{1,1,K}$	$Y_{1,2,K}$	...	$Y_{1,J,K}$	$Y_{2,1,K}$	$Y_{2,2,K}$	...	$Y_{2,J,K}$

Similarly, the hypothesis test of *equality among all specimens* (i.e., the choice of specimen C_k has no effect on the measurement)

$K_0: C_1 = C_2 = \cdots = C_K$ *(Null hypothesis)*

$K_1: C_k \neq C_\ell$ for some $k \neq \ell$ *(Alternate hypothesis)*

is determined via the p-value $p_C = 1 - \Phi_F(F_C; df_C, df_W)$ with $df_C = K - 1$ and $df_W = (I - 1) \bullet (K - 1)$.

How can this model be adapted for clustered (i.e., multi-reader) data? Table 6.7 provides the formal mechanism. Here $Y_{i,j,k}$ can be the measured signal or diagnostic classification of the kth sample by the jth reader using the ith test.

The full three-way model associated with Table 6.7 is

$$Y_{ijk} = \mu + \tau_i + R_j + C_k + (\tau R)_{ij} + (RC)_{jk} + (\tau C)_{ik} + (\tau RC)_{ijk} + \varepsilon_{ijk}. \quad (6.16)$$

In parallel to the two-way model, μ is the overall mean value, τ_i is the effect of the ith test, R_j is the effect of the jth reader, C_k is the effect of the kth specimen, $(\tau R)_{ij}$ is the interaction of test i and reader j, $(RC)_{jk}$ is the interaction of reader j and specimen k, $(\tau C)_{ik}$ is the interaction of test i and specimen k, and ε_{ijk} is the random error within the ith measurement by the jth reader for the k observations; that is, ε_{ijk} are i.i.d. $\mathcal{N}(0, 1)$. As stated earlier, K is the number of specimens, I is the number of tests, and J is the number of readers. Before establishing the three-way model ANOVA table, a comment about employing the information in Table 6.7 to render decisions about diagnostic assessment metrics is offered. That is, how can the variations in *sensitivity*, *specificity*, or *diagnostic likelihood ratio* be assessed from Table 6.7? The answer is by producing estimates of the *pseudo-value* of an assessment metric via the *QT-jackknife method* of Quenoille [11] and Tukey [14]. For the remainder of this section, let θ represent an assessment metric (e.g., *positive diagnostic likelihood ratio*). Then the following definitions are required.

$\widehat{\theta}_{ij}$ = the estimate of the assessment metric based on *all* of the specimen data for *test i* and *reader j*.

$\widehat{\theta}_{ij[k]}$ = the estimate of the assessment metric based on *all* data except the kth specimen for *test i* and *reader j*. This is the QT-jackknife estimate of the kth value of the assessment metric.

The *pseudo-value* of the assessment metric θ with respect to test i, reader j, and specimen k is

$$Y_{ijk} = K \cdot \widehat{\theta}_{ij} - (K-1) \cdot \widehat{\theta}_{ij[k]}. \tag{6.17}$$

Expanding the "dot – bar" definitions from the two-way model yields

$$\left. \begin{aligned} \bar{Y}_{ij\bullet} &= \frac{1}{K} \sum_{k=1}^{K} Y_{ijk} = \widehat{\theta}_{ij} - \left(\frac{K-1}{K}\right) \sum_{k=1}^{K} \widehat{\theta}_{ij[k]} \\ \bar{Y}_{i\bullet k} &= \frac{1}{J} \sum_{j=1}^{J} Y_{ijk} \\ \bar{Y}_{\bullet jk} &= \frac{1}{I} \sum_{i=1}^{I} Y_{ijk} \end{aligned} \right\} \tag{6.18a}$$

$$\left. \begin{aligned} \bar{Y}_{i\bullet\bullet} &= \frac{1}{J} \sum_{j=1}^{J} \bar{Y}_{ij\bullet} = \frac{1}{J\cdot K} \sum_{j=1}^{J}\sum_{k=1}^{K} Y_{ijk} \\ \bar{Y}_{\bullet j\bullet} &= \frac{1}{I} \sum_{i=1}^{I} \bar{Y}_{ij\bullet} = \frac{1}{I\cdot K} \sum_{i=1}^{I}\sum_{k=1}^{K} Y_{ijk} \\ \bar{Y}_{\bullet\bullet k} &= \frac{1}{J} \sum_{j=1}^{J} \bar{Y}_{\bullet jk} = \frac{1}{I\cdot J} \sum_{j=1}^{J}\sum_{i=1}^{I} Y_{ijk} \end{aligned} \right\} \tag{6.18b}$$

$$\bar{Y}_{\bullet\bullet\bullet} = \frac{1}{I} \sum_{i=1}^{I} \bar{Y}_{i\bullet\bullet} = \frac{1}{I\cdot J\cdot K} \sum_{i=1}^{I}\sum_{j=1}^{J}\sum_{k=1}^{K} Y_{ijk}. \tag{6.18c}$$

From formulae (6.18a)–(6.18c), the ANOVA tables for the full three-way model are provided in Tables 6.8a–6.8c.

There are two basic hypothesis tests associated with the full three-way model: Equality with factors and uncorrelated factors. These are expressed as follows.

Equality Within Factors

$H_0(\phi)$: $\phi_1 = \phi_2 = \cdots = \phi_N$ (*Null hypothesis*)

$H_1(\phi)$: $\phi_i \neq \phi_\ell$ for some $i \neq \ell$ (*Alternate hypothesis*)

Here ϕ is one of the factors such as *test* (τ), *reader* (R), or *specimen* (C), and N is the number of recurrences of the factor in the model. As indicated in Table 6.8c, a p-value less than the prescribed significance level indicates a rejection of the null hypothesis $H_0(\phi)$ in favor of the alternate hypothesis $H_1(\phi)$. For example, $\phi = R \Rightarrow N = J$ and $p_R < \alpha \Rightarrow$ reject $H_0(R)$ in favor of $H_1(R)$.

TABLE 6.8a   Sum of squares/residual squared errors and degrees of freedom for full three-way model

Factor	Sum of squares/residual squared errors	Degrees of freedom
Test (τ)	$SS_\tau = J \cdot K \sum_{i=1}^{I} (\bar{Y}_{i\bullet\bullet} - \bar{Y}_{\bullet\bullet\bullet})^2$	$df_\tau = I - 1$
Reader (R)	$SS_R = I \cdot K \sum_{j=1}^{J} (\bar{Y}_{\bullet j\bullet} - \bar{Y}_{\bullet\bullet\bullet})^2$	$df_R = J - 1$
Specimen (C)	$SS_C = I \cdot J \sum_{k=1}^{K} (\bar{Y}_{\bullet\bullet k} - \bar{Y}_{\bullet\bullet\bullet})^2$	$df_C = K - 1$
(τ,R)	$SS_{(\tau,R)} = K \sum_{i=1}^{I} \sum_{j=1}^{J} (\bar{Y}_{ij\bullet} - \bar{Y}_{i\bullet\bullet} - \bar{Y}_{\bullet j\bullet} + \bar{Y}_{\bullet\bullet\bullet})^2$	$df_{(\tau,R)} = (I-1)\bullet(J-1)$
(τ,C)	$SS_{(\tau,C)} = J \sum_{i=1}^{I} \sum_{k=1}^{K} (\bar{Y}_{i\bullet k} - \bar{Y}_{i\bullet\bullet} - \bar{Y}_{\bullet\bullet k} + \bar{Y}_{\bullet\bullet\bullet})^2$	$df_{(\tau,C)} = (I-1)\bullet(K-1)$
(R,C)	$SS_{(R,C)} = I \sum_{j=1}^{J} \sum_{k=1}^{K} (\bar{Y}_{\bullet jk} - \bar{Y}_{\bullet j\bullet} - \bar{Y}_{\bullet\bullet k} + \bar{Y}_{\bullet\bullet\bullet})^2$	$df_{(R,C)} = (J-1)\bullet(K-1)$
(τ,R,C)	$SS_{(\tau,R,C)} = \sum_{i=1}^{I} \sum_{j=1}^{J} \sum_{k=1}^{K} (Y_{ijk} - \bar{Y}_{ij\bullet} - \bar{Y}_{\bullet jk} - \bar{Y}_{i\bullet k} + 2\bar{Y}_{\bullet\bullet\bullet})^2$	$df_{(\tau,R,C)} = (I-1)\bullet(J-1)\bullet(K-1)$
Residual error (τ,R)	$RS_{(\tau,R)} = \sum_{i=1}^{I} \sum_{j=1}^{J} \sum_{k=1}^{K} (Y_{ijk} - \bar{Y}_{ij\bullet})^2$	$df_R(\tau,R) = I\bullet J\bullet(K-1)$
Residual error (τ,C)	$RS_{(\tau,C)} = \sum_{i=1}^{I} \sum_{j=1}^{J} \sum_{k=1}^{K} (Y_{ijk} - \bar{Y}_{i\bullet k})^2$	$df_R(\tau,C) = I\bullet K\bullet(J-1)$
Residual error (R,C)	$RS_{(R,C)} = \sum_{i=1}^{I} \sum_{j=1}^{J} \sum_{k=1}^{K} (Y_{ijk} - \bar{Y}_{\bullet jk})^2$	$df_R(R,C) = J\bullet K\bullet(I-1)$
Mean error	$SS(\bar{Y}) = I \cdot J \cdot K \cdot \bar{Y}_{\bullet\bullet\bullet}$	$df_{\bar{Y}} = 1$
Total error	$SS(Total) = \sum_{i=1}^{I} \sum_{j=1}^{J} \sum_{k=1}^{K} (Y_{ijk} - \bar{Y}_{\bullet\bullet\bullet})^2$	$df_{Total} = I\bullet J\bullet K$

TABLE 6.8b   Mean squared errors for full three-way model

Factor	Mean squares	F-statistic
Test (τ)	$MS_\tau = SS_\tau/df_\tau$	$F_\tau = \dfrac{MS_\tau}{MS_{(\tau,R,C)}} \sim F_{df_\tau, df_{(\tau,R,C)}}$
Reader (R)	$MS_R = SS_R/df_R$	$F_R = \dfrac{MS_R}{MS_{(\tau,R,C)}} \sim F_{df_R, df_{(\tau,R,C)}}$
Specimen (C)	$MS(C) = SS_C/df_C$	$F_C = \dfrac{MS_C}{MS_{(\tau,R,C)}} \sim F_{df_C, df_{(\tau,R,C)}}$
(τ,R)	$MS_{(\tau,R)} = SS_{(\tau,R)}/df_{(\tau,R)}$	$F_{\tau,R} = \dfrac{MS_{(\tau,R)}}{MS_{(\tau,R,C)}} \sim F_{df_{(\tau,R)}, df_{(\tau,R,C)}}$
(τ,C)	$MS_{(\tau,C)} = SS_{(\tau,C)}/df_{(\tau,C)}$	$F_{\tau,C} = \dfrac{MS_{(\tau,C)}}{MS_{(\tau,R,C)}} \sim F_{df_{(\tau,C)}, df_{(\tau,R,C)}}$
(R,C)	$MS_{(R,C)} = SS_{(R,C)}/df_{(R,C)}$	$F_{R,C} = \dfrac{MS_{(R,C)}}{MS_{(\tau,R,C)}} \sim F_{df_{(R,C)}, df_{(\tau,R,C)}}$
(τ,R,C)	$MS_{(\tau,R,C)} = SS_{(\tau,R,C)}/df_{(\tau,R,C)}$	
Residual error (τ,R)	$MR_{(\tau,R)} = RS_{(\tau,R)}/df_R(\tau,R)$	$\hat{F}_{\tau,R} = \dfrac{MS_{(\tau,R)}}{MR_{(\tau,R)}} \sim F_{df_{(\tau,R)}, df_R(\tau,R)}$
Residual error (τ,C)	$MR_{(\tau,C)} = RS_{(\tau,C)}/df_R(\tau,C)$	$\hat{F}_{\tau,C} = \dfrac{MS_{(\tau,C)}}{MR_{(\tau,C)}} \sim F_{df_{(\tau,C)}, df_R(\tau,C)}$
Residual error (R,C)	$MR_{(R,C)} = RS_{(R,C)}/df_R(R,C)$	$\hat{F}_{R,C} = \dfrac{MS_{(R,C)}}{MR_{(R,C)}} \sim F_{df_{(R,C)}, df_R(R,C)}$
Mean error	$MS(\bar{Y}) = SS(\bar{Y})$	
Total error	$MS(Total) = SS(Total)/df_{Total}$	

TABLE 6.8c   p-Values and hypothesis test decision criterion

Source	Test statistic	p-Value	Hypothesis	Decision
Test	F_τ	$p_\tau = 1 - \Phi_F(F_\tau; df_\tau, df_{(\tau,R,C)})$	$H_0(\tau)$	$p_\tau < \alpha \Rightarrow$ reject $H_0(\tau)$
Reader	F_R	$p_R = 1 - \Phi_F(F_R; df_R, df_{(\tau,R,C)})$	$H_0(R)$	$p_R < \alpha \Rightarrow$ reject $H_0(R)$
Specimen	F_C	$p_C = 1 - \Phi_F(F_C; df_C, df_{(\tau,R,C)})$	$H_0(C)$	$p_C < \alpha \Rightarrow$ reject $H_0(C)$
(τ,R)	$F_{\tau,R}$	$p_{\tau,R} = 1 - \Phi_F(F_{\tau,R}; df_{(\tau,R)}, df_{(\tau,R,C)})$	$K_0(\tau,R)$	$p_{\tau,R} < \alpha \Rightarrow$ reject $K_0(\tau,R)$
(τ,C)	$F_{\tau,C}$	$p_{\tau,C} = 1 - \Phi_F(F_{\tau,C}; df_{(\tau,C)}, df_{(\tau,R,C)})$	$K_0(\tau,C)$	$p_{\tau,C} < \alpha \Rightarrow$ reject $K_0(\tau,C)$
(R,C)	$F_{R,C}$	$p_{R,C} = 1 - \Phi_F(F_{R,C}; df_{(R,C)}, df_{(\tau,R,C)})$	$K_0(R,C)$	$p_{R,C} < \alpha \Rightarrow$ reject $K_0(R,C)$
$\Phi_F(x; df_1, df_2)$ is the *cdf* of the F-distribution with df_1 and df_2 degrees of freedom				

Uncorrelated Factors (No Interaction)

$$K_0(\phi_1, \phi_2): \sigma_{\phi_1,\phi_2} = 0 \hspace{3cm} \textit{(Null hypothesis)}$$
$$K_1(\phi_1, \phi_2): \sigma_{\phi_1,\phi_2} \neq 0 \hspace{3cm} \textit{(Alternate hypothesis)}$$

In this case, σ_{ϕ_1,ϕ_2} is the correlation in-between the two factors ϕ_1 and ϕ_2. Since the general linear model assumes each factor is normally distributed, then a zero correlation implies that the factors are independent. Again, the usual p-value criterion dictates the decision with respect to the null hypothesis $K_0(\phi_1, \phi_2)$. As an example from Table 6.8c, $p_{R,C} < \alpha$ implies that $K_0(R,C)$ is *rejected* so that the factors R and C are not independent.

The order in which these hypotheses are tested is a function of the test statistic. Satterthwaite [13] suggests a form of the F-statistic (F^*) and degrees of freedom as per equations (6.19a)–(6.19c). If these test statistics are utilized, then Satterthwaite recommends, for example, that $K_0(\tau,R)$ vs. $K_1(\tau,R)$ should be tested prior to $H_0(\tau)$ vs. $H_1(\tau)$.

$$
\left. \begin{aligned}
F^*_\tau &= \frac{MS_\tau}{MS_{(\tau,R)} + MS_{(\tau,C)} + MS_{(\tau,R,C)}} \sim F_{df_1,df_2} \\[2mm]
df_1 &= I-1, \ df_2 = \left[\frac{[MS_{(\tau,R)} + MS_{(\tau,C)} - MS_{(\tau,R,C)}]^2}{\frac{MS^2_{(\tau,R)}}{(I-1)(J-1)} + \frac{MS^2_{(\tau,C)}}{(I-1)(K-1)} + \frac{MS^2_{(\tau,R,C)}}{(I-1)(J-1)(K-1)}} \right]
\end{aligned} \right\} \quad (6.19a)
$$

$$
\left. \begin{aligned}
F^*_R &= \frac{MS_R}{MS_{(\tau,R)} + MS_{(R,C)} + MS_{(\tau,R,C)}} \sim F_{df_1,df_2} \\[2mm]
df_1 &= J-1, \ df_2 = \left[\frac{[MS_{(\tau,R)} + MS_{(R,C)} - MS_{(\tau,R,C)}]^2}{\frac{MS^2_{(\tau,R)}}{(I-1)(J-1)} + \frac{MS^2_{(R,C)}}{(J-1)(K-1)} + \frac{MS^2_{(\tau,R,C)}}{(I-1)(J-1)(K-1)}} \right]
\end{aligned} \right\} \quad (6.19b)
$$

$$
\left. \begin{aligned}
F^*_C &= \frac{MS_C}{MS_{(\tau,C)} + MS_{(R,C)} + MS_{(\tau,R,C)}} \sim F_{df_1,df_2} \\[2mm]
df_1 &= K-1, \ df_2 = \left[\frac{[MS_{(\tau,C)} + MS_{(R,C)} - MS_{(\tau,R,C)}]^2}{\frac{MS^2_{(\tau,C)}}{(I-1)(K-1)} + \frac{MS^2_{(R,C)}}{(J-1)(K-1)} + \frac{MS^2_{(\tau,R,C)}}{(I-1)(J-1)(K-1)}} \right]
\end{aligned} \right\} \quad (6.19c)
$$

Which test statistic then should be used to exercise the hypothesis tests $H_0(\phi)$ vs. $H_1(\phi)$ and $K_0(\phi,\psi)$ vs. $K_1(\phi,\psi)$? Hillis et al. [4] recommend the following strategy. If there is no evidence of (*test, reader*) or (*test, specimen*) independence; that is, if $K_0(\tau,R)$ and $K_0(\tau,C)$ are both rejected, then use the Satterthwaite F-statistic F^*_τ to test $H_0(\tau)$.

The details of this strategy are listed by factor.

Test τ

If $K_0(\tau,R)$: $\sigma_{\tau R} = 0$ and $K_0(\tau,C)$: $\sigma_{\tau C} = 0$ are both rejected, then $H_0(\tau)$: $\tau_1 = \tau_2 = \cdots = \tau_I$ has Satterthwaite test statistic F^*_{τ} and degrees of freedom $df = (df_1, df_2)$ as defined via (6.19a). The null decision criterion is $p^*_{\tau} \equiv 1 - \Phi_F(F^*_{\tau}; df_1, df_2) < \alpha$, implies reject $H_0(\tau)$.

Reader R

If $K_0(\tau,R)$: $\sigma_{\tau R} = 0$ and $K_0(R,C)$: $\sigma_{RC} = 0$ are both rejected, then $H_0(R)$: $R_1 = R_2 = \cdots = R_J$ has Satterthwaite test statistic F^*_{R} and degrees of freedom $df = (df_1, df_2)$ as defined via (6.19b). The null decision criterion is $p^*_{R} \equiv 1 - \Phi_F(F^*_{R}; df_1, df_2) < \alpha$, implies reject $H_0(R)$.

Specimen C

If $K_0(\tau,C)$: $\sigma_{\tau C} = 0$ and $K_0(R,C)$: $\sigma_{RC} = 0$ are both rejected, then $H_0(C)$: $C_1 = C_2 = \cdots = C_K$ has Satterthwaite test statistic F^*_{C} and degrees of freedom $df = (df_1, df_2)$ as defined via (6.19c). The null decision criterion is $p^*_{C} \equiv 1 - \Phi_F(F^*_{C}; df_1, df_2) < \alpha$, implies reject $H_0(C)$.

Dorfman et al. [2], however, present a different test statistic to assess the *equality* H_0 vs. H_1 and *correlation* K_0 vs. K_1 hypotheses listed earlier. In particular, the *DBM method* employs an F-statistic based on the signs of the paired factor covariances

$$\left.\begin{array}{l} \hat{\sigma}_{\tau R} = \dfrac{1}{K}(MS_{(\tau,R)} - MS_{(\tau,R,C)}) \\[2mm] \hat{\sigma}_{\tau C} = \dfrac{1}{J}(MS_{(\tau,C)} - MS_{(\tau,R,C)}) \\[2mm] \hat{\sigma}_{RC} = \dfrac{1}{I}(MS_{(R,C)} - MS_{(\tau,R,C)}) \end{array}\right\}. \tag{6.20}$$

The corresponding *DBM* test statistics and degrees of freedom for each of the factors, test (τ), reader (R), and specimen (C), are as below.

$$\left.\begin{array}{l} F^{DBM}_{\tau} = \dfrac{MS_{\tau}}{MS_{(\tau,R)} + \max\{MS_{(\tau,C)} - MS_{(\tau,R,C)}, 0\}} \\[3mm] df_1 = I - 1 \\[3mm] df_2 = \begin{cases} \text{Satterthwaite formula for } df_2 \text{ as per (6.19a)} & \text{for } \hat{\sigma}_{\tau C} > 0 \\ (I-1)\cdot(J-1) & \text{for } \hat{\sigma}_{\tau C} \leq 0 \end{cases} \end{array}\right\} \tag{6.21a}$$

$$\left.\begin{array}{l} F^{DBM}_{R} = \dfrac{MS_{R}}{MS_{(R,C)} + \max\{MS_{(\tau,R)} - MS_{(\tau,R,C)}, 0\}} \\[3mm] df_1 = J - 1 \\[3mm] df_2 = \begin{cases} \text{Satterthwaite formula for } df_2 \text{ as per (6.19b)} & \text{for } \hat{\sigma}_{\tau R} > 0 \\ (J-1)\cdot(K-1) & \text{for } \hat{\sigma}_{\tau R} \leq 0 \end{cases} \end{array}\right\} \tag{6.21b}$$

$$
\left.
\begin{aligned}
F_C^{DBM} &= \frac{MS_C}{MS_{(\tau,C)} + \max\{MS_{(R,C)} - MS_{(\tau,R,C)}, 0\}} \\
df_1 &= K - 1 \\
df_2 &= \begin{cases} \text{Satterthwaite formula for } df_2 \text{ as per (6.19c)} & \text{for } \hat{\sigma}_{RC} > 0 \\ (I-1)\cdot(K-1) & \text{for } \hat{\sigma}_{RC} \le 0 \end{cases}
\end{aligned}
\right\}
\quad (6.21c)
$$

Let ϕ represent a factor with respect to the hypothesis test $H_0(\phi)$: $\phi_1 = \phi_2 = \cdots$ $= \phi_N$ vs. $H_1(\phi)$: $\phi_i \ne \phi_\ell$ for some $i \ne \ell$. The p-value corresponding to the DBM method for this hypothesis test is $p_\phi^{DBM} = 1 - \Phi_F(F_\phi^{DBM}; df_1(\phi), df_2(\phi))$ with respect to the factors $\phi = \tau, R, C$. The correlation hypothesis test $K_0(\phi_1, \phi_2)$: $\sigma_{\phi_1,\phi_2} = 0$ vs. $K_1(\phi_1, \phi_2)$: $\sigma_{\phi_1,\phi_2} \ne 0$ is exercised via the standard method identified in Table 6.8c. Table 6.9 summarizes the factor equality hypothesis test decision criteria.

Finally, Hillis et al. [4] use an alternate form for the denominator degree of freedom $df_2(\phi)$ for the DBM method. This is done to correct the "conservative" nature of the DBM method [2]. The alternate form for the second degree of freedom associated with the factor ϕ is provided via (6.22a)–(6.22c) and identified by the symbol $ddf_H(\phi)$. By combining F_ϕ^{DBM} with $ddf_H(\phi)$, the new simplification method or NSM is composed. The p-value is defined via $p_\phi^{NSM} = 1 - \Phi_F(F_\phi^{DBM}; df_1(\phi), ddf_H(\phi))$ and null hypothesis decision criteria for the equal factor hypothesis test is summarized in Table 6.10.

$$
ddf_H(\tau) = (I-1)\bullet(J-1) \cdot \left[\!\!\left[\frac{(MS_{(\tau,R)} + \max\{MS_{(\tau,C)} - MS_{(\tau,R,C)}, 0\})^2}{MS_{(\tau,R)}^2} \right]\!\!\right]
$$
$$(6.22a)$$

$$
ddf_H(R) = (J-1) \cdot (K-1) \cdot \left[\!\!\left[\frac{(MS_{(R,C)} + \max\{MS_{(\tau,R)} - MS_{(\tau,R,C)}, 0\})^2}{MS_{(R,C)}^2} \right]\!\!\right]
$$
$$(6.22b)$$

$$
ddf_H(C) = (I-1) \cdot (K-1) \cdot \left[\!\!\left[\frac{(MS_{(\tau,C)} + \max\{MS_{(R,C)} - MS_{(\tau,R,C)}, 0\})^2}{MS_{(\tau,C)}^2} \right]\!\!\right]
$$
$$(6.22c)$$

Some examples are provided in the next section.

TABLE 6.9 Hypothesis decision criteria for the DBM method, $\phi = \tau, R, C$

Factor	Test statistic	$df_1(\phi)$	$df_2(\phi)$	p-value	Decision
ϕ	F_ϕ^{DBM}	$N_\phi - 1$	(6.21a)–(6.21c)	p_ϕ^{DBM}	$p_\phi^{DBM} < a \Rightarrow$ reject $H_0(\phi)$

TABLE 6.10 Hypothesis decision criteria for the NSM, $\phi = \tau, R, C$

Factor	Test statistic	$df_1(\phi)$	$df_2(\phi)$	p-value	Decision
ϕ	F_ϕ^{DBM}	$N_\phi - 1$	$ddf_H(\phi)$	p_ϕ^{NSM}	$p_\phi^{NSM} < a \Rightarrow$ reject $H_0(\phi)$

TABLE 6.11   ANOVA for the reduced three-way layout

Factor	Sum of Squares	df	MS	F-statistic
Test	$SS_\tau = (JK) \cdot \sum_{i=1}^{I} (\bar{Y}_{i\cdot\cdot} - \bar{Y}_{\cdots})^2$	$I-1$	$MS_\tau = \dfrac{SS_\tau}{df_\tau}$	$F_\tau = \dfrac{MS_\tau}{MS_{(\tau,R)}}$
Reader	$SS_R = (IK) \cdot \sum_{i=1}^{I} (\bar{Y}_{\cdot j\cdot} - \bar{Y}_{\cdots})^2$	$J-1$	$MS_R = \dfrac{SS_R}{df_R}$	$F_R = \dfrac{MS_R}{MS_{(\tau,R)}}$
Specimen	$SS_C = (IJ) \cdot \sum_{k=1}^{K} (\bar{Y}_{\cdot\cdot k} - \bar{Y}_{\cdots})^2$	$K-1$	$MS_C = \dfrac{SS_C}{df_C}$	$F_C = \dfrac{MS_C}{MS_{(\tau,R)}}$
Residual	$SS_{(\tau,R)} = K \sum_{i=1}^{I} \sum_{j=1}^{J} (\bar{Y}_{ij\cdot} - \bar{Y}_{i\cdot\cdot} - \bar{Y}_{\cdot j\cdot} + \bar{Y}_{\cdots})^2$	$(I-1) \cdot$ $(J-1)$	$MS_{(\tau,R)} = \dfrac{SS_{(\tau,R)}}{df_{(\tau,R)}}$	
Total	$SS_T = \sum_{i=1}^{I} \sum_{k=1}^{K} (Y_{ijk} - \bar{Y}_{\cdots})^2$	$IJK-1$		

EXERCISES

6.8 Suppose that the full three-way model of equation (6.16) has *a priori* information that the test (τ_i) and reader (R_j) factors are independent of the specimens
(C_k). Then one obtains the reduced three-way model $Y_{ijk} = \mu + \tau_i + R_j + C_k + (\tau R)_{ij} + \varepsilon_{ijk}$. Using the notation developed in Section 6.4, show that the ANOVA
table for the reduced three-way model is Table 6.11.

6.9 Using the pseudo-values of Y_{ijk} as defined in (6.17), show that $\bar{Y}_{ij\bullet} = \hat{\theta}_{ij} - \left(\frac{K-1}{K}\right) \sum_{k=1}^{K} \hat{\theta}_{ij[k]}$. That is, verify the first formula in (6.18a).

6.5   EXAMPLES FOR ANOVA

Next, the theory developed in Section 6.4 is put to use for two distinct assessment
metrics. The data presented contain diagnostic information for 1000 specimens as
graded by three readers using two different systems (*Control* and *Test*). With respect
to the full three-way model (6.16), there are $K = 1000$ specimens, $J = 3$ readers, and
$I = 2$ tests (*Control* and *Test*). Moreover, the diagnostic information was divided into
eight categories as illustrated in Table 6.12.

TABLE 6.12   Diagnostic levels

Clinical D_x	Indeterminate	Level⁻	$Level_1^+$	$Level_2^+$	$Level_3^+$	$Level_4^+$	$Level_5^+$	$Level_6^+$
Recorded level	1	2	3	4	5	6	7	8

The notation $Level^-$ means that the specimen is *negative* for disease, while $Level_j^+$ indicates that the specimen is *positive* for a level j state of the disease. For example, cancers are generally delineated in four stages: Stage I (small and localized cancers), Stage II (locally advanced cancers with some penetration of the local lymph nodes), Stage III (more advanced than Stage II with larger and more disorganized tumors and greater infiltration into the lymph nodes), and Stage IV (inoperable or metastatic cancer). In the case of the aforementioned data, $Level_1^+$ represents the *indication* of a precancerous condition. As the positive levels increase, so does the severity of the disease state. Finally, $Level_6^+$ represents the diagnosis of cancer. The standard of care for a $Level_1^+$ specimen does not indicate immediate treatment. Instead, follow-up testing after a short period of time (3–9 months) can be recommended. Consequently, some therapeutic protocols place $Level_1^+$ specimens in the negative diagnostic category. Thus, depending upon which level of disease is required for a treatment regimen, the *negative* versus *positive* partition of the diagnostic set will vary.

There are three MATLAB data files (MAT-files) which contain pseudo-values as per equation (6.17) of *sensitivity, positive diagnostic likelihood ratio,* and *negative diagnostic likelihood ratio* for two different negative/positive partitions of the diagnostic separation space. These files and their associated conditions are summarized in Table 6.13.

An analysis using the data contained in `Y_3waymodel_sen_Level3_1000.mat` and formulae from Tables 6.8a–c for sensitivity pseudo-values Y_{ijk} as modeled by equation (6.17) are presented in Table 6.14a. Here, it is seen when the standard methods of Tables 6.8a–c are applied to the *sensitivity metric*, then the null hypotheses $K_0(\phi_1, \phi_2)$ of zero correlations in-between the factors and zero difference $H_0(\phi)$ within the factors are *not* rejected. If the *DBM* and *NSM* methods of equations (6.21a)–(6.21c) and (6.22a)–(6.22c), respectively, are employed, then the results of the equal factors hypothesis tests change. Indeed, for the tests of *reader* and *specimen* equality, both the *DBM* and *NSM* methods *reject* the null hypotheses of equality. Table 6.14b summarizes these comments.

Three things are evident from the results listed in Tables 6.14a and 6.14b. First, none of the "uncorrelated effects" null hypotheses $K_0(\tau,R)$, $K_0(\tau,C)$, or $K_0(R,C)$ are rejected. Thus, the strategy of Hillis et al. [4] can be implemented. Second, there is no apparent difference in the outcomes (degrees of freedom, F-statistic, or p-value) in-between the *DBM* and *NSM* methods. Third, the *standard methods* are overly optimistic; that is, *anti-conservative*. This can be seen from the small values of the

TABLE 6.13 Three-way model data files and associated diagnostic partitions

File name	D_x partition	Pseudo-value
`Y_3waymodel_sen_Level3_1000.mat`	$Neg = \{Level^-\},$ $Pos = \{Level_1^+,\dots, Level_6^+\}$	Sensitivity
`Y_3waymodel_DLRplus_Level4_1000.mat`	$Neg = \{Level^-, Level_1^+\},$ $Pos = \{Level_2^+,\dots, Level_6^+\}$	DLR^+
`Y_3waymodel_DLRminus_Level4_1000.mat`	$Neg = \{Level^-, Level_1^+\},$ $Pos = \{Level_2^+,\dots, Level_6^+\}$	DLR^-

TABLE 6.14a Standard ANOVA for three-way layout and *sensitivity* pseudo-values

Factor	Null hypothesis	p-Value	Null decision
Test (τ)	$H_0(\tau)$: $\tau_1 = \tau_2 = \cdots = \tau_I$	0.760	Do not reject
$SS_\tau = 0.267$, $MS_\tau = 0.267$, $df = [1, 1998]$, $F_\tau = 0.093$			
Reader (R)	$H_0(R)$: $R_1 = R_2 = \cdots = R_J$	0.899	Do not reject
$SS_R = 0.613$, $MS_R = 0.307$, $df = [2, 1998]$, $F_R = 0.107$			
Specimen (C)	$H_0(C)$: $C_1 = C_2 = \cdots = C_K$	1	Do not reject
$SS_C = 956.5$, $MS_C = 0.9575$, $df = [999, 1998]$, $F_C = 0.334$			
(τ, R)	$K_0(\tau, R)$: $\sigma_{\tau R} = 0$	0.973	Do not reject
$SS_{(\tau, R)} = 0.155$, $MS_{(\tau, R)} = 0.077$, $df = [2, 1998]$, $F_{(\tau, R)} = 0.027$			
(τ, C)	$K_0(\tau, C)$: $\sigma_{\tau C} = 0$	1	Do not reject
$SS_{(\tau, C)} = 151.6$, $MS_{(\tau, C)} = 0.152$, $df = [999, 1998]$, $F_{(\tau, C)} = 0.053$			
(R, C)	$K_0(R, C)$: $\sigma_{RC} = 0$	1	Do not reject
$SS_{(R,C)} = 304.4$, $MS_{(R,C)} = 0.1524$, $df = [1998, 1998]$, $F_{(R,C)} = 0.053$			
$SS(\bar{Y}) = 5245$, $MS(\bar{Y}) = 5245$, $df_{\bar{Y}} = 1$			
$SS_{Total} = 1596$, $MS_{Total} = 0.266$, $df_{Total} = 6000$			

TABLE 6.14b *DBM* and *NSM* ANOVA for three-way layout and *sensitivity* pseudo-values

NSM method			
Factor	Null hypothesis	p-Value	Null decision
Test (τ)	$H_0(\tau)$: $\tau_1 = \tau_2 = \cdots = \tau_I$	0.2044	Do not reject
$df = [1, 2]$, $F_\tau = 3.45$			
Reader (R)	$H_0(R)$: $R_1 = R_2 = \cdots = R_J$	0.134	Do not reject
$df = [2, 1998]$, $F_R = 2.01$			
Specimen (C)	$H_0(C)$: $C_1 = C_2 = \cdots = C_K$	0	Reject
$df = [999, 999]$, $F_C = 6.31$			
DMB method			
Test (τ)	$H_0(\tau)$: $\tau_1 = \tau_2 = \cdots = \tau_I$	0.204	Do not reject
$df = [1, 2]$, $F_\tau = 3.45$			
Reader (R)	$H_0(R)$: $R_1 = R_2 = \cdots = R_J$	0.134	Do not reject
$df = [2, 1998]$, $F_R = 2.01$			
Specimen (C)	$H_0(C)$: $C_1 = C_2 = \cdots = C_K$	0	Reject
$df = [999, 999]$, $F_C = 6.31$			

F-statistics F_τ, F_R, and F_C and the large degrees of freedom in the denominator term df_2. These small F-values, in turn, result in large p-values. The large p-values in Table 6.14a indicate that null hypotheses $H_0(\tau)$, $H_0(R)$ and $H_0(C)$ should not be rejected.

The M-file used to generate these tables is clusteranova.m on the data file Y_3waymodel_sen_Level3_1000.mat. See MATLAB Commands below.

MATLAB Commands
(*Tables 6.14a and 6.14b*)

```
% Data directory
Ddir = 'C:\PJC\Math_Biology\Chapter_6\Data';
```
% *Partition 1*: Positive $\geq Level_1{}^+$, Negative $= Level^-$, and Indeterminate $= Level_1$; see Table 6.12
```
load(fullfile(Ddir,'Y_3waymodel_sen_Level3_1000.mat'));
```
% Perform the ANOVA analysis
```
S = clusteranova(Y1);
```
% Which returns (recover the tables from the Standard, DBM, and NSM fields)
```
S =
           Standard: [1x1 struct]
                DBM: [1x1 struct]
                NSM: [1x1 struct]
      Satterthwaite: [1x1 struct]
```

The *DBM and NSM methods*, however, provide *p-values* that are *considerably smaller than the standard method*. While neither of these alternate methods reject either $H_0(\tau)$ or $H_0(R)$, they both indicate that $H_0(C)$ should be rejected. How can these differences be reconciled? Obuchowski and Rockette [10] recommend examining the two-way model

$$\theta_{ij} = \mu + \tau_i + R_j + (\tau R)_{ij} + \varepsilon_{ij}, \qquad (6.23)$$

where the θ_{ij} are estimates of the appropriate assessment metric for the ith test by the jth reader. This approach is referred to as the *OR* method and has the following test statistics for the hypotheses of equal test and equal reader effects.

$$F_{OR}(\tau) = \frac{MS_\tau}{MS_{OR}^\tau(\tau, R)} \qquad (6.24a)$$

$$F_{OR}(R) = \frac{MS_R}{MS_{OR}^R(\tau, R)} \qquad (6.24b)$$

TABLE 6.15 ANOVA for the *OR* two-way layout

Factor	Sum of squares	df	MS	F-statistic
Test (τ)	$SS_\tau = J \cdot \sum\limits_{i=1}^{I}(\bar{\theta}_{i\cdot} - \bar{\theta}_{\cdot\cdot})^2$	$df_\tau = I - 1$	$MS_\tau = \dfrac{SS_\tau}{df_\tau}$	$F_{OR}(\tau)$
Reader (R)	$SS_R = I \cdot \sum\limits_{k=1}^{K}(\bar{\theta}_{\cdot k} - \bar{\theta}_{\cdot\cdot})^2$	$df_R = J - 1$	$MS_R = \dfrac{SS_R}{df_R}$	$F_{OR}(R)$
(τ,R)	$SS_{(\tau,R)} =$ $\sum\limits_{i=1}^{I}\sum\limits_{j=1}^{J}(\theta_{ij} - \bar{\theta}_{i\cdot} - \bar{\theta}_{\cdot j} + \bar{\theta}_{\cdot\cdot})^2$	$df_{(\tau,R)} = (I-1)$ $\cdot (J-1)$	$MS_{(\tau,R)} = \dfrac{SS_{(\tau,R)}}{df_{(\tau,R)}}$	
OR Mean Squares	$MS_{OR}^{\tau}(\tau, R) = MS_{(\tau,R)} +$ $\max\{J \cdot (Cov_2 - Cov_3), 0\}$	$ddf_H(\tau)$ from (6.25a)		
	$MS_{OR}^{R}(\tau, R) = MS_{(\tau,R)} +$ $\max\{I \cdot (Cov_1 - Cov_3), 0\}$	$ddf_H(R)$ from (6.25b)		

These statistics are applied to the hypotheses

$H_0(\tau){:}\tau_1 = \tau_2 = \cdots = \tau_I$ (Null hypothesis)

$H_1(\tau){:}\tau_j \neq \tau_k$ for some $j \neq k$ (Alternate hypothesis)

and

$H_0(R){:} R_1 = R_2 = \cdots = R_J$ (Null hypothesis)

$H_1(R){:} R_j \neq R_k$ for some $j \neq k$ (Alternate hypothesis)

respectively.

The mean squares and other ANOVA measures are provided in Table 6.15. The previously undefined symbols used within Table 6.15 are provided below.

$$\bar{\theta}_{i\cdot} = \frac{1}{J}\sum_{j=1}^{J}\theta_{ij},\ \bar{\theta}_{\cdot j} = \frac{1}{I}\sum_{i=1}^{I}\theta_{ij},\ \bar{\theta}_{\cdot\cdot} = \frac{1}{I\cdot J}\sum_{i=1}^{I}\sum_{j=1}^{J}\theta_{ij}$$

$$Cov_1 = \frac{2}{I\cdot J\cdot(J-1)}\sum_{i<i'}\sum_{j=1}^{J}(\theta_{ij} - \bar{\theta}_{\cdot\cdot})(\theta_{i'j} - \bar{\theta}_{\cdot\cdot})$$

$$Cov_2 = \frac{2}{I\cdot J\cdot(J-1)}\sum_{i=1}^{I}\sum_{j<j'}(\theta_{ij} - \bar{\theta}_{\cdot\cdot})(\theta_{ij'} - \bar{\theta}_{\cdot\cdot})$$

$$Cov_3 = \frac{2}{I\cdot J\cdot(J-1)}\sum_{i<i'}\sum_{j<j'}(\theta_{ij} - \bar{\theta}_{\cdot\cdot})(\theta_{i'j'} - \bar{\theta}_{\cdot\cdot})$$

TABLE 6.16   Three-way model data files and associated diagnostic partitions

File name	D_x Partition	Value
`Y_ORmodel_sen_Level3_1000.mat`	$Neg = \{Level^-\}$, $Pos = \{Level_1^+, \ldots, Level_6^+\}$	Sensitivity
`Y_ORmodel_DLRminus_Level4_1000.mat`	$Neg = \{Level^-, Level_1^+\}$, $Pos = \{Level_2^+, \ldots, Level_6^+\}$	DLR$^+$
`Y_ORmodel_DLRplus_Level4_1000.mat`	$Neg = \{Level^-, Level_1^+\}$, $Pos = \{Level_2^+, \ldots, Level_6^+\}$	DLR$^-$

The OR denominator degrees of freedom are

$$ddf_H(\tau) = (I-1)\cdot(J-1)\left[\!\!\left[\frac{(MS_{(\tau,R)} + \max\{J\cdot(Cov_2 - Cov_3), 0\})^2}{MS_{(\tau,R)}^2}\right]\!\!\right]$$

$$= (I-1)\cdot(J-1)\left[\!\!\left[\left(1 + \frac{\max\{J\cdot(Cov_2 - Cov_3), 0\}}{MS_{(\tau,R)}}\right)^2\right]\!\!\right] \quad (6.25a)$$

$$ddf_H(R) = (I-1)\cdot(J-1)\left[\!\!\left[\frac{(MS_{(\tau,R)} + \max\{I\cdot(Cov_1 - Cov_3), 0\})^2}{MS_{(\tau,R)}^2}\right]\!\!\right]$$

$$= (I-1)\cdot(J-1)\left[\!\!\left[\left(1 + \frac{\max\{I\cdot(Cov_1 - Cov_3), 0\}}{MS_{(\tau,R)}}\right)^2\right]\!\!\right]. \quad (6.25b)$$

In this setting, $F_{OR}(\tau) \sim \mathcal{F}_{df_\tau}$ where $df_\tau = [df_\tau, ddf_H]$, and $F_{OR}(R) \sim \mathcal{F}_{df_R}$ with $df_R = [df_R, ddf_H]$. If $\psi = \tau$ or R is a factor within the OR two-way model, then the associated hypothesis test $H_0(\psi)$ vs. $H_1(\psi)$ has p-value $p_\psi = 1 - \Phi_F(F_{OR}(\psi); df_\psi, ddf_H)$. Thus, $p_\psi < \alpha$ means that the null $H_0(\psi)$ is *rejected* in favor of the alternate $H_1(\psi)$. Otherwise, *do not reject* $H_0(\psi)$.

For the genuine (i.e., not pseudo-values) values of the assessment metrics as estimated according to the OR two-way model and equation (6.23), Table 6.16 contains the associated partitions and data files. Now the OR method is exercised on the first data set which measures *sensitivity* with respect to the positive $\geq Level_1^+$ partition. The results are summarized in Table 6.17 via the MATLAB Commands listed afterward.

TABLE 6.17   *OR* ANOVA for two-way layout and *sensitivity* values

Factor	Null hypothesis	p-Value	Null decision
Test (τ)	$H_0(\tau)$: $\tau_1 = \tau_2 = \cdots = \tau_I$	0.2044	Do not reject
$df = [1, 2]$, $F_{OR}(\tau) = 3.45$			
Reader (R)	$H_0(R)$: $R_1 = R_2 = \cdots = R_J$	0.1468	Do not reject
$df = [2, 4]$, $F_{OR}(\tau) = 3.22$			

MATLAB Commands
(Table 6.17, *OR ANOVA Method*)

% Data directory
```
Ddir = 'C:\PJC\Math_Biology\Chapter_6\Data';
```
% *Partition 1*: Positive $\geq Level_1^+$, Negative $= Level^-$, and Indeterminate $= Level_1$;
see Table 6.12
```
load(fullfile(Ddir, 'Y_ORmodel_sen_Level3_1000.mat'));
```
% Significance level
```
alpha = 0.05;
```
% Perform the ANOVA analysis
```
S = anovaor(Yor1,alpha);
```
% Which returns (recover the tables from the `Test` and `Reader` fields)
```
H =
      Test: [1x1 struct]
    Reader: [1x1 struct]
```

The results from the *OR method* are consistent with all of the methods with respect to the hypothesis of equal test and reader effects. Note that the standard ANOVA method has a substantially larger *p*-value than any of the other (i.e., *DBM*, *NSM*, or *OR*) methods. Figure 6.3 illustrates how the *p*-values for the *Test* $H_0(\tau)$ vs. $H_1(\tau)$ and *Reader* $H_0(R)$ vs. $H_1(R)$ hypotheses evolve as a function of method. This reinforces the comment that the standard method is overly optimistic (i.e., *anti-conservative*). The methods (standard, *DBM*, and *NSM*) diverge, however, with respect to the hypothesis of equal *specimen* effects. This may be explained, in part, by the disparity in the denominator degree of freedom, $df_2 = 1998$ for standard ANOVA, while $df_2 = 999$ for *DBM* and *NSM*.

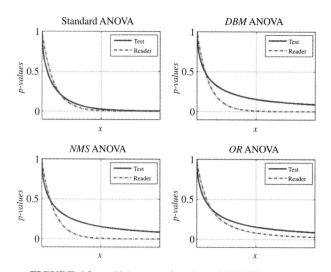

FIGURE 6.3 *p*-Values as a function of ANOVA method.

TABLE 6.18a   Standard ANOVA for three-way layout and DLR^+ pseudo-values

Factor	Null hypothesis	p-Value	Null decision
Test (τ)	$H_0(\tau)$: $\tau_1 = \tau_2 = \cdots = \tau_I$	0.892	Do not reject
$SS_\tau = 99.2$, $MS_\tau = 99.2$, $df = [1, 1998]$, $F_\tau = 0.0185$			
Reader (R)	$H_0(R)$: $R_1 = R_2 = \cdots = R_J$	0.2673	Do not reject
$SS_R = 1.4 \bullet 10^4$, $MS_R = 7066$, $df = [2, 1998]$, $F_R = 1.32$			
Specimen (C)	$H_0(C)$: $C_1 = C_2 = \cdots = C_K$	$<10^{-2}$	Reject
$SS_C = 7.7 \bullet 10^6$, $MS_C = 7717$, $df = [999, 1998]$, $F_C = 1.44$			
(τ,R)	$K_0(\tau,R)$: $\sigma_{\tau R} = 0$	0.874	Do not reject
$SS_{(\tau,R)} = 1444$, $MS_{(\tau,R)} = 722$, $df = [2, 1998]$, $F_{(\tau,R)} = 0.135$			
(τ,C)	$K_0(\tau,C)$: $\sigma_{\tau C} = 0$	1	Do not reject
$SS_{(\tau,C)} = 1.3 \bullet 10^6$, $MS_{(\tau,C)} = 1345$, $df = [999, 1998]$, $F_{(\tau,C)} = 0.251$			
(R,C)	$K_0(R,C)$: $\sigma_{RC} = 0$	1	Do not reject
$SS_{(R,C)} = 4 \bullet 10^6$, $MS_{(R,C)} = 2007$, $df = [1998, 1998]$, $F_{(R,C)} = 0.375$			
$SS(\bar{Y}) = 6.4 \bullet 10^4$, $MS(\bar{Y}) = 6.4 \bullet 10^4$, $df_{\bar{Y}} = 1$			
$SS_{Total} = 1.5 \bullet 10^7$, $MS_{Total} = 2562$, $df_{Total} = 6000$			

An examination of a different partition and assessment metric yields disharmonious results with respect to the *reader* effects. Suppose the second partition, described in Table 6.13, row two, is utilized with respect to the *positive diagnostic likelihood ratio DLR^+*. The three alternate methods *DBM*, *NSM*, and *OR* disagree with respect to the hypothesis of equal reader effect. The first two (*DBM* and *NSM*) *reject* the null while the *OR* method *does not reject*. Tables 6.18a–6.18d summarize the computations. One reason for the discord is the disparity in degrees of freedom. Indeed, denominator degrees of freedom for the test and reader hypotheses $H_0(\tau)$ and $H_0(R)$ with respect to the *DBM* and *NSM* are $df_2 = 2, 1998$, whereas the *OR* method has $df_2 = 2, 106$, respectively. Hillis et al. [4] indicate that the *DBM* method is "*theoretically unsatisfactory since it assumes that the pseudo-values are independent and normally distributed—but they are neither.*" Since the *NSM method* also used pseudo-values, it appears most sensible to employ the *OR* methodology on the test and reader hypotheses.

What should be done with respect to the third hypothesis of equal specimen effect? The recommended strategy of Hillis et al. [4] from Section 6.4 cannot be applied as *none* of the correlation hypotheses $K_0(\tau,R)$, $K_0(\tau,C)$, $K_0(R,C)$ are rejected. Ignoring the caution of testing the correlation hypotheses first and directly employing the Satterthwaite [13] F^*-statistics from equations (6.19a)–(6.19c) of Section 6.4 yields the results that parallel the standard ANOVA method and are summarized in Table 6.18c. In particular, the Satterthwaite approach *does not reject* the null hypothesis $H_0(C)$ of no specimen effect (contradicting the *DBM* and *NSM* results).

TABLE 6.18b *DBM* and *NSM* ANOVA for three-way layout and *DLR*+ pseudo-values

NSM method			
Factor	Null hypothesis	*p*-Value	Null decision
Test (τ)	$H_0(\tau): \tau_1 = \tau_2 = \cdots = \tau_I$	0.7465	Do not reject
df = [1, 2], $F_\tau = 0.1374$			
Reader (R)	$H_0(R): R_1 = R_2 = \cdots = R_J$	0.03	Reject
df = [2,1998], $F_R = 3.52$			
Specimen (C)	$H_0(C): C_1 = C_2 = \cdots = C_K$	0	Reject
df = [999,999], $F_C = 5.74$			
DMB method			
Test (τ)	$H_0(\tau): \tau_1 = \tau_2 = \cdots = \tau_I$	0.7465	Do not reject
df = [1, 2], $F_\tau = 0.1374$			
Reader (R)	$H_0(R): R_1 = R_2 = \cdots = R_J$	0.03	Reject
df = [2,1998], $F_R = 3.52$			
Specimen (C)	$H_0(C): C_1 = C_2 = \cdots = C_K$	0	Reject
df = [999,999], $F_C = 5.74$			

TABLE 6.18c *Satterthwaite* method for three-way layout and *DLR*+ pseudo-values

Satterthwaite Method			
Test (τ)	$H_0(\tau): \tau_1 = \tau_2 = \cdots = \tau_I$	0.9085	Do not reject
df = [1, 39], $F_\tau^* = 0.0134$			
Reader (R)	$H_0(R): R_1 = R_2 = \cdots = R_J$	0.4295	Do not reject
df = [2, 25], $F_R^* = 0.8743$			
Specimen (C)	$H_0(C): C_1 = C_2 = \cdots = C_K$	0.8812	Do not reject
df = [999, 220], $F_C^* = 0.8865$			

TABLE 6.18d *OR* ANOVA for two-way layout and *DLR*+ values

Factor	Null hypothesis	*p*-Value	Null decision
Test (τ)	$H_0(\tau): \tau_1 = \tau_2 = \cdots = \tau_I$	0.7523	Do not reject
df = [1, 2], $F_{OR}(\tau) = 0.131$			
Reader (R)	$H_0(R): R_1 = R_2 = \cdots = R_J$	0.2589	Do not reject
df = [2, 106], $F_{OR}(\tau) = 1.37$			

The MATLAB code used to produce these computations summarized in Tables 6.18a, 6.18b, 6.18c, and 6.18d is located below.

An alternate to these aforementioned methods would be to extend the *OR* two-way model to a reduced three-way model in which the test (τ_i) and reader (R_j) effects are

---

MATLAB Commands
(Tables 6.18a, 6.18b, and 6.18c)

% Data directory
```
Ddir = 'C:\PJC\Math_Biology\Chapter_6\Data';
```
% *Partition 2*: Positive $\geq Level_2{}^+$, Negative $= \{Level_1{}^+, Level^-\}$, and
Indeterminate $= Level_1$; see Table 6.12
```
load(fullfile(Ddir,'Y_3waymodel_DLRplus_Level4_1000.mat'));
```
% Perform the ANOVA analysis
```
S = clusteranova(Y2);
```
% Which returns (recover the tables from the `Standard`, `DBM`, `NSM`, and
`Satterthwaite` fields)
```
S =
          Standard: [1x1 struct]
               DBM: [1x1 struct]
               NSM: [1x1 struct]
     Satterthwaite: [1x1 struct]
```

---

assumed to be independent of the specimens (C_k). This is achieved via the three-way model

$$\theta_{ijk} = \mu + \tau_i + R_j + C_k + (\tau R)_{ij} + \varepsilon_{ijk} \qquad (6.26)$$

for $i = 1, 2,\ldots, I$; $j = 1, 2,\ldots, J$; and $k = 1, 2,\ldots, K$. To apply this model, however, the pseudo-values from equation (6.17) must be used. This, in turn, leads back to the dilemma of Hillis noted earlier. This conundrum is not readily rectified. For the sake of completeness, the equations corresponding to the extended *OR* ANOVA method are provided in Table 6.19.

MATLAB Commands
(Table 6.18d)

% Data directory
```
Ddir = 'C:\PJC\Math_Biology\Chapter_6\Data';
```
% *Partition 2*: Positive $\geq Level_2{}^+$, Negative $= \{Level_1{}^+, Level^-\}$, and
Indeterminate $= Level_1$; see Table 6.12
```
load(fullfile(Ddir,'Y_ORmodel_DLRplus_Level4_1000.mat'));
```
% Perform the ANOVA analysis
```
H = anovaor(Yor2,alpha)
```
% Which returns (recover the tables from the `Test` and `Reader` fields)
```
H =
       Test: [1x1 struct]
     Reader: [1x1 struct]
```

TABLE 6.19   ANOVA table for the reduced three-way model

Factor	SS	df	MS	F-statistic
Test (τ)	$SS_\tau = J \cdot K \sum_{i=1}^{I} (\bar{\theta}_{i\bullet\bullet} - \bar{\theta}_{\bullet\bullet\bullet})^2$	$df_\tau = I - 1$	$MS_\tau = \dfrac{SS_\tau}{df_\tau}$	$F_\tau = \dfrac{MS_\tau}{MS_{OR}^\tau(\tau, R)}$
Reader (R)	$SS_R = I \cdot K \sum_{j=1}^{J} (\bar{\theta}_{\bullet j\bullet} - \bar{\theta}_{\bullet\bullet\bullet})^2$	$df_R = J - 1$	$MS_R = \dfrac{SS_R}{df_R}$	$F_R = \dfrac{MS_R}{MS_{OR}^R(\tau, R)}$
Specimen (C)	$SS_C = I \cdot J \sum_{k=1}^{K} (\bar{\theta}_{\bullet\bullet k} - \bar{\theta}_{\bullet\bullet\bullet})^2$	$df_C = K - 1$	$MS_C = \dfrac{SS_C}{df_C}$	$F_C = \dfrac{MS_C}{MS_{OR}^C(\tau, R)}$
(τ,R)	$SS_{(\tau,R)} = K \sum_{i=1}^{I} \sum_{j=1}^{J} (\bar{\theta}_{ij\bullet} - \bar{\theta}_{i\bullet\bullet} - \bar{\theta}_{\bullet j\bullet} + \bar{\theta}_{\bullet\bullet\bullet})^2$	$df_{(\tau,R)} = (I - 1)\bullet(J - 1)$	$MS_{(\tau,R)} = \dfrac{SS_{(\tau,R)}}{df_{(\tau,R)}}$	
Mean Error	$SS(\bar{\bar{\theta}}) = I \cdot J \cdot K \cdot \bar{\bar{\theta}}_{\bullet\bullet\bullet}$	$df_{\bar{\bar{\theta}}} = 1$		
Total	$SS(\text{Total}) = \sum_{i=1}^{I} \sum_{j=1}^{J} \sum_{k=1}^{K} (\theta_{ijk} - \bar{\theta}_{\bullet\bullet\bullet})^2$	$df_{Total} = I\bullet J\bullet K$		

Extended or Three-Way Method

This model, based on equation (6.26), has parallel definitions for $MS^{\tau}_{OR}(\tau, R)$, $MS^{R}_{OR}(\tau, R)$, $ddf_H(\tau)$, and $ddf_H(R)$ as found in Table 6.15 and equations (6.25a) and (6.25b), respectively. The variation in the degrees of freedom is the replacement of the factor $(I - 1)(J - 1)$ by $(I - 1)(J - 1)(K - 1)$. The definitions of $MS^{C}_{OR}(\tau, R)$ and $ddf_H(C)$ follow from Table 6.19; the ANOVA table associated with (6.26). See Exercise 6.8 of Section 6.4 as a reference.

$$\left. \begin{array}{l} MS^{C}_{OR}(\tau, R) = MS_{(\tau,R)} + \max\{K \cdot (Cov_1 - Cov_2), 0\} \\[4mm] ddf_H(C) = (I - 1)(J - 1)(K - 1) \left[\!\!\left[\dfrac{(MS_{(\tau,R)} + \max\{K \cdot (Cov_1 - Cov_2), 0\})^2}{MS^2_{(\tau,R)}} \right]\!\!\right] \end{array} \right\} \tag{6.27}$$

The formulae for Cov_{ℓ} change in a corresponding fashion.

$$\bar{\theta}_{\cdots} = \frac{1}{I \cdot J \cdot K} \sum_{i=1}^{I} \sum_{j=1}^{J} \sum_{k=1}^{K} \theta_{ijk}$$

$$Cov_1 = \frac{2}{I \cdot J \cdot K \cdot (J-1)} \sum_{i<i'} \sum_{j=1}^{J} \sum_{k=1}^{K} (\theta_{ijk} - \bar{\theta}_{\cdots})(\theta_{i'jk} - \bar{\theta}_{\cdots})$$

$$Cov_2 = \frac{2}{I \cdot J \cdot K \cdot (J-1)} \sum_{i=1}^{I} \sum_{j<j'} \sum_{k=1}^{K} (\theta_{ijk} - \bar{\theta}_{\cdots})(\theta_{ij'k} - \bar{\theta}_{\cdots})$$

$$Cov_3 = \frac{2}{I \cdot J \cdot K \cdot (J-1)} \sum_{i<i'} \sum_{j<j'} \sum_{k=1}^{K} (\theta_{ijk} - \bar{\theta}_{\cdots})(\theta_{i'j'k} - \bar{\theta}_{\cdots})$$

The final portion of this section will address the issue of *nested* effects. Suppose that a chemical assay which acts as a biomarker (indicating the presence or absence of disease) is tested over a set of days, a particular number of runs per day, and a number of replicates (of each assay specimen) per run. The effect of the runs can be seen as a function of the day so that these effects are *nested*. The resulting three-way nested model, which follows the development from the *National Institute of Standards and Technology* [8], is

$$Y_{ijk} = \mu + d_i + r_{j(i)} + \rho_k + \varepsilon_{ijk}, \tag{6.28}$$

where μ is the mean signal response, d_i is the effect of the ith day, $r_{j(i)}$ is the effect of the jth run on the ith day, ρ_k is the effect of the kth replicate, and ε_{ijk} are $\mathcal{N}(0, 1)$ effect errors. The usual notation is maintained so that $i = 1, 2, \ldots, I$; $j = 1, 2, \ldots, J$; and $k = 1, 2, \ldots, K$. The ANOVA table for this model is presented as Table 6.20. The p-values

TABLE 6.20 Three-way nested ANOVA

Source	Sum of squares	df	Mean SS	F-statistic	p-Value
Day	$SS_d = J \cdot K \sum_{i=1}^{I} (\bar{Y}_{i\bullet\bullet} - \bar{Y}_{\bullet\bullet\bullet})^2$	$df_d = I - 1$	$MS_d = \dfrac{SS_d}{df_d}$	$F_d = \dfrac{MS_d}{MS_{r(d)}}$	p_d
Run (Day)	$SS_{r(d)} = K \sum_{i=1}^{I} \sum_{j=1}^{J} (\bar{Y}_{ij\bullet} - \bar{Y}_{i\bullet\bullet})^2$	$df_{r(d)} = I \cdot (J - 1)$	$MS_{r(d)} = \dfrac{SS_{r(d)}}{df_{r(d)}}$	$F_{r(d)} = \dfrac{MS_{r(d)}}{MS_w}$	$p_{r(d)}$
Replicate	$SS_\rho = I \cdot J \sum_{k=1}^{K} (\bar{Y}_{\bullet\bullet k} - \bar{Y}_{\bullet\bullet\bullet})^2$	$df_\rho = K - 1$	$MS_\rho = \dfrac{SS_\rho}{df_\rho}$	$F_\rho = \dfrac{MS_\rho}{MS_w}$	p_ρ
Residual	$SS_w = \sum_{i=1}^{I} \sum_{j=1}^{J} \sum_{k=1}^{K} (Y_{ijk} - \bar{Y}_{i\bullet\bullet} - \bar{Y}_{ij\bullet} - \bar{Y}_{\bullet\bullet k} + 2\bar{Y}_{\bullet\bullet\bullet})^2$	$df_w = I \cdot J \cdot (K - 1)$	$MS_w = \dfrac{SS_w}{df_w}$		
Total	$SS_t = \sum_{i=1}^{I} \sum_{j=1}^{J} \sum_{k=1}^{K} (Y_{ijk} - \bar{Y}_{\bullet\bullet\bullet})^2$	$df_t = I \cdot J \cdot K - 1$			

TABLE 6.21   Three-way nested ANOVA for a 20 (day) × 2 (run) × 3 (replicate) data cube

Source	Sum of squares	df	Mean SS	F-statistic	p-Value
Day	0.0017	19	$9.106 \cdot 10^{-5}$	1.5576	0.1666
Run (day)	0.0012	20	$5.846 \cdot 10^{-5}$	0.7878	0.7202
Replicate	$8.55 \cdot 10^{-6}$	2	$4.275 \cdot 10^{-6}$	0.0576	0.9441
Residual	0.0059	80	$7.420 \cdot 10^{-5}$		
Total	0.0053	119			

have the usual forms $p_d = 1 - \Phi_F(F_d, df_d, df_{r(d)})$, $p_{r(d)} = 1 - \Phi_F(F_{r(d)}, df_{r(d)}, df_w)$, and $p_\rho = 1 - \Phi_F(F_\rho, df_\rho, df_w)$.

Using the data from Y_3way_nested.mat yields the corresponding numeric Table 6.21. Since the p-values for the tests of equal day, runs, and replicates are all greater than the usual significance level of 5% (i.e., $p_d = 0.1666 > \alpha = 0.05$, $p_{r(d)} = 0.72 > \alpha = 0.05$, and $p_\rho = 0.94 > \alpha = 0.05$), then none of the null hypotheses are rejected. That is, *do not reject* $H_0(d)$: $d_1 = d_2 = \cdots = d_I$, $H_0(r)$: $r_1 = r_2 = \cdots = r_J$, or $H_0(\rho)$: $\rho_1 = \rho_2 = \cdots = \rho_K$.

MATLAB Commands
(Table 6.21)

```
% Data directory
Ddir = 'C:\PJC\Math_Biology\Chapter_6\Data';
% Three-way nested ANOVA
load(fullfile(Ddir,'Y_3way_nested.mat'));
% Perform the ANOVA analysis
H = anova3nest(Y)
% Which returns (recover the tables from the Test and Reader fields)
H =
```

SSd: 0.0017	DFd: 19	MSd: 9.1057e-05	Fd: 1.5576	PvalueD: 0.1666
SSrd: 0.0012	DFrd: 20	MSrd: 5.8458e-05	Frd: 0.7878	PvalueRd: 0.7202
SSp: 8.5500e-06	DFp: 2	MSp: 4.2750e-06	Fp: 0.0576	PvalueP: 0.9441
SSw: 0.0059	DFw: 80	MSw: 7.4204e-05		
SSt: 0.0053	DFt: 119			

EXERCISE

6.10 Apply the *extended Obuchowski–Rockette method* to any of the pseudo-value data sets described in Table 6.16. How does the result compare to the *DBM* or *NSM* methods reported in the tables corresponding to the choice of pseudo-value?

6.6 BOOTSTRAPPING AND CONFIDENCE INTERVALS

Formulae for calculating the confidence intervals of the various assessment metrics of *sensitivity*, *specificity*, *positive/negative diagnostic likelihood ratios*, and *relative diagnostic ratios* for clustered data have been provided in Sections 6.2 and 6.3. Moreover, the M-file `dlrtest.m` contains the coded versions of the aforementioned formulae. For such complicated expressions as the variance of the natural logarithm of the positive relative diagnostic likelihood ratio presented in equations (6.13a) and (6.13b), the task of coding these correctly is non-trivial; nor is the risk of errant formulae entry. Therefore, as an alternative to the analytic approach of Sections 6.2 and 6.3, a sampling method called *bootstrapping* will be described. This technique, along with the previously described jackknife method of pseudo-values, was first formalized by Efron [3]. Rutter [12] implemented the approach for clinical samples.

Consider the relatively straightforward matter of computing a confidence interval for a selected assessment metric when there are N_s samples, two distinct diagnostic devices, and a "gold standard" diagnostic for each sample. Furthermore, let the diagnostic partition be dichotomous. That is, either positive (1) or negative (-1). Table 6.22 gives an example of these data.

For each measurement, there are three diagnoses. This can be represented as an $N_s \times 4$ matrix, where the first column contains the sample identifications $[ID_1, ID_2,\dots, ID_{N_s}]^T$ and the remaining three columns contain the diagnostic determination made on each sample by device$_1$, device$_2$, and the gold standard, respectively. With respect to clustered data, $N_s = \sum_{k=1}^{K} n_k$, where K is the number of clusters and n_k is the number of samples in cluster k. The bootstrapping method for the $(1 - \alpha) \cdot 100\%$ confidence interval of the assessment metric θ is defined sequentially.

The Bootstrap

1. Form the matrices $A = \begin{bmatrix} ID_1 & D_x(1,1) & Truth_1 \\ \vdots & \vdots & \vdots \\ ID_{N_s} & D_x(N_s,1) & Truth_{N_s} \end{bmatrix}$ and $B =$

$\begin{bmatrix} ID_1 & D_x(1,2) & Truth_1 \\ \vdots & \vdots & \vdots \\ ID_{N_s} & D_x(N_s,2) & Truth_{N_s} \end{bmatrix}$ of the sample identifications (ID_j), device

TABLE 6.22 Diagnostic tally for bootstrap

Sample *ID*	D_x on device 1	D_x on device 2	Gold standard D_x
ID_1	1	1	1
ID_1	1	-1	-1
ID_1	-1	-1	-1
$\vdots$	$\vdots$	$\vdots$	$\vdots$
ID_{N_s}	1	1	1

TABLE 6.23   Contingency for assessment metric estimation on device$_j$

Device$_j$ D_x		Gold standard D_x	
		Positive	Negative
	Positive	$n_{1,1}$	$n_{1,2}$
	Negative	$n_{2,1}$	$n_{2,2}$

diagnoses $(D_x(j, \ell))$, and gold standard diagnoses (Truth$_j$) for *device*$_1$ ($\ell = 1$) and *device*$_2$ ($\ell = 2$), respectively.

2. Set the number of resamplings (N_{boot}).
3. For each resampling (i.e., for each $j = 1, 2,..., N_{boot}$),
   a) Create a random sample $i_{sample} = \{i_1, i_2,..., i_{N_s}\}$ of the indices $\{1, 2,..., N_s\}$ of size N_s *with replacement*. That is, the elements $\{i_1, i_2,..., i_{N_s}\}$ of i_{sample} need not be unique.
   b) Form a 2×2 contingency matrix for each device based on extracting the i_{sample} rows of A and B. Specifically, $A(i_{sample}, :) =$
   $$\begin{bmatrix} ID_{i_1} & D_x(i_1, 1) & Truth(i_1) \\ ID_{i_2} & D_x(i_2, 1) & Truth(i_2) \\ \vdots & \vdots & \vdots \\ ID_{i_{N_s}} & D_x(i_{N_s}, 1) & Truth(i_{N_s}) \end{bmatrix}$$ with a similar definition for $B(i_{sample}, :)$.
   The corresponding contingency table with respect to device$_j$ is modeled by Table 6.23.
   c) Based on the contingency Table 6.23, estimate the assessment metric for each device: $\theta_{j,1}$ and $\theta_{j,2}$ for devices 1 and 2, respectively.
4. Following the completion of step 3 for each j, record the device-dependent assessment metric vectors $\theta_1 = [\theta_{1,1}, \theta_{2,1},..., \theta_{N_s,1}]$ and $\theta_2 = [\theta_{1,2}, \theta_{2,2},..., \theta_{N_s,2}]$. These two vectors provide a sample distribution (i.e., histogram) of the assessment metric θ with respect to each device.
5. For each sample distribution computed in step 4, calculate the $(\frac{1}{2}\alpha) \cdot 100\%$ and $(1 - \frac{1}{2}\alpha) \cdot 100\%$ quantiles $q_{\alpha/2}$ and $q_{1-\alpha/2}$. The desired confidence interval (for the respective device) is $CI(\theta) = [q_{\alpha/2}, q_{1-\alpha/2}]$.

EXERCISE

6.11 Write a MATLAB program to compute a 95% bootstrap confidence interval for the positive relative diagnostic likelihood ratio for clustered data. Use the data contained in the MAT–files BootA.mat and BootB.mat in the Chapter 6 Data folder to exercise the program.

REFERENCES

[1] Agresti, A., *Categorical Data Analysis*, Second Edition, Wiley–Interscience Books, Hoboken, NJ, 2002.

[2] Dorfman, D. D., K. S. Berbaum, and C. E. Metz, "Receiver operating characteristic rating analysis: generalization to the population of readers and patients with the Jackknife method," *Investigative Radiology*, vol. 26, pp. 723–731, 1992.

[3] Efron, B., *The Jackknife, the Bootstrap and Other Resampling Plans*, SIAM Books, Philadelphia, PA, 1982.

[4] Hillis, S. L., K. S. Berbaum, and C. E. Metz, "Recent developments in the Dorfman–Berbaum–Metz procedure for multireader ROC study analysis," *Academic Radiology*, vol. 15, no. 5, pp. 647–661, May 2008.

[5] Liu, J–p, H–m Hsueh, E. Hsieh, and J. J. Chen, "Tests for equivalence or non–inferiority for paired binary data," *Statistics in Medicine*, vol. 21, pp. 231–245, 2002.

[6] Nam, J–m and D. Kwon, "Non–inferiority tests for clustered matched–pair data," *Statistics in Medicine*, vol. 28, pp. 1668–1679, 2008.

[7] Nam, J–m, "Power and sample size requirements for non–inferiority in studies comparing two matched proportions where the events are correlated," *Computational Statistics and Data Analysis*, vol. 55, pp. 2880–2887, 2011.

[8] National Institute of Standards and Technology, *Engineering Statistics Handbook*, http://itl.nist.gov/div898/handbook/ppc/section2/ppc233.htm.

[9] Obuchowski, N. A., "On the comparison of correlated proportions for clustered data," *Statistics in Medicine*, vol. 17, pp. 1495–1507, 1998.

[10] Obuchowski, N. A. and H. E. Rockette, "Hypothesis testing of diagnostic accuracy for multiple readers and multiple tests: an ANOVA approach with dependent observations," *Communications in Statistics Simulation and Computation*, vol. 24, pp. 285–308, 1995.

[11] Quenoille, M. H., "Notes on bias in estimation," *Biometrika*, vol. 43, pp. 353–360, 1956.

[12] Rutter, C. M., "Bootstrap estimation of diagnostic accuracy with patient–clustered data," *Academic Radiology*, vol. 7, pp. 413–419, 2000.

[13] Satterthwaite, F. E., "Synthesis of variance," *Psychometrika*, vol. 6, pp. 309–316, 1941.

[14] Tukey, J. W., "Bias and confidence in not quite large samples," *Annals of Mathematical Statistics*, vol. 29, p. 614, 1958 (Abstract).

[15] Zhou, X–H, N. A. Obuchowski, and D. K. McClish, *Statistical Methods in Diagnostic Medicine*, Wiley–Interscience, New York, 2002.

FURTHER READING

Bickel, P. J., and K. A. Doksum, *Mathematical Statistics: Basic Ideas and Selected Topics*, Holden–Day, Incorporated, 1977, San Francisco, CA.

Draper, N. and H. Smith, *Applied Regression Analysis*, Second Edition, Wiley–Interscience, New York, 1981.

Pepe, M. S., *The Statistical Evaluation of Medical Tests for Classification and Prediction*, Oxford Statistical Science Series, vol. 31, Oxford University Press, 2003.

APPENDIX: MATHEMATICAL MATTERS

This appendix contains the details of the mathematical formulations and computations used in the main body of the book.

A.1 LINEAR LEAST SQUARES FIT

It is noted in Chapter 4, Section 4.2, that the correlation of two random variables measures how closely they are linearly related. This, in turn, led to the idea of a "best fit" of two matched samples $X = \{x_1, x_2, \ldots, x_n\}$ and $Y = \{y_1, y_2, \ldots, y_n\}$. To determine the "best" line $\mathcal{L}$ that fits these data (x_j, y_j), criteria required to find the slope m and intercept b that *minimize* the deviation of the points from the line must be established.

The *least squares error* of the slope m and intercept b corresponding to the line

$$\mathcal{L} : y_j = m \cdot x_j + b \tag{A.1a}$$

is

$$E(m, b) = \sum_{j=1}^{n} (y_j - m \cdot x_j - b)^2. \tag{A.1b}$$

The error E is a function of m and b. Recalling a basic result of multivariate calculus, the minimum of the error function is attained when the gradient of E, with respect to the model variables, is equal to the zero vector. That is,

Applied Mathematics for the Analysis of Biomedical Data: Models, Methods, and MATLAB®, First Edition. Peter J. Costa.
© 2017 Peter J. Costa. Published 2017 by John Wiley & Sons, Inc.
Companion website: www.wiley.com/go/costa/appmaths_biomedical_data

for $v = [m, b]$, $E(v_0) = \min\limits_v E(v)$ provided $\nabla E(v_0) = 0$. Notice that $\nabla E(v_0) = \left(\frac{\partial E}{\partial m}(v_0), \frac{\partial E}{\partial b}(v_0) \right) = \left(\frac{\partial E}{\partial m}(m_0, b_0), \frac{\partial E}{\partial b}(m_0, b_0) \right)$. For the least squares error (A.1b), $\frac{\partial E}{\partial m} = 2 \sum_{j=1}^{n}(-x_j \cdot y_j + m \cdot x_j^2 + b \cdot x_j)$ and $\frac{\partial E}{\partial b} = 2 \sum_{j=1}^{n}(-y_j + m \cdot x_j + b)$. Therefore,

$$\nabla E(v) = 2 \left(\sum_{j=1}^{n} \left(-x_j \cdot y_j + m \cdot x_j^2 + b \cdot x_j \right), \sum_{j=1}^{n}(-y_j + m \cdot x_j + b) \right). \quad \text{(A.2)}$$

To find the critical points of the error function, set $\nabla E(v) = 0$. This gives rise to the linear equation

$$\begin{bmatrix} \frac{1}{n} \sum\limits_{j=1}^{n} x_j^2 & \bar{x} \\ \bar{x} & 1 \end{bmatrix} \cdot \begin{bmatrix} m \\ b \end{bmatrix} = \begin{bmatrix} \frac{1}{n} \sum\limits_{j=1}^{n} x_j y_j \\ \bar{y} \end{bmatrix} \quad \text{(A.3)}$$

where $\bar{x} = \frac{1}{n} \sum_{j=1}^{n} x_j$ and $\bar{y} = \frac{1}{n} \sum_{j=1}^{n} y_j$ are the sample means of the measurements. Recall that the inverse of the 2×2 matrix $A = \begin{bmatrix} a_{11} & a_{12} \\ a_{21} & a_{22} \end{bmatrix}$ is $A^{-1} = \frac{1}{a_{11}a_{22} - a_{12}a_{21}} \begin{bmatrix} a_{22} & -a_{12} \\ -a_{21} & a_{11} \end{bmatrix}$ provided $a_{11}a_{22} - a_{12}a_{21} \neq 0$. Therefore, the solution of (A.3) is

$$\begin{bmatrix} m \\ b \end{bmatrix} = \frac{1}{\frac{1}{n} \sum\limits_{j=1}^{n} x_j^2 - (\bar{x})^2} \begin{bmatrix} 1 & -\bar{x} \\ -\bar{x} & \frac{1}{n} \sum\limits_{j=1}^{n} x_j^2 \end{bmatrix} \cdot \begin{bmatrix} \frac{1}{n} \sum\limits_{j=1}^{n} x_j y_j \\ \bar{y} \end{bmatrix}.$$

Some algebraic simplification leads to

$$\left. \begin{array}{c} m = \dfrac{\sum\limits_{j=1}^{n} x_j y_j - n \cdot \bar{x} \cdot \bar{y}}{\sum\limits_{j=1}^{n} x_j^2 - n \cdot (\bar{x})^2} \\[4mm] b = \dfrac{\bar{y} \cdot \sum\limits_{j=1}^{n} x_j^2 - \bar{x} \cdot \sum\limits_{j=1}^{n} x_j y_j}{\sum\limits_{j=1}^{n} x_j^2 - n \cdot (\bar{x})^2} \end{array} \right\}. \quad \text{(A.4)}$$

Figure A.1 illustrates a linear least squares fit of data.

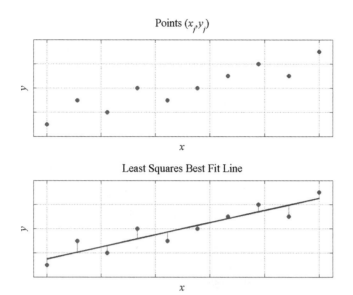

FIGURE A.1   Linear least squares fit.

**MATLAB Commands
(Linear Least Squares Fit)**

```
% Set the data
x = linspace(0,5,10);
y = [2.5,3.5,3,4,3.5,4,4.5,5,4.5,5.5];
% Calculate the slope m and intercept b via the least squares formulae
mx = mean(x); my = mean(y); n = numel(x);
m = (sum(x.*y) - n*mx*my)/(sum(x.^2) - n*(mx^2));
b = (my*sum(x.^2) - mx*sum(x.*y))/(sum(x.^2) - n*(mx^2));
% Compute the least squares line
Ls = m*x + b;

% MATLAB has a function which can be used to calculate the least squares line.
% Fit a polynomial of order 1 to the data wrt the independent variable x
p = polyfit(x,y,1);
% Compute the line. In this case L will give the same results as Ls above.
L = polyval(p,x);
```

A.2   ELEMENTARY MATRIX THEORY

While it is assumed that the reader has some familiarity with linear algebra, this section contains a concise review of select fundamental concepts. The matrix

$A \in \mathcal{Mat}_{n \times m}(\mathbb{R})$ is an array of real numbers $a_{jk} \in \mathbb{R}$ of the form

$$A = \begin{bmatrix} a_{11} & a_{12} & \cdots & a_{1m} \\ a_{21} & a_{22} & \cdots & a_{2m} \\ \vdots & \vdots & \ddots & \vdots \\ a_{n1} & a_{n2} & \cdots & a_{nm} \end{bmatrix}. \tag{A.5}$$

The matrix A can be thought of as a collection of n vectors $a_k = [a_{1k}, a_{2k}, \ldots, a_{nk}]^T$, $k = 1, 2, \ldots, m$. In this case, $A = [a_1, a_2, \ldots, a_m]$. The symbol a^T is the *transpose* of the vector a. If $a = [a_1, a_2, \ldots, a_n]$, then $a^T = \begin{bmatrix} a_1 \\ a_2 \\ \vdots \\ a_n \end{bmatrix}$. Similarly, for $b = \begin{bmatrix} b_1 \\ b_2 \\ \vdots \\ b_n \end{bmatrix}$, the transpose

is $b^T = [b_1, b_2, \ldots, b_n]$. The transpose converts rows into columns and columns into rows. The *transpose* operator essentially turns a matrix on its side so that the transpose of the matrix A from (A.5) makes the first row of A to be the first column of A^T, the second row of A becomes the second column of A^T, etc. Equation (A.6) gives the mathematical description.

$$A^T = \begin{bmatrix} a_{11} & a_{21} & \cdots & a_{n1} \\ a_{12} & a_{22} & \cdots & a_{n2} \\ \vdots & \vdots & \ddots & \vdots \\ a_{1m} & a_{2m} & \cdots & a_{nm} \end{bmatrix} \in \mathcal{Mat}_{m \times n}(\mathbb{R}) \tag{A.6}$$

A *square matrix* has an equal number of rows and columns, $A \in \mathcal{Mat}_{n \times n}(\mathbb{R})$. To simplify matters, consider the case of two dimensions $n = 2$. The matrix $A = \begin{bmatrix} a_{11} & a_{12} \\ a_{21} & a_{22} \end{bmatrix}$ represents a parallelogram in $\mathbb{R}^2$. The vectors $a_1 = [a_{11}, a_{21}]^T$ and $a_2 = [a_{12}, a_{22}]^T$ define the adjacent sides of the parallelogram. The *determinant* of the square matrix A is a number $(\in \mathbb{R})$ which is a measure of certain properties of A. It is denoted by the symbol $\det(A)$. In particular, the absolute value of the determinant is the area subtended by the parallelogram defined by the columns of A: $area(A) = |\det(A)|$. For the 2×2 matrix A, the determinant is defined as $\det(A) = a_{11}a_{22} - a_{12}a_{21}$. Figure A.2 illustrates these ideas for the matrix $A = \begin{bmatrix} 1 & 1 \\ 1 & 3 \end{bmatrix}$. For this example, it is seen that $a_1 = [1, 1]^T$ and $a_2 = [1, 3]^T$. The determinant of the matrix is $\det(A) = 1 \cdot 3 - 1 \cdot 1 = 2$ so that the area subtended by the corresponding parallelogram is $|\det(A)| = 2$.

The multi-dimensional equation $Ax = b$ can be solved in much the same manner as the one-dimensional equation $ax = b$. That is, $x = a^{-1} \cdot b$. What does A^{-1} mean for a matrix? This is the symbol for the *inverse of the matrix A*. For a square matrix A, A^{-1} is the matrix whose multiplication by A results in the identity matrix I: $A \cdot A^{-1} = I = A^{-1} \cdot A$. Again, for the 2×2 case, $A^{-1} = \frac{1}{\det(A)} \begin{bmatrix} a_{22} & -a_{12} \\ -a_{21} & a_{11} \end{bmatrix}$. This

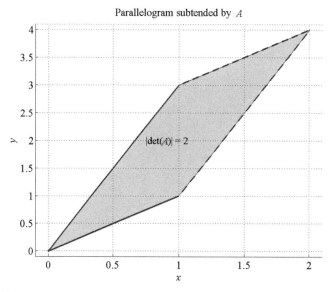

FIGURE A.2 The vectors of the 2 × 2 matrix A and corresponding area.

formula reveals the reason why the determinant of a matrix determines its "invertibility." Plainly, if $\det(A) = 0$, then A^{-1} cannot be calculated. Using the example above,

$$A^{-1} = \frac{1}{2}\begin{bmatrix} 3 & -1 \\ -1 & 1 \end{bmatrix} = \begin{bmatrix} \frac{3}{2} & -\frac{1}{2} \\ -\frac{1}{2} & \frac{1}{2} \end{bmatrix}.$$

The notions of determinant and inverse can be extended to dimensions greater than 2. A full treatment of linear algebra can be read from myriad texts, most notably Strang [10]. Fortunately, MATLAB contains commands for these matrix operations; det for determinant and inv for matrix inverse. These can be seen in the command insert at the end of this section.

Another vital topic in linear algebra is the decomposition of a matrix into its characteristic direction and magnitude. This can be encapsulated in the notion of *eigenvalues* and *eigenvectors*. The approximate translation of the German words *eigenvalue* and *eigenvector* are "characteristic value" and "characteristic vector." The *eigenvalues* and *eigenvectors* of a square matrix $A \in \mathcal{Mat}_{n\times n}(\mathbb{R})$ are the numbers λ and vectors $\boldsymbol{v}$ that are parallel to the image of A. More specifically, the number $\lambda \neq 0$ is an eigenvalue with associated (nonzero) eigenvector $\boldsymbol{v}$ of the matrix A, provided

$$A\boldsymbol{v} = \lambda\boldsymbol{v}. \tag{A.7}$$

Observe that equation (A.7) gives way to

$$(A - \lambda I)\boldsymbol{v} = \boldsymbol{0} \tag{A.8}$$

where $I = \begin{bmatrix} 1 & 0 & \cdots & 0 \\ 0 & 1 & \cdots & 0 \\ \vdots & \vdots & \ddots & \vdots \\ 0 & 0 & \cdots & 1 \end{bmatrix}$ is the identity matrix and $\mathbf{0} = [0, 0, \ldots, 0]^T$ is the zero vec-

tor. This is referred to as the *characteristic equation* for the square matrix A. Again, since eigenvalues and eigenvectors are nonzero, then the only way (A.8) can be satisfied is for $\det(A - \lambda I) = 0$. Indeed, if $\det(A - \lambda I) \neq 0$ then $(A - \lambda I)^{-1}$ exists and the solution of (A.8) would be $\mathbf{v} = \mathbf{0}$. Therefore, it must be the case that $\det(A - \lambda I) = 0$.

In the case of a *symmetric matrix* (i.e., $A = A^T$), the eigenvalues of A are all real. Moreover, the eigenvectors are mutually orthogonal so that $A\mathbf{v}_j = \lambda_j \mathbf{v}_j$ and $A\mathbf{v}_k = \lambda_k \mathbf{v}_k$ implies that $\mathbf{v}_j^T \cdot \mathbf{v}_k = 0$ for $j \neq k$. Form the matrix $V = [\mathbf{v}_1, \mathbf{v}_2, \ldots, \mathbf{v}_n]$ whose

columns are the eigenvectors of A and the diagonal matrix $\Lambda = \begin{bmatrix} \lambda_1 & 0 & \cdots & 0 \\ 0 & \lambda_2 & \cdots & 0 \\ \vdots & \vdots & \ddots & \vdots \\ 0 & 0 & \cdots & \lambda_n \end{bmatrix} \equiv$

$\text{diag}(\lambda_1, \lambda_2, \ldots, \lambda_n)$ whose *nonzero* elements are the eigenvalues. This leads to one of the most significant results in linear algebra.

The Spectral Theorem: If $A = A^T$, V is the matrix of eigenvectors, and Λ is the diagonal matrix of eigenvalues, then $A = V\Lambda V^T$ and $VV^T = I = V^T V$. Moreover, all the eigenvalues are *real valued*.

The second-order equation $\mathbf{x}^T A \mathbf{x} = 1$ describes an n-dimensional geometric form. If A is a symmetric matrix, then the quadratic equation is written as $1 = \mathbf{x}^T A \mathbf{x} = \mathbf{x}^T (V\Lambda V^T)\mathbf{x} = (V^T\mathbf{x})^T \Lambda (V^T\mathbf{x})$. By setting $\mathbf{y} = V^T\mathbf{x}$, the quadratic equation is simply $1 = \mathbf{y}^T \Lambda \mathbf{y} = \lambda_1 y_1^2 + \lambda_2 y_2^2 + \cdots + \lambda_n y_n^2 = \frac{y_1^2}{a_1^2} + \frac{y_2^2}{a_2^2} + \cdots + \frac{y_n^2}{a_n^2}$. In two dimensions ($n = 2$), this is the canonical equation for conic sections (i.e., circle, parabola, ellipse, or hyperbola). In this milieu, the eigenvectors are along the semi-major and semi-minor axes while the *lengths* of these axes are $a_1 = 1 \big/ \sqrt{\lambda_1}$ and $a_2 = 1 \big/ \sqrt{\lambda_2}$.

Example: $A = \begin{bmatrix} 1 & 1 \\ 1 & 3 \end{bmatrix}$. The quadratic equation $1 = \mathbf{x}^T A \mathbf{x}$ becomes $x_1^2 + 2x_1 x_2 + 3x_2^2 = 1$. This describes an ellipse as is exhibited in Figure A.3. The characteristic equation $0 = \det(A - \lambda I)$ produces $0 = \det \begin{pmatrix} 1 - \lambda & 1 \\ 1 & 3 - \lambda \end{pmatrix} = \lambda^2 - 4\lambda + 2$ which has roots $\lambda_1 = 2 - \sqrt{2}$ and $\lambda_2 = 2 + \sqrt{2}$. To calculate the eigenvalue $\mathbf{v}_1 = [x_1, x_2]^T$ associated with λ_1, solve $A\mathbf{v}_1 = \lambda_1 \mathbf{v}_1$. This produces the set of equations $x_1 = -(1 + \sqrt{2})x_2$ and $x_2 = (1 - \sqrt{2})x_1$. Selecting $x_2 = 1$ yields $x_1 = -(1 + \sqrt{2})$. To give the eigenvector unit length, set $\mathbf{v}_1 = \frac{1}{\sqrt{x_1^2 + x_2^2}} [x_1, x_2]^T = \frac{1}{\sqrt{4 + 2\sqrt{2}}} \begin{bmatrix} -(1 + \sqrt{2}) \\ 1 \end{bmatrix} \approx$

$\begin{bmatrix} -0.924 \\ 0.383 \end{bmatrix}$. It is left to the reader to show that the eigenvalue $\lambda_2 = 2 + \sqrt{2}$ has the

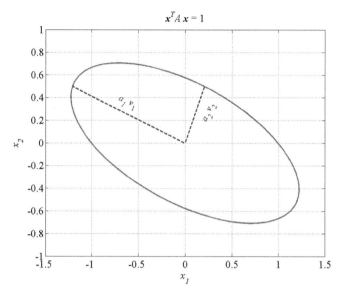

FIGURE A.3   Ellipse for the quadratic form $x^T A x = 1$.

corresponding eigenvector $v_2 = \dfrac{1}{\sqrt{4+2\sqrt{2}}} \begin{bmatrix} 1 \\ 1+\sqrt{2} \end{bmatrix} \approx \begin{bmatrix} 0.383 \\ 0.924 \end{bmatrix}$. The semi-major and semi-minor axes of the ellipse are defined by the two vectors $a_1 v_1$ and $a_2 v_2$ with $a_j = 1/\sqrt{\lambda_j}, j = 1, 2$.

The example above is a special case of the fundamental decomposition theorem of linear algebra. Specifically, if a matrix has n distinct eigenvalues and n linearly independent corresponding eigenvalues, then the matrix can be written as the similarity product of the diagonal matrix of eigenvalues, the matrix of eigenvectors, and its inverse. A *similarity product* or *similarity transformation* on a matrix A is of the form $A = SBS^{-1}$. If $A = SBS^{-1}$, then A is *similar* to B.

The fundamental decomposition theorem says that (under the appropriate circumstances), the [square] matrix A is similar to the diagonal matrix of its eigenvalues. In the special case of a symmetric matrix, the *mapping matrix* $S = V$ is orthogonal.

Suggestion: Using the MATLAB command `eigshow`, see the geometric interplay of the image of the linear transformation A on the vector x against the rotation of the vector x in the plane $\mathbb{R}^2$.

Fundamental Decomposition Theorem: If $A \in \mathcal{Mat}_{n \times n}(\mathbb{R})$ has n linearly independent eigenvectors $v_1, v_2, \ldots, v_n$ and associated eigenvalues $\lambda_1, \lambda_2, \ldots, \lambda_n$ which form the matrices $\Lambda = \text{diag}(\lambda_1, \lambda_2, \ldots, \lambda_n)$ and $V = [v_1, v_2, \ldots, v_n]$, then A is similar to Λ via $A = V\Lambda V^{-1}$.

This decomposition is of great utility in ordinary differential equations. For example, the *system* of linear equations $\frac{dx}{dt} = Ax, x(t_0) = x_0$ has solution $x(t) = e^{A(t-t_0)}x_0$.

If $A \in \mathcal{M}at_{n \times n}(\mathbb{R})$ has n linearly independent eigenvectors, then $e^{A(t-t_0)} = Ve^{\Lambda(t-t_0)}V^{-1}$ where $e^{\Lambda(t-t_0)} = \text{diag}(e^{\lambda_1(t-t_0)}, e^{\lambda_2(t-t_0)}, \ldots, e^{\lambda_n(t-t_0)})$.

Thus far, the presentation has been concerned with square matrices. What happens in the case of a *rectangular* matrix $A \in \mathcal{M}at_{m \times n}(\mathbb{R})$, $m \neq n$? The concept of eigenvalues and eigenvectors no longer applies. A more general methodology, called the *singular value decomposition*, is implemented. This approach was examined by Beltrami and Jordan as early as 1873 (Singular Value Decomposition, https://en.wikipedia.org/wiki/Singular_value_decomposition). It was approximately 90 years later that Golub and Kahan [5] developed a viable and efficient numerical approach to computing this decomposition. First, the statement of the result.

Singular Value Decomposition: Every rectangular matrix $A \in \mathcal{M}at_{m \times n}(\mathbb{R})$ can be factored as $A = U\Sigma V^T$ where U and V are orthogonal matrices and Σ is block-diagonal. More precisely, $A \in \mathcal{M}at_{m \times n}(\mathbb{R})$ implies there exists $U \in \mathcal{M}at_{m \times m}(\mathbb{R})$, $V \in \mathcal{M}at_{n \times n}(\mathbb{R})$, and $\Sigma \in \mathcal{M}at_{m \times n}(\mathbb{R})$ so that $U^T U = UU^T = I_{m \times m}$, $V^T V = VV^T = I_{n \times n}$, and Σ is a rectangular matrix with the *singular values* $\sigma_1, \sigma_2, \ldots, \sigma_r$, along its diagonal and zeroes in all other elements. Note that $r < \min(m, n)$.

Remark: By *block-diagonal* it is meant that the matrix Σ is of the form $\Sigma = \begin{bmatrix} D_r & O_{r \times n - r} \\ O_{m - r \times r} & O_{m - r \times n - r} \end{bmatrix}$, where $D_r = \begin{bmatrix} \sigma_1 & 0 & \cdots & 0 \\ 0 & \sigma_2 & \cdots & 0 \\ \vdots & \vdots & \ddots & \vdots \\ 0 & 0 & \cdots & \sigma_r \end{bmatrix}$ and $O_{p \times q}$ is $p \times q$ matrix of zeros.

Example: $A = \begin{bmatrix} 0 & 2 \\ 1 & 1 \\ 2 & 0 \end{bmatrix}$. An application of the MATLAB command svd produces the following (approximate) result: $A = U\Sigma V^T$, $U = \begin{bmatrix} -\frac{\sqrt{3}}{3} & \frac{\sqrt{2}}{2} & \frac{1}{\sqrt{6}} \\ -\frac{\sqrt{3}}{3} & 0 & -\frac{2}{\sqrt{6}} \\ -\frac{\sqrt{3}}{3} & \frac{\sqrt{2}}{2} & \frac{1}{\sqrt{6}} \end{bmatrix}$, $\Sigma = \begin{bmatrix} 2.45 & 0 \\ 0 & 2 \\ 0 & 0 \end{bmatrix}$, and $V = \begin{bmatrix} -\frac{\sqrt{2}}{2} & -\frac{\sqrt{2}}{2} \\ -\frac{\sqrt{2}}{2} & \frac{\sqrt{2}}{2} \end{bmatrix}$. In this case, the matrix A has two singular values $\sigma_1 = 2.45$ and $\sigma_2 = 2$. It is left to the reader to verify that $UU^T = I_{3 \times 3} = U^T U$ and $VV^T = I_{2 \times 2} = V^T V$.

Remark: The singular values of a matrix A are *not* its eigenvalues. If σ_j is a singular value of the matrix A, then $\lambda_j = \sigma_j^2$ is an eigenvalue of the matrix $B = AA^T$. The number $\lambda_j = \sigma_j^2$ is also an eigenvalue of the matrix $C = A^T A$. Observe that, by construction, B and C are symmetric matrices. Therefore, the eigenvectors λ_j will be real numbers.

The importance of linear algebra and matrix theory to applied mathematics, statistics, and data analysis *cannot be overstated*. As a consequence, it is impossible to list all of the vital topics within this discipline in an appendix. Instead, the reader is urged to look at the references [4, 9, 10, 12] and other sources for insight and inspiration.

**MATLAB Commands
(Linear Algebra)**

```
% Example of a 2×2 matrix
A = [1,1;1,3]
A =
       1       1
       1       3
% Determinant of the matrix A
det(A) =
       2
% Inverse of the matrix A⁻¹
```
% Inverse of the matrix A^{-1}
```
inv(A) =
     1.5000   -0.5000
    -0.5000    0.5000
% Compute the eigenvalues and eigenvectors of A
[V,D] = eig(A);
% The columns of V are the eigenvectors
V =
    -0.9239    0.3827
     0.3827    0.9239
% The diagonal elements of D are the eigenvalues
D =
     0.5858         0
          0    3.4142
% Geometry of eigenvalues and eigenvectors
eigshow(A)

% Singular value decomposition
A = [0,2;1,1;2,0];
[U,S,V] = svd(A)

U =                                      V =

    -0.5774    0.7071    0.4082            -0.7071   -0.7071
    -0.5774   -0.0000   -0.8165            -0.7071    0.7071
    -0.5774   -0.7071    0.4082

S =
```

$$
\begin{vmatrix}
2.4495 & 0 \\
0 & 2.0000 \\
0 & 0
\end{vmatrix}
$$

A.3 ELEMENTARY PROBABILITY THEORY

Probability theory is the basis for the development of the objects used to define the distributions in mathematical statistics. In particular, *probability density functions* and *cumulative density functions* play crucial roles in the development of *test statistics* for hypothesis testing. Therefore, a brief summary of the key components of probability theory are presented.

A *random variable* is a mathematical variable whose values have a distribution. More precisely, a *continuous random variable* X is a mapping from a sample space Ω into the real line $\mathbb{R}$ with the following properties.

(i) $X: \Omega \to \mathbb{R}$ maps the domain of definition into the real line.

(ii) The values of X are governed by a *probability density function* $f_X(x)$ and its associated *cumulative density function* $F_X(x)$.

(iii) The *probability density function* (*pdf*) $f_X(x)$ of the random variable X is a non-negative function on the real line $\mathbb{R}$ whose cumulative values define the probability that X takes a certain range of values.

(iv) The *cumulative density function* (*cdf*) $F_X(x)$ of the random variable X is a non-decreasing function on the real line into the unit interval $[0, 1]$ so that $F_X(-\infty) = 0$ and $F_X(\infty) = 1$.

There are two kinds of random variables: Discrete and continuous. For a *continuous random variable*, the *pdf* $f: \mathbb{R} \to \mathbb{R}^+$ so that $f(x) \geq 0$ for all $x \in \mathbb{R}$ and $\int_{-\infty}^{\infty} f(x)dx = 1$. The associated *cdf* $F: \mathbb{R}^+ \to [0, 1]$ has the added properties $x \geq y \Rightarrow F(x) \geq F(y)$ (i.e., F is non-decreasing), $F_X(-\infty) = 0$, $F_X(\infty) = 1$, and $Prob(X \leq x_0) \equiv F(x_0) = \int_{-\infty}^{x_0} f(x)dx$. Moreover, $Prob(a \leq X \leq b) = F(b) - F(a) = \int_a^b f(x)dx$.

If X is a discrete random variable, then it has a finite number of nonzero values. For example, the event of selecting a single card from a deck of (52) playing cards is a discrete random variable. The probability of selecting a Queen is $4/52 = 1/13$. The *pdf* for the discrete random variable X is a real-valued non-negative function $f: \mathbb{R} \to \mathbb{R}^+$ so that $f(x) = Prob(X = x)$. The associated *cdf* is defined by $F(x|\mathscr{A}) = Prob(X \in \mathscr{A}) = \sum_{x \in \mathscr{A}} f(x)$. In particular, if $\mathscr{A}$ is the set of discrete points $\mathscr{A} = \{x_1, x_2, \ldots, x_n\}$, then $F(x|\mathscr{A}) = Prob(X \in \mathscr{A}) = \sum_{j=1}^{n} f(x_j)$. If $\Omega = \{x_1, x_2, \ldots\}$ is the countably infinite or finite set on which the discrete random variable X is defined, then $\sum_{x_j \in \Omega} f(x_j) = 1$.

A *probability space* $(\Omega, \mathscr{A}, P)$ is a sample space Ω, a collection of subsets $\mathscr{A}$ of Ω, and a probability measure P. As seen earlier, the probability measure P is the *cdf* while the sample space Ω is the set on which the random variable X is defined. The collection $\mathscr{A}$ of subsets of Ω is called a *σ-field* or *σ-algebra*. $\mathscr{A}$ is, essentially, the set of all *subsets* closed under countable unions, intersections, and complement. For more details on σ-algebras, see Bickel and Doksum [2], Hoel et al. [6], Hogg and Craig [7], or Lamperti [8].

One of the applications of the *pdf* of a random variable is to compute its *moments*. In particular, the first and second moments of a random variable help define its corresponding *mean* and *variance*. The nth moment of a *continuous random variable X* with *pdf* $\phi_X(x)$ and sample space Ω is

$$E[X^n] = \int_\Omega x^n \cdot \phi_X(x)dx \text{ for } n \in \mathbb{Z}^+. \tag{A.9a}$$

If X is a *discrete random variable* with *pdf* $f_X(x)$ and sample space $\Omega = \{x_1, x_2, \ldots, x_n\}$, then the nth moment is

$$E[X^n] = \sum_{j=1}^n x_j^n \cdot f_X(x_j) \text{ for } n \in \mathbb{Z}^+. \tag{A.9b}$$

In either case (continuous or discrete), the *mean* μ and *variance* σ^2 of a random variable X are directly related to the first and second moments. Specifically,

$$\mu = E[X] \tag{A.10a}$$

$$\sigma^2 = E[X^2] - E[X]^2 = E[X^2] - \mu^2. \tag{A.10b}$$

As the focus of this book is on *continuous random variables* only such random variables will be discussed further. Four of the most popular probability density functions will be listed next.

A.3.1   Normal Distribution

A random variable X is *normally distributed with mean* μ and *variance* σ^2 provided it has *pdf* $\phi(x; \mu, \sigma)$ and *cdf* $\Phi(x; \mu, \sigma)$ defined as follows. That is, $X \sim \mathcal{N}(\mu, \sigma^2)$ implies that (A.11a) is the *pdf* and (A.11b) is the *cdf*.

$$\phi(x; \mu, \sigma^2) = \frac{1}{\sigma\sqrt{2\pi}} e^{-\frac{(x-\mu)^2}{2\sigma^2}}, \text{ for } -\infty < x < \infty \tag{A.11a}$$

$$\Phi(x; \mu, \sigma) = \frac{1}{\sigma\sqrt{2\pi}} \int_{-\infty}^x \phi(t; \mu, \sigma)dt = \frac{1}{\sigma\sqrt{2\pi}} \int_{-\infty}^x e^{-\frac{(t-\mu)^2}{2\sigma^2}} dt \tag{A.11b}$$

As can be seen from the definitions of the *pdf*, the normal distribution is completely specified by its first and second moments. More specifically, it is determined by the mean μ and variance σ^2 of the random variable.

The *standard normal distribution* refers to a normal random variable with mean 0 and variance 1. This has the common notation $Z \sim \mathcal{N}(0, 1)$, $\phi(x, 0, 1) \equiv \phi(x)$, and $\Phi(x; 0, 1) = \Phi(x)$. Figure A.4 provides examples of normal distributions for a variety of means and variances.

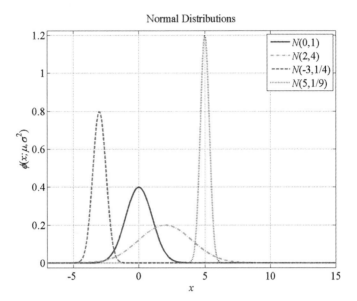

FIGURE A.4   Normal *probability density functions.*

A.3.2 χ^2-Distribution

Suppose $X = \{X_1, X_2, \ldots, X_n\}$ is an independent and identically distributed random sample from a standard normal distribution. That is, each $X_j \sim \mathcal{N}(0, 1)$, X_j and X_k are *independent* whenever $j \neq k$. Then $W = \sum_{j=1}^{n} X_j^2$ has a χ^2-*distribution* with $df = n$ degrees of freedom. Such random variables have the *pdf*

$$\phi_{\chi^2}(x; n) = \frac{x^{\frac{1}{2}(n-2)} e^{-\frac{1}{2}x}}{2^{\frac{n}{2}} \cdot \Gamma\left(\frac{n}{2}\right)} \text{ for } x \geq 0. \tag{A.12}$$

The *cdf* of any continuous random variable is the integral of the corresponding *pdf*. In this case, the *cdf* of a χ^2-random variable is directly related to the *incomplete gamma function* $G(z; a) = \frac{1}{\Gamma(a)} \int_0^z x^{a-1} e^{-x} dx$. That is,

$$\Phi_{\chi^2}(x; df) = \int_0^x \phi_{\chi^2}(\xi; df) d\xi = \frac{1}{2^{\frac{df}{2}} \cdot \Gamma\left(\frac{df}{2}\right)} \int_0^x \xi^{\frac{1}{2}(df-2)} e^{-\frac{1}{2}\xi} d\xi.$$

Now, the substitutions $z = 2\xi$ followed by $\zeta = \frac{1}{2}z$ result in

$$\frac{1}{2^{\frac{df}{2}} \cdot \Gamma\left(\frac{df}{2}\right)} \int_0^x \xi^{\frac{1}{2}(df-2)} e^{-\frac{1}{2}\xi} d\xi = \frac{1}{\Gamma\left(\frac{df}{2}\right)} \int_0^{2x} \left(\frac{1}{2}z\right)^{\frac{1}{2}df-1} e^{-\frac{1}{2}z} dz$$

$$= \frac{1}{\Gamma\left(\frac{df}{2}\right)} \int_0^x \zeta^{\frac{1}{2}df-1} e^{-\zeta} d\zeta \equiv G\left(x; \frac{1}{2} df\right).$$

χ^2-Distributions

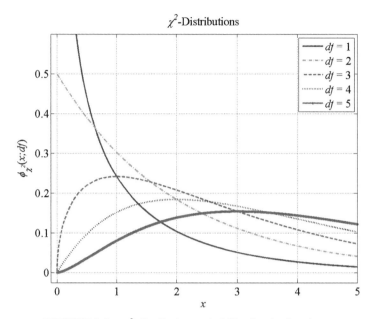

FIGURE A.5 χ^2-distribution probability density functions.

For more details about the *incomplete gamma function*, see Abramowitz and Stegun [1].

Consequently, the *cdf* for a χ^2-random variable with *df* degrees of freedom is

$$\Phi_{\chi^2}(x; df) = G\left(x; \tfrac{1}{2}df\right). \tag{A.13}$$

A selection of χ^2-distribution *pdfs* for varying degrees of freedom is illustrated in Figure A.5.

A.3.3 $\mathcal{T}$-Distribution

A random variable T has a *Student t-distribution* with *df* degrees of freedom provided it has *pdf* $\phi_T(t; df)$ defined as

$$\phi_T(t; df) = \frac{\Gamma\left(\tfrac{1}{2}(df + 1)\right)}{\sqrt{\pi \cdot df}\,\Gamma\left(\tfrac{1}{2}df\right)} \cdot \frac{1}{\left(1 + \tfrac{1}{df}t\right)^{(df+1)/2}} \quad \text{for } -\infty < t < \infty. \tag{A.14}$$

This distribution was popularized in the English-speaking literature by Sealy Gosset whose 1908 paper [11] was written under the pseudonym "Student." Gosset used the *nom de plume* "Student" in keeping with his employer's (the Guinness Brewery of Dublin, Ireland) policy directives. See the Wikipedia article https://en.wikipedia.org/wiki/Student's_t-distribution for more details. The distribution now is generally called the $\mathcal{T}$-distribution or Student t-distribution.

Remark: If $Z \sim \mathcal{N}(0, 1)$ and $W \sim \chi^2_{df}$ (a χ^2-random variable with df degrees of freedom), then $T = \dfrac{Z}{\sqrt{\frac{1}{df}W}}$ is a $\mathcal{T}$-distributed random variable with df degrees of freedom. This is a fundamental result of mathematical statistics as can be found in Bickel and Doksum [2] or Hogg and Craig [7]. The *cdf* for the $\mathcal{T}$-distribution is significantly more complicated. By definition, the *cdf* of a continuous random variable is the integral of its corresponding *pdf*. Thus, $\Phi_T(t; df) = \int_{-\infty}^{t} \phi_T(\tau; df) d\tau$. For positive values of t, the *cdf* relies upon the regularized *incomplete beta function* $I_x(a, b)$. For negative values of t, the symmetric nature of the $\mathcal{T}$-distribution defines the *cdf*. That is, for $t < 0$, $\Phi_T(t; df) = 1 - \Phi_T(|t|; df)$.

$$\Phi_T(t; df) = 1 - I_{v(t)} \left(\frac{1}{2}df, \frac{1}{2}\right) \text{ for } t > 0 \text{ and } v(t) = \frac{df}{df + t^2} \qquad \text{(A.15)}$$

Here $I_t(a, b) = \dfrac{\int_0^t \tau^{a-1}(1-\tau)^{b-1} d\tau}{B(a,b)}$ is the *regularized incomplete beta function* and $B(a, b) = \dfrac{\Gamma(a) \cdot \Gamma(b)}{\Gamma(a+b)}$ is the *beta function*. $\Gamma(z)$ is the *gamma function*. These formulae and special functions are presented in great detail by Abramowitz and Stegun [1].

Figure A.6 illustrates the $\mathcal{T}$-distribution *pdf* for various degrees of freedom.

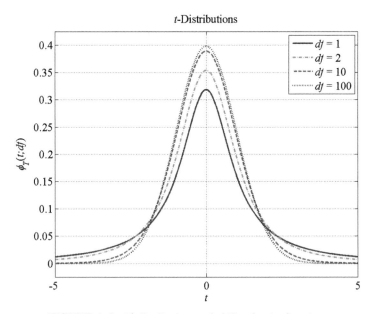

FIGURE A.6 $\mathcal{T}$-distribution *probability density functions*.

A.3.4 $\mathcal{F}$-Distribution

Let V_1 and V_2 be independent χ^2-random variables with df_1 and df_2 degrees of freedom, respectively. Then $F = \dfrac{\frac{1}{df_1}V_1}{\frac{1}{df_2}V_2} = \dfrac{df_2}{df_1}\dfrac{V_1}{V_2}$ is an $\mathcal{F}$-distributed random variable with $df = [df_1, df_2]$ degrees of freedom and *pdf*

$$\phi_F(x; df) = \frac{\left(\frac{df_1}{df_2}\right)x^{\frac{1}{2}(df_1-2)}}{B\left(\frac{1}{2}df_1, \frac{1}{2}df_2\right) \cdot \left(1 + \left(\frac{df_1}{df_2}\right) \cdot x\right)^{\frac{1}{2}(df_1+df_2)}}, \quad x \geq 0. \qquad (A.16)$$

As in Section A.3.3, $B(a,b) = \frac{\Gamma(a)\cdot\Gamma(b)}{\Gamma(a+b)}$ is the *beta function*. Similar to the $\mathcal{T}$-distributed *cdf*, the *cdf* of F relies upon the *regularized incomplete beta function*, which is defined in equation (A.15).

$$\Phi_F(x; df) = I_{v(x)}\left(\frac{1}{2}df_2, \frac{1}{2}df_1\right), \quad x \geq 0 \text{ and } v(x) = \frac{df_2}{df_2 + df_1 \cdot x} \qquad (A.17)$$

For a variety of $df = $ [numerator, denominator] pairs of degrees of freedom, Figure A.7 provides examples of $\mathcal{F}$-distribution *pdf*s. Observe that, as the number of degrees of freedom increases, the $\mathcal{F}$-*pdf* mimics a $\mathcal{N}(1,1)$ random variable. Indeed, this is a common theme for these distributions. As the degrees of freedom increase,

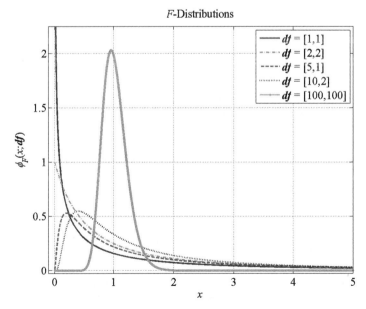

FIGURE A.7 $\mathcal{F}$-distribution probability density functions.

TABLE A.1 Normal vs. $\mathcal{T}$-, χ^2-, $\mathcal{F}$-distributions with large df

Distribution (X)	Degrees of freedom	Normal distribution	RMS error
$\mathcal{T}$	$df = 200$	$\mathcal{N}(0, 1.005^2)$	0.0005
χ^2	$df = 200$	$\mathcal{N}(200, 20^2)$	0.0003
$\mathcal{F}$	$df = [200, 200]$	$\mathcal{N}(1.01, 0.144^2)$	0.0617

the more the distribution resembles a normally distributed random variable. Table A.1 summarizes the fit of a normal distribution to a $\mathcal{T}$-, χ^2-, or $\mathcal{F}$-distribution for large degrees of freedom. The *root mean square error* in-between the aforementioned distributions $\phi_X(x; df)$ and a normal *pdf* $\phi(x; \mu, \sigma^2)$ is defined via (A.18).

$$RMS_X = \sqrt{\frac{1}{n} \sum_{j=1}^{n} (\phi_X(x_j; df) - \phi(x_j; \mu, \sigma^2))^2} \qquad (A.18)$$

The fits are illustrated in Figure A.8 with the $\mathcal{T}$-, χ^2-, or $\mathcal{F}$-distribution marked by a solid line (——) and the normal approximation denoted by a dotted line (–.–.–). These results demonstrate a consequence of the *Central Limit Theorem* which implies that if $X = \{X_1, X_2, \ldots, X_n\}$ is an independent and identically distributed sample from a distribution $\Phi_X(x)$, then the ratio $\frac{X_1 - \bar{X}}{S}$ converges to $Z \sim \mathcal{N}(0, 1)$, where

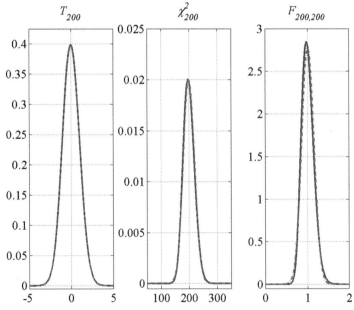

FIGURE A.8 Normal *pdfs* versus $\mathcal{T}$-, χ^2-, $\mathcal{F}$-distributions for large df.

$\bar{X} = \frac{1}{n} \sum_{j=1}^{n} X_j$ is the *sample mean* and $S = \sqrt{\frac{1}{n-1} \sum_{j=1}^{n} (X_j - \bar{X})^2}$ is the *sample standard deviation*.

Central Limit Theorem: Let $\{X_1, X_2, \ldots, X_n\}$ be a random sample from a distribution with mean μ and finite variance σ^2. If $S_n = \sum_{j=1}^{n} X_j$ and $\bar{X}_n = \frac{1}{n} S_n$, then $\frac{S_n - n\mu}{\sigma \sqrt{n}} = \frac{\sqrt{n}}{\sigma}(\bar{X}_n - \mu)$ has a limiting distribution equal to the standard normal. That is, $\lim_{n \to \infty} (\bar{X}_n - \mu) = Z \sim \mathcal{N}(0, 1)$.

The four distributions listed above, normal, χ^2, $\mathcal{T}$, and $\mathcal{F}$, are among the most popular. They are well understood and can all be derived as functions of the standard normal distribution. The MATLAB statistics toolbox provides numerical access to *many* other distributions. Christiansen [3] has catalogued an encyclopedic collection.

The intention of this section of the appendix is to give the reader a notion of elementary probability theory and mathematical statistics germane to the contents of this book. For a deeper view, consult any of the references below or at the end of Chapter 5.

REFERENCES

[1] Abramowitz, M., and I. A. Stegun, *Handbook of Mathematical Functions: With Formulas, Graphs, and Mathematical Tables*, Dover Books, 1972, New York.

[2] Bickel, P. J., and K. A. Doksum, *Mathematical Statistics: Basic Ideas and Selected Topics*, Holden–Day, Incorporated, 1977, San Francisco, CA.

[3] Christiansen, R., *Data Distributions*, Second Edition, Entropy Limited, 1989, Lincoln, MA.

[4] Demmel, J. W., *Applied Numerical Linear Algebra*, SIAM Books, Philadelphia, PA, 1997.

[5] Golub, G. and W. Kahan, *Calculating the Singular Values and Pseudo–Inverse of a Matrix*, SIAM Journal of Numerical Analysis, Series B, Volume 2, Number 2, pp. 205–224.

[6] Hoel, P. G., S. C. Port, and C. J. Stone, *Introduction to Probability Theory*, Houghton Mifflin Company, Boston, MA, 1971.

[7] Hogg, R. V. and A. T. Craig, *Introduction to Mathematical Statistics*, Fourth Edition, Macmillan, New York, 1978.

[8] Lamperti, J., *Probability*, Benjamin–Cummings Publishing, 1966, Reading, MA.

[9] Robert, A. M., *Linear Algebra, Examples and Applications*, World Scientific Press, 2005, Hackensack, NJ.

[10] Strang, G., *Introduction to Linear Algebra*, Wellesley–Cambridge Press, 1993, Wellesley, MA.

[11] Student (Sealy Gosset), *The Probable Error of a Mean*, Biometrika, Volume 6, 1908, pp. 1–25.

[12] Trefethen, L. N., and D. Bau III, *Numerical Linear Algebra*, SIAM Books, 1997, Philadelphia, PA.

FURTHER READING

Levine, A., *Theory of Probability*, Addison–Wesley Publishing Company, 1971, Reading, MA.

Lingren, B. W. and G. W. McElrath, *Introduction to Probability and Statistics*, Second Edition, MacMillan Company, 1966, New York.

GLOSSARY OF MATLAB FUNCTIONS

The following is a partial list of the MATLAB functions developed for this text. The functions are listed alphabetically. Not all of the M–files are listed since their inclusion would unnecessarily add to this book's length. The majority of the files are described below. All M–files are included in the web page developed for this project. To expedite reader access, a list of M–files and MAT–files used, by Chapter, are provided below.

It is *strongly recommended* that the user download these and other related M– and MAT–files which are included in this text in the following manner. Under the MATLAB folder create a subfolder named Math_Biology. Then under this file, create the following subfolders.

```
Math_Biology
      M_files
      Chapter_1
         Data
      Chapter_2
         Data
      Chapter_3
         Data
      Chapter_4
         Data
```

Applied Mathematics for the Analysis of Biomedical Data: Models, Methods, and MATLAB®, First Edition. Peter J. Costa.
© 2017 Peter J. Costa. Published 2017 by John Wiley & Sons, Inc.
Companion website: www.wiley.com/go/costa/appmaths_biomedical_data

```
Chapter_5
    Data
Chapter_6
    Data
```

In each of the "data" subfolders, place the MAT–files indicated in the table below. Many of these M–files are used in conjunction with the MATLAB® Statistic Toolbox; particularly in Chapters 5 and 6. Several other M–files are included in the software written for this book but are not listed as they are merely "helper functions" for the main M–files. All other M–files are part of the base MATLAB software package.

Section	M–files	MAT–file (data file)
Introduction	deming.m	
Chapter 1	chindex.m	BasketballData.mat
	dafeplot.m	Clustered_Data_Classes1_2.mat
	kcluster.m	HockeyData.mat
	pcamb.m	SEIR_Simulated_Data.mat
	pcacompare.m	SEIR_Time.mat
	rmsstd.m	X_2D_outlier.mat
	scorecompare.m	Y_2D_outlier.mat
	screeplot.m	
	stdnorm.m	
	toutlier.m	
	toutlier2.m	
Chapter 2	rtpcr.m	Ktrpcr.mat
Chapter 3	addnoise.m	SEIR_Simulated_Data.mat
	histplot.m	SEIR_Time.mat
	hivseirplot.m	
	seirekbf.m	
	seriekbfout.m	
	seirekbfout.m	
Chapter 4	corrvars.m	Mahal_Example_X.mat
	dafe.m	Mahal_Example_Y.mat
	deming.m	Multivariate_Multicalss_Data.mat
	fried.m	NegClass.mat
	mahalonbis.m	PosClass.mat
	pooledcov.m	
	rda.m	
	stdnorm.m	
	svmbdry.m	
	svmclass.m	
	svmkernel.m	

Section	M–files	MAT–file (data file)
Chapter 5	cvtest.m	Basketball_Data.xlsx
	dlratios.m	BasketballData.mat
	ftest.m	Hockey_Data
	levene.m	HockeyData.mat
	mwutest.m	WilcononA.mat
	niftiest.m	WilcoxonB.mat
	nimean.m	Xcv.mat
	niprop.m	Ycv.mat
	nipropi.m	
	nisamplesize.m	
	smprop.m	
	smprofdiff.m	
	tmean.m	
Chapter 6	anovaor.m	DataCube.mat
	anova2nest.m	DataCube3.mat
	anova3nest.m	DataCube4.mat
	anova4nest.m	Y_3way_nested.mat
	anovawoi.m	Y_3way_woi.mat
	clusteranova.m	Y_3waymodel_DLRminus_Level4_1000.mat
	clustmatchedpair.m	Y_3waymodel_DLRplus_Level4_1000.mat
	dlrtest.m	Y_3waymodel_sen_Level3_1000.mat
	modobuchowski.m	Y_4way_nested.mat
		Y_ORmodel_DLRminus_Level4_1000.mat
		Y_ORmodel_DLRplus_Level4_1000.mat
		Y_ORmodel_sen_Level3_1000.mat

ADDNOISE

ADDNOISE   Adds noise to a matrix of measurements.

A = addnoise(M,p) adds noise proportional to p to the matrix
of measurements M.   The noise is taken from a uniform
distribution on the interval [-1/2,1/2]: Noise ~ U([-1/2,1/2]).

A = addnoise(M,p,'norm') adds noise proportional to p to the
matrix of measurements M.   The noise is taken from a normal
distribution with mean 0 and standard deviation p.

Note: p is the percent of the original signal added as noise
to the system.   Thus, p should be in the interval [0,1].
p = 0.01 means adding 1% noise to the signal, etc.

A = addnoise(M,p,'unif') is equivalent to addnoise(M,p).

See also seirekbf.

ANOVAOR

ANOVAOR   The Obuchowski-Rockette ANOVA method for MRMC data.

```
H = anovaor(Yor,alpha) returns the structure H whose fields
contain the F-statistic (Fstat), the assocaited degrees of
freedom (df), the decision of the null hypothesis (Ho), and a
p-value of the test for the two-way Obuchowski-Rockette model
```

```
 Theta(i,j) = mu + t(i) + R(j) + (tR)(i,j) + e(i,j)
```

```
where t(i) is the effect of the ith test and R(j) is the
effect of the jth reader. The term (tR) is the interaction
of the ith test on the jth reader and mu is the mean value of
the assessment metric as specified by the input array Theta.
```

```
More specifically, H has the fields TEST and READER each of
which contain the subfields
```

```
        Ho: Decision with respect to the null hypothesis
     Fstat: Value of the Obuchoski-Rockette test F-statistic
        df: Degrees of freedom of the F-statistic
    Pvalue: p-value of the test
```

```
The two hypotheses are
```

```
Equal test effects
Ho: t(1) = t(2) = ... t(I)
H1: t(j) ~= t(k) for some j~= k
```

```
Equal reader effects
Ho: R(1) = R(2) = ... R(J)
H1: R(j) ~= R(k) for some j~= k
```

```
Example: Ddir = 'C:\PJC\Math_Biology\Chapter_6\Data';
         load(fullfile(Ddir,'Y_ORmodel_sen_Level3_1000.mat'));
         alpha = 0.05;
         H = anovaor(Yor1,alpha)    returns
```

```
H.Test =                        H.Reader =

       Ho: 'do not reject'             Ho: 'do not reject'
    Fstat: 3.45                     Fstat: 3.219
       df: [1 2]                       df: [2 4]
   Pvalue: 0.2044                  Pvalue: 0.1468
```

See also <u>clusteranova</u>.

Reference: S.L. Hillis, K.S. Berbaum, and C.E. Metz
          Recent Developments in the Dorfman-Berbaum-Metz
          Procedure for Multireader ROC Study Analysis
          Academic Radiology, 2008 May, Volume 15, Number 5,
          pages 647-661

ANOVA2NEST

ANOVA2NEST   Two–way Nested ANOVA Model.

S = anova2nest(Y) returns the structure S whose fields contain
the elements of the ANOVA table listed below for the input 3-D
array Y indexed by days, runs, and replicates. Y = Y(Ndays,
Nruns,Nreps). The elements of the table are the factors
(e.g., Day, Run, Replicate), the sum of residual squares (SS),
the degrees of freedom, the mean sum of squares, the F-values,
and P-values as listed below. Notice the assumption of nested
factors Run with Day.

```
ANOVA Table
----------------------------------------------------------------
Factor        SS        Df        MS        F-value
----------------------------------------------------------------
Day           SSd       DFd       MSd       MSd/MSw
----------------------------------------------------------------
Run (Day)     SSr       DFr       MSr       MSr/MSw
----------------------------------------------------------------
Replicate     SSp       DFp       MSp       MSp/MSw
----------------------------------------------------------------
Residual      SSw       DFw       MSw
----------------------------------------------------------------
Total         SSt       DFt
```

If Yo is the overall mean of the 3-D array Y, I = Ndays,
J = Nruns, and K = Nreps, then the following formulae hold.

```
N = I*J*K;
SSd = J*K*sum(Y(i,.,.)-Yo).^2,1);    DFd = I-1;
SSr = K*sum(Y(.,j,.)-Yo).^2,2);      DFr = I*(J-1);
SSp = I*J*sum(Y(.,.,k)-Yo).^2,3);    DFp = K-1;
```

```
SSw = sum(sum(sum((Y(i,j,k)-Y(i,.,.)-Y(i,j,.)-Y(.,.,k)+2Yo).
       ^2,3),2),1);                                    DFw = I*J*(K-1)
SSt = sum(sum(sum( (Y(i,j,k)-Yo).^2, 3),2),1);  DFt = N - 1;
```

$S = ANOVA2N(Y,1)$ returns a form of the table listed above (see example below).

Example: Ddir = 'C:\PJC\Math_Biology\Chapter_6\Data';
         load(fullfile(Ddir,'Y_2way_nested.mat'));

         S = anova2nest(Y,1);   returns

SSd	DFd	MSd	Fd	PvalueD
0.00173	19	9.106e-05	1.779	0.03991
SSr	DFr	MSr	Fr	PvalueR
0.002899	20	0.000145	2.833	0.0005266
SSp	DFp	MSp	Fp	PvalueP
8.55e-06	2	4.275e-06	0.08353	0.9199
SSw	DFw	MSw		
0.004094	80	5.118e-05		
SSt	DFt			
0.005272	119			

See also, anovaor, clusteranova, anova3nest, anova4nest.

ANOVA3NEST

ANOVA3NEST   Three–way Nested ANOVA Model.

$S = ANOVA3NESY(Y)$ returns the structure S whose fields contain the elements of the ANOVA table listed below for the input 3-D array Y indexed by days, runs, and replicates. $Y = Y(Ndays, Nruns,Nreps)$. The elements of the table are the factors (e.g., Day, Run, Replicate), the sum of residual squares (SS), the degrees of freedom, the mean sum of squares, the F-values, and P-values as listed below. The runs are a function of the day and hence the nested factor

```
ANOVA Table
----------------------------------------------------
```

Factor	SS	Df	MS	F-value
Day	SSd	DFd	MSd	MSd/MSr(d)
Run(Day)	SSr	DFr(d)	MSr(d)	MSr(d)/MSw

```
------------------------------------------------
Replicate    SSp      DFp        MSp       MSp/MSw
------------------------------------------------
Residual     SSw      DFw        MSw
------------------------------------------------
Total        SSt      DFt
------------------------------------------------
```

If Yo is the overall mean of the 3-D array Y, a = Ndays,
b = Nruns, and c = Nreps, then the following formulae hold.

```
N = a*b*c;
SSd = b*c*sum(Y(i,.,.)-Yo).^2,1);                DFd = a-1;
SSr(d) = a*c*sum(sum(Y(i,j,.)-Yo).^2„1)2);       DFr(d) = a*(b-1);
SSp = a*b*sum(Y(.,.,k)-Yo).^2,3);                DFp = c-1;
SSw = sum(sum(sum( (Y(i,j,k)-Y(i,.,.)-Y(.,j,.)-Y(.,.,k)-
      Yo).^2,3),2),1);                           DFw = a*b*(c-1)
SSt = sum(sum(sum( (Y(i,j,k)-Yo).^2, 3),2),1);   DFt = N - 1;
```

S = anova3nest(Y,1) returns a form of the table listed above
(see example below).

Example: Ddir = 'C:\PJC\Math_Biology\Chapter_6\Data';
       load(fullfile(Ddir,'Y_3way_nested.mat');
       S = anova3nest(Y,1)   returns

SSd	DFd	MSd	Fd	PvalueD
0.00173	19	9.106e-05	1.558	0.1666
SSr(d)	DFr(d)	MSr(d)	Fr(d)	PvalueR(d)
0.001169	20	5.846e-05	0.7878	0.7202
SSp	DFp	MSp	Fp	PvalueP
8.55e-06	2	4.275e-06	0.05761	0.9441
SSw	DFw	MSw		
0.005936	80	7.42e-05		
SSt	DFt			
0.005272	11			

See also, <u>anova2nest</u>.m, <u>anova4nest</u>.m, <u>anovaor</u>.m, <u>anova3woi</u>.m

ANOVA4NEST

ANOVA4NEST   Four–way Nested ANOVA Model.

S = anova4nest(Y) returns the structure S whose fields contain
the elements of the nested ANOVA table listed below for the

4-D input array Y. The input Y has the form
    Y(Ntesters,Ndays,Nruns,Nreplicates).
The nesting occurs with respect to runs (that is, runs/day)
and replicates (replicates/run/day). The model is

$$Y(j,k,l,m) = mu + t(j) + D(k) + R(l(k)) + P(m(l,k))$$
$$+ e(j,k,l,m)$$

The corresponding nested ANOVA table is

Factor	SS	DF	MS	F-value
Tester	SSt	Nt-1	MSt = SSt/DFt	MSt/MSw
Day	SSd	Nd-1	MSd = SSd/DFd	MSd/MSr(d)
Run(Day)	SSr(d)	Nd(Nr-1)	MSr(d) = SSr(d)/DFd	MSr(d)/MSp(d,r)
Rep(Day,Run)	SSp(d,r)	Nd(Nr-1)(Np-1)	MSp(d,r) = SSp(d,r)/DFp(d,r)	MSp(rd,r)/MSw
Residual	SSw	NtNd(Nr-1)(Np-1)	MSw = SSw/DFw	
Total	SStot	NtNdNrNp - 2		

Here SS = sum of squares, DF = degrees of freedom, MS = mean
squares, Nt = number of testers, Nd = number of days, Nr =
number of runs, and Np = number of replicates.

The output structure has the fields
S =
            Test: [1x1 struct]
             Day: [1x1 struct]
          RunDay: [1x1 struct]
           RunDa: [1x1 struct]
      RepDayRun: [1x1 struct]
       Residual: [1x1 struct]
          Total: [1x1 struct]

and each subfield has the additional fields

S.Test =
        SS: Test factor sum of squares
        DF: degrees of freedom
        MS: Mean Square value
     Fstat: F-statistic
     Pvalue: p-value

Each p-value detemined the disposition of the null hypothesis of equal values across the associated factor. That is, p-value < alpha implies reject Ho.

S = anova4nest(Y,1) returns a form of the table listed above (see example below).

Example: Ddir = 'C:\PJC\Math_Biology\Chapter_6\Data';
        load(fullfile(Ddir,'Y_4way_nested.mat'));
        S = anova4nest(Y4,1)   returns

SSt	DFt	MSt	Ft	Pvalue Test
0.005191	2	0.002595	23.73	6.607e-07
SSd	DFd	MSd	Fd	Pvalue Day
0.0002286	4	5.715e-05	1.329	0.3743
SSr(d)	DFr(d)	MSr(d)	Fr(d)	Pvalue Run(Day)
0.0002151	5	4.301e-05	0.4709	0.79
SSp(d,r)	DFp(d,r)	MSp(d,r)	Fp(d,r)	Pvalue Rep(Day,Run)
0.0009133	10	9.133e-05	0.8352	0.5994
SSw	DFw	MSw		
0.003281	30	0.0001094		
SStotal	DFtotal			
0.009828	88			

See also <u>anova2nest</u>.m, <u>anova3nest</u>.m, <u>anova3woi</u>.m, <u>anovaor</u>.m

ANOVA3WOI

ANOVA3WOI   Three–way ANOVA Model without factor interactions.

S = anova3woi(Y) returns the structure S whose fields contain the elements of the ANOVA table listed below for the input 3-D array Y indexed by days, runs, and replicates. Y = Y(Ndays, Nruns,Nreps). The elements of the table are the factors (e.g., Day, Run, Replicate), the sum of residual squares (SS), the degrees of freedom, the mean sum of squares, the F-values, and P-values as listed below.

ANOVA Table

Factor	SS	Df	MS	F-value
Day	SSd	DFd	MSd	MSd/MSw
Run	SSr	DFr	MSr	MSr/MSw

Replicate	SSp	DFp	MSp	MSp/MSw
Residual	SSw	DFw	MSw	
Total	SSt	DFt		

If Yo is the overall mean of the 3-D array Y, a = Ndays,
b = Nruns, and c = Nreps, then the following formulae hold.

```
N = a*b*c;
SSd = b*c*sum(Y(i,.,.)-Yo).^2,1);     DFd = a-1;
SSr = a*c*sum(Y(.,j,.)-Yo).^2,2);     DFr = b-1;
SSp = a*b*sum(Y(.,.,k)-Yo).^2,3);     DFp = c-1;
SSw = sum(sum(sum((Y(i,j,k)-Y(i,.,.)-Y(.,j,.)-Y(.,.,k)-
      Yo).^2,3),2),1);                DFw = N-a-b-c-1
SSt = sum(sum(sum( (Y(i,j,k)-Yo).^2, 3),2),1); DFt = N - 1;
```

S = anova3woi(Y,1) returns a form of the table listed above
(see example below).

Example: Ddir = 'C:\PJC\Math_Biology\Chapter_6\Data';
load(fullfile(Ddir,'Y_3way_woi.mat'));
S = anova3woi(Y3woi,1)   returns

SSd	DFd	MSd	Fd	PvalueD
0.00173	19	9.106e-05	2.592	0.001239
SSr	DFr	MSr	Fr	PvalueR
0.0001261	1	0.0001261	3.589	0.06113
SSp	DFp	MSp	Fp	PvalueP
8.55e-06	2	4.275e-06	0.1217	0.8855
SSw	DFw	MSw		
0.003407	97	3.513e-05		
SSt	DFt			
0.005272	119			

See also, <u>anova2nest</u>.m, <u>anova3nest</u>.m, <u>anova4nest</u>.m, <u>anovaor</u>.m

BLANDALTMAN

BLANDALTMAN   Bland Altman (Allowable Total Difference) Plot.

BLANDATLMAN(X,Y,alpha,mu,s) returns a Bland-Altman "trumpet"
plot displaying the (1-alpha)*100% confidence Allowable Total
Difference region. That is, BLANDALTMAN returns the scatter

plot of (X+Y)/2 versus Y-X along with the Allowable Total
Difference (ATD) Zone.
mu is the "average" measurement value at which the ATD zone
transitions between a constant width interval and a slope.
s is the standard deviation to be used in calculating the
ATD zone.
It is typically derived from the measurement precision from
a separate study.

p = BLANDATLMAN(X,Y,alpha,mu,s) returns the plots as above and
returns the percentage of the difference Y-X contained within
the ATD Zone.

[p,H] = BLANDATLMAN(X,Y,alpha,mu,s) returns the plots as
above, the percentage of the difference Y-X contained within
the ATD Zone, and the (1-alpha)*100% standard and Newcombe
Confidence intervals in the structure H.

Example: N = 1000; x = 5 + randn(N,1); y = 6 + (1/sqrt(2))
         *rand(N,1);
       mu = 5.5; s = sqrt(3/2);
      [p,H] = blandaltman(x,y,alpha,mu,s)   returns

         p = 98.5000

         H =
             p: 0.9850
            CI: [0.9775 0.9925]
           NCI: [0.9754 0.9909]

BWTNGRPS

BWTNGRPS   Between Groups Covariance Matrix.

Cb = bwtngrps(S) returns the Between-Groups Covariance Matrix
from the cell array of data classes S = {S1,S2,...,Sg}.

Example: Ddir = 'C:\PJC\Math_Biology\Chapter_4\Data';
         load(fullfile(Ddir,'Multivariate_MultiClass_Data
          .mat'));
         M = vertcat(D.X{:}); Z = stdnorm(M,D.X{:});

         B = bwtngrps(Z);

See also <u>dafe</u>, <u>pooledcov</u>, <u>crimcoords</u>, <u>fried</u>, <u>rda</u>.

CHI2TEST

CHI2TEST   Chi-squared Test of Independence

The Chi-squared test of independence produces a test statistic
whose value determines whether the input variables are
independent. The determination is achieved at the (1-alpha)
*100% confidence level. This method tests the hypothesis that
the probability of occurrence of one of the classes is the
same across all populations. That is,

```
  Ho: p1 = p2 = p3 = ... = pc
  H1: pj ~= pk     for some distinct j and k.
```

H = CHI2TEST(A,alpha) returns the structure H whose fields
contain the decision of the test with respect to the null
hypothesis (Ho), the value of the test statistic (T), the
(1-alpha)*100% confidence interval CI, the number of degrees
of freedom of the input contingency table A, the p-value for
the test, and the achieved effect size (w) of the data.

More precisely, the structure H has the fields

```
      Ho: Null hypothesis status (i.e., 'do not reject' or
          'reject')
       T: Test statistic
      CI: (1-alpha)*100% confidence interval
      df: Degrees of freedom.
  Pvalue: The p-value of the test (that is, the significance
          probability = the smallest level of significance at
          which the null hypothesis is rejected on the basis
          of the observed data)
  Effect: The Chi-squared (df = (r-1)x(c-1)) effect size
          of the data.
```

Example: A = [76,78,16,7;19,29,10,13;2,6,7,8]; alpha = 0.05;
         H = chi2test(A,alpha)   returns
```
      H =
          Ho: 'reject'
           T: 43.0439
          CI: [0 12.5916]
          df: 6
      Pvalue: 1.1433e-007
      Effect: 0.3985
```

See also <u>ftest</u>.

Remark: CHI2TEST can be used for χ^2 Tests of Independence and $r \times c$ contingency table hypothesis testing.

CHINDEX

CHINDEX   The Calinski-Harabasz index.

```
ch = chindex(S) returns the Calinski-Harabasz index with
respect to the data matrix partitioned into k-clusters via
KCLUSTER.
```

Example: `Ddir = 'C:\PJC\Math_Biology\Chapter_5\Data';`
`                load(fullfile(Ddir,'Clustered_Data_Classes1_2.mat'));`
`                X = C{1}; Y = C{2}; S = kcluster(Y,2)`

`                ch = chindex(S)    returns    2.2981`

```
Reference:    Data Clustering:  Theory, Algorithms, adn
              Applications
              G. Gan, C. Ma, and J. Wu
              ASA-SIAM, Copyright (c) 2007
              ISBN: 978-0-898716-23-8
```

```
See also mahalanobis, kcluster, kmeans, mdist, rmsstd.
```

CIMV

CIMV   Confidence Intervals for the Mean and Variance of a Sample.

```
CI = CIMV(X,alpha) returns the structure CI whose fields
contain the (1 - alpha)*100% confidence intervals for the mean
and variance of the distribution sampled by the input
vector X.  The structure CI has the form
```

```
  CI =

    mu: The estimate of the mean of X
   Cmu: (1-alpha)*100% confidence interval for the mean of X
   var: The estimate of the variance of X
  Cvar: (1-alpha)*100% confidence interval for the variance
        of X
```

Examples:

```
X = [20888,14858,17444,21573,13013,12427,12540,11882,12191,...
          11102,8697,15696,24745,30311,20239,24023,16143,...
          3919,9331,4653];
Y = [4273,2077,3837,3756,1587,2013,1714,2208,490,705,539,...
          423,6188,7284,11314,10046,25799,23384,24003,22917];

CI = cimv(X,0.05)     returns

    CI =
         mu: 15283.75
        Cmu: [12110.54 18456.96]
        var: 45970599.99
       Cvar: [26586896.06 98067679.04]

CI = cimv(Y,0.05)     returns

    CI =
         mu: 7727.85
        Cmu: [3560.28 11895.42]
        var: 79295215.29
       Cvar: [45860042.03 169158064.61]
```

See also <u>ftest</u>, <u>tmean</u>.

CLUSTERANOVA

CLUSTERANOVA   Analysis of Variance for Clustered Data.

```
S = CLUSTERANOVA(Y) returns the structure whose fields contain
the Analysis of Variance (ANOVA) measures for the clustered
data Y and associated assessment metric as specified via
CLUSTERPREP.  There are three methods used to compute the
ANOVA measures:  Standard, DBM (Dorfman, Berbaum, Metz) and
NSM (Hillis).

All subfields ("Standard","DBM", and "NSM") contain decisions
on the hypotheses

  Ho(phi): phi(1) = phi(2) = ... = phi(Nphi) vs.
  H1(phi): phi(j) ~= phi(k) for some j ~= k
```

while only the "Standard" subfield contains information on the hypotheses

    Ko(phi,psi): sigma(phi,psi) = 0 vs.
    K1(phi,psi): sigma(phi,psi) ~= 0

where phi and psi are distinct factors (that is, variables within the clustered data) and Nphi is the number of measurements of the factor phi.

Also, the P-value of each hypothesis test (with respect to each factor), the F-statistic used to test the null hypothesis, and the corresponding degrees of freedom df1 (numerator) and df2 (denominator) for the F-statistic are returned.

Finally, the "Standard" subfield contains the sum of squared errors (SS), the mean squared errors (MS), the residual sum of squares (SR), the mean residuals (MR), and the corresponding degrees of freedom are returned.

The subfields are as follows.

        S.Standard =
                      Test: [1x1 struct]
                    Reader: [1x1 struct]
                  Specimen: [1x1 struct]
                TestReader: [1x1 struct]
              TestSpecimen: [1x1 struct]
            ReaderSpecimen: [1x1 struct]
                      Mean: [1x1 struct]
                     Total: [1x1 struct]
                  Residual: [1x1 struct]

        S.Standard.Test =
              Ho: Decision on Ho: T1 = T2 = ... = Tn
          Pvalue: p-value of the test statistic
              SS: Sum of Squares errors
              MS: Means sum of squared errors
           Fstat: The (F-) Test statistic
             df1: degrees of freedom numerator
             df2: degrees of freedom denominator

        S.Standard.TestReader =
              Ko: Decision on Ko: sigma(Test,Reader) = 0

```
Pvalue: p-value of the test statistic
    SS: Sum of Squares errors
    MS: Means sum of squared errors
 Fstat: The (F-) Test statistic
   df1: degrees of freedom numerator
   df2: degrees of freedom denominator
```

with similar values for S.Standard.Reader, S.Standard.Specimen, S.Standard.TestSpecimen, S.Standard.ReaderSpecimen, S.DBM.Test, S.DBM.Reader, S.DBM.Specimen, S.NSM.Test, S.NSM.Reader, and S.NSM.Specimen.

Examples:
```
Ddir = 'C:\PJC\Math_Biology\Chapter_6\Data';
load(fullfile(Ddir,'Y_3waymodel_sen_Level3_1000
    .mat'));
S = clusteranova(Y1);
load(fullfile(Ddir,'Y_3waymodel_DLRplus_Level4_1000
    .mat'));
S = clusteranova(Y2);
```

CLUSTMATCHEDPAIR

CLUSTMATCHEDPAIR   Hypothesis Test for Clustered matched-pair data.

H = CLUSMATCHEDPAIR(Dcube,delta,alpha) returns the structure H whose fields contain the null hypothesis decision (Ho), the modified Obuchowski test statistic (Zo), the margin-adjusted difference in proportions Dp = (p1.,. - p.1,. - delta) and the corresponding (1-alpha)*100% confidence interval (CI) of Dp for the following hypothesis test with respect to a (negative) non-inferiority margin delta.

```
Ho:  p10,. - p01,. = delta(Null hypothesis)
H1:  p10,. - p01,. > delta(Alterate hypothesis)
```

Notice that these hypotheses are equivalent to

```
Ho:  pt(+) - pc(+) = delta(Null hypothesis)
H1:  pt(+) - pc(+) > delta(Alterate hypothesis)
```

where pt(+) = p11,. + p10,. = p1.,. and
      pc(+) = p11,. + p01,. = p.1,.

The input data cube Dcube is modelled from the K-by-2-by-2 set
of contingency tables

```
                         Control
                    Yes             No
          Yes  |   X(k,1,1)   |   X(k,1,2)
  Test    No   |   X(k,2,1)   |   X(k,2,2)


  k = 1, 2, ..., K
```

The fields of the output structure H are

```
        Zo: Value of the Modified Obuchowski statistic
        Ho: Null hypothesis decision
        Dp: p10,. - p01,. = pt(+) - pc(+)
        CI: (1-alpha)*100% confidence interval for Dp
    Pvalue: p-value of the test
     PTest: pt(+)
  PControl: pc(+)
```

Example: alpha = 0.05; delta = -0.03;
        Ddir = 'C:\PJC\Math_Biology\Chapter_6\Data';
        D = load(fullfile(Ddir,'DataCube.mat'));

```
  H = clustmatchedpair(D.Cube,delta,alpha)      returns
  H =
        Zo: 4.9199
        Ho: 'reject'
        Dp: -0.0014
        CI: [-0.0128 0.0100]
    Pvalue: 5.0608e-07
     PTest: 0.4259
  PControl: 0.4273
```

See also <u>modobuchowski</u>.

Reference: Jun-mo Nam and Deukwoo Kwon
           Non-inferiority test for clustered matched-pair
           data
           Statistics in Medicine
           Volume 28, 2009, pages 1668-1679

CORRTEST

CORRTEST   Correlation hypothesis test.

H = CORRTEST(X1,X2,alpha,Tflag) returns the structure H whose fields contain the decision of the hypothesis test (with respect to the null hypothesis) value of the Pearson Correlation coefficient (rho), the (1-alpha)*100% confidence interval for the correlation rho (CI), the number of degrees of freedom (df), the hypothesis test description (test), and the p-value of the test (Pvalue).

The input vectors X1 and X2 are correlated at the significance level alpha. The hypothesis test is indicated by Tflag as below.

Tflag	Ho	H1	
2	rho = 0	rho ~= 0	(two-tail)
1	rho >= 0	rho < 0	(left-tail)
0	rho <= 0	rho > 0	(right-tail)

Example: X1 = [9.8,7.4,7.9,8.3,8.3,9.0,9.7,8.8,7.6,6.9];
X2 = [9.5,6.7,7.0,8.6,6.7,9.5,9.0,7.6,8.5,8.6];
alpha = 0.05;

H = corrtest(X1,X2,alpha,2)   returns

H =
      Ho: 'do not reject'
     rho: 0.5041
      CI: [-0.7041 0.7041]
      df: 8
    Test: 'two-sided, equal'
  Pvalue: 0.1374

See also <u>regressci</u>, <u>tmean</u>.

CORRVARS

CORVARS   Correlated Variables

[Cv,Names] = corrvars(C,Fnames,thres) returns the matrix Cv of variable indices which are thres*100% correlated with variables 1, 2, ..., p, respectively, the cell array Names of associated variable indices and names as derived from the input cell array of data matrices C = {C{1},C{2},...,C{g}} and variable names (Fnames).

The input cell array C = {C{1},...C{g}} is the collection of data classes with p variables and THRES is the correlation threshold.

[Cv,Names] = corrvars(C,Fnames,thres,plevel) returns the matrix Cv and cell array Names as above using the probability of rejecting the null hypothesis (that variables have 0 correlation) to the significance PLEVEL. That is, p < plevel => reject Ho: uncorrelated.

The default value of PLEVEL is 0.05.

NOTE: This function operates under the [exceedingly optimistic] assumption that the underlying distributions for each variable is normal and hence correlation is associative.

Example: Ddir = 'C:\PJC\Math_Biology\Chapter_4\Data';
                load(fullfile(Ddir,'\Multivariate_MultiClass_Data
                   .mat'));
                w = D.w; p = numel(w); alpha = 0.05; rho = 0.80;
                for j = 1:p; Feature{j} = ['Phi_' num2str(j)]; end

                [R,F] = corrvars(D.X,Feature,rho,alpha)   returns

```
R =
    1    3    4    5    6
    3    4    5    6    0
   13   14   15   16    0
   14   15   16    0    0
   26   27   28   29    0
   28   29    0    0    0
```

See also stdnorm, pca, corr, dafe.

CVCI

CVCI   Confidence interval for the coefficient of variation (CV)

CI = CVCI(CV,n,alpha) returns the structure whose fields contain the Miller, McKay, and Vangel (1 - alpha)*100% confidence intervals for the input coefficient of variation CV and sample size n.

The fields of CI are

>   MH: The Mahmoudvand and Hassani CI (also valid for
>   small n)
>   Miller: The Miller CI
>   McKay: The McKay CI
>   Vangel: The Vangel CI

Example: X = rand(1,100); [cv,N] = coefvar(X);
                CI = cvci(cv,N,0.05)   returns

>   CI =
>
>   MH: [0.5019 0.6636]
>   Miller: [0.4703 0.6757]
>   McKay: [0.4852 0.7074]
>   Vangel: [0.4840 0.7039]

See also coefvar

> Reference: R. Mahmoudvand and H. Hassani
> "Two new confidence intervals for the coefficient of variation
> in a normal distribution"
> Journal of Applied Statistics
> Volume 36, Number 4, April 2009, pps. 429 - 442

See also cvtest.

CVPDF

CVPDF   Coefficient of Variation Probability Density Function.

f = cvpdf(x,df,ncp) returns the probability density function
for the coefficient of variation CV = std/mean. The second
input argument is the degrees of freedom (from the estimate of
the CV) and the third input argument is the (t-distribution)
non-centrality parameter.

Example: nu = [6.3,7.2]; df = 3; x = linspace(0,1,1000);
                f = cvpdf(x,df,nu(1)); g = cvpdf(x,df,nu(2));

See also nctpdf, cvci, cvcvcoef.

CVTEST

CVTEST   Hypothesis test for Coefficient of Variance

H = cvtest(X,Y,alpha) returns the structure H whose fields contain the null hypothesis decision, an estimate of the test statistic, the corresponding p-value, and the (1-alpha)*100% confidence interval for the difference in CVs.

That is, the two-sided hypothesis test

        Ho: CVx = CVy
        H1:   CVx ~= CVy

is exercised for the input vectors of samples X = [x1,x2, ..., xn] and Y = [y1,y2, ... , ym], significance level alpha, with respect to four distinct approaches: The Exact method, the method of Bennett, the Modified Bennett method, and the method of Miller. The structure H contains the fields

            CVx: The CV of the input vector X
            CVy: The CV of the input vector Y
          CICVx: The (1-alpha)*100% Mahmoudvand-Hassani CI for
                 CVx
          CICVy: The (1-alpha)*100% Mahmoudvand-Hassani CI for
                 CVy

        Exact:
            Ho: Decision with respect to the null hypothesis
        Pvalue: The exact test p-value
            T: The estimate [exact method] test statistic
                T = CVx - CVy
            CI: The exact method (1-alpha)*100% confidence
                interval for CVx - CVy

        Bennett:
            Ho: Decision with respect to the null hypothesis
        Pvalue: The Bennett test p-value
            T: The estimate [Bennett method] test statistic
                B
        ModBen:
            Ho: Decision with respect to the null hypothesis
        Pvalue: The Modified Bennett test p-value
            T: The estimate [Modified Bennett method] test
                statistic MB

```
    Miller:
        Ho: Decision with respect to the null hypothesis
    Pvalue:  The Miller test p-value
        T:  The estimate [Miller method] test statistic
            M2
```

Example: Sdir = 'C:\PJC\Math_Biology\Chapter_5\Data';
            load(fullfile(Sdir,'Xcv.mat'));
            load(fullfile(Sdir,'Ycv.mat'));
            alpha = 0.05;

    H = cvtest(X,Y,alpha);   returns
  H =

        CVx: 6.1067
        CVy: 26.6319
      CICVx: [5.1118 7.4948]
      CICVy: [22.8071 31.7299]

      Exact:
                Ho: 'reject'
            Pvalue: 0.0156
                T: -20.5252
                CI: [-24.7672 -16.2832]

      Bennett:
                Ho: 'reject'
            Pvalue: 0
                T: 84.6454

      ModBen:
                Ho: 'reject'
            Pvalue: 0
                T: 84.6454

      Miller:
                Ho: 'do not reject'
            Pvalue: 0.6395
                T: -0.3572

See also cvci, coefvar

DAFE

DAFE   Discriminant Analysis Feature Extraction

x = dafe(S) returns the cell array of feature-coordinates of
the input input cell array S with respect to the weighting
matrix. The number of coordinates correponds to the number
of singular values which constitute 98% of the sum of those
singular values. The weighting matrix is the inverse of the
pooled covariance matrix W and the between groups covariance
across all data classes in the cell array S

x = dafe(S,p) returns a cell array each of whose elements
contain p*100% of the sum of the singular values (i.e., p*100%
of the discrimination information) resulting in N(p) feature-
coordinates of the corresponding element of the input cell
array S.
Note: To use this option, p MUST be in the open interval
(0,1).

x = dafe(S,n) returns a cell array each of whose elements
contain n feature-coordinates of the corresponding element of
the input cell array S.

Example: Ddir = 'C:\PJC\Math_Biology\Chapter_4\Data';
        load(fullfile(Ddir,'Multivariate_MultiClass_Data
          .mat'));
        x = dafe(D.X,0.99);

See also dafeplot, pooledcov, bwtngrps.

DAFEPLOT

DAFEPLOT   Plot of Discriminant Analysis Feature Extraction coordinates

dafeplot(dF,Fnames) returns the 2- or 3- dimensional plot of
the Discriminant Analysis Feature Extraction data contained in
the input cell array dF = [dF{1},dF{2},...,dF{g}} with respect
to the color map associated with the cell array of Feature
Names Fnames.

If dF{k} is an n(k)-by-2 matrix, then a 2-dimensionnal plot is
returned. If, conversely, dF{k} is an n(k)-by-3 matrix, then
a 3-dimensionnal plot is returned. Finally, if dF{k} is an
n(k)-by-p matrix with p > 3, then only the first three
dimensions are plotted and a warning is returned.

Example: Ddir = 'C:\PJC\Math_Biology\Chapter_4\Data';
load(fullfile(Ddir,'Multivariate_MultiClass_Data
.mat'));
M = vertcat(D.X{:}); Z = stdnorm(M,D.X{:});
F = dafe(Z,3);

dafeplot(F,{'Level_{0}','Level_{1}','Level_{2}',
'Level_{3}'});

See also dafe.

DEMING

DEMING   Deming Regression.

S = deming(X,Y,alpha) returns the structure S whose fields
contain the estimated slope (m) and intercept (b) for the
linear regression model y = m*x + b with respect to Deming
weighting. It is assumed that the ratio of variance errors is
one: delta = 1. The output structure also contains an
estimate of Pearson's correlation coefficient (rho) along with
the (1-alpha)*100% confidence intervals for the slope (CIm),
intercept (CIb), and correlation (CIrho).

S = deming(X,Y,alpha,'plot') returns the structure S above
along with a scatterplot of the data, the Deming regrssion
line (red), and the identity line (dotted green)

S = deming(X,Y,alpha,[],delta) returns the structure S above
with the Deming variance ratio of delta. If a plot is also
desired, input 'plot' in the fourth argument rather than the
empty string [].

Reference: P. J. Cornbleet and N. Cochman
              "Incorrect Least-Squares Regression Coefficients in
              Method-Comparison Analysis"
              Clinical Chemistry, Volume 25, Number 3, 1979,
              pp 432-438

Example:
x = ...
[0.978,6.87,7.2,1.15,9.41,2.74,2.31,3.83,6.61,8.57,8.36,2.02,
1.01,6.08,5.25,7.18];

```
y = ...
[1.04,8.24,5.46,1.01,9.41,2.46,2.04,5.26,7.23,7.5,8.53,6.4,
1.09,4.61,7.65,6.96];
alpha = 0.05;

    S = deming(x,y,alpha)    returns

    S =
            m: 0.9893
            b: 0.3858
          rho: 0.8677
          CIm: [0.7883 1.1903]
          CIb: [-0.6136 1.3852]
        CIrho: [0.6527 0.9533]
```

See also regressci.

DLRATIOS

DLRATIOS   Diagnostic Likelihood Ratios, Confidence Intervals, and Equivalence
Test for Two Diagnostic systems.

```
S = dlratios(Acomp,alpha) returns the structure S whose fields
contain the positive DLR+(j) and negative DLR-(j) diagnostic
likelihood ratios for each system (j = 1,2) and their
corresponding (1-alpha)*100% confidence intervals; the positive
(rho+) and negative (rho-) relative diagnostic ratios (rho+ =
DLR+(1)/DLR+(2)) and their corresponding (1-alpha)*100%
confidence intervals; and the standard deviations of the
natural logarithm of the relative diagnostic ratios.

The input comparison matrix Acomp is a 2-by-4 matrix of the
form
```

		T1 = 1		T1 = 0		Totals
	T2 = 1	T2 = 0	T2 = 1	T2 = 0		
Dx = 1	s11	s10	s01	s00		s
Dx = 0	r11	r10	r01	r00		r
Totals	n11	n10	n01	n00		n

where only s11, s01, s01, s00, r11, r10, r01, and r00 are
input. The symbols s and r are the sum of the first and second
rows, respectively, while njk = (sjk + rjk) and n = s + r =
n11 + n10 + n01 + n00.

The fields of S contain the positive and negative diagnostic
likelihood ratios DLRplus, DLRminus) for each test (T1 and
T2), their individual (1-alpha)*100% confidence intervals (CI),
estimates of the natural logarithm of the DLRs (Wplus =
ln(DLRplus(1)/DLRplus(2)) with a similar defintion for Wminus),
the standard deviations of Wplus and Wminus (Std), the
confidence intervals of Wplus and Wminus (LnCI), and the
confidence intervals of the ratios RHOplus = DLRplus(1)/
DLRplus(2), RHOminus = DLRminus(1)/DLRminus(2). Finally, the
decisions with respect to the hypotheses

  Ho: DLRplus(1) = DLRplus(2)   Ko: DLRminus(1) = DLRminus(2)
  H1: DLRplus(1) ~= DLRplus(2)   Ko: DLRminus(1) ~= DLRminus(2)

are provided in the fields HoPlus and HoMinus along with their
corresponding p-values.

Example: A = [786, 29, 183, 25;69, 46, 176, 151];
        alpha = 0.05;
        S = dlratios(A,alpha)   returns

```
  T1                        T2
    DLRplus                   DLRplus
       3.1                       1.7
    DLRminus                  DLRminus
       0.27                      0.12
    CI                        CI
       DLRplus                   DLRplus
         [2.6 3.6]                 [1.6 1.9]
       DLRminus                  DLRminus
         [0.24 0.31]               [0.09 0.16]

  Ratios
    Std
       Wplus
          0.11
       Wminus
          0.15
                    LnCI
    Wplus                 Wplus
       0.58                  [0.37 0.8]
```

```
Wminus                Wminus
    0.84                  [0.55 1.1]
                   CI
RHOplus               RHOplus
    1.8                   [1.4 2.2]
RHOminus              RHOminus
    2.3                   [1.7 3.1]


 HoPlus               HoMinus
    reject                reject


 Pplus                Pminus
   9.142e-08             1.756e-08
```

```
Reference:   J. A. Nofuentes and J de Dios Luna del Castillo
             Comparison of the likelihood ratios of two binary
             diagnostic tests in paired designs
             Statistics in Medicine. Volume 26, pages 4179-4201
             March 2007
```

```
See also dlrtest.
```

DLRTEST

DLRTEST   Hypothesis test of equality of Diagnostic Likelihood Ratios for clustered data.

```
H = DLRTEST(Cube4,alpha) returns the structure H whose fields
contain estimates of the (clustered) sensitivity, specificity,
DLR+, DLR- and their corresponding (1-alpha)*100% confidence
intervals for the Control and Test diagnostic tests (as
compared against the gold standard truth Dx) as contained in
the input K-by-2-by-4 data cube of the form below.
```

	Control = 1		Control = 0		Totals
	Test = 1	Test = 0	Test = 1	Test = 0	
Dx = 1	s11(k)	s10(k)	s01(k)	s00(k)	s(k)
Dx = 0	r11(k)	r10(k)	r01(k)	r00(k)	r(k)
Totals	n11(k)	n10(k)	n01(k)	n00(k)	n(k)

Recall that DLR+ = Sen/(1-Spec) and DLR- = (1-Sen)/Spec. The relative diagnostic ratios rDLR+ and rDLR- are ratios of the postive and negative DLRs with respect to the Control and Test. That is,

rDLR+ = DLR+(Control)/DLR+(Test) and rDLR- = DLR-(Control)/ DLR-(Test)

These quantities along with their associated (1-alpha)*100% confidence intervals are contained in the field rDLR.

Concisely, H has the fields

```
H.Control
  Sen
    Value: Estimate of Sen(Control)
       CI: Confidence interval for Sen(Control)
  Spec
    Value: Estimate of Spec(Control)
       CI: Confidence interval for Spec(Control)
  DLRplus
    Value: Estimate of DLR+(Control)
       CI: Confidence interval for DLR+(Control)
  DLRminus
    Value: Estimate of DLR-(Control)
       CI: Confidence interval for DLR-(Control)

H.Test
  Sen
    Value: Estimate of Sen(Test)
       CI: Confidence interval for Sen(Test)
  Spec
    Value: Estimate of Spec(Test)
       CI: Confidence interval for Spec(Test)
  DLRplus
    Value: Estimate of DLR+(Test)
       CI: Confidence interval for DLR+(Test)
  DLRminus
    Value: Estimate of DLR-(Test)
       CI: Confidence interval for DLR-(Test)

H.Hypothesis
  Sen
       Ho: Decision with respect to Ho: Sen(T1) = Sen(T2)
   Pvalue: P-value associated with the hypothesis test
     dSen: Sen(T1) - Sen(T2)
```

CIdSen: (1-alpha)*100% Confidence Interval for dSen
Spec
    Ho: Decision with respect to Ho: Spec(T1) = Spec(T2)
  Pvalue: P-value associated with the hypothesis test
   dSpec: Spec(T1) - Spec(T2)
 CIdSpec: (1-alpha)*100% Confidence Interval for dSpec
rDLRp
    Ho: Decision with respect to Ho: DLR+(T1) = DLR+(T2)
  Pvalue: P-value associated with the hypothesis test
   rDLRp: DLR+(T1)/DLR+(T2)
 CIrDLRp: (1-alpha)*100% Confidence Interval for rDLR+
rDLRm
    Ho: Decision with respect to Ho: DLR-(T1) = DLR-(T2)
  Pvalue: P-value associated with the hypothesis test
   rDLRm: DLR-(T1)/DLR-(T2)
 CIrDLRm: (1-alpha)*100% Confidence Interval for rDLR-

Example: Ddir = 'C:\PJC\Math_Biology\Chapter_6\Data';
      E = load(fullfile(Ddir,'DataCube4.mat')); alpha = 0.05;
      H = dlrtest(E.Cube4,alpha);

See also clustmatchedpair.m, dlratios.m

FRIED

FRIED   The Normalized Friedman Matrix for Regularized Discriminant Analysis.

F = fried(A,B,gam,lam) returns the normalized Friedman matrix
of the n-by-p input matrices A, B with weighting parameters
gam and lam as defined by

  S = (1-gam)*[(1-lam)*A + lam*B] + (gam/p)*trace((1-lam)*A
    + lam*B)*I.

Here I is the p-by-p identity matrix and p = column_size(A).

Example: Ddir = 'C:\PJC\Math_Biology\Chapter_4\Data';
      load(fullfile(Ddir,'Multivariate_MultiClass_Data
        .mat'));
      M = vertcat(D.X{:}); Z = stdnorm(M,D.X{:});
      Cp = pooledcov(Z);

      F = fried(cov(Z{1}),Cp,0.4,0.9);

See also rda, pooledcov, bwtngrps.

FTEST

FTEST   The F–test for the comparison of variances.

```
H = FTEST(X,Y,alpha) returns the structure H whose fields
contain the decision with respect to the null hypothesis (Ho),
the values f the test statistic (F), the (1-alpha)*100%
confidence interval for the test statistic, the degrees of
freedom (df), the p-value of the test, and the effect size for
the indicated test.

    Ho:  Sx^2 = Sy^2    (equal variances)
    H1:  Sx^2 ~= Sy^2   (unequal variances)

    Sx = std(X) and Sy = std(Y).

H = FTEST(X,Y,alpha,Tflag) computes a left-tailed, right-
tailed, or two-sized test as indicated by the test flag (Tflag)
below

    Tflag              Ho                 H1
    -----            -----              -----
      2            Sx^2 = Sy^2        Sx^2 ~= Sy^2   (two-tail)
      1            Sx^2 >= Sy^2       Sx^2 < Sy^2    (left-tail)
      0            Sx^2 <= Sy^2       Sx^2 > Sy^2    (right-tail)

The structure H contains the fields

       Ho: Null hypothesis decision state
        F: Value of the F-statistic
       CI: (1-alpha)*100% confidence interval
       df: Degrees of freedom
   Pvalue: The p-value of the test (i.e., the significance
           probability = the smallest level of significance at
           which the null hypothesis is rejected on the basis
           of the observed data
   Effect: The F-test effect size of the data with respect to
           the signficance level alpha.
```

Example:
```
X = [20888,14858,17444,21573,13013,12427,12540,11882,12191, ...
       11102,8697,15696,24745,30311,20239,24023,16143,3919, ...
       9331,4653];
Y = [4273,2077,3837,3756,1587,2013,1714,2208,490,705,539, ...
       423,6188,7284,11314,10046,25799,23384,24003,22917];
```

```
H = ftest(X,Y,0.05)    returns

H =

        Ho: 'do not reject'
         F: 0.5797
        CI: [0.3958 2.5265]
        df: [19 19]
    Pvalue: 0.1219
    Effect: 0.4774
```

See also, wilcoxonsr, tmean.

HISTPLOT

HISTPLOT   Histogram Distribution Plot.

histplot(Data,Nh) plots a histogram of the input data (Data) divided into Nh bins along with the corresponding normal probability density function based upon the mean and standard deviation of the data. Note that the histogram is normalized so that it sums to 1 and therefore is plotted on the same scale as the normal distribution. Moreover, the number of bins Nh can be replaced by a vector of bin centers (Bcen) so that the centers of a histogram can be established.

histplot(Data,Nh,colour) returns the same plot as above using the color COLOOUR to display the results.

histplot(Data,Nh,colour,Tstr) returns the same plot as above using the color COLOOUR and title Tstr to display the results.

[x,y] = histplot(Data,Nh,...) returns the plot above along with the corresponding x- and y-coordinates from the histogram.

See also distplot.

HIVSEIRPLOT

HIVSEIRPLOT   Plotting function for the HIV/AIDS SEIR Model.

hivseirplot(Cflag,T) returns the graphic containing the population curves for the susceptible S(t), HIV+ X(t), HIV+

but no longer infectious Xq(t), AIDS+ Y(T), and AIDS+ but no
longer infectious Yq(t) with respect to the contact profile
specified by Cflag and the time partition T.

hivseirplot(Cflag,T,P) returns the graphic as above with
respect to the vector P = {p1,p2,...,pm] of "not-quarantined"
proportions. The default value for P is P = [0.95, 0.70,
0.50].

Cflag	Definition	Time Partition
'single'	c(t) = 1 for t >= to	T = to
'compound'	c(t) = 0 for to <= t < t1	
	= 1 for t1 <= t < t2	
	= 0 for t2 <= t < t3	T = [to,t1,t2,t3,t4,t5]
	= 2 for t3 <= t < t4	
	= 0 for t4 <= t < t5	
	= 3 for t >= t5	
'multiple'	c(t) = m for t >= to	T = to

Examples: hivseirplot('single',0);
         hivseirplot('compound',[0,1,2,3,5,7]);
         hivseirplot('multiple',0);

See also <u>probhiv</u>, <u>serohiv</u>, <u>gexpcdf</u>, <u>gexppdf</u>, <u>aidsrate</u>,
       <u>hivseirsys</u>, <u>hivseirsim</u>, <u>Cprofile3</u>, <u>Cprofile</u>

KCLUSTER

KCLUSTER   Plotting function for the HIV/AIDS SEIR Model.

S = KCLUSTER(X,k) returns the structure S whose fields
contain the cell array C whose elements are the k-means
partition of the input n-by-p data matrix X along with the
partition centroids and standard deviations. That is, S
contains the fields

   C = {C{1},C{2},   ..., C{k}} where each C{j} is an nj-by-p
       matrix with n1 + n2 + ... + nk = n,
   mu = {mu{1},mu{2},...,mu{k}} where each mu{j} is the 1-by-p
       centroid of the partition element C{j},
   sig = {sig{1},sig{2},...,sig{k}} where each sig{j} is the
       1-by-p standard deviation of the partition element C{j},

```
xindx = {xindx{1},xindx{2},...,xindx{k}} where each xindx{j}
    is the index into X which yields the partition C{j}:
    C{j} = X(xindx{j},:),
N = [n(1),n(2),...,n(k)] where n(j) is the number of samples
    in the partition element C{j}.
```

Example: Ddir = 'C:\PJC\Math_Biology\Chapter_1\Data';
        load(fullfile(Ddir,'Clustered_Data_Classes1_2.mat'));
        X = C{1}; Y = C{2};

        S = kcluster(Y,4)   returns
  S =

           C: {[99x2 double] [101x2 double]}
         mu: {[-5.0994 1.1119] [-2.8774 -0.2291]}
       sig: {[0.7453 0.9430] [0.9142 0.8061]}
    xindx: {[99x1 double] [101x1 double]}
          N: [99 101]

See also <u>mahalanobis</u>, sdindex, <u>kmeans</u>, <u>mdist</u>, sdbwindex,
<u>rmsstd</u>, <u>chindex</u>.

LEVENE

LEVENE   Levene's Test for the Statistical Equivalence of Variances.

```
H = LEVENE(X,Y,alpha) returns the structure whose fields
contain the null hypothesis decision, the test statistic (T),
the (1-alpha)*100% confidence interval CI, and the p-value
for the  null hypothesis test of equal variances between the
distributions.
```

The null hypothesis is that the sample variances are equal.

```
Ho:  Sx^2 = Sy^2
H1:  Sx^2 ~= Sy^2.
```

```
D = LEVENE(X,Y,alpha,'plot') returns structure as above along
with a plot of the confidence interval and test statistic
value.
```

Example: alpha = 0.05;
        X = [4.7,3.7,5.2,6.3,6.2,6.7,2.8,4.8,6.1,3.9];

```
Y = [10.1,8.6,10.9,9.7,9.7,10,9.4,10.1,9.9,10,10.8,
     8.7];
H = levene(X,Y,alpha,'plot')    returns

H =
  Ho: 'reject'
   T: 2.3428
  CI: [-2.0860 2.0860]
Pvalue: 0.0296
```

along with the plot of the corresponding T–distribution, acceptance region, and value of the test statistic.

```
Reference:  "Applied Statistics", Second Edition
            J. P. Marques de Sá
            Springer, 2007
            Especially, pages 130 - 131.
```

See also <u>regressci</u>, <u>ftest</u>.

MAHALANOBIS

MAHALANOBIS   The Mahalanobis distance.

```
d = mahalanobis(X,mu,C) returns the Mahalanobis distance of
the input matrix X to data class represented by the group
covariance matrix C and group mean mu:

   d = sqrt( abs( (X-mu)*inv(C)*((X-mu).') ) )
```

Example: Ddir = 'C:\PJC\Math_Biology\Chapter_4\Data';
```
            load(fullfile(Ddir,'Multivariate_MultiClass_Data
              .mat'));
            M = vertcat(D.X{:}); Z = stdnorm(M,D.X{:});

            d = mahalanobis(Z{2},mean(Z{4}),pooledcov(Z));
```

See also <u>qda</u>, <u>lda</u>, <u>rda</u>, <u>bhattacharyya</u>.

MCNEMAR

MCNEMAR   McNemar's Test for a 2–by–2 Contingency Table.

This method tests the hypothesis that nominal data have equal column and row frequencies. It is applied to 2-by-2 contingency tables with dichotomous traits on matched pairs.

H = MCNEMAR(A,alpha) returns the structure H whose fields contain the decision with respect to the null hypothesis (Ho), the value of the test statistic (T), the (1-alpha)*100% confidence interval (CI) for T, and the p-value for the test.

Two separate methods are used to compute the hypothesis test: McNemar, and Corrected McNemar. The structure H has the fields

```
    H =
          McNemar: [1x1 struct]
        Corrected: [1x1 struct]
```

and each field contains the null hypothesis decision (Ho), test statistic value (T), and confidence interval (CI).

Example: A = [49,21;8,82]; alpha = 0.05;
            H = mcnemar(A,alpha)   returns

```
    H.McNemar =
          Ho: 'reject'
           T: 5.8276
          CI: [9.8207e-004 5.0239]
      Pvalue: 0.0158
        CIdp: [-0.1460 -0.0165]

    H.Corrected =
          Ho: 'reject'
           T: 4.9655
          CI: [9.8207e-004 5.0239]
      Pvalue: 0.0259
```

See also chi2test.

MEANCI

MEANCI   Confidence interval for the mean.

CI = MEANCI(data,alpha) returns a (1 - alpha)*100% confidence interval CI for the mean of the input data.

CI = MEANCI(data,alpha,'Bootstrap') returns a (1 - alpha)*100% confidence interval CI for the bootstrapped mean mean of the input data using m = 10000 resamplings of the data.

CI = MEANCI(data,alpha,'Bootstrap',m) returns a (1 - alpha) *100% confidence interval CI for the bootstrapped mean of the input data using m resamplings of the data.

[CI,mu] = MEANCI(data,alpha, ...) returns the (1 - alpha)*100% confidence interval for the computed mean mu. Note that in the case of the Bootstrap method, mu is the bootstrapped mean.

For details, see
  (1) "Applied Statistics (Second Edition)" by J. P. Marques de Sá, Springer-Verlag, Berlin, 2007
  (2) "Probability and Statistical Inference" by R. V. Hogg and E. A. Tanis, Macmillian Publishing Company, New York, 1977
  (3) "Mathematical Statistics: Basic Ideas and Selected Topics" by P. J. Bickel and K. A. Doksum, Holden-Day, San Francisco, 1977

Examples: data = [1.02,1.21,1.49,1.18]; alpha = 0.05;

  CI = meanci(data,alpha)   returns [0.9141, 1.5359]

  CI = meanci(data,alpha,'Bootstrap')   returns [1.0914, 1.3613]

  CI = meanci(data,alpha,'Bootstrap',1e+06)   returns [1.0905, 1.3596]

See also <u>tmean</u>.

MODOBUCHOWSKI

MODOBUCHOWSKI   The modified Obuchowski statistic.

[Zo,Vz] = MODOBUCHOWSKI(Dcube,delta) returns the modified Obuchowski statistic Zo and the variance of that statistic (Vz) used in non-inferiority tests on clusters of matched pairs with correlations. The second input is the non-inferiority margin (delta).

[Zo,Vz,P] = MODOBUCHOWSKI (Dcube,delta) returns the modified Obuchowski statistic Zo and its associate variance Vz as above along with the vector P = [Pt(+),Pc(+)] the proportions of "Yes" diagnoses contained in the contingency latter X(k;i,j) described below.

NOTE: The non-inferiority margin is assumed to be negative: delta < 0

The first input (Dcube) is the K-by-2-by-2 data cube (Dcube) which contains the 2-by-2 Control versus Test contingency tables for each (cluster) sample k = 1,2, ..., K.

The contingency tables are represented as below.

```
                        Control
                 Yes            No              Total
        Yes  |   X(k,1,1)   |   X(k,1,2)   |  sum(X(k,1,:))
Test    No   |   X(k,2,1)   |   X(k,2,2)   |  sum(X(k,2,:))
        Total|sum(X(k,:,1)) | sum(X(k,:,2)) |  sum(X(k,:))
```

for k = 1, 2, ..., K (= number of clusters)

Example: delta = -0.03;
          Ddir = 'C:\PJC\Math_Biology\Chapter_6\Data';
          D = load(fullfile(Ddir, 'DataCube.mat'));

          [Zo,Vz,P] = modobuchowski(D.Pcube,delta)   returns

          Zo = 4.9199, Vz = 3.3774e-05, and P = [0.4259,
              0.4273]

Reference: Jun-mo Nam and Deukwoo Kwon
            Non-inferiority test for clustered matched-pair
            data
            Statistics in Medicine
            Volume 28, 2009, pages 1668-1679

See also clustmatchedpair.m

MWUTEST

MWUTEST   Mann–Whitney *U*–test.

H = mwutest(X,Y,alpha,Tflag) returns the result of the Mann-Whitney U-Test (also called rank-sum test or Wilcoxon-Mann-Whitney test) of stochastic equality for samples taken from two independent populations at a significance level alpha.

More specifically, let X = [x(1),x(2),...,x(n)] and Y = [y(1), y(2),...,y(m)] be samples of size n and m of the independent random variables X and Y with distributions Fx and Fy. To determine whether the distributions are equal (greater or less than) the Mann-Whitney U-test is performed on the input data with respect to the significance level alpha and test flag Tflag (as below).

Tflag	Ho	H1	
----	------	------	
2	Fx = Fy	Fx ~= Fy	(two-tail)
1	Fx >= Fy	Fx < Fy	(left-tail)
0	Fx <= Fy	Fx > Fy	(right-tail)

Example: X = [4.3, 5.9, 4.9, 3.1, 5.3, 6.4, 6.2, 3.8, 7.5, 5.8];
Y = [5.5, 7.9, 6.8, 9.0, 5.6, 6.3, 8.5, 4.6, 7.1];
alpha = 0.05;
H = mwutest(X,Y,alpha,2)   returns

H =

Ho: 'do not reject'
 U: 24
CR: [20.9954 69.0046]
Pvalue: 0.0864

Note: If the sample size for X and Y is less than or equal to 8, then an exact p-value is returned

Reference: Mathematical Statistics: Basic Ideas and Selected Topics
Peter J. Bickel and Kjell A. Doksum
Holden-Day, San Francisco Copyright (c) 1977
(Especially section 9.1)

See also tmean, wilcoxonsr, wilcLookup

NIFTEST

NIFTEST Non-inferiority, equivalence, and superiority hypothesis tests for distinct variances.

```
H = niftest(X,Y,alpha,delta) returns the structurs H whose
fields contain the disposition of the null hypothesis, the
two-sided confidence interval for Var(X)/Var(Y), the p-value,
the type of test administered, and an echo of the clinical
margin.  The fields of the structure H are

        Pvalue: The p-value of the test
            Ho: Null hypothesis decision state
          VarX: Estimate of the variance from the sample X
          CIVx: Two-sided (1-alpha)*100% confidence interval
                for VarX
          VarY: Estimate of the variance from the sample Y
          CIVy: Two-sided (1-alpha)*100% confidence interval
                for VarY
             F: Value of the F-statistic F = Sx^2/Sy^2
            CI: (1-alpha)*100% confidence interval for F
      TestType: The type of hypothesis test (i.e., Equivalence,
                Non-inferiority)
        margin: The clincial margin

    The tests exercised by niftest are defined by the test flag
Tflag

Tflag       Hypothesis Test                      Type
____        _____                        ____
  0         Ho: Var(X)/Var(Y) >= 1 + delta       Non-inferiority
            H1: Var(X)/Var(Y) <  1 + delta

  1         Ho: Var(X)/Var(Y) >= 1 - delta       Superiority
            H1: Var(X)/Var(Y) <  1 - delta

  2         Ho: |Var(x)/Var(Y) - 1| >= delta     Equivalence
            H1: |Var(x)/Var(Y) - 1| <  delta

Example:    Ddir = 'C:\PJC\Math_Biology\Chapter_5\Data';
            load(fullfile(Ddir,'HockeyData.mat'))
            load(fullfile(Ddir,'BasketballData.mat'))
            X = D.Wt; Y = B.Wt;
            alpha = 0.05; delta = 0.025;
```

```
H = niftest(X,Y,alpha,delta,2)    returns

H =
      Pvalue: 1
          Ho: 'do not reject'
        VarX: 229.7623
        CIVx: [207.2159 256.2195]
        VarY: 684.1271
        CIVy: [601.7618 784.7357]
           F: 0.3358
          CI: [0.2908 0.3974]
    TestType: 'Equivalence'
      margin: 0.0250
See also nimean, niprop, nipropi
```

NIMEAN

NIMEAN   Non–inferiority, equivalence, and superiority hypothesis tests for distinct means.

```
H = nimean(X,Y,alpha,delta) returns the structure H whose
fields contain the decision with respect to the null hypothesis
(Ho), the value of the test statistic (T), the estimates of
the means of the input data vectors X and Y (Mx and My), their
respective confidence intervals (CIx and CIy), the difference
in means (dM) and confidence interval (CIdM), the number of
degrees of freedom for the test statistic (df), the hypothesis
test conducted (Test), and the p-value of the test (Pvalue).
More directly, the fields of H are

          Ho: Null hypothesis decision
           T: Value of the test statistic
          Mx: The mean of the X data
          My: The mean of the Y data
         CIx: Confidence interval for Mx
         CIy: Confidence interval for My
          dM: dM = Mx - My
        CIdM: CI(dM)
          df: Degrees of freedom for T
        Test: Type of hypothesis
      Pvalue: p-value
```

The type of hypothesis test (equivalence, non-inferiority, superiority) is identified via the test flag Tflag. The default value of the test flag is Tflag = 2.

Tflag	Ho	H1	
2	|My - Mx| >= delta	|My - Mx| < delta	(Equivalence)
1	My - Mx <= delta	My - Mx > delta	(Superiority)
0	My - Mx <= -delta	My - Mx > -delta	(Non-inferiority)

Remark: nimean uses the Levene test to determine whether the variances are equal or unequal.

Examples: x = 10*rand(1,400); y = 2*rand(1,250);
          alpha = 0.05; delta = 1;

    H = nimean(x,y,alpha,delta,2)   returns
    H =
          Ho: 'reject'
           T: 19.7393
          Mx: 4.9455
          My: 1.0230
         CIx: [4.6623 5.2287]
         CIy: [0.9558 1.0903]
          dM: 3.9225
        CIdM: [3.5603 4.2846]
          df: 444
        Test: 'two-sided, equivalence'
      Pvalue: 1.0084e-62

    H = nimean(x,y,alpha,delta,1)   returns
    H =
          Ho: 'do not reject'
           T: 19.7393
          Mx: 4.9455
          My: 1.0230
         CIx: [4.6623 5.2287]
         CIy: [0.9558 1.0903]
          dM: 3.9225
        CIdM: [-Inf 4.2262]
          df: 444
        Test: 'one-sided, superiority'
      Pvalue: 1

    H = nimean(x,y,alpha,delta,0)   returns

```
     H =
           Ho: 'reject'
            T: 33.2481
           Mx: 4.9455
           My: 1.0230
          CIx: [4.6623 5.2287]
          CIy: [0.9558 1.0903]
           dM: 3.9225
         CIdM: [3.6187 Inf]
           df: 444
         Test: 'one-sided, non-inferiority'
       Pvalue: 7.0887e-123
```

See also niprop, tmean

NIPROP

NIPROP   Equivalence hypothesis test for paired binary data and non–inferiority (or clinical) margin.

H = niprop(A,alpha,delta) returns the structure H whose fields contain the decision with respect to the (equivalence test) null hypothesis

    Ho: |Ptest - Pcontrol| >= -delta

for a non-inferiority/clinical margin delta.

Also, a (1-alpha)*100% confidence interval for the difference in proportions Dp = Ptest - Pcontrol is a field within H.

H = niprop(A,alpha,delta,0) returns the structure H whose fields contain the decision with respect to the (non-inferiority test) null hypothesis

    Ho: Ptest - Pcontrol <= -delta   (delta > 0)

H = niprop(A,alpha,delta,1) returns the structure H whose fields contain the decision with respect to the (superiority test) null hypothesis

    Ho: Ptest - Pcontrol <= delta

The computations are done with respect to the Wald and Restricted Maximum Likelihood Estimation (RMLE) methods.

Tlower and Tupper are the lower and upper test statistics, respectively.

The equivalence test can also be called via H = niprop (A,alpha,delta,2)

The input matrix A is a 2-by-2 Control/Test contingency matrix of the form

```
                         Control

                  Yes   |  No   |  Totals
               _____|_____|_____
                        |       |
       Test |   Yes  |  X11  |  X10  |  X1.
            |_____|_____|_____|_____
            |        |       |       |
            |   No   |  X01  |  X00  |  X0.
            |_____|_____|_____|_____
            |        |       |       |
            | Totals |  X.1  |  X.0  |  n
```

and A is the matrix A = [X11,X10;X01,X00].

Reference: J-p Liu, H-m Hsueh, E. Hsieh, and J.J. Chen
            Tests for equivalence or non-inferiority for
            paired binary data
            Statistics in Medicine, Volume 21, pages 231-245
            Copyright (c) 2002

Example: A = [36,5;4,2]; alpha = 0.05; delta = 0.10;
            H = niprop(A,alpha,delta); returns

```
Wald                         RMLE
   Ho                           Ho
      do not reject               do not reject
   Tstat                        Tstat
      [1.902 -1.235]               [1.647 -1.176]
   Pvalue                       Pvalue
      0.1085                       0.1198
   TestType                     TestType
      Equivalence                  Equivalence
CI
   Wald
      [-0.1037 0.1462]
   RMLE
      [-0.09987 0.1314]
```

See also <u>nisamplesize</u>, <u>clustmatchedpair</u>, <u>modobuchowski</u>.

NIPROPI

NIPROPI   Non–inferiority, equivalence, superiority hypothesis tests for independent (i.e., uncorrelated) portions.


```
H = nipropi(X,Y,alpha,delta) returns the structure whose fields
contain the disposition of the null hypothesis, the p-value of
the hypothesis test, estimates of the portions corresponding
to the dichotomous input vectors X = [x(1),x(2),...,x(n1)] and
Y = [y(1),y(2),...,y(n2)], the difference of portions, and the
corresponding (1-alpha)*100% Newcombe score method confidence
intervals.  The constant delta is the CLINICAL MARGIN defining
the equivalence/non-inferiority/superiority of the
proportions.
```

```
The tests exercised by nipropi are defined by the test flag Tflag
```

Tflag	Hypothesis Test	Type
0	Ho: py - px <= -delta H1: py - px > -delta	Non-inferiority
1	Ho: py - px <= delta H1: py - px > delta	Superiority
2	Ho: |py - px| >= delta H1: |py - px| < delta	Equivalence

```
The output struct H has the fields
```

```
      Pvalue: p-value of the hypothesis test Ho vs. H1
          Ho: Decision with respect to the null hypothesis Ho
          p1: Estimate of px
        NCI1: Newcombe (1-alpha)*100% confidence interval for p1
          p2: Estimate of py
        NCI2: Newcombe (1-alpha)*100% confidence interval for p2
          dp: Difference in proportions dp = pi - p2
         NCI: Newcombe (1-alpha)*100% confidence interval for dp
    TestType: Type of hypothesis test
      margin: Clinical margin used in the hypothesis test
```

```
Example: Ddir = 'C:\PJC\Math_Biology\Chapter_5\Data';
         load(fullfile(Ddir,'HockeyData.mat'));
         load(fullfile(Ddir,'BasketballData.mat'))
```

```
    X = D.Wt; Y = B.Wt; Xh = (X >= 200); Yb = (Y >= 200);
    alpha = 0.05; delta = 0.025;

    H = nipropi(Xh,Yb,alpha,delta,1)    returns
H =

    Pvalue: 1.5573e-22
        Ho: 'reject'
        p1: 0.5716
      NCI1: [0.5343 0.6082]
        p2: 0.8059
      NCI2: [0.7663 0.8402]
        dp: -0.2343
       NCI: [-Inf -0.1804]
  TestType: 'Superiority'
    margin: 0.0250
```

See also <u>niprop</u>, <u>nimean</u>, <u>niftest</u>

NISAMPLESIZE

NISAMPLESIZE   Non–inferiority, Equivalence, Superiority Hypothesis tests for independent (i.e., uncorrelated) portions.

N = nisamplesize(Dcube,alpha,beta,delta) returns the required sample (that is, cluster) size for a non-inferiority test comparing two matched proportions with correlated samples. The input parameters alpha, beta, and delta are the significance level (alpha), the compliment of the power (pwr = 1 - beta) and non-inferiority margins (delta = [delta(1), delta(2)]). Two margins are required to determine the sensitivity of the sample size estimate. The second margin (delta(2)) is applied to the alternate hypothesis and is larger or equal to the null hypothesis margin (delta(1)).

The first input is the K-by-2-by-2 data cube (Dcube) which contains the 2-by-2 Control versus Test continency tables for each (cluster) sample k = 1,2, ..., K. The K contingency tables are represented as below.

```
                      Control
                  Yes            No           Total
          Yes  |   X(1,1)   |   X(1,2)   |  sum(X(1,:))
  Test    No   |   X(2,1)   |   X(2,2)   |  sum(X(2,:))
          Total |sum(X(:,1)) | sum(X(:,2)) |  sum(X(:))
```

If K = 1, then the output Ns is a structure which contains the Wald and RMLE estimates for the equivalence hypothesis test with non-inferiority margin delta.

Examples: alpha = 0.05; beta = 0.2; delta = [0.1, 0.12];
                 Ddir = ' C:\PJC\Math_Biology\Chapter_6\Data';
                 Dcube = load([dirchk(Ddir) 'DataCube3.mat']);

                 Ns = nisamplesize(Dcube.Pcube,alpha,beta,delta)
                     returns
                 Ns =
                     505

                 A = [36,5;4,2];
                 Ns = nisamplesize(A,alpha,beta,delta(1))   returns
                 Ns =
                     Wald: 134
                     RMLE: 168

References: (1) Jun-mo Nam
                 Power and sample size requirements for
                 non-inferiority in studies comparing two
                 matched proportions where the events are
                 correlated
                 Computational Statistics and Data Analysis
                 Volume 55, Copyright (c) 2011, pages 2880-2887

            (2) J-p Liu, H-m Hsueh, E. Hsieh, and J.J. Chen
                 Tests for equivalence or non-inferiority for
                 paired binary data
                 Statistics in Medicine, Volume 21,
                 pages 231-245
                 Copyright (c) 2002

See also modobuchowski, niprop.

QQPLOTMB

QQPLOTMB   Display an empirical quantile-quantile plot modified for mathematical biology project.

qqplotmb(X) makes an empirical QQ-plot of the quantiles of the data set X versus the quantiles of a standard Normal distribution.

qqplotmb(X,Y) makes an empirical QQ-plot of the quantiles of
the data set X versus the quantiles of the data set Y.

H = qqplotmb(X,Y,PVEC) allows you to specify the plotted
quantiles in the vector PVEC. H is a handle to the plotted
lines.

[H,rms] = qqplotmb(X,...) returns the plot handle H and the
average root mean square error.

When both X and Y are input, the default quantiles are those
of the smaller data set.

The purpose of the quantile-quantile plot is to determine
whether the sample in X is drawn from a Normal (i.e.,
Gaussian) distribution, or whether the samples in X and Y come
from the same distribution type. If the samples do come from
the same distribution (same shape), even if one distribution
is shifted and re-scaled from the other (different location
and scale parameters), the plot will be linear.

PCAMB

PCAMB   Principal Component Analysis Transformation and Scores (written for
mathematical biology project).

[Score,Load,Lambda] = pca(X) returns the n-by-p matrix of
Principal Component Scores (Score), the the n-by-p matrix of
Principal Component Loadings (Load), and the n weighted
eigenvalues (Lambda) of the sample covariance matrix S of the
n-by-p input data matrix X.
The method is to compute the SVD of (X - Xave) where Xave is
the 1-by-p vector of column averages of X. That is,

U*D*V.' = svd([X - Xave]) where V is a p-by-p orthogonal
matrix whose columns are the eigenvectors of S = cov(X); D is
the n-by-p matrix of singular values so that Lambda(j,j) =
D(j,j)^2/(n - 1) are the eigenvectors of S; and U is the n-by-
n matrix whose columns are the eigenvectors of [X - Xave].

Notes: (1) n = number of measurements while p = number of
variables.
(2) This method uses Group Mean Centering

The eigenvalues (Lambda) of the covariance matrix (S) for X
are sorted and the first N values are taken so that

$$
\sum_{j=1}^{N} Lambda(j,j) \Bigg/ \sum_{j=1}^{p} Lambda(j,j) \geq 0.98
$$

where Lambda(j,j) = D(j,j)^2/(n - 1).

[Score,Load,Lambda] = pca(X,q) for 0 < q < 1 returns the
Principal Component scores, loadings, and variance totals
(Lambda) which contain q*100% of the variance.

[Score,Load,Lambda] = pca(X,m) for m > 1 an integer, returns
the first m Principal Component scores, loadings, and variance
totals.

[Score,Load,Lambda,Z] = pca(X,...) returns the m-by-n Principal
Component scores (Score), the p-by-m Principal Component
Loadings (Load), the m-by-1 variance measures (Lambda), and
the p-by-m Principal Component transformation matrix (Z).
Here m is either the number of specified Principal Components
or m is the number of Principal Components required to obtain
q*100% of the variance.

Example: Ddir = 'C:\PJC\Math_Biology\Chapter_4\Data';
            load(fullfile(Ddir,'Multivariate_MultiClass_Data
               .mat'));
            X = vertcat(D.X{:});

            [Sc,Ld,sig,PI] = pcamb(X,0.99);

See also <u>scorecompare</u>, <u>pcacompare</u>.

PCACOMPARE

PCACOMPARE   Compare the PCA Decomposition.

pcacompare(Ld,Colours,Shapes) Plots the PCA Decompositions
from the cell array of Principal Components loadings Ld with
respect to the input colors and shapes.

Example: Ddir = 'C:\PJC\Math_Biology\Chapter_4\Data';
               load(fullfile(Ddir,'Multivariate_MultiClass_Data
               .mat');
               X = vertcat(D.X{:}); M = vertcat(D.X{:});
               Z = stdnorm(M,D.X{:});
               for j = 1:numel(Z)
                 [Sc{j},Ld{j},Lam{j}] = pcamb(D.X{j},0.99);
               end
               colours = {'b','g','r','k'}; shapes = {'o','d','s',
               'h'};
               pcacompare(Ld,colours,shapes);

See also <u>pcamb</u>, <u>scorecompare</u>.

POOLEDCOV

POOLEDCOV   Pooled Within-Groups Covariance Matrix.

Cp = pooledcov(S) returns the pooled within-groups covariance matrix formed as the weighted sum of the input cell array of data matrices S = {T1,T2,...Tg}.

Example: Ddir = 'C:\PJC\Math_Biology\Chapter_4\Data';
             load(fullfile(Ddir,'Multivariate_MultiClass_Data
             .mat'));
             M = vertcat(D.X{:}); Z = stdnorm(M,D.X{:});

             B = pooledcov(Z);

See also <u>dafe</u>, <u>bwtngrps</u>, <u>crimcoords</u>, <u>fried</u>, <u>rda</u>.

RDA

RDA   Regularized Discriminant Analysis.

[cf,indx] = rda(x,gam,lam,r,A1,A2, ... ,Ag) returns the rda Classification Score (cf) and Group classification (indx) of the test object x with respect to the g-data groups A1, A2,..., Ag and the weighting parameters gam (to favor individual Group Covariance matrices and Quadratic Discriminant Analysis) and lam (to favor Pooled Within-Group Covariance matrix and Linear Discriminant Analysis).   The vector r = [r(1),r(2),...,r(g)]

are the prior probabilities that the measurement x belongs to
class k. The data class matrices A1, A2, etc. are n(k)-by-p.
If the input vector r <= 0, then the default value r(k) =
n(k)/N where N = sum(n) is used.

NOTE: The input Group data Aj must be n(j)-by-p matrices.
Moreover, 0 <= gam <= 1 and 0 <= lam <= 1.

Example: Ddir = 'C:\PJC\Math_Biology\Chapter_4\Data';
        load(fullfile(Ddir,'Multivariate_MultiClass_Data
          .mat'));
        M = vertcat(D.X{:}); Z = stdnorm(M,D.X{:});

        [bs,ind] = rda(Z{2},0.2,0.75,-1,Z);

See also <u>qda</u>, <u>lda</u>, <u>mahalanobis2</u>, <u>pooledcov</u>, <u>fried</u>.

REGRESSCI

REGRESSCI   Confidence Intervals for Linear Regression Models

S = REGRESSCI(X,Y,alpha) returns the (1-alpha)*100% confidence
intervals for the linear regression model Y = b + m*X. That
is, S is a structure whose field CIb contains the confidence
interval for the linear offset b and CIm contains the
confidence interval for the slope m. The output S is a
structure with the fields

S =

      CIb: (1-alpha)*100% Confidence interval for the intercept
         b
      CIm: (1-alpha)*100% Confidence interval for the slope m
        b: Estimated (regression) Intercept
        m: Estimated (regression) Slope
    rhoXY: The correlation coefficient between X and Y
      RMS: Root mean square error
  coefdet: Coefficient of determination
    Ypred: (1-a)*100% confidence bounds on the predicted
        regression points
     CIbD: 1-alpha)*100% Confidence interval for the intercept
        b of the difference D = Y - X
     CImD: (1-alpha)*100% Confidence interval for the slope m
        of the difference D = Y - X

```
        bD: Estimated (regression) intercept of the difference
        mD: Estimated (regression) slope of the difference
      rhoXD: The correlation coefficient between X and D = Y - X
```

S = REGRESSCI(X,Y,alpha,'plot') returns a plot of the data, estimated regression line, and (1-alpha)*100% confidence bounds on the estimated regression parameters.

```
Reference:  Applied Statistics (Second Edition)
            Using SPSS, STATISTICA, MATLAB, and R
            J. P. Marques de Sá
            Springer-Verlag, 2007
```

Example: X = [91,104,107,107,106,100,92,92,105,108];
           Y = [9.8,7.4,7.9,8.3,8.3,9.0,9.7,8.8,7.6,6.9];
           alpha = 0.05;

           S = regressci(X,Y,alpha)   returns

```
       CIb: [14.0290 26.3492]
       CIm: [-0.1775 -0.0560]
         b: 20.1891
         m: -0.1168
     rhoXY: -0.8431
   CIrhoXY: [-0.9620 -0.4549]
       RMS: 0.5485
   coefdet: 0.7108
     Ypred: [10x2 double]
      CIbD: [14.0290 26.3492]
      CImD: [-1.1775 -1.0560]
        bD: 20.1891
        mD: -1.1168
     rhoXD: -0.9978
```

See also levene, deming.

RMSSTD

RMSSTD   Root-mean-squared standard deviation index.

rms = rmsstd(S) returns the RMS standard deviation index with respect to the data matrix partitioned into k-clusters via KCLUSTER.

Example: Ddir = 'C:\PJC\Math_Biology\Chapter_1\Data';
                load(fullfile(Ddir,'Clustered_Data_Classes1_2.mat'));
                X = C{1}; Y = C{2};
                S = kcluster(Y,2);

                rms = rmsstd(S)   returns   0.8560

Reference: Data Clustering: Theory, Algorithms, adn
                Applications
                G. Gan, C. Ma, and J. Wu
                ASA-SIAM, Copyright (c) 2007
                ISBN: 978-0-898716-23-8

See also <u>mahalanobis</u>, <u>kcluster</u>, <u>kmeans</u>, <u>chindex</u>.

RTPCR

RTPCR   Real Time Polymerase Chain Reaction Model.

F = rtpcr(Ncycle,K,C) returns the structure whose fields
contain the fluorescence, efficiency, and the list of cycles
with respect to the input number of cycles (Ncycle), the input
structure of initial conditions, and initial concentration C.
The output structure F has the fields

        fluorescence: the fluorescence level at each cycle
                Eff:
                        den: efficiency at cycle n during
                             denaturation
                        ann: efficiency at cycle n during annealing
                        ext: extension efficiency
                        pcr: overall PCR efficiency at cycle n
              cycles: the cycle values

Example: P = load('C:\PJC\Math_Biology\Chapter_2\Data\Ktrpcr
                .mat');
                K = P.K;
                F = rtpcr(40,K,1)   returns

   F =
     fluorescence: [1x40 double]
              Eff: [1x1 struct]
           cycles: [1x40 double]

See also <u>rtpcrode</u>.

RTPCRODE

RTPCRODE   System of nonlinear ordinary differential equations for the real time polymerase chain reaction model.

```
f = rtpcrode(x,K) returns the right-hand-side of the system of
ODEs which model the real time PCR model AND is compatible
with MATLABs ODE solver suite (in particular, ODE45).

The parameters K listed below will be set within this
function.
Observe that K is a 1-by-8 vector K = [k(1),k(2), ..., k(8)]
containing the association (sense) and dissociation
(anti-sense) rates for the hybrids (H1, H2), template hybrid
(U), and primer-dimer (D) as follows.

    k(1) = kaH1, k(2) = kdH1, k(3) = kaH2, k(4) = kdH2
    k(5) = kaU, k(6) = kdU, k(7) = kaD, k(8) = kdD

The equilibrium constants are constructed from the rates.

  KH1 = kdH1/kaH1, KH2 = kdH2/kaH2, KU = kdU/kaU, KD = kdD/kaD
```

Example: P = load('C:\PJC\Math_Biology\Chapter_2\Data\Ktrpcr
             .mat');
          k = ...
  [P.K.kaH1,P.K.kdH1,P.K.kaH2,P.K.kdH2,P.K.kaU,P.K.kdU,P.K.kaD,
  P.K.kdD];
             Xinit = [ones(1,2)*1e-6, ones(1,2)*1e-10, ones(1,4)*
                   1e-08];
             t,Y] = ode45(@(t,y) rtpcrode(y,k),[0,P.K.tann],
                   Xinit);

See also rtpcr.

Reference: J. L. Gevertz, S. M. Dunn, and C. M. Roth
             Mathematical Model of Real-Time PCR Kinetics
             Biotechnology and Bioengineering, 5 November 2005
             Volume 92, Number 3, pp 346-355

SAMPLESIZE

SAMPLESIZE   The sample size for a specified statistical test significance level, power, and effect size.

N = SAMPLESIZE(Test,alpha,pwr,ES) returns the number of
samples (N) required for the statistical test (Test),
significance level (alpha), power (pwr), and effect size.

The statistical tests the SAMPLESIZE can process are listed
below.

```
      Test                 Hypothesis Test
      ----                 ---------------
      'T'                  Difference of means
      'Tcorr'              Difference of correlations
      'F'                  Difference of variances
      'Chi2'               Goodness of Fit/Contingency
      'Prop'               Difference of proportions
      'Wilx'               Matched-pairs Wilcoxon test
```

The input options vary with respect to the statistical test.
These options are spelled out as below.

```
% One-Sided Tests (means)
% Ho:  mu1 >= mu2 & H1: mu1 < mu2  or  Ho: mu1 <= mu2 & H1:
      mu1 > mu2
N = SAMPLESIZE('T',alpha,pwr,es,1);

% Two-Sided Test (means)
% Ho: mu1 = mu2 & H1: mu1 ~= mu2
N = SAMPLESIZE('T',alpha,pwr,es,2);

% One-Sided Tests (correlation)
% Ho:  rho = 0 & H1: rho < 0  or  Ho: rho = 0 & H1: rho > 0
N = SAMPLESIZE('Tcorr',alpha,pwr,es,1);

% Two-Sided Test (correlation)
% Ho: rho = 0 & H1: rho ~= 0
N = SAMPLESIZE('Tcorr',alpha,pwr,es,2);

% One-Sided Tests (proportions)
% Ho:  p1 >= p2 & H1: p1 < p2  or  Ho: p1 <= p2 & H1: p1 > p2
N = SAMPLESIZE('Prop',alpha,pwr,p2,p1,N2/N1,1);

% Two-Sided Test (proportions)
% Ho: p1 = p2 & H1: p1 ~= p2
N = SAMPLESIZE('Prop',alpha,pwr,p2,p1,N2/N1,2);

% One-Sided Test (matched-pairs)
% Ho: Control = Treatment & H1: Control > Treatment    OR
```

```
% Ho: Control = Treatment & H1: Control < Treatment
N = SAMPLESIZE('Wilx',alpha,pwr,es,1);

% Two-Sided Test (matched-pairs)
% Ho: Control = Treatment & H1: Control ~= Treatment
N = SAMPLESIZE('Wilx',alpha,pwr,es);

% Goodness-of-fit/Contingency Table (chi-squared) Test
% Ho: p(j) = p(j) for all j and k vs. H1: p(j) ~= p(k) for
     some j and k
% w = effect size and df = degreees of freedom
[N,ncp] = SAMPLESIZE('Chi2',alpha,pwr,w,df);
% Returns the sample size N and the non-centrality parameter
  ncp

% F-Test (equivalence of variance)
% Ho: Var1 = Var2 vs. H1: Var1 ~= Var2
% w = effect size and df = [df1,df2] = degreees of freedom
[N,ncp] = SAMPLESIZE('F',alpha,pwr,w,df);
% Returns the sample size N and the non-centrality parameter
  ncp

[N,ncp] = SAMPLESIZE('F',alpha,pwr,w,df,1);
% Returns the sample size N and the non-centrality parameter
  ncp
% for the one-sided tests
% Ho: Var1 >= Var2 vs. H1: Var1 < Var2    OR
% Ho: Var1 <= Var2 vs. H1: Var1 > Var2
```

See also teffectsize, feffectsize, chi2effectsize,
propeffectsize.

SCORECOMPARE

SCORECOMPARE   Compare the PCA Decomposition.

```
scorecompare(score,colours) Plots the PCA scores of pairs of
PCA components with respect to the input colors.
```

Example: Ddir = 'C:\PJC\Math_Biology\Chapter_4\Data';
         load(fullfile(Ddir,'Multivariate_MultiClass_Data
            .mat'));
         X = vertcat(D.X{:}); M = vertcat(D.X{:});
         Z = stdnorm(M,D.X{:});

```
        for j = 1:numel(Z)
          [Sc{j},Ld{j},Lam{j}] = pcamb(D.X{j},0.99);
        end
        colours = {'b','g','r','k'};
        scorecompare(Sc,colours);
```

See also pcamb, pcacompare

SEIREKBF

SEIREKBF   Extended Kalman-Bucy Filter for SEIR Epidemiology model
            SEIR = susceptible, exposed, infectious, recovered

F = seirekbf(IC,Sap,D,T) returns the structure S whose fields
contain the Kalman-filter estimates of the susceptibles (S),
exposed (E), infectious (I), and recovereds (R) obtained from
the vector of initial conditions IC = [So,Eo,Io,Ro], structure
of system adjustable parameters Sap, time series data D, and
vector of times T.

The input stucture of parameters P has the form

 Sap =
        B: immigration rate into the susceptible population
        v: rate of sero conversion
     beta: probability of disease transmission
        a: recovery delay rate
    kappa: loss of immunity rate
        d: death rate due to disease
       nd: death rate due to natural causes
       bo: process (plant) noise fraction

 The output structure F contains the fields

        S: susceptibles
        E: exposed/infected
        I: infectious
        R: recovered
     Ecov: cell array of Error covariance matrices.

Example: % Load the simulated data and time vectors
         Ddir = 'C:\PJC\Math_Biology\Chapter_3\Data';
         data = load(fullfile(Ddir,'SEIR_Simulated_Data.mat'));

```
D = data.D;
t = load(fullfile(Ddir,'SEIR_Time.mat')); T = t.T;
% Clear extraneous data structures
clear data t
% Initial conditions for the nonlinear ODE (state
dynamics)
So = 6.5e+04; IC = [So,So*0.05,D(1),D(1)*0.5];
% Structure of system adjustable parameters (SAP)
dy = 365.25;
Sap.v = 1.5/dy; Sap.beta = 0.0056; Sap.a = 0.7222;
Sap.d = 0.0003224/dy; Sap.kappa = 0.05; Sap.nd =
0.008724/dy;
Sap.bo = 0.03*365.25*IC(3); Sap.Mnoise = 0.1;
```

F = seirekbf(IC,Sap,D,T) returns the structure F containing
   the fields

```
   dm: Vector of measurement residuals
 Info: The information matrix inv(H*P(k+1)*H.' + Mnoise)
   In: Vector of Innovation measurement residuals
  Iup: Vector of estimated infections upper bound
 Ilow: Vector of stimated infections lower bound
    S: Vector of estimated susceptibles
    E: Vector of estimated Exposed
    I: Vector of estimated Infecteds
    R: Vector of estimated Recovereds
 Ecov: Celll arrays of 4-by-4 Covariance matrices at each
       time t
```

```
   seirekbfplot(F,T,D)
```

See also addnoise, seirekbfplot.

SEIREKBFPLOT

SEIREKBFPLOT   Plot of the Extended Kalman-Bucy Filter solution of the SEIR
                model


```
seirekbfplot(F,t,m) returns the 2-by-2 plots of the
susceptibles S(t),exposed E(t), infecteds I(t), and Recovered
R(t) populations from the extended Kalman-Bucy filter solution
of the SEIR model.  Also returned are the innovations dm =
m(t) - mp(t) and the overlay of the measurements m(t) against
```

the filtered solution I(t) +/- std(I(t)). The second input t
is the vector of times over which the Extended Kalman-Bucy
filter is run. The third input m is the vector of times-series
measurements used in the filter.

The input structure F is the output of SEIREKBF.

See also seirekbf.

SEIREKBFOUT

SEIREKBFOUT   Batch run for the Extended Kalman-Bucy Filter estimates of the
               SEIR model for disease outbreak.

S = seirekbfout(IC,Sap,D,T,Mtype,Nout) returns the structure
S whose fields contain the probability of outbreak detection
(Pd), the sensitivity (Sen) and specificity (Spec) of the
algorithm, the times at which the outbreaks were simulated
(Tout), and the duration (d) and rate (m) of the outbreak with
respect to the input vector of initial conditions (IC), the
structure of system adjustable parameters (Sap), the
measurement data (D), and the times (T). The type of model
outbreak is identified via Mtype (as noted below). The number
of outbreak simulations is identified via the positive integer
Nout while the number of consecutive days required for
successful outbreak detection is designated by Ndays.

The possible values for Mtype are

```
    Mtype          Model
    -----          -----

    'alin'         f(t,to) = m*(t-to) for to <= t <= to + d
                           = 0         otherwise

    'lramp'        f(t,to) = m*(t-to) for to <= t <= to + d
                           = m*d       for t > to + d
                           = 0         otherwise

    'hat'          f(t,to) = m*(t-to) for to <= t <= to + d
                           = m*d - m*(t-to-d) for to+d <= t <= to+2*d
                           = 0         otherwise

    'Gauss'        f(t,to) = m*d*exp(-[t-to-d]^2/(2*s^2))
                         s = d/1.96
```

The nominmal values of d and m are d = 7 (days) and m = 1.

S = seirekbfout(IC,Sap,D,T,Mtype,Nout,d) returns True
Positives, True Negative, False Positives, and False Negatives
as above for an outbreak value d (number of days).

S = seirekbfout(IC,Sap,D,T,Mtype,Nout,d,m) returns True
Positives, True Negative, False Positives, and False Negatives
as above for an outbreak value d (number of days) and m (rate
of the outbreak).

Example: % Load the simulated data and time vectors
```
Ddir = 'C:\PJC\Math_Biology\Chapter_3\Data';
data = load(fullfile(Ddir,'SEIR_Simulated_Data.mat'));
D = data.D;
t = load(fullfile(Ddir,'SEIR_Time.mat')); T = t.T;
% Clear extraneous data structures
clear data t
% Initial conditions for the nonlinear ODE (state
dynamics)
So = 6.5e+04; IC = [So,So*0.05,D(1),D(1)*0.5];
% Structure of system adjustable parameters (SAP)
dy = 365.25;
Sap.v = 1.5/dy; Sap.beta = 0.0056; Sap.a = 0.7222;
Sap.d = 0.0003224/dy; Sap.kappa = 0.95; Sap.nd =
0.008724/dy;
Sap.bo = 0.03*dy*IC(3); Sap.Ndays = 3; Sap.thres =
0.05;

S = seirekbfout(IC,Sap,D,T,'Lramp',100,7,4)    returns
```
S =

```
        Pd: 1
       Sen: 0.8081
      Spec: 1
      Tout: [1x100 double]
  duration: 7
      rate: 4
     Ndays: 3
```

See also seirekbf, outbreaks.

SMPROP

SMPROP   Score Method Confidence Interval for a single proportion.

H = SMPROP(k,n,alpha) returns the structure H whose fields
contain an estimate of the proportion (p), the conventional
(CI) and Newcombe score method (NCI) (1-alpha)*100% confidence
intervals for the proportion estimated from the number of
successes (k) and trials (n).

Example: H = smprop(96,100,0.05) returns

```
    H =
          p: 0.9600
         CI: [0.9216 0.9984]
        NCI: [0.9016 0.9843]
```

See also <u>smpropdiff</u>, <u>pooledvar</u>.

SMPROPDIFF

SMPROPDIFF   Score Method Confidence Interval and Hypothesis Test for the
             difference in proportions.

H = SMPROPDIFF(k,n,alpha) returns the structure H whose fields
contain the null hypothesis decision (Ho), the hypothesis test
p-value, the conventional (CI) and Newcombe (NCI) (1-alpha)*
100% confidence intervals for the difference in proportions
estimated from the two-dimensional vector of successes (k) and
trials (n). The structure also contains estimated proportions
(p) and corresponding two-sided (1-alpha/2) confidence
intervals (CI1,NCI1,CI2,NCI2).

```
    Ho: p1 = p2          %  p1 = k(1)/n(1), p2 = k(2)/n(2)
    H1: p1 ~= p2         %  dp = p1 - p2
```

H = SMPROPDIFF(k,n,alpha,Tflag) returns the structure H above
for the two-sided, right-tailed, or left-tailed hypothesis
test as indicated by the input test flag Tflag.

```
    Tflag                 Null Hypothesis
    ----                  ----------------
      0                   Ho: p1 <= p2 (right-tailed)
      1                   Ho: p1 >= p2 (left-tailed)
      2                   Ho: p1 = p2 (two-tailed)
```

The default value of Tflag is 2 (two-tailed test).

H = SMPROPDIFF(k,n,alpha,Tflag,A) returns the structure H
above using the Newcombe-Wilson confidence interval for the

difference of CORRELATED proportions. The input A is the 2-by-2 contingency matrix for the matched pairs associated with the correlated proportions. The matrix A has the form below.

```
                          Method 2
                    Negative      Positive
          Negative   A(1,1)        A(1,2)
Method 1  Positive   A(2,1)        A(2,2)
```

In additional to the fields noted above, the correlated coefficient for paired proportions (rho) is also returned. If the proportions are not paired (that is, are independent), then rho is set to 0.

Examples: H = smpropdiff([12,97],[15,228],0.05) returns

```
    H =
          Ho: 'reject'
          dp: 0.3746
           Z: 3.4571
          CI: [0.1622 0.5869]
         NCI: [0.1145 0.5183]
      Pvalue: 5.4603e-004
           p: [0.8000 0.4254]
         CI1: [0.5976 1.0024]
        NCI1: [0.5481 0.9295]
         CI2: [0.3613 0.4896]
        NCI2: [0.3630 0.4903]

          alpha = 0.05;
          k = [23,12]; n = [395, 432]; A = [10,15;25,345];
          H = smpropdiff(k,n,alpha,2)   returns
    H =
          Ho: 'reject'
          dp: 0.0305
           Z: 2.1460
          CI: [0.0026 0.0583]
         NCI: [0.0027 0.0605]
      Pvalue: 0.0319
           p: [0.0582 0.0278]
         CI1: [0.0351 0.0813]
        NCI1: [0.0391 0.0859]
         CI2: [0.0123 0.0433]
        NCI2: [0.0160 0.0479]
         rho: 0
```

```
        H = smpropdiff(k,n,alpha,2,A)      returns
  H =
        Ho: 'reject'
        dp: 0.0305
         Z: 2.1460
        CI: [0.0026 0.0583]
       NCI: [0.0067 0.0574]
    Pvalue: 0.0319
         p: [0.0582 0.0278]
       CI1: [0.0351 0.0813]
      NCI1: [0.0391 0.0859]
       CI2: [0.0123 0.0433]
      NCI2: [0.0160 0.0479]
       rho: 0.2665
```

See also <u>smprop</u>.

STDNORM

STDNORM   Standard Normalization.

`Z = stdnorm(M,T1,T2,...Tg)` returns the cell array of mean and standard deviation normalized data matrices `Z = {Z1,Z2,...,Zg}` obtained by subtracting from the input cell array of matrices `T = {T{1},T{2},...,T{g}}` the mean and dividing by the standard deviation of the input matrix M. That is,

$$Zj = (T\{j\} - mean(M))/std(M).$$

`Z = stdnorm(T,T1,T2,...Tg)` returns the cell array of matrices Zk as above using `M = vertcat(T{:})` whenever the first input is a cell array.

`[Z,mu,s] = stdnorm(M,T1,...,Tg)` returns the overall mean (mu) and standard deviation vectors.

Example: `Ddir = 'C:\PJC\Math_Biology\Chapter_4\Data';`
        `load(fullfile(Ddir,'Multivariate_MultiClass_Data`
           `.mat'));`
        `M = vertcat(D.X{:});`

        `Z = stdnorm(M,D.X{:});`

See also <u>meanorm</u>.

SVMBDRY

SVMBDRY   Support Vector Machine Boundary

F = svmbdry(X,y,kernel,alpha,nsv) returns the string which
represents the function that is the SVM boundary associated
with the kernel, Lagrange multipliers (alpha), and indices of
non-zero support vectors associated with the data matrix X and
class identification vector y.

Example: X = [1,1;3,3;1,3;3,1;2,2.5;3,2.5;4,3];
         y = [-1,1,1,-1,1,-1,-1];
         K.kernel = 'polynomial'; K.parameter = 3;
         S = svmclass(X,y,K);

   F = svmbdry(X,y,K,S.alpha,S.nsv)   returns

   F = + -0.0060628*(1 + x1*1)^3 + -0.0060628*(1 + x2*1)^3 +...
       0.00079635*(1 + x1*3)^3 + 0.00079635*(1 + x2*3)^3 + ...
       0.012736*(1 + x1*2)^3 + 0.012736*(1 + x2*2.5)^3 + ...
       -0.0074698*(1 + x1*3)^3 + -0.0074698*(1 + x2*2.5)^3 + ...
       -1.1771

See also, qpip, svmkernel.

Warning: The support vector machine functions use the QPC packaged created by
the School of Electrical Engineering and Computer Science at the University of Newcastle, Callaghan, NSW, 2308, Australia. It is used by permission of Professor Brett
Ninness and is not included within this text. To gain access, contact the University of
Newcastle directly.

SVMCLASS

SVMCLASS   Support Vector Machine Classifier

S = svmclass(X,y,ker) returns the structure S whose fields
contain the vector of Lagrange multipliers (alpha), the index
of non-zero support vectors (nsv), the optimal weight vector
(w), and the boundary offset (b) which separates the two
classes identified via the input vector y. The input arguments
are

     X = n-by-p matrix of positive/negative class measurements
     y = n-by-1 vector of class identifiers (-1 -> negative,
         1 -> positive)

ker = structure for SVM kernel. ker.kernel = kernel name
and
ker.parameter = kernel parameters.

The kernel name, parameter, and mathematical formuation is as
below.

```
Kernel Name      Parameter    Mathematical Form
----------       -------      ----------------
'linear'            []        K(i,j) = <x(i),x(j)>
'polynomial'        d         K(i,j) = (1 + <x(i),x(j)>).^d
'neural'         [a(1),a(2)]  K(i,j) = tanh(a(1)*<x(i),x(j)> + a(2))
'radial'            g         K(i,j) = exp(-g*||x(i)-x(j)||^2)
'exponential'       g         K(i,j) = exp(-g*||x(i)-x(j)||)
```

[alpha,nsv,w] = svmclass(X,y,ker,Ub) returns the output
described above with respect to the upper bound Ub placed on
the Lagrange multipliers.

The Lagrange multipliers alpha = [alpha(1),alpha(2),...,alpha
(n)] are the solution of the optimization problem

    LD[alpha] = min [1/2 alpha*H*alpha.' - f.'*alpha]

subject to the constraints

    alpha*y = 0 and alpha(k) >= 0 for all k = 1, 2, ... , n
    f = -ones(1,n) = [-1,-1, ... ,-1], H = Y*K*Y, K = the
      selected kernel
    Y = diag(y).

Example: X = [1,1;3,3;1,3;3,1;2,2.5;3,2.5;4,3];
        y = [-1,1,1,-1,1,-1,-1];
        K.kernel = 'polynomial'; K.parameter = 3;

        S = svmclass(X,y,K)   returns

    S =
      alpha: [7x1 double]
        nsv: [4x1 double]
          w: [2x1 double]
          b: -1.0089

Nota Bene: svmclass uses the QPC package created by the
School of Electrical Engineering & Computer Science at the
University of Newcastle, Callaghan, NSW, 2308, AUSTRALIA

See also, qpip, svmkernel.

Warning: The support vector machine functions use the QPC packaged created by the School of Electrical Engineering and Computer Science at the University of Newcastle, Callaghan, NSW, 2308, Australia. It is used by permission of Professor Brett Ninness and is not included within this text. To gain access, contact the University of Newcastle directly.

SVMKERNEL

SVMKERNEL   Kernel generator for a Support Vector Machine

```
K = svmkernel(X,Y,Ktype,Kparameter) returns the n-by-n Kernel
matrix K(i,j) = kernel(X(i,:),X(j,:)) with respect to the
n-by-p input matrix X, the n-dimensional vector of class
identification Y, the Kernel flag Ktype, and the associated
kernel parameters Kparameter. The kernels and associated
parameters which can be returned by this M-file are as follows.
```

Kernel Type	Parameter	Mathematical Form
'linear'	[]	K(i,j) = <x(i),x(j)>
'polynomial'	d	K(x,y) = (1 + <x(i),x(j)>).^d
'neural'	[a(1),a(2)]	K(x,y) = tanh(a(1)*<x(i),x(j)> + a(2))
'radial'	g	K(x,y) = exp(-g*||x(i)-x(j)||^2)
'exponential'	g	K(x,y) = exp(-g*||x(i)-x(j)||)

```
Examples:   K = svmkernel(X,'polynomial',3)   returns the
              polynomial
            kernel whose (i,j) element is
            K(i,j) = (1 + X(i,:)*X(j,:).').^3

            K = svmkernel(X,'radial',[2,4])   returns the neural
            network kernel whose (i,j) element is
            K(i,j) = tanh(2*X(i,:)*X(j,:).' + 4)
```

```
See also svmclass, svmbdry.
```

Warning: The support vector machine functions use the QPC packaged created by the School of Electrical Engineering and Computer Science at the University of Newcastle, Callaghan, NSW, 2308, Australia. It is used by permission of Professor Brett Ninness and is not included within this text. To gain access, contact the University of Newcastle directly.

TMEAN

TMEAN   The T–test comparing two sample means.

H = TMEAN(X,Y,alpha,Tflag) returns the structure H whose fields
contain the decision with respect to the null hypothesis (Ho),
the values of the test T-statistic (T), the (1-alpha)*100%
confidence interval/level for the null hypothesis, the degrees
of freedom (df), the type of hypothesis test as indicated by
the input Tflag, the p-value of the test, and the effect size
for the indicated test.

Tflag	Ho	H1	
2	Mx = My	Mx ~= My	(two-tail)
1	Mx >= My	Mx < My	(left-tail)
0	Mx <= My	Mx > My	(right-tail)

Mx = mean(X) and My = mean(Y).

H = TMEAN(X,Y,alpha,Tflag,Pwr) returns the structure H as above
for a specified power Pwr. The default value of the power is
90%: Pwr = 0.9

The structure H contains the fields

      Ho: Null hypothesis decision state
       T: Value of the T-statistic
      CI: (1-alpha)*100% confidence interval
      df: Degrees of freedom
    Test: Test computed (i.e., two-sided, one-side greater
          than, etc.)
  Pvalue: The p-value of the test (i.e., the significance
          probability = the smallest level of significance at
          which the null hypothesis is rejected on the basis
          of the observed data)
  Effect: The t-test effect size of the data for a power = Pwr.

Example: X = 10*rand(1,400); Y = 2*rand(1,250);
          H = tmean(X,Y,0.05,0)   returns
    H =
          Ho: 'reject'
           T: 16.8864
          CI: [-Inf 1.6483]
          df: 444

```
    Test: 'one-sided, right tail'
  Pvalue: 4.4969e-050
  Effect: 1.7167


      H = tmean(X,Y,0.05,1)    returns
H =
      Ho: 'do not reject'
       T: 16.8864
      CI: [1.6483 Inf]
      df: 444
    Test: 'one-sided, left tail'
  Pvalue: 1
  Effect: 1.7167


      H = tmean(X,Y,0.05,2)    returns
H =
      Ho: 'reject'
       T: 16.8864
      CI: [-1.9653 1.9653]
      df: 444
    Test: 'two-sided, equal'
  Pvalue: 8.9937e-050
  Effect: 1.7167
```

See also, wilcoxonsr.

WILCOXONSR

WILCOXONSR   The Wilcox Signed Rank Test statistic

H = WILCOXONSR(Con,Treat,alpha,Tflag) returns the structure H whose fields contain the decision with respect to the null hypothesis (Ho), the values of Wilcoxon Signed Rank Statistic (W), the critical region for the Wilcoxon statistic W (CR), the p-value of the test, the number of ties (Nties), the number of non-zero differences (n), and the test type as defined by the flag Tflag.

```
  Ho:  Con = Treat
  H1:  Con ~= Treat
```

Here Con = Control and Treat = Treatment.

```
Tflag              Ho                  H1
----           ---------           --------
  2            Con = Treat         Con ~= Treat    (Two-tail)
  1            Con >= Treat        Con < Treat     (Left-tail)
  0            Con <= Treat        Con > Treat     (Right-tail)
```

H = WILCOXONSR(X,Y,alpha,Tflag,Rflag) returns the structure H as above using the tied-rank replacement option Rflag.

```
Rflag                  Action
----                   ------
mean                   tied ranks are replaced by mean rank
median                 tied ranks are replaced by median ranks
```

The default value for Rflag is 'med'.

The structure H contains the fields

```
        Ho: The null hypothesis decision
         W: The Wilcoxon Signed Rank Statistic (W+)
        CR: The (1-alpha)*100% critical region for W+
    Pvalue: The p-value of the test
     normT: The normalized test statistic Z = (W*-E[W])/
            var(W*)
            W* = -Wminus is sufficient as W = W+ - Wminus.
         n: Number of non-zero differences between paired
            samples
     Nties: The number of samples in which Control =
            Treatment
  TestType: Right-, Left-, or Two-tailed
```

Example 1:
```
X = [0,0,0,0,2,1,1,1,1,1,0,0,0,0,0,0,0,0,0,0,0,0,0,0,0,1,0,0, ...
     0,0,0,0,0,0,0,0,0,1,0,0,0,0,0,0,0,0,0,0,0,0,0,0,0,0,0,0, ...
     0,0,0,0,1,0,0,0,1,1,1,1,0,0,0,2,1,0,0,0,1,1,0,0,0,0,0, ...
     0,0,0,1,1,0,0,0,0,2,0,0,0,0,1,1,0,0,0,0,0,0,0,1,0,0,1, ...
     0,0,1,0,0,0,0,0,0,1,0,1,0,0,0,0,0,0,0,0,0,0,1,0,0,0,0, ...
     1,0,0,0,0,1,0,0,1,0,1,1,0,0,0,0,0,0,0,1,0,2,0,0,0,0,0, ...
     2,0,0,3,0,0,1,0,0,0,0,2,0,1,1,0,0,1,1,0,0,0,0,0,0,0,0, ...
     0,0,0,0,1,0];

Y = [1,0,1,0,0,0,0,0,0,0,0,0,0,0,0,0,0,0,0,0,0,0,0,0,1,1,0, ...
     0,0,0,0,1,0,1,0,0,0,0,0,0,0,1,1,0,1,0,1,0,0,1,2,2,0,0, ...
     0,0,0,0,1,0,0,0,0,0,0,0,1,0,0,0,0,1,1,1,1,0,0,0,0,1,2, ...
```

```
      0,0,0,0,0,2,0,0,0,0,0,1,0,0,0,1,0,0,0,0,0,0,0,2,0,0,1, ...
      2,0,0,2,0,0,1,0,0,0,0,0,0,0,0,0,0,1,0,0,1,0,0,0,1,1,0,0, ...
      0,0,0,0,1,0,0,0,0,0,0,1,0,0,0,0,0,0,0,1,1,0,0,0,1,1,0, ...
      1,0,0,3,1,3,0,1,0,0,0,0,0,0,1,0,1,0,3,0,0,0,0,0,0,1,0, ...
      1,0,0,0,0,2];
```

```
H = wilcoxonsr(X,Y,0.05,2)    returns
```

```
   H =
           Ho: 'do not reject'
            W: 1531
           CR: [993 1708]
       Pvalue: 0.2872
        normT: -1.0643
            n: 73
        Nties: 122
     TestType: 'Two-tailed'
```

Example 2: Sdir = 'C:\PJC\Math_Biology\Chapter_3\Data';
              load(fullfile(Sdir,'WilcoxonA.mat'));
              load(fullfile(Sdir,'WilcoxonB.mat'));

```
              H = wilcoxonsr(A,B,alpha,2)    returns
   H =
           Ho: 'reject'
            W: 950
           CR: [434 841]
       Pvalue: 0.0019
        normT: -3.1124
            n: 50
        Nties: 4
     TestType: 'Two-tailed'
```

Example 3: H = wilcoxonsr(A,B,alpha,1) returns

```
   H =
           Ho: 'reject'
            W: 950
           CR: [-Inf 808]
       Pvalue: 9.4370e-04
        normT: -3.1124
            n: 50
        Nties: 4
     TestType: 'Left-tailed'
```

Example 4: H = wilcoxonsr(A,B,alpha,0) returns

```
    H =
            Ho: 'do not reject'
             W: 950
            CR: [467 Inf]
        Pvalue: 0.9991
         normT: -3.1124
             n: 50
         Nties: 4
      TestType: 'Right-tailed'
```

See also signrank.

WSMEAN

WSRMEAN   Alternate Hypothesis mean for the Wilcoxon Signed–Rank Test.

m = WSRMEAN(n,theta) returns the mean of the Wilcoxon Signed-Rank Test under the alternate hypothesis that the control and treatment variables are stochastically different (theta ~= 0).

    m = p1*n +p2*(n+1)/2

where p1 and p2 are weights determined by the standard Normal probability density.

See also wsrpower, wsrtest, wsrmeano.

WSRSAMPLE

WSRSAMPLE   Sample size required to attain a power and significance level for Wilcoxon Signed–Rank Test.

N = wsrsample(theta,alpha,beta) returns the structure N whose fields contain the number of samples (via the Lehmann, Noether, and exact methods) required to attain an input power (1 - beta) and significance level (alpha) for the Wilcoxon Signed-Rank Test.

Example: N = wsrsample(0.1,0.05,0.2) returns

```
    N =
      Noether: 652
      Lehmann: 648
        Exact: 649
```

See also <u>wsrpower</u>, <u>wsrtest</u>, <u>wsrmean</u>, <u>wsrmeano</u>, <u>wsrvaro</u>.

WSRVAR

WSRVAR   Alternate Hypothesis variance for the Wilcoxon Signed–Rank Test

v = WSRVAR(n,theta) returns the mean of the Wilcoxon Signed-
Rank Test under the alternate hypothesis that the control and
treatment variables are stochastically different (theta ~= 0).

```
  v = n*p(1)*(1 - p(1)) + n*(n - 1)*p(2)*(1 - p(2))/2 + ...
    2*n*(n - 1)*(p(3) - p(1)*p(2)) + n*(n - 1)*(n - 2)* ...
    (p(4) - p(2)^2);
```

See also <u>wsrpower</u>, <u>wsrtest</u>, <u>wsrmean</u>, <u>wsrmeano</u>, <u>wsrvaro</u>.

INDEX

Applied Mathematics for the Analysis of Biomedical Data: Models, Methods, and MATLAB®, First Edition. Peter J. Costa.
© 2017 Peter J. Costa. Published 2017 by John Wiley & Sons, Inc.
Companion website: www.wiley.com/go/costa/appmaths_biomedical_data